Keep Track of Physics and Astronomy All Year Long!

The editors at Wadsworth Publishing Company hope you will enjoy Margaret Greenwood's PHYSICS: THE EXCITEMENT OF DISCOVERY. To accompany this exciting book, we have produced a handsome physics and astronomy wall calendar, beautifully illustrated with dramatic full-color photographs and paintings and extensively annotated with important physical and astronomical dates and events occuring throughout the year.

We are pleased to make available to you a complimentary copy of this useful, attractive calendar. To receive your free calendar, simply fill out and remove the coupon below, and mail to: Wadsworth Publishing Company, Attn: Sales Services Department, Box BK-CAL-PE1, Ten Davis Drive, Belmont, California 94002.

Note: If your request is received before May 1st, you will be sent a calendar for the current year, if your request is received after May 1st, you will be sent a calendar for the coming year. (Calendars for each year are published in October of the preceding year.)

You must use the *original* coupon printed below to receive your complimentary calendar.

Mail to:

Wadsworth Publishing Company
Attn: Sales Service Department
Box BK-CAL-PE1
Ten Davis Drive
Belmont, CA 94002

Please send me my complimentary copy of Wadsworth's Physics and Astronomy Calendar

(print clearly as this is your mailing label)

_____
Name

_____
Street Address

_____
City

_____
State   *Zip Code*

**Keep Track of Physics and Astronomy
All Year Long!**

(To receive your complimentary
calendar, see preceding page.)

PHYSICS
THE EXCITEMENT OF DISCOVERY

MARGARET STAUTBERG GREENWOOD
DePaul University

WADSWORTH PUBLISHING COMPANY
Belmont, California

A Division of Wadsworth, Inc.

Physics Editor: Thomas P. Nerney
Designer: Janet Bollow
Copy Editor: Stuart Kenter
Technical Illustrations: Ayxa Art Studio
Drawings: Tom Barnett
Part- and chapter-opening art: Heather Preston

Cover photograph: A computer-enhanced, false-color photo shows the visible light from a passing 20-nanosecond-long (about 20-billionths of a second) 1500-amp pulse of electrons produced by Lawrence Livermore National Laboratory's Experimental Test Accelerator. The pulse is traveling through nitrogen gas at atmospheric pressure. As the electrons pass, they heat nitrogen molecules, causing them to glow, which in effect describes the size and path of the beam. The various colors indicate the intensity of the light, with blue being the most intense and red the least intense. Note that the beam is not diverging. This indicates that the like-charged electrons that would normally repel each other are being held together by their own magnetic field. The apparent segmenting is not due to any breakup of the beam, but rather to obstruction in the camera's field of view. (Lawrence Livermore National Laboratory photo)

Printed in the United States of America

2 3 4 5 6 7 8 9 10—87 86 85 84 83

ISBN 0-534-01260-4

Library of Congress Cataloging in Publication Data

Greenwood, Margaret Stautberg.
  Physics, the excitement of discovery.

  Includes index.
  1. Physics. I. Title.
QC23.G793 1983   530   82−24780
ISBN 0−534−01260−4

To Larry, Brian, and Susan

Contents

CHAPTER 8

Waves, Sound, and Light 168

CHAPTER 9

Relativity 196

CHAPTER 10

The Dual Nature of Light 230

CHAPTER 11

Radioactivity 254

CONTENTS

CONTENTS

CONTENTS

Preface

A few years ago, our family received a carton of books as a gift. One of the books was the biography *Madame Curie,* written by her daughter Eve Curie. I had read it in college, but decided to read it again. While her life story is fascinating and compelling, I eagerly awaited the chapters dealing with her research. However, while these chapters certainly presented the overall trend of her research, they did not answer my rather detailed questions: What method did she use to measure radioactivity? How did she separate radium from the pitchblende mineral? How did she determine the atomic mass of radium? Standard textbooks didn't answer these questions either.

Reading her doctoral thesis finally satisfied my curiosity. The method she used for measuring radioactivity was the forerunner of the Geiger counter (invented in 1908), which is widely used today to detect radiation. Pierre Curie, then a well-known physicist for his work on magnetism, joined his wife in the study of radioactivity. Their purified radium sample (one-tenth of a gram from one ton of pitchblende) remained above room temperature and this was the first observation of energy stored in the atom—nuclear energy we call it today. The Curies' story is such a singular one that I decided to include an episode of their lives in the chapter on radioactivity and weave the physics into it. To capture the excitement of their great discoveries, the text includes data and firsthand accounts from their research papers and from Madame Curie's thesis, as well as quotations from her biography. The result is Chapter 11.

I became convinced that the historical approach—showing physicists at work, so to speak—was a fascinating way to present physics and decided to continue it. In this way, the student sees, often from firsthand accounts, how a researcher got an idea for an experiment in the first place, and how numbers collected in a data book can reveal properties of the atom. Chapter 18, which deals with the discovery of fission, probably contains the most quotations from journals and other books, including an account by a physicist who witnessed the first self-sustaining chain reaction at the University of Chicago on December 2, 1942.

Originally, this text was started because, like so many instructors, I could not find one that exactly suited my needs as a teacher for nonscience majors. The course I taught had a modern physics theme. Most texts treated modern physics only at the end of the book and, furthermore, did not have enough chapters on this subject. There was somewhat of a dilemma in teaching this course. On one hand, classical physics was important, and I did want to discuss it but not to the extent that other textbooks did. One certainly needs to grasp the concepts of force and conservation of energy, for example, to comprehend modern physics. Yet, on the other hand, I wanted to get to modern physics right away. The solution evolved with increased teaching experience and continued revision of chapters. In the final analysis, I chose a traditional format—an introductory chapter, classical physics, modern physics—and then superimposed a modern physics theme *throughout*.

Chapter 1 presents an overview of modern physics through the discovery of the neutron in 1932. Chapters 2 through 6 (Part II) deal with the concepts of classical physics, such as acceleration, force, gravity, momentum, and energy. Since the student knows some modern physics from Chapter 1, illustrations from this area can now be included in Part II. Easily visualized examples from everyday experience are presented side by side with the more abstract concepts of the atom. There is, after all, not so much difference between a car coming to a stop when the brakes are applied and an alpha particle slowing down as it enters a gold foil. Nor, basically, does a car traveling in a circle differ much in principle from an electron traveling in a circular path in Bohr's model of the hydrogen atom. In Chapter 6 the principle of conservation of energy is applied to a rock falling freely to the ground and to a newly forming star in a nebula.

One of the unique features of Part II is the discussion of Rutherford's nuclear model of the atom and Bohr's model of the hydrogen atom. These two topics are used as examples each time a basic concept is introduced. In summary, Part II presents a parallel discussion of classical and modern physics, where the "line" so often separating these branches has been erased.

Gravitation (Chapter 5) is an exception to the modern physics theme. This is such an important topic in classical physics that many instructors like to include it in their courses. To meet this need, this text covers gravitation, as well as sections on heat and sound, to round out the discussion of classical physics. The text includes all of the major topics: mechanics, heat, waves, sound, light, electricity and magnetism, relativity, and modern physics. However, all of the topics are not given equal weight, and modern physics *is* the theme of the text.

There are, in fact, good reasons for this balancing of topics with the modern physics theme. Without doubt, the thirty-year period beginning with the discovery of X-rays in 1895 is one of the most productive and exciting in physics. Since nonscience majors may take only one course in this discipline, it makes good sense to expose them to the most dramatic material. Surely, the unraveling of the structure of the atom and nucleus is equal to any Sherlock Holmes mystery. In addition, since modern physics is the work of physicists today, students gain some insight into what contemporary researchers are trying to do, what their goals are. Finally, students have already read and heard much about the applications of modern physics: lasers, reactors, the electron microscope, television, the X-ray CT-scan, to name only a few. The study of modern physics will allow students to understand these applications and the concepts behind them.

Capitalizing on this student awareness, the text discusses a newsworthy application such as the X-ray CT scan in Chapter 1, so that readers may be drawn in right away.

Of interest to both student and instructor alike is the mathematical level of the text. In my view, a physics course should not avoid mathematics.

Yet, the approach taken in a general physics course, which involves extensive use of algebraic derivations, is *not* appropriate for a course for nonmajors. *Physics: The Excitement of Discovery* occupies a middle ground—it uses analogies and plausibility arguments augmented with numerical, rather than algebraic, calculations. For example, when discussing a car traveling in a circle, the student is asked: What happens to the force on a passenger, if the speed increases? If the radius increases? By considering the units of force, we arrive at the formula relating the force to the passenger's mass, speed, and the radius of the circle. This formula is then used to carry out *arithmetical* calculations for the force on a car and for the force on an electron in Bohr's model of the hydrogen atom. The discussion of motion in Chapter 2 uses formulas and arithmetical calculations, and yet I would not say that it's mathematical. The emphasis is on the physics, not the math. Sometimes, simple notions of algebra are used: A term may be transferred from one side of an equation to the other, or an equation may be simplified. Overwhelmingly, the calculations are arithmetical, rather than algebraic.

Thus, the major features of this text include:

• An overview of modern physics in Chapter 1
• A parallel discussion of classical physics and modern physics in Part II
• An historical approach with data and quotations from original papers, Nobel Prize lectures, the *Scientific American*, and so on
• The inclusion of technological applications
• Extensive use of plausibility arguments and analogies coupled with an appropriate level of mathematics

Acknowledgements

The following reviewers have made many valuable comments:

Professor Ronald Stoner
Physics and Astronomy Department
Bowling Green State University
Bowling Green, OH 43401

Professor Robert O. Garrett
Physics Department
Beloit College
Beloit, WI 53511

Professor Wayne W. Sukow
Physics Department
University of Wisconsin-River Falls
River Falls, WI 54022

Professor Alex F. Burr
Physics Department
New Mexico State University
Box 3D
Las Cruces, NM 88003

Professor Dietrich Schroeer
Physics Department
University of North Carolina
Chapel Hill, NC 27514

Professor Richard Obermyer
Physics Department
Pennsylvania State University-McKeesport
University Dr.
McKeesport, PA 15132

Professor Stanley J. Shepherd
Physics Department
Pennsylvania State University
104 Davey Laboratory
University Park, PA 16802

Professor D. Lee Rutledge
Physics Department
Oklahoma State University
Stillwater, OK 74074

Professor P. N. Swamy
Physics Department
Southern Illinois University
Edwardsville, IL 62026

Professor William J. Mullin
Physics Department
University of Massachusetts
Amherst, MA 01003

Professor Joseph W. Connolly
Physics Department
St. Bonaventure University
St. Bonaventure, NY 14778

Professor William W. Pratt
Physics Department
Pennsylvania State University
University Park, PA 16802

Professor Rexford Adelberger
Physics Department
Guilford College
Greensboro, NY 27410

Professor Russell N. Coverdale
Physics Department
Purdue University
West Lafayette, IN 47907

Professor Eugene D. Jacobson
Physics Department
Suffolk County Community College
533 College Rd.
Selden, NY 11784

Professor Victor J. Stenger
Physics and Astronomy Department
University of Hawaii-Manoa Campus
2505 Correa Rd.
Honolulu, HI 96822

Professor Alan K. Miller
Physical Science Department
Pasadena City College
Pasadena, CA 91106

Professor William B. Good
Physics Department
New Mexico State University
Box 3D
Las Cruces, NM 88003

Professor Richard S. Masada
Physics Department
Santa Monica College
1815 Pearl St.
Santa Monica, CA 90405

James Merkel
Physics Department
University of Wi-Eau Claire
Eau Claire, WI 54701

My thanks to several people at Wadsworth for their helpful suggestions: Autumn Stanley, Joan Garbutt, Mary Arbogast, Sandra Craig, and Tom Nerney. I would also like to acknowledge the help of my typist, Sally Hupert.

I also would like to acknowledge very helpful suggestions from my colleagues at DePaul University, Thomas G. Stinchcomb and the late James J. Vasa.

I especially want to thank my husband, Larry, for his wholehearted support and encouragement during this long project and our two children, Brian and Susan, who provided inspiration.

Margaret Stautberg Greenwood

Physics: The Excitement of Discovery

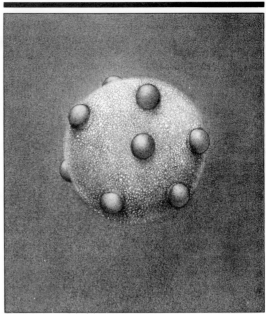

Thomson's model of the atom.

PART I

Introduction to Modern Physics

Homogenous sphere
model of the
atom (1808)

J. J. Thompson, who
discovered the
electron (1897)

Marie Curie, who
investigated
radioactivity (1898)

James Chadwick, who
discovered the
neutron (1932)

Planetary model
of the atom

Ernest Rutherford, who
discovered the nucleus
of the atom (1913)
and the proton (1919)

H. G. J. Moseley, who found how
to determine the number of electrons
and protons within an atom (1913)

Neils Bohr, who proposed
a model of the
hydrogen atom (1913)

FIG. 1–1   Some of the physicists who investigated the structure of the atom. (Photos: Thomson, courtesy of Argonne
National Laboratory; Curie, AIP Niels Bohr Library; Rutherford, AIP Niels Bohr Library; Bohr, AIP Niels Bohr Library,
W. F. Meggers Collection; Moseley, AIP Niels Bohr Library, Burndy Library; Chadwick, AIP Niels Bohr Library, Meggers
Gallery of Nobel Laureates)

Introduction

The discovery of X-rays in 1895 by the German physicist Wilhelm Roentgen marked the beginning of 30 years or so of monumental developments unparalleled in the history of science. "Here the great age opens. Physics becomes in those years the greatest collective work of science—no, more than that, the great collective work of art of the twentieth century."[1] One revolutionary discovery or theory rapidly followed another during that time. In 1896, the French physicist Henri Becquerel, looking elsewhere for the X-ray phenomenon, discovered that a uranium salt emitted penetrating rays of some kind. Becquerel had found radioactivity. Within the next several years, the extraordinary Polish-born scientist Marie Curie and her husband Pierre isolated two new radioactive elements. In 1897, an English researcher, J. J. Thomson, discovered the electron and showed that atoms contain electrons.

The turn of the century yielded a more complete understanding of the nature of light (1900) as well as Albert Einstein's theory of relativity (1905). Researchers made three major advances in 1913. Up to then, the atom was thought to be a homogenous sphere of matter. The English physicist Ernest Rutherford showed that, in fact, an atom is mostly empty space, that most of its mass is concentrated in its nucleus, a sphere only one ten-thousandth its size with electrons occupying the region that surrounds the nucleus. Next, Niels Bohr, a Danish physicist, proposed a planetary model of the hydrogen atom in which a single electron travels in a circular path around the tiny nucleus. Finally, Henry G. J. Moseley, carrying out experiments with X-rays, determined a way to find the number of electrons in an atom. In 1919, Rutherford made two major discoveries in a single experiment: He found (1) that the nucleus of the atom has structure and that it contains particles called protons, and (2) that he could change atoms of nitrogen into atoms of oxygen, a process called transmutation of the elements. In 1932, James Chadwick, one of Rutherford's collaborators, showed that the

Wilhelm Roentgen

CHAPTER 1

The Structure of the Atom

nucleus contains yet other particles called neutrons.

This listing, by no means complete, centers on a number of key discoveries that revealed the submicroscopic world of the atom and its nucleus. Figure 1–1 shows some of the physicists responsible for the excitement generated during this 30-year period. In this chapter, we shall be looking at their work to see how the concept of the atom evolved.

Elements and Atoms

Hundreds of thousands of varieties of matter exist on this earth. Most of these substances can be broken down into several components. However, 88 substances, called **chemical elements,** resist any attempts by chemists to reduce them to simpler components. All of the other substances can be broken down into their respective chemical elements. For example, water consists of the elements hydrogen and oxygen. If you have a container of hydrogen and oxygen gases, and you apply a small flame or spark to this mixture, you will cause the hydrogen and oxygen gases to combine to form water (and you will initiate an explosion as well). The metals iron, copper, and zinc are also chemical elements. The rust on a piece of iron is caused by oxygen from the air combining with the iron to form a compound called *iron oxide.* You may have seen a copper church dome that has turned green due to the formation of *copper oxide* (oxygen combined with copper). Brass is also a metal, but it consists of the alloyed elements copper and zinc.

Consider now a sample of an element, a piece of copper, say, and imagine that we can somehow cut it into smaller and smaller pieces. We cannot continue this subdivision indefinitely, for eventually the copper pieces will become so incredibly tiny that they absolutely cannot be

cut in half. These indivisible units of copper would be copper atoms. An **atom** of an element is the smallest unit of matter that retains the properties of that element.

When one or more elements combine to form a chemical compound, their atoms join to form a larger unit called a **molecule.** Thus, the smallest unit of a chemical compound is a molecule. For example, the familiar symbol H_2O, which represents the chemical compound water, indicates that 2 atoms of hydrogen (H) combine with 1 atom of oxygen (O) to form a molecule of water. In ordinary table salt, NaCl, 1 atom of sodium (Na) combines with 1 atom of chlorine (Cl) to form a molecule of sodium chloride. A molecule of common table sugar, $C_{12}H_{22}O_{11}$, has 12 atoms of carbon (C), 22 of hydrogen, and 11 of oxygen. Molecules of hydrogen gas H_2 contain 2 atoms of hydrogen.

We owe this connection between atoms and molecules to John Dalton (1766–1844), an English schoolteacher. In 1808, he proposed the following atomic theory, which gave great insight into the results of many chemical experiments:

1. The most distinctive feature of an atom is its mass.
2. The atoms of a given chemical element all have the same mass.
3. Atoms of different elements have different characteristic masses.

With this theory, Dalton was able to determine the relative masses of atoms of different elements from the results of chemical experiments. For example, he knew from the work of others that 1 gram (abbreviation: g) of hydrogen combines completely with 8 of oxygen to form 9 g of water, with no atoms of either element left over after the combining took place. Similarly, 0.1 g of hydrogen combines with 0.8 g of oxygen. He asked, Would not one-millionth of a gram of hydrogen combine with eight-millionths of a gram of oxygen? Going one step further, he raised the question, Would not 1 unit of hydrogen combine with 1 unit of oxygen, if the unit of oxygen were 8 times more massive?

Dalton argued that these basic units were atoms of the elements. Using the simplest pos-

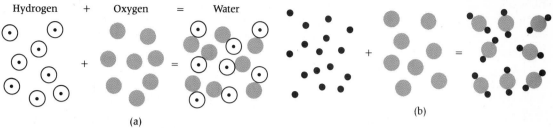

FIG. 1–2 Atoms of hydrogen combine with atoms of oxygen to form molecules of water. No atoms of hydrogen or oxygen are left over after the water forms. (a) How John Dalton viewed this process. Dalton assumed that 8 hydrogen atoms combined with 8 oxygen atoms to form 8 molecules of water. He thought—incorrectly as we now know—that a molecule of water contains 1 atom each of hydrogen and oxygen. (b) The correct way in which 16 atoms of hydrogen combine with 8 atoms of oxygen to form 8 water molecules. The chemical formula H_2O shows that a water molecule contains 2 atoms of hydrogen and 1 atom of oxygen.

sible combination, he assumed (incorrectly, as it turned out) that molecules of water consist of 1 atom of hydrogen and 1 atom of oxygen. Dalton used a circle with a dot in the center to symbolize hydrogen and an open circle for oxygen. Figure 1–2(a) shows how hydrogen and oxygen atoms combine under Dalton's assumption. There are, naturally, the same number of molecules of water as there were initially atoms of hydrogen and of oxygen. If the hydrogen sample had a mass of 1 g and the oxygen a mass of 8 g, then 1 atom of oxygen would have had a mass 8 times that of 1 atom of hydrogen.

Actually, of course, 2 atoms of hydrogen and 1 atom of oxygen make up a molecule of water. Figure 1–2(b) shows the correct way in which atoms of hydrogen and oxygen combine to form water. Here, we see that 1 atom of oxygen is 8 times as massive as 2 atoms of hydrogen, or 1 atom of oxygen is 16 times more massive than 1 atom of hydrogen. To specify the *relative* masses of atoms of elements, we compare them to hydrogen, which is the lightest atom. The mass of 1 hydrogen atom is denoted as 1 **atomic mass unit** (abbreviation: amu). Thus, the mass of an oxygen atom is 16 amu.

To cite another example: Suppose that 84 g of nitrogen combine with 96 g of oxygen to form a chemical compound in which the molecules have 1 atom of nitrogen and 1 atom of oxygen. Since the nitrogen and the oxygen samples have the same number of atoms, 1 atom of nitrogen is less massive than 1 atom of oxygen. In fact, an atom of nitrogen has a mass equal to 84/96 times that of oxygen. Since the mass of 1 oxygen is 16 amu, then nitrogen's atomic mass must be 84/96 × 16 amu, or 14 amu.

Many of Dalton's atomic masses were in error because he assumed incorrect chemical formulas for chemical compounds. However, he pointed in the right direction by showing that the relative atomic masses of elements could be determined, and, in the years that ensued, chemists followed his lead and did just that.

THE PERIODIC TABLE AND RELATIVE MASS

By about 1860, the number of elements that had been identified stood at 63. Some of these seemed to fall into families with similar chemical and physical properties. For example, fluorine, chlorine, bromine, and iodine form one family called the *halogens*. When halogens combine with hydrogen, compounds called *acids* are formed. Lithium, sodium, and potassium form another family called *alkali metals*. All alkali metals react violently with water, and their solutions are said to be *alkaline*.

THE STRUCTURE OF THE ATOM

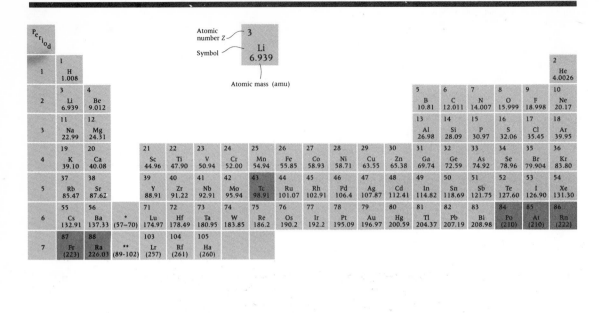

FIG. 1–3 Periodic table of the elements. Elements with atomic numbers equal to 84, 86, 88, 89, 90, 91, and 92 are radioactive, but they do occur naturally. Those with atomic numbers equal to 43, 61, 85, 87, and greater than 92 are also radioactive, but they must be produced artificially. Shaded squares indicate radioactive elements.

A Russian chemist, Dimitri Mendeleev (1834–1907), related the chemical and physical properties of these families of elements to their atomic masses. In 1871, he arranged the elements in a table so that those having similar chemical properties appeared in the same column. To do this, he had to leave gaps in his table for elements that had not yet been discovered, and he even predicted their properties (quite correctly, as later research substantiated). Figure 1–3 shows a modern version of the **periodic table of the elements.** The elements are listed in order of increasing atomic mass, and the **atomic number,** which is represented by the symbol Z, indicates an element's place in this listing. Table 1–1 lists the element's name, chemical symbol, and atomic number Z.

Figure 1–3, the periodic table, shows that the alkali metals—lithium (Li), sodium (Na), potassium (K), and so forth—appear in the first column, and that the halogens—fluorine (F), chlorine (Cl), and so forth—appear in the second-to-last column. The last column contains the inert gases—helium (He), neon (Ne), argon (Ar), and so on. These are called *inert* because they refuse to form chemical compounds with other elements.

The periodic table also reveals that most of the atomic masses are very close to the whole number: 1.008 amu for hydrogen, 15.999 amu

TABLE 1–1   THE ELEMENTS IN ALPHABETICAL ORDER

ELEMENT	SYMBOL	ATOMIC NUMBER (Z)	ELEMENT	SYMBOL	ATOMIC NUMBER (Z)
Actinium	Ac	89	Mercury	Hg	80
Aluminum	Al	13	Molybdenum	Mo	42
Americium	Am	95	Neodymium	Nd	60
Antimony	Sb	51	Neon	Ne	10
Argon	Ar	18	Neptunium	Np	93
Arsenic	As	33	Nickel	Ni	28
Astatine	At	85	Niobium	Nb	41
Barium	Ba	56	Nitrogen	N	7
Berkelium	Bk	97	Nobelium	No	102
Beryllium	Be	4	Osmium	Os	76
Bismuth	Bi	83	Oxygen	O	8
Boron	B	5	Palladium	Pd	46
Bromine	Br	35	Phosphorus	P	15
Cadmium	Cd	48	Platinum	Pt	78
Calcium	Ca	20	Plutonium	Pu	94
Californium	Cf	98	Polonium	Po	84
Carbon	C	6	Potassium	K	19
Cerium	Ce	58	Praseodymium	Pr	59
Cesium	Cs	55	Promethium	Pm	61
Chlorine	Cl	17	Protactinium	Pa	91
Chromium	Cr	24	Radium	Ra	88
Cobalt	Co	27	Radon	Rn	86
Copper	Cu	29	Rhenium	Re	75
Curium	Cm	96	Rhodium	Rh	45
Dysprosium	Dy	66	Rubidium	Rb	37
Einsteinium	Es	99	Ruthenium	Ru	44
Erbium	Er	68	Rutherfordium	Rf	104
Europium	Eu	63	Samarium	Sm	62
Fermium	Fm	100	Scandium	Sc	21
Fluorine	F	9	Selenium	Se	34
Francium	Fr	87	Silicon	Si	14
Gadolinium	Gd	64	Silver	Ag	47
Gallium	Ga	31	Sodium	Na	11
Germanium	Ge	32	Strontium	Sr	38
Gold	Au	79	Sulfur	S	16
Hafnium	Hf	72	Tantalum	Ta	73
Hahnium	Ha	105	Technetium	Tc	43
Helium	He	2	Tellurium	Te	52
Holmium	Ho	67	Terbium	Tb	65
Hydrogen	H	1	Thallium	Tl	81
Indium	In	49	Thorium	Th	90
Iodine	I	53	Thulium	Tm	69
Iridium	Ir	77	Tin	Sn	50
Iron	Fe	26	Titanium	Ti	22
Krypton	Kr	36	Tungsten	W	74
Lanthanum	La	57	Uranium	U	92
Lawrencium	Lw	103	Vanadium	V	23
Lead	Pb	82	Xenon	Xe	54
Lithium	Li	3	Ytterbium	Yb	70
Lutetium	Lu	71	Yttrium	Y	39
Magnesium	Mg	12	Zinc	Zn	30
Manganese	Mn	25	Zirconium	Zr	40
Mendelevium	Md	101			

for oxygen, 22.99 amu for sodium, and so forth. The integer obtained by rounding off the atomic mass to a whole number is called the **atomic mass number,** which is represented by the symbol A. For hydrogen, the atomic mass number is equal to 1 ($A = 1$); for oxygen, A equals 16; and, for sodium, A equals 23. There are some notable exceptions in which the atomic masses of a few elements are not as close to a whole number as are the majority of atomic masses. For example, neon has an atomic mass of 20.18 amu, and chlorine has an atomic mass of 35.45 amu. Near the end of the chapter, we shall consider these exceptions again and also apply the concept of the atomic mass number A.

AVOGADRO'S NUMBER

A very important principle is hidden in the periodic table of the elements. To uncover it, let's consider a concrete example first.

Suppose we have two boxes, one containing marbles (mass equal to 5 g each) with a total mass of 500 g, and the other containing golf balls (mass equal to 50 g each) with a total mass of 5000 g. Each box must contain 100 balls, since the masses of a marble and a golf ball are different by a factor of 10, and the masses of the boxes (neglecting the mass of the box itself) are different by the *same factor,* 10.

Now, let's apply this situation to elements and their atoms. Suppose we have 12 g of carbon, 24 g of magnesium, and 48 g of titanium. The atomic masses, rounded off to the nearest integer for simplicity, tell us that the mass of a magnesium atom is 2 times that of a carbon atom, and the mass of a titanium atom is 4 times that of carbon. Since the masses of the samples differ by the same factor as the masses of the atoms, the samples must all contain the same number of atoms. We can extend this argument to all

elements in the periodic table and arrive at an extremely important principle that was first realized by the Italian chemist and physicist Count Amedeo Avogadro (1776–1856): *A sample of an element with a mass in grams equal to its atomic mass (amu) contains N_A atoms,* where N_A is called **Avogadro's number.** The value of Avogadro's number is 6.02×10^{23}. (See Appendix A for a discussion of power-of-ten notation and Appendix B for the metric system of units.) For example, 1.008 g of hydrogen has N_A atoms, and so does 15.999 g of oxygen.

CALCULATING THE MASS AND DIAMETER OF AN ATOM

We can now easily find the masses of atoms. Since 1.008 g of hydrogen contains N_A atoms, each individual hydrogen atom has a mass given by 1.008 g divided by N_A (1.008 g/N_A), or 1.66×10^{-24} g. The mass of an oxygen atom, for example, is about 16 times that of a hydrogen atom, and so on.

Our next step is to calculate the diameter of an atom, but, to do that, we first have to consider the concept of **density.** Density, a characteristic of all substances, is defined as the mass per unit volume. Figure 1–4 shows that two aluminum blocks of differing mass and volume have the same density. Indeed, each element has a characteristic density. For example, the density of iron is 2.9 times that of aluminum; copper, 3.3 times; lead, 4.2 times; and gold, 7.2 times. (Recall the scene from the film *Goldfinger* in which James Bond lifts a gold vase effortlessly. If that vase had really been gold, even Agent 007 would have had to flex a muscle or two to lift it.)

Rather than in the random way that marbles fill a box, the aluminum atoms in the blocks of Fig. 1–4 are arranged in a regular pattern. (See Fig. 1–5.) In fact, most solids are *crystalline,* meaning that the atoms are arranged in a regular fashion. The cubic shape of the sodium chloride crystal (Fig. 1–6) surely points to the orderly arrangement of its atoms.

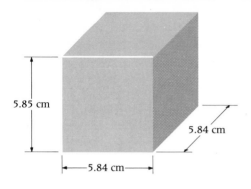

5.85 cm

5.84 cm

5.84 cm

Mass = 540 g

Volume = 5.85 cm × 5.84 cm × 5.84 cm
= 200 cm^3

Density = $\dfrac{\text{mass}}{\text{volume}}$ = $\dfrac{540\text{ g}}{200\text{ cm}^3}$
= 2.70 g/cm^3

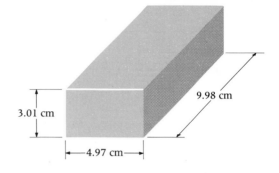

9.98 cm

3.01 cm

4.97 cm

Mass = 405 g

Volume = 3.01 cm × 4.97 cm × 9.98 cm
= 149 cm^3

Density = $\dfrac{405\text{ g}}{149\text{ cm}^3}$
= 2.72 g/cm^3

FIG. 1–4   If we are given the mass and the dimensions of two aluminum blocks, we can determine the density of aluminum. (Density is defined as mass per unit volume.) The slightly different density values derived for each block are due to uncertainties in the measurements of mass and volume. The density of aluminum is 2.70 g/cm^3. Shape makes no difference—the density of a substance is always the same.

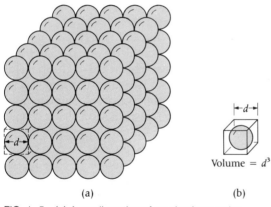

$\leftarrow d \rightarrow$

Volume = d^3

(a)   (b)

FIG. 1–5   (a) A small section of an aluminum cube, greatly magnified. (b) The volume occupied by the atom-cube, which accounts for the space occupied by the atom itself as well as its share of the empty space surrounding the atom. The symbol d represents the diameter of the aluminum atom. The 125 aluminum atoms in (a) occupy a volume of 125 d^3.

FIG. 1–6   Sodium chloride crystals. (Courtesy of Institute of Geological Sciences, London)

Figure 1–5 shows that an aluminum cube can be divided into atom-cubes. Thus, the volume of the aluminum cube is given by:

$$\text{Volume of cube} = \text{number of atoms in cube} \times \text{volume of atom-cube.} \quad (1)$$

We know the volume of the aluminum cube from Fig. 1–4, and we can calculate the number of atoms in it. Since the aluminum cube has a mass of 540 g (and we know that 27 g contains N_A atoms), the aluminum cube contains 20 N_A atoms, or 1.204×10^{25} atoms. From Equation 1, we find the volume of the atom-cube, represented by d^3, to be

$$d^3 = \frac{\text{volume of cube}}{\text{number of atoms in cube}}$$
$$= \frac{200\,\text{cm}^3}{1.204 \times 10^{25}}$$
$$= 16.6 \times 10^{-24}\,\text{cm}^3 \quad (2)$$
$$d = 2.55 \times 10^{-8}\,\text{cm,}$$
$$= 2.55 \times 10^{-10}\,\text{m.}$$

We can carry out similar calculations for other elements and will always find that the size of the atom is on the order of 10^{-10} meters (abbreviation: m). The sizes of atoms, drawn to scale in Fig. 1–7, show the results of such calculations.

Atoms and Electric Charge

We are all familiar with the static electricity that occurs when two different materials are rubbed together: the static cling of laundry placed in a clothes dryer, the shock you feel at turning on a light switch after just walking across a carpeted floor, and so on. To illustrate some properties of **electric charge,** let us consider another common situation, a girl combing her hair (see Fig. 1–8a). Although the comb and hair were initially neutral, the friction between them separates the positive and negative charges: The comb becomes negatively charged and the hair positively charged by the same amount. "Unlike charges attract and like charges repel" is an echo from childhood, and we see these effects here. The fly-away hair is due to the repulsion between the positive charges on the hair. The positively charged hair is likewise attracted to the negative charge on the comb.

Figure 1–8b shows yet another effect: The negatively charged comb is able to pick up a small piece of paper, even though the paper is neutral. When the comb is brought near the paper, the negative charge in the paper, repelled by that on the comb, moves away. This results in a positive charge nearer to the comb, which is attracted to the negative charge on the comb.

Thus, although the net charge on matter is zero (it is neutral), matter is composed of equal amounts of positive and negative charge. This fact means that, since matter is composed of atoms, an atom has equal amounts of positive and negative charge, while having no net charge. However, atoms of different elements contain different amounts of positive and negative charge. For example, an atom of hydrogen contains 1 positive charge and 1 negative charge, and an atom of oxygen contains 8 positive charges and 8 negative charges. Later in this chapter, we shall see how the charges in an atom are determined.

The nature of electric charge was finally understood when J. J. Thomson discovered the *electron* in 1897. Thomson showed that the electron has a negative charge and that atoms contain electrons (as well as the complementary amount of positive charge). The electron carries the smallest unit of charge, and we shall frequently denote this charge by -1. (During the last 10 years or so, physicists have searched exhaustively for particles called *quarks*, which are expected to carry charges only one-third of

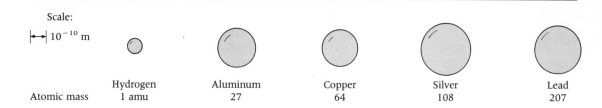

Scale:

$\longmapsto$ 10^{-10} m

	Hydrogen	Aluminum	Copper	Silver	Lead
Atomic mass	1 amu	27	64	108	207

that on the electron. So far, they have not been successful in finding quarks.)

The charge on other particles is compared to that on the electron. A particle having a +2 charge has twice the charge as that of an electron, but, since it has the opposite charge as an electron, it is attracted to the electron. In fact, the **sign** of a charged particle is really a bookkeeping device to indicate whether a given particle is attracted to or repelled by an electron. The amount of charge specifies the strength of this attraction or repulsion. A neutral particle is simply unaffected by an electron.

That atoms are neutral, but composed of positive and negative charges, leads to a better understanding of what's going on in Fig. 1-8(a). When the comb is pulled through the hair, electrons are pulled off the atoms in the hair and travel onto the comb, leaving the hair with the remaining positive charge of the atoms. In Fig. 1-8(b), electrons travel away from the negative charge on the comb, leaving behind the positive charge.

We know from our review so far that, throughout the nineteenth century, scientists had learned a great deal about the nature of the atom. We'll now focus on some of the experiments begun at the end of that century. These extended knowledge about the atom appreciably.

Elements and Light

During the late 1800s, many scientists were investigating the light emitted by various elements. What especially intrigued them was that each element emitted a characteristic color, or, more accurately, a characteristic *set* of colors.

FIG. 1–7 The diameters of several atoms drawn to scale. These diameters were calculated in the same way as that of aluminum mentioned in the text. Even though a copper atom is more massive than an aluminum atom, its size is smaller than aluminum, which shows that copper is denser than aluminum. The atomic masses are rounded off to the nearest integer.

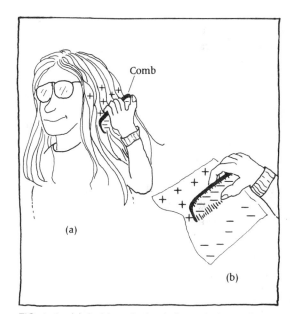

FIG. 1–8 (a) A girl combs her hair, producing static electricity. The friction between the comb and the hair separates the positive and negative charges, so that the comb becomes charged negatively and the hair positively by the same amount. (b) The negatively charged comb can pick up a small piece of paper, even though the paper is neutral.

THE STRUCTURE OF THE ATOM

11

TABLE 1–2   THE COLORS OF THE SPECTRUM, AND THE WAVELENGTHS (IN NANOMETERS) ASSOCIATED WITH EACH COLOR

COLOR	RANGE OF WAVELENGTHS (nm)
Ultraviolet	less than 400
Violet	400–450
Blue	450–500
Green	500–570
Yellow	570–590
Orange	590–610
Red	610–700
Infrared	greater than 700

(Visible spectrum brackets Violet through Red)

The red glow of a neon sign, the violet hue of a mercury-vapor streetlight, and the yellow glow of a sodium-vapor streetlight are familiar examples of this phenomenon. How can one explain these effects? At that time, the only known difference between atoms of, say, neon and mercury was that an atom of mercury was about 10 times more massive and, of course, larger as well. Why should neon atoms emit red light, while mercury atoms emit violet light?

The color of light, red or violet, is a physiological response of our eyes to light of differing **wavelengths.** Light travels as a wave, much like a water wave travels over the surface of the water. The wavelength is defined as the distance between two adjacent peaks or troughs of a wave (Fig. 1–9). Unlike water waves, the wavelength of light is extremely small, being only about 5000 times larger than the size of an atom. The wavelength of light is given in nanometers (abbreviation: nm), where 1 nanometer is equal to 10^{-9} meters (1 nm = 10^{-9} m). The ranges of the wavelength associated with each color are listed in Table 1–2.

To study the light emitted by elements, the researchers placed a gas inside a specially designed glass tube called a **cathode ray tube.** They then connected the electrodes (the metal pieces that pass through the glass) to a high

voltage source. The electrode connected to the negative terminal of the high voltage is called the **cathode,** and the one connected to the positive terminal is called the **anode.** When the gas in the tube is at atmospheric pressure, nothing happens. Interestingly, when the workers pumped some of the gas out to the point where the internal pressure was only one one-hundredth of the **atmospheric pressure,*** they found that the gas began to glow (Fig. 1–10). When the pressure was decreased to one one-hundredth of atmospheric pressure, only one one-hundredth of the gas atoms originally in the tube still remained.

The researchers naturally asked, Why is light emitted by connecting a high voltage to the electrodes? They knew that the cathode becomes negatively charged and the anode positively charged. When the electrodes, initially uncharged, are connected to the high voltage, negative charges (electrons) leave the anode and flow onto the cathode. Hence, the cathode becomes negatively charged. The deficit of negative charge on the anode means that it is positively charged. But how this charge on the electrodes affects the gas atoms was unknown, nor did the investigators have any idea of how atoms go about emitting light. When confronted with many questions, the first step scientists take is to obtain data. They then examine the data to see if any patterns or regularities

---

*Gravity is responsible for retaining the atmosphere that surrounds the earth and for the density of air molecules at the earth's surface. This density (number/cm^3) of air molecules in turn governs what we call *atmospheric pressure.* Atmospheric pressure is due to air molecules striking and rebounding from an object immersed in the atmosphere, much like a ball striking a wall and rebounding. When we say a vessel is filled with a gas (not necessarily air) at atmospheric pressure, we mean that the pressure generated by the gas molecules striking the walls of the vessel is equal to the pressure on objects in the atmosphere. For example, when the vessel is evacuated so that only 10 percent of the gas molecules remain, the internal pressure is one-tenth of atmospheric pressure due to correspondingly fewer gas molecules striking the walls. The pressure is proportional to the number of molecules in the vessel. Since scientists have instruments to measure the pressure, it is usually the pressure that is specified, rather than the density of molecules remaining in the vessel.

FIG. 1–9 Section of a wave. The wavelength λ is the distance between two adjacent peaks, or between two adjacent troughs.

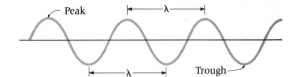

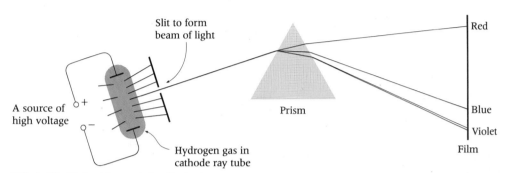

FIG. 1–10 Method for capturing the hydrogen spectrum on film.

emerge. If so, then perhaps some insight (with subsequent interpretation) will be provided into what's going on.

In their study of light, the scientists directed a beam onto a glass prism. After passing through the prism and emerging from the other side, the light broke up into a spectrum of colors. A beam of sunlight, for instance, will show a **continuous spectrum** of colors, ranging from red at one end, through orange, yellow, green, and blue, to violet at the other end. However, the spectrum of light from an element placed in a cathode ray tube shows quite a different behavior: It is not continuous, but consists only of certain very definite colors. We call this phenomenon a **line spectrum.** The line spectrum for hydrogen (Fig. 1–10) shows only four colors: red, blue, and two slightly different shades of violet. (The spectrum can be captured by exposing photographic film.) Figure 1–11 shows the line spectra of hydrogen, helium, sodium, and mercury. In each case, only very narrow regions—lines—on the film are exposed. No light reaches the rest of the film. Since each position on the film corresponds to a certain wavelength, the wavelengths of the spectral lines can be determined. The hydrogen line spectrum has

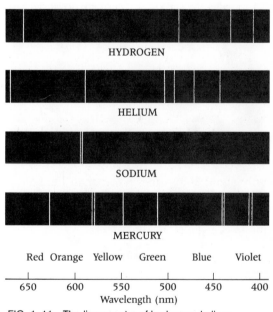

FIG. 1–11 The line spectra of hydrogen, helium, sodium, and mercury.

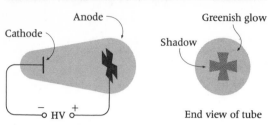

FIG. 1–12   A cathode ray tube showing the shadow cast by cathode rays striking the anode. The glass fluoresces with an apple-green color when struck by cathode rays. HV stands for *high voltage*.

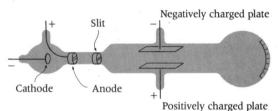

FIG. 1–13   Cathode ray tube used by J. J. Thomson to discover the electron.

the following wavelengths: 656.2 nm (red), 486.1 nm (blue), 434.0 nm (violet), and 410.1 nm (violet).

In 1885, the German scientist Johann Balmer (1825–1898) found a mathematical formula that accounts for all four wavelengths with great accuracy. This formula is

$$\lambda = 364.56\,nm \times \frac{n^2}{n^2 - 2^2},\qquad (3)$$

where λ (symbol for the lowercase Greek letter lambda) is the wavelength of a spectral line. The four wavelengths are found using n equals 3 for red, n equals 4 for blue, and n equals 5 and 6 for the two violet lines. For example, when n equals 3, the wavelength is 364.56 nm × 9/5, which equals 656.2 nm. The fact that one equation can describe all four wavelengths indicates that there is some regularity, or some relationship, among the four wavelengths. Yet, this formula by itself does not help us with the basic question, What happens when a hydrogen atom emits light?

The answer to this question, which takes us deeper into atomic structure, has an interesting history that we'll examine shortly. First, though, let's continue to stay with the development of cathode ray experiments and see what more this area revealed about the structure of the atom.

The Discovery of the Electron

In the 1890s, scientists began studying what happens when the pressure in a cathode ray tube is reduced to one ten-thousandth of atmospheric pressure. Here, a gas ceases to emit light (or, at least, we cannot detect it because so few atoms are left at this pressure). Researchers found that the end of the tube near the anode started to glow or fluoresce with an apple-green color, except where a shadow of the cross-shaped anode appeared (Fig. 1–12). Apparently, the cathode emits rays, called **cathode rays,** that travel toward the anode. Some hit the glass, causing it to emit green light, but others strike the anode. Therefore, a shadow appears on the end of the cathode ray tube.

What are these cathode rays? Are they particles or waves? Are they charged or neutral? If charged, are they positive or negative? These questions plagued J. J. Thomson (1856–1940), who set out to find answers, using the apparatus shown in Fig. 1–13.

When the tube is operating, the cathode rays travel from the cathode to the anode, where some pass through a slit in the anode and then through another slit, forming a narrow beam that travels the length of the tube. Thomson could observe the behavior of the beam because it hit a fluorescent screen at the end, causing a bright spot to appear there.

When Thomson passed the beam between two plates, one charged negatively and the other positively, he found that the bright spot shifted downward from the center. That is, the cathode ray beam deflected downward because it was attracted to the positive charge and was repelled by the negative charge. This meant that cathode rays—now called **electrons**—are negatively charged particles. From these observations and similar ones using a magnetic field, Thomson was also able to show that electrons have a mass less than one one-thousandth that of a hydrogen atom. *The hydrogen atom could no longer be considered the smallest unit of matter.* (Today, we know that an electron has a mass only 1/1833 that of a hydrogen atom.)

But why does the cathode emit rays in the first place? To find out, Thomson placed different gases in the cathode ray tube. He also used a variety of metals for the cathode itself. Despite these variations, his results were always the same. Thomson concluded that *all* atoms contain electrons. In addition, since an atom as a whole is neutral and the electrons have a negative charge, it followed that the atom must also have a positive part. The cathode rays result when gas atoms near the cathode are ripped apart. The electron is strongly repelled by the negatively charged cathode, while the positive part of the atom is attracted to the cathode. The electron then travels to the anode, and the positive part, called a *positive ion,* goes to the cathode (Fig. 1–14). Since

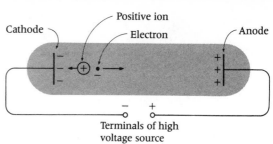

FIG. 1–14 Schematic diagram of a cathode ray tube. The electrodes at each end of the cathode ray tube are initially uncharged. When the high voltage is connected between them, negative charges leave the anode and flow onto the cathode. As a result, the anode is positively charged, and the cathode is negatively charged. The diagram shows a gas atom being ripped apart because the electron in the atom is strongly repelled by the negative charge on the cathode, while the positive charge is attracted to it. The freed electron travels to the anode. This process is responsible for the production of cathode rays (now called electrons) in the cathode ray tube.

THE STRUCTURE OF THE ATOM

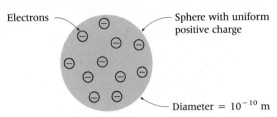

FIG. 1–15 Thomson's model of the atom.

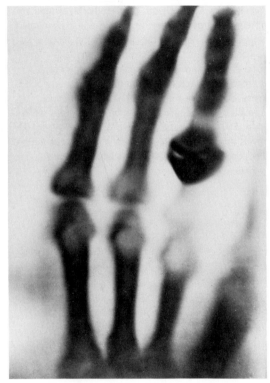

FIG. 1–16 Roentgen's first X-ray photograph. (Courtesy of Deutsches Museum)

all atoms contain electrons, it didn't make any difference if the cathode ray tube contained air, neon, or hydrogen.

To account for his observations, Thomson proposed a model of the atom in which the tiny lightweight electrons were imbedded in a sphere of positively charged matter (see Fig. 1–15). According to Thomson, most of the mass of the atom was due to this positive sphere, which had a diameter of about 10^{-10} m (the known size of the atom). Usually, Thomson said, the electrons were motionless and did not emit light. But when the electrons oscillated about their normal position, light should be emitted. Thomson, however, could *not* account for the wavelengths of hydrogen's lines, even though the hydrogen atom contained only one electron.

The discovery of the electron's existence was a major advance along the difficult road that led to the uncovering of the atom's structure. Yet other significant information came from those working with cathode ray tubes: more knowledge about the nature of the atom itself, and a spin-off by-product that fundamentally altered the field of medicine.

X-Rays and the Atom

When Wilhelm Roentgen (1845–1923) reported his discovery of rays that could penetrate human flesh to show the bones within, his audience at the Würzburg Medical Society saw a whole new world of diagnostic possibilities open up. For the first time, physicians would be able to see inside the body without surgery. Within months, doctors were using these rays routinely for diagnosis.

Roentgen discovered **X-rays** quite accidentally, while studying the mysterious cathode rays (electrons). In his darkened laboratory, he noticed that a fluorescent screen, lying on a nearby bench, glowed when the cathode ray tube was on and stopped glowing when the tube was switched off. This phenomenon happened even after he covered the tube with black cardboard. Roentgen knew that a fluorescent material (such as that which comprises the coating of a picture tube in a TV set) emits light when struck by radiation or by a stream of particles. So, he concluded that some kind of ray was coming from the glass tube, passing through the cardboard, and traveling to the fluorescent screen. He called these rays X-rays after the mathematical symbol *"x,"* which stands for an unknown quantity. Further experiments showed that the *X-rays were produced when electrons struck the anode or the glass wall of the tube.* Thus, X-rays and cathode rays were completely different.

Roentgen found that X-rays are extremely penetrating. Even with a piece of wood an inch thick between the switched-on tube and the screen, the screen continued to glow. A piece of aluminum five-eights of an inch thick dimmed the fluorescence because the metal absorbed some of the rays, but did not cut them off altogether. Roentgen's spectacular observation was that, "If the hand be held between the discharge tube and the screen, the darker shadow of the bones is seen within the slightly dark shadow image of the hand itself"[2] Figure 1–16 shows Roentgen's first X-ray photograph.

Roentgen found that X-rays exposed a photographic plate just as ordinary light rays do. When he placed his hand in front of the photographic plate, the bones absorbed more X-rays than the surrounding flesh. This difference in absorption produced a contrast in exposure and showed a shadow of the bones in the photograph.

Interestingly, some scientists before Roentgen had observed that photographic plates placed near a cathode ray tube became fogged, but instead of trying to analyze the problem, they complained to the plate manufacturer, who could not offer an explanation. Any one of these people might have discovered X-rays, but did not.

X-rays, then, are produced when energetic electrons strike a solid target. When the electrons are stopped, their energy is converted into X-rays, which are very similar to ordinary visible light, only about 5000 times more energetic. X-rays have a wavelength on the order of 0.1 nm.

The news of Roentgen's work amazed the public as no scientific discovery had before. Between the first announcement of the discovery on January 16, 1896, and June of that year, *The New York Times* alone published over 40 articles on X-rays (see Fig. 1–17). In such an atmosphere of excitement, fears abounded, as did speculations of impossible benefits. People feared, for instance, that, since X-rays could penetrate not only human flesh but brick walls, the privacy of their homes would be invaded. Some bought supposedly X-ray-proof clothing to protect their modesty. Among the wildest of the proposed benefits was that X-rays might bring the dead back to life!

Once found, X-rays became the subject of much experimentation. In 1913, H. G. J. Moseley (1887–1915) carried out experiments with X-rays using many different elements for the anode of the cathode ray tube. He found that the energy of X-rays depended upon the anode material. (We can think of the energy of X-rays as the ability to penetrate matter: the greater the energy, the greater the penetration.) Moseley's systematic analysis showed that the greater the atomic number Z of the anode material, the greater the energy of the X-rays it produced. Given this fact, Moseley deduced that *the atomic number Z corresponded to the number of electrons in the atom.* And since the atom is neutral, it must likewise contain Z positive charges as well.

The Discovery of the Nucleus

RADIOACTIVITY

Quickly following the discovery of X-rays was the discovery of radioactivity by Henri Becquerel (1852–1908). Becquerel found that ura-

January 16, 1896

HIDDEN SOLIDS REVEALED

Prof. Routgen's Experiments with Crookes's Vacuum Tube.

BULLETS FOUND BY USING LIGHT

Opaque Bodies Covered by Other Bodies Photographed — Views of Profs. O. N. Rood and Halleck of Columbia.

Men of science in this city are awaiting with the utmost impatience the arrival of European technical journals, which will give them the full particulars of Prof. Routgen's great discovery of a method of photographing opaque bodies covered by other bodies, hitherto regarded as wholly impenetrable by light rays of any kind.

Prof. Routgen of Würzburg University has recently discovered a light which, for purposes of photography, will penetrate wood, paper, flesh, and nearly all other organic substances. Thus, the bones of the human frame can be photographed in relief without the flesh which covers them, or metals inclosed in a box covered with a woolen cloth can be photographed as if the cloth and the wood did not exist.

In one sense, it is a misnomer to call the process photography, as now understood, because no lens is employed to project the image. It also seems, from the brief accounts of the process which have already been sent by cable, that the new images of concealed bodies resemble rather the old-fashioned daguerreotypes than the modern finished photographs, inasmuch as they appear only in outline.

Briefly, the new images are obtained by the energy given out in a Crookes's vacuum tube. The object to be photographed is placed behind the tube, and a dry plate is placed behind the object. If the object be, say, a hand, the image on the dry plate will be the bones in it, without any flesh covering whatever.

Prof. Routgen has already used his process to detect the exact location of bullets in gunshot wounds, and one of its first practical uses is expected to be a transformation of modern surgery by enabling the surgeon to detect the presence of foreign bodies of whatever kind in any part of the human body.

April 28, 1896

THE X RAY IN MEDICINE

SOME EXPERIMENTS MADE BEFORE DOCTORS OF THIS CITY.

Needle Plainly Seen in the Hand of a Woman Patient Who Had Suspected Its Presence—Shadowgraph of a Man's Hand Shows the Point of Fracture in a Finger—Stereopticon Views of Tests Made with Various Substances.

At the meeting of the Medical Society of the County of New-York, at the Academy of Medicine last night, Dr. William J. Morton gave a lecture of interest on "The X Ray and Some of Its Relations to Medicine." There were also demonstrations of apparatus at work and stereopticon views.

President Edward D. Fisher was in the chair, and the attendance was so large that many had to stand.

Dr. Morton, who was assisted by several experts, said that from time immemorial there has been a desire to fully explore the mysteries of the human body, and told of the various aids to such research up to the present time. It was no wonder, then, that the X ray has excited general interest, as no greater auxiliary to diagnosis has been given.

FIG. 1–17 Clippings about X-rays from *The New York Times*. (© 1896 by The New York Times Company. Reprinted by permission.)

nium ($Z = 92$) spontaneously emits penetrating rays; that is, uranium is **radioactive.** Marie Curie (1867–1934), who was searching for a suitable thesis topic for her doctorate, decided to investigate this curious phenomenon. Systematically, she tested all known elements and immediately found that thorium ($Z = 90$) was also radioactive. She found no others. Next, she tested minerals and found that pitchblende (an ore that contains uranium) was far more radioactive than either uranium or thorium. She reasoned that there must be as-yet-undiscovered radioactive elements in pitchblende. Using chemical separation procedures, she found two new elements, which she named radium and polonium (after her native Poland). After four years of work, Marie Curie finally obtained one-tenth of a gram of pure radium from one ton of pitchblende! She used this sample to determine the atomic mass of radium. Pierre Curie (Marie's husband) discovered that the rays from radium could be used to treat cancer. The Curies became famous as a result, but not wealthy. They did not seek a patent on the separation process, but published these procedures openly. Soon, **radiation therapy,** or **Curie-therapy** as it was called, was used all over the world in treating cancer.

ALPHA PARTICLES

Many scientists besides the Curies investigated the rays emitted by radioactive sources. Among them was Ernest Rutherford (1871–1937), who had studied with J. J. Thomson. Rutherford placed thin sheets of aluminum foil in front of a radioactive source and observed how many rays were able to pass through these sheets. One group of rays, which he called the **alpha rays,** was stopped completely by 4 sheets, while another group, the **beta rays,** could pass through 100 sheets. Rutherford was able to show that the alpha rays were positively charged particles. Then, in 1909, he and T. D. Royds captured alpha particles in a cathode ray tube, and the resulting spectrum was the same as helium. The

question was finally settled: *Alpha particles are helium atoms with two electrons missing so that they carry two positive charges.* Beta rays turned out to be electrons.

Gamma rays, a third emission from radioactive sources, have the same nature as light and X-rays and are characterized by their wavelength. Gamma rays have a wavelength even smaller than X-rays, and they are also more penetrating than X-rays. Today, the gamma rays from cobalt-60 are often used to destroy cancerous tumors.

The term *ray* was used initially to refer to cathode rays and X-rays, and then for alpha, beta, and gamma rays. Lest there be any confusion, let us clarify this term. The word *ray* was most likely borrowed from the familiar *light ray,* and anything whose nature was unknown was called a ray. For example, Thomson showed that a cathode ray was a negatively charged particle that is now called an electron. In this context, one cathode ray means one particle; that is, one electron. Both alpha rays and beta rays are particles, and each ray refers to one particle. In fact, an alpha ray is most commonly called an alpha particle, which is the phrase that we shall use. Convention also dictates that beta rays are still called beta rays, and are not called beta particles.

Before long, Rutherford realized that alpha particles could be a marvelous tool for probing the atom. No one had thought that it might be possible to test Thomson's model of the atom, but now Rutherford realized it could be done. He thought that if he scattered alpha particles from the atoms in a gold foil, the scattering pattern would give information (rather indirectly, of course) about the shape of the gold atom. In order to understand how Rutherford could learn about an invisible atom, consider the analogy generated by the computer program SCAT-

BOX 1-1   X-RAY CT-SCAN

In 1979, Godfrey N. Hounsfield and Allan M. Cormack shared the Nobel Prize in physiology and medicine for their imaginative development of the X-ray CT-scanner, which revolutionized the use of X-rays for medical diagnosis. Before the scanner, X-ray photographs differed little from those obtained by Roentgen himself. In his Nobel lecture, Hounsfield relates:

Some time ago I investigated the possibility that a computer might be able to reconstruct a picture from sets of very accurate X-ray measurements taken through the body at a multitude of different angles. Many hundreds of thousands of measurements would have to be taken, and reconstructing a picture from them seemed to be a mammoth task. . . When I investigated the advantages over conventional X-ray techniques, however, it became apparent that the conventional methods were not making full use of all the information X-rays could give. [3]

For example, a conventional X-ray photograph compresses a three-dimensional object into a two-dimensional picture and "cannot distinguish between soft tissues. . . Variations in soft tissues such as the liver and the pancreas are not discernible at all, and certain other organs

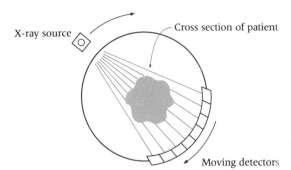

FIG. 1   A third generation X-ray CT-scanner. Many narrow beams from the X-ray source strike the patient. The X-ray source and X-ray detectors, one for each beam, rotate around the patient, while the source is turned on and off at regular intervals to get readings at desired angles.

may be rendered visible only through the use of radiopaque dyes." [4]

Figure 1 shows a schematic diagram of a third-generation X-ray CT-scan machine. Each detector records the intensity of X-rays reaching it, which indicates how much has been absorbed in traveling along *that* straight line path through the patient. This in turn indicates the average

---

TER,[5] using an Apple II computer (Fig. 1–18). The TV monitor shows how balls, aimed at an invisible object, are scattered. The program asks you to identify the shape and size of the obstacle.

Before he began his experiment, Rutherford determined how the alpha particles should be scattered by gold atoms, assuming that Thomson's model was correct. An alpha particle striking such a gold atom would change its direction by only about one-thirtieth of a degree,

because of the repulsion between the alpha particle and the positive sphere of the atom. Since an alpha particle would encounter many Thomson-model gold atoms, its direction upon leaving the foil would reflect this accumulated effect. Most alpha particles would be scattered only by a few degrees, but a few might be scattered to an angle as large as twenty degrees.

The first scattering experiment was carried out by Rutherford's inventive associate, Hans Geiger (1882–1945), the same person who invented the Geiger counter. Geiger used the setup shown in Fig. 1–19. He aimed a narrow beam of alpha particles at the gold foil and watched for flashes of light on the fluorescent screen behind the foil. These flashes marked density of matter along that straight line. Such a system would actually have from 300 to 500 detectors rather than the 8 shown in the illutration. After the data at one angle have been obtained, the X-ray source and detector array are advanced by about one-third of a degree, and the measurements are repeated. If the source and the 500-detector array are rotated through 180°, there are 540 angular positions and 500 absorption measurements at each angle, yielding a total of 270,000 absorption measurements.

All this data is fed into the computer. Solving a complex mathematical problem, the computer calculates (from the average densities) what the density at *each point* in the cross section must be in order to account for all these absorption measurements. Cormack showed how to solve this problem.

Figure 2 shows a computerized image of a section through a patient's abdomen, where the brightness at each point in the image corresponds to the density. The CT in CT-scan stands for *computer-assisted tomography*, which is what this process is called.

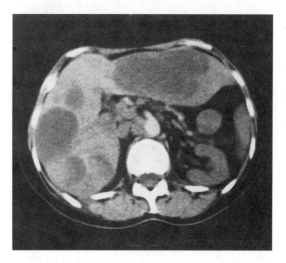

FIG. 2 X-ray CT-scan showing a cross section of the abdomen. The white area is the spine. The large colon is at the top, and the stomach is to the left. The darker areas, dark gray or black, are air- or fluid-filled cavities. This photograph shows the subtle differences in tissue densities that can be observed using X-ray CT-scans. (Courtesy of General Electric Medical Systems Division, Milwaukee)

particles striking the screen. He tallied the number of alpha particles scattered to various angles and found that the number fell off rapidly as the scattering angle increased. Still, it was not *exactly* the pattern predicted by Thomson's model.

Rutherford recalls what happened next:

One day Geiger came to me and said, 'Don't you think that young Marsden,* whom I am training in radioactive methods, ought to begin a small research?' Now I had thought that too, so I said, 'Why not let him see if any alpha particles can be scattered through a large angle?' I may tell you in

*Ernest Marsden (1884–1970), a student from New Zealand who had come to work with Rutherford.

confidence that I did not believe that there would be, since we knew that the alpha particle was a very fast massive particle, with a great deal of energy, and you could show that if the scattering was due to the accumulated effect of a number of small scatterings the chance of an alpha particle being scattered backwards was very small. Then I remember two or three days later Geiger coming to me in great excitement saying, 'We have been able to get some alpha particles coming backwards. . .' It was quite the most incredible event that has ever happened to

(a)

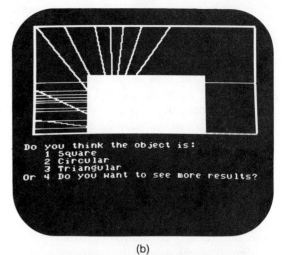

(b)

FIG. 1–18 The SCATTER program. (a) The TV monitor shows three different objects: a square, a circle, and a triangle. As the program proceeds, balls coming from the left will strike one of these objects. The size of the objects can vary. (b) The monitor now shows the scattering of the balls from an invisible object. The program asks you to identify the object from this scattering and then to estimate its size. (c) The object is revealed, and you can see how the object scatters the balls.

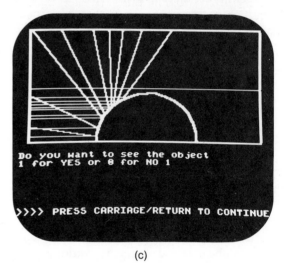

(c)

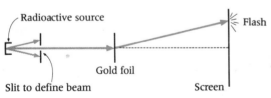

FIG. 1–19 An experimental setup for observing the scattering of alpha particles by a thin gold foil.

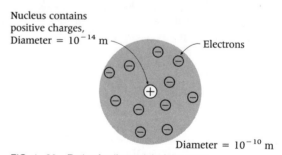

FIG. 1–20 Rutherford's model of the atom.

me in my life. It was almost as incredible as if you fired a 15-inch shell at a piece of tissue paper and it came back and hit you.[6]

To be sure, the number of alpha particles that scattered backwards (like a ball bounding off a wall) was very small: only 1 in 8000. Yet, the fact that *any* came back at all meant that Thomson's model must be wrong. The question then became, What must the atom look like to account for backward-scattering?

This question nagged Rutherford for two years, while he vigorously pursued many other lines of research, before he came up with an answer (the model in Fig. 1–20). Rutherford realized that the positive charge of the atom must be concentrated in a sphere only about 10^{-14} m in diameter—10,000 times smaller than in Thomson's model. This sphere of positive charge is called the **nucleus.** The electrons were contained in a larger surrounding sphere 10^{-10} m in diameter, conforming to the known size of the atom. This meant that an atom is mostly empty space! Thus, when alpha particles strike a gold foil, most of them pass through the empty spaces; only a very small fraction hit a nucleus head on. These few particles experience such a large repulsive force (due to the concentrated positive charge of the nucleus) that they will come to a stop and then rebound backwards.

Rutherford hadn't described the behavior of the electrons very well simply because the scattering of alpha particles gave him no information about them. The reason for this is that the mass of the alpha particle is about 8000 times that of an electron. When an alpha particle collides with an electron, the *electron's* direction changes, not the direction of the massive alpha particle. (For example, when a golf ball sails through the air, its direction doesn't change if it hits a bee, but the golf ball certainly will change course if it hits a tree. Obviously, the golfer will know if the ball hits the tree, but won't if it hits the bee.) When two objects collide, the less massive one suffers the greater effect. Since the alpha particle's direction does not change when it hits an electron, Rutherford could infer nothing about electron behavior.

A Model of the Hydrogen Atom

Niels Bohr (1885–1962), studying with Rutherford, was on the scene in England while the alpha scattering experiments were being carried out. Fascinated with the nuclear model, Bohr began searching for a connection between it and the light emitted by a hydrogen atom.

Shortly after returning to his native city of Copenhagen, Denmark, Bohr proposed that the single electron of the hydrogen atom revolved in a circular orbit around the positively charged nucleus. He further proposed that these orbits had certain "allowed" radii, with no other orbit sizes being permitted. He calculated that the smallest one, the so-called first allowed orbit, had a radius of 0.528×10^{-10} m. This seems reasonable because the size of the atom was known to be approximately 10^{-10} m. The second allowed orbit had a radius four times that of the first; the third, nine times. These three, of many **allowed orbits,** are shown in Fig. 1–21. According to Bohr's description, the electron can occupy any one of the allowed orbits, but tends to jump to an orbit with a smaller radius. The atom emits light each time the electron jumps between orbits.

Recall that researchers studied the light emitted by hydrogen by placing hydrogen in a cathode ray tube and connecting the electrodes to a source of high voltage. We can now link Thomson's discovery of the electron and Bohr's model. Hydrogen atoms near the cathode are torn apart due to the electron's repulsion to the (negative) cathode and the attraction of the nucleus to it. This freed electron, now attracted to the (positive) anode, will collide with hydrogen atoms as it travels toward the anode. As a result of the collision, the electron in the first allowed orbit of the hydrogen atom will jump to a larger orbit. Then it will start jumping to smaller allowed orbits until it again reaches the first allowed orbit, where it will remain until

THE STRUCTURE OF THE ATOM

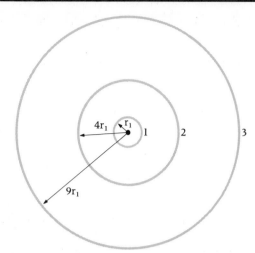

FIG. 1–21 Bohr's model of the hydrogen atom, showing three of many allowed orbits. The first allowed orbit has a radius r_1 of 0.528×10^{-10} m.

TABLE 1–3 WAVELENGTH OF LIGHT EMITTED BY THE HYDROGEN ATOM WHEN AN ELECTRON JUMPS FROM AN INITIAL ORBIT TO A SMALLER FINAL ORBIT

WAVELENGTH (nm)	COLOR	INITIAL ORBIT	FINAL ORBIT
656.2	Red	3	2
486.1	Blue	4	2
434.0	Violet	5	2
410.1	Violet	6	2

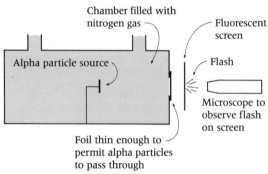

FIG. 1–22 The apparatus Rutherford was using when he discovered the proton.

the hydrogen atom is struck by another freed electron. Each time the electron jumps to a smaller orbit, the hydrogen atom emits light.

The most essential (and revolutionary) aspect of Bohr's theory was the criterion for selecting the radii of the allowed orbits. While Bohr's model with its allowed orbits may bring to mind our solar system with its planets orbiting the sun, there is one extremely important difference. With the right thrust, we can put an artificial satellite in *any* orbit around the sun; that is, *all* satellite orbits are allowed. In Bohr's model, however, *only certain* electron orbits are permitted. Using basic concepts of physics, Bohr calculated the radii of the allowed orbits and also the wavelength of light emitted when the electron jumps between allowed orbits. He thus derived the Balmer formula (Equation 3, p. 14) exactly! Table 1–3 shows some of Bohr's predictions, where the electron jumps from the larger or initial orbit to the smaller or final orbit. Bohr's theory solved the long-baffling problem of how atoms emit light.

Chapters 2, 3, 4, and 6 deal with the fundamental concepts that Bohr used in his model of the hydrogen atom. In each of these chapters, we shall use Bohr's model as an example to gain a deeper understanding of both the concepts and their important applications.

The Structure of the Nucleus: Protons and Neutrons

Rutherford had used alpha particles to discover the nucleus. A few years later, in 1919, that genius of experimental nuclear physics used them once again, in a single experiment, to discover the proton and transmutation of the elements. He placed an alpha particle source in a chamber containing nitrogen gas (Fig. 1–22). He could observe the particles that had passed through the very thin foil at the other end because they produced a flash of light on the fluorescent screen. In the course of these experiments, Rutherford placed aluminum foils in front of the fluorescent screen and increased

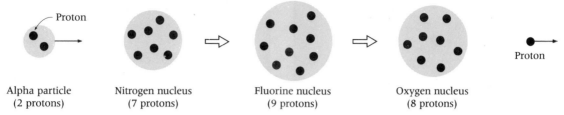

Alpha particle
(2 protons)

Nitrogen nucleus
(7 protons)

Fluorine nucleus
(9 protons)

Oxygen nucleus
(8 protons)

Proton

FIG. 1–23   The results of an alpha particle colliding with a nitrogen nucleus.

the number of foils until he couldn't see any flashes at all. What had happened was that the particles, after colliding with many aluminum atoms in the foil, had come to a stop or had been absorbed by the foil.

Rutherford was very surprised by the results: The thickness of aluminum was about *three times thicker* than that needed to stop alpha particles or nitrogen nuclei (set into motion by being hit by an alpha particle). From these results, he knew that these not-so-easily-absorbed particles must have less charge and less mass than an alpha particle because, in that case, they would interact less with the positive and negative charges in aluminum atoms. He found that these particles—now called **protons**—were nuclei of hydrogen atoms having a mass of about 1 amu and 1 unit of positive charge. (Recall that the alpha particle has a mass of about 4 amu and 2 units of positive charge.)

Rutherford reasoned that all nuclei must contain protons. For example, since the atomic number Z for nitrogen is 7, the nitrogen atom must have 7 electrons orbiting the nitrogen nucleus, which has 7 positive charges. Therefore, the nitrogen nucleus would contain 7 protons and, similarly, the alpha particle, 2 protons.

When an alpha particle collides with a nitrogen nucleus, a composite particle forms that contains 9 protons. But 1 of the protons breaks away from it, and a nucleus containing 8 protons, an oxygen nucleus, remains (Fig. 1–23).

This experiment was also the first instance of **the transmutation of elements.** By knocking loose that proton, Rutherford had changed nitrogen into oxygen. As these experiments

continued in collaboration with James Chadwick (1891–1974), these colleagues changed boron into carbon, fluorine into neon, and so on. The ancient alchemists, who dreamed of turning lead into gold, were merely born a few hundred years too soon.

It was thought that there must be other particles in the nucleus besides protons. The nitrogen nucleus, for example, has a mass of about 14 amu, but its 7 protons only account for about 7 amu. What particles account for the remaining mass of 7 amu?

In 1932, Chadwick identified these particles. They are called **neutrons,** neutral particles having about the same mass as a proton, 1 amu. Chadwick shot alpha particles at a beryllium target. As before, a composite particle formed, but, in this case, a neutron broke loose. He concluded that the alpha particle with its mass of 4 amu must contain 2 protons, to account for two units of positive charge and 2 amu, and 2 neutrons (to account for the remaining 2 amu). Similarly, the beryllium nucleus, having a mass of about 9 amu and 4 units of positive charge, should have 4 protons and 5 neutrons. These 9 particles, each one having a mass of 1 amu, would account for the mass and charge of the beryllium nucleus. The composite particle would then have 6 protons and 7 neutrons. After 1 neutron left, the resulting nucleus would have 6 protons and 6 neutrons.

THE STRUCTURE OF THE ATOM

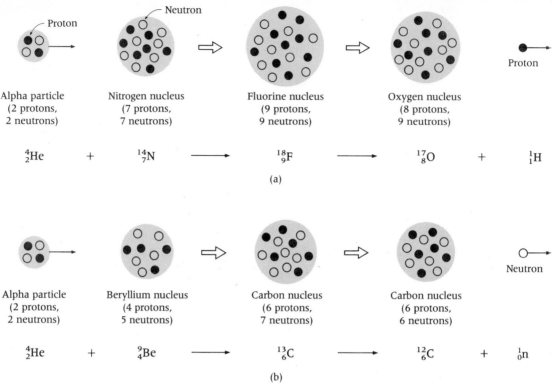

FIG. 1-24 (a) Nuclear reactions producing protons. (b) Nuclear reactions producing neutrons. The symbol for a neutron is $_0^1n$, where the 1 indicates its atomic mass number and 0 shows that it is neutral.

To indicate a nucleus, we use the chemical symbol for the element. The subscript on the left of this symbol indicates the number of protons, which equals the atomic number Z; the superscript on the left indicates the total number of particles in the nucleus, which equals the atomic mass number A (the atomic mass rounded off to a whole number). The number of neutrons in a nucleus is equal to $A - Z$. For example, $_4^9Be$ shows that this nucleus of a beryllium atom contains 4 protons and 5 neutrons. The 9 also indicates that the mass of the nucleus is, to a nearest whole number, 9 amu.

Figure 1-24 displays diagrams for the **nuclear reactions** (as these collisions among particles are called) that produce protons and neutrons. Also shown are the nuclear reaction equations symbolizing them.

Isotopes and the Modern Periodic Table

We have already learned about Dalton's atomic theory: Atoms of a given element all have the same characteristic mass. Now, this theory has to be modified. Experiments show, for example, that atoms of neon have three different masses: 19.99 amu (90.51 percent), 20.99 amu (0.27 percent), and 21.99 amu (9.22 percent). The average mass of all neon atoms is 20.17

amu, and this average value is listed as the chemical atomic mass in Fig. 1–3. Atoms of the same element having different atomic masses are called **isotopes** of the element. The atomic number Z of neon is 10. Atoms of neon-20* (20 is the atomic mass number A of the isotope) have 10 electrons orbiting the nucleus, and 10 protons and 10 neutrons in the nucleus, while atoms of neon-22 have the same number of electrons and protons, but 12 neutrons in the nucleus. Thus, isotopes of an element are atoms with different numbers of neutrons in their nuclei, and, as a result, they have different atomic masses.

Little did Mendeleev realize that his periodic table of the elements contained a code that, when deciphered, would provide information about the structure of the atom and its nucleus. The atomic number Z, which indicates the place of an element in the table, also indicates the number of electrons and protons in the atom. The atomic mass number A, which is the atomic mass rounded off to a whole number, indicates the number of particles in the nucleus. $A - Z$ gives the number of neutrons in the nucleus.

Summary

Chemical elements are substances that cannot be broken down by the chemist into simpler components. All substances on earth are composed of one or more of the 88 chemical elements. An atom of an element is the smallest unit of matter that retains the properties of that element, while a molecule is a larger unit that results when atoms combine. A molecule is the smallest unit of a chemical compound that retains the properties of that compound.

Dalton's atomic theory of 1808 states that the atoms of each element have a characteristic *mass.*

*Isotopes may also be expressed by using the symbol for the element with the atomic mass number of the isotope appearing to the left of the symbol in the superscript position. Thus, neon-20 may appear as ^{20}Ne, and carbon-14 may appear as ^{14}C. You will see this form of isotope notation frequently.

The atomic mass, relative to hydrogen, can be determined from chemical reactions. Avogadro concluded that a sample of an element having a mass in grams equal to the atomic mass (amu) contains N_A atoms, where $N_A = 6.02 \times 10^{23}$. Knowledge of N_A leads directly to the mass of the atom and, further, from the density (mass per unit volume), the diameters of atoms are found to be about 10^{-10} m.

Mendeleev devised a periodic table of the elements, in which each column contains elements with similar chemical properties. The elements are placed in order of increasing atomic mass. The atomic number Z indicates the location of an element in this table.

At the end of the nineteenth century, researchers made great advances from experiments conducted with cathode ray tubes. When the electrodes are connected to a high-voltage source and the gas pressure is one one-hundredth of atmospheric pressure, light is emitted by the gas. When this light is sent through a prism, its spectrum is a line spectrum, consisting of only certain definite colors, rather than a continuous spectrum. When the pressure is reduced to one ten-thousandth of atmospheric pressure, cathode rays travel from the cathode to the anode. In 1895, Roentgen found that X-rays are produced when cathode rays strike the anode. In 1897, J. J. Thomson showed that cathode rays are negatively charged particles; they have a mass 1/1833 times that of a hydrogen atom. This subatomic particle (the first found) is produced when the electron (modern term for cathode ray) is torn away from a gas atom in the tube by the large repulsive force of the cathode.

In 1898, Becquerel discovered that uranium is radioactive (that it spontaneously emitted rays). Marie Curie found that thorium is also radioactive and discovered two new radioactive elements, radium and polonium, in pitchblende. Physicists showed that rays consist of three types:

THE STRUCTURE OF THE ATOM

alpha (nucleus of helium atom), beta (electron), and gamma rays (more penetrating than X-rays, but similar in nature). Rutherford and his colleagues scattered alpha particles from a gold foil. The startling result that 1 in 8000 were scattered backward led Rutherford to propose a nuclear model of the atom: The positive charge is concentrated in a sphere having a diameter of 10^{-14} m, while the electrons occupy a sphere of 10^{-10} m diameter.

Bohr proposed a planetary model of the hydrogen atom, in which the electron traveled in allowed circular orbits. When it jumped to a smaller orbit, light was emitted. Bohr triumphantly predicted the wavelength of lines in the hydrogen spectrum. Each element has a characteristic line spectrum.

In 1919, Rutherford showed that, when an alpha particle hits a nitrogen nucleus, a proton (nucleus of a hydrogen atom) is knocked loose during the collision. Similarly, Chadwick found that a neutron (same mass as proton, but no charge) is released when an alpha particle collides with a beryllium nucleus.

In his experiments with X-rays, Moseley showed that Z electrons orbit around a nucleus containing Z protons.

The existence of isotopes shows that Dalton's theory had to be modified. Atoms of the same element having different atomic masses are called isotopes of the element. The mass listed in the periodic table of the elements is the *average* atomic mass. Atoms of isotopes of an element have different numbers of neutrons in their nuclei. An atom having an atomic mass number A and atomic number Z contains Z electrons, Z protons, and $A - Z$ neutrons.

True or False Questions

Indicate whether the following statements are true or false. Change all of the false statements so that they read correctly.

1–1 Equal masses of two different elements contain the same number of atoms.

1–2 From the density of an element and Avogadro's number, you can estimate the diameter of the atoms of that element.

1–3 If it were possible to evacuate *all* of the gas atoms from a cathode ray tube, cathode rays would not travel from the cathode to the anode.

1–4 X-rays are cathode rays that bounce off the anode of a cathode ray tube.

1–5 Becquerel discovered radioactivity when he found that uranium spontaneously emitted penetrating rays.

1–6 Madame Curie concluded that pitchblende contained undiscovered radioactive elements because it was more radioactive than uranium or thorium.

1–7 An alpha particle is the nucleus of a helium atom.

1–8 When Rutherford aimed alpha particles at a thin gold foil, he was astounded to find that most of them were not scattered at all.

1–9 While the diameter of the earth's orbit around the sun is known to be about 200 times the diameter of the sun, the diameter of the atom is nearly 10,000 times the diameter of the nucleus.

1–10 An atom with atomic mass number A and atomic number Z has the same number of electron, protons, and neutrons only when Z equals $A/2$.

Questions for Thought

1–1 Match the name or term with its description by inserting the correct letter from the lettered column into each blank in the descriptive column. Entries from the lettered column may be used more than once.

a. Henri Becquerel   **l.** James Chadwick
b. John Dalton   **m.** electron
c. J. J. Thomson   **n.** gamma rays
d. Johann Balmer   **o.** atomic mass
e. Dimitri Mendeleev **p.** line spectra
f. Wilhelm Roentgen **q.** strong yellow lines
g. Marie Curie   **r.** red, blue, and two
h. Pierre Curie   violet lines
i. Ernest Rutherford **s.** density
j. Hans Geiger and   **t.** Avogadro's number
   Ernest Marsden   **u.** 10^{-10} m
k. Niels Bohr   **v.** 10^{-14} m

___ discovered X-rays in 1895
___ discovered radioactivity in 1896
___ discovered the electron in 1897
___ modern term for cathode ray
___ proposed an atomic theory in which all atoms of an element have the same mass
___ first scientist to determine atomic mass of elements
___ relative mass of atom compared to hydrogen
___ determined atomic mass of radium
___ mass per unit volume
___ devised the periodic table of the elements
___ predicted the properties of undiscovered elements
___ size of the nucleus
___ size of the atom
___ 6.02×10^{23}
___ characteristic of the light emitted by elements
___ spectrum of sodium
___ spectrum of hydrogen
___ found a mathematical formula that predicted the wavelengths of hydrogen spectral lines
___ discovered the elements polonium and radium

___ obtained one-tenth of a gram of radium from one ton of pitchblende
___ discovered a treatment for cancer
___ determined that the alpha ray was a helium atom with two electrons missing
___ another name for a beta ray
___ more penetrating than X-rays
___ proposed a model of the atom as a sphere of positive charge with electrons embedded inside
___ observed the scattering of alpha particles by a thin gold foil and found 1 in 8000 scattered backwards
___ proposed a nuclear model of the atom
___ developed a model of the hydrogen atom that accounted for the hydrogen spectrum
___ discovered the proton
___ transformed nitrogen into oxygen
___ discovered the neutron

1–2 Define the following terms: element, atom, chemical compound, and molecule.

1–3 a. Briefly describe Dalton's atomic theory.
   b. What was the significance of the atomic mass?
   c. Explain how Dalton found the atomic masses of elements.

1–4 According to Dalton's theory, all atoms that have an atomic mass of 16 amu are oxygen atoms.
   a. How has Dalton's theory been modified?
   b. Could atoms with an atomic mass of 17 amu be oxygen atoms?
   c. Could atoms of nitrogen and fluorine have an atomic mass of 16 amu?

1-5 In the periodic table of the elements in Fig. 1–3 (p. 6), most of the atomic masses are very close to a whole number. Oxygen, for example, has an atomic mass of 15.999 amu and helium, 4.0026 amu.

    a. Look through the periodic table and list the elements whose atomic masses have neither a 9 nor a 0 after the decimal point.

    b. What is the connection between the elements found in **a** and the existence of isotopes?

1-6 a. The element strontium has similar chemical properties to which of the following: titanium, calcium, potassium, or magnesium? Explain how you arrived at your answer.

    b. One by-product from atmospheric testing of atomic bombs is the production of the isotope strontium-90, which is radioactive. Strontium-90 eventually is found in the _____ of human beings, due to the similarity in the chemical properties of the elements strontium and _____ .

1-7 When Madame Curie separated radium from the pitchblende mineral using chemical means, she found that barium and radium came off together in the same sample.

    a. What does this suggest about the chemical properties of radium and barium?

    b. Locate barium and radium in the periodic table. What do their locations indicate?

    c. If a molecule of barium chloride ($BaCl_2$) contains 1 atom of barium and 2 atoms of chlorine, what does a molecule of radium chloride contain?

1-8 With the nuclear model of the atom, we can no longer think of the atom as having a definite surface—a sphere—that encloses the atom. Make a sketch similar to Fig. 1–5 (p. 9), but indicate the nucleus in each atom. Now erase

the circles indicating the atom boundary. What does the distance d (used in Fig. 1–5) now represent?

1-9 What is the difference between a continuous spectrum of light and a line spectrum? How is each one obtained?

1-10 A cathode ray tube consists of _____

_____

The anode is connected to the _____ terminal of the high voltage, and the cathode to the _____ terminal of the high voltage. The air in the tube emits light when the pressure is _____ . When the pressure is reduced to _____ , the glass glows with an apple-green color. At this reduced pressure, cathode rays travel from the _____ toward the _____ . Cathode rays are (*particles* or *waves*) _____ that have a mass about (*equal to*, 1/10, 1/100, or 1/2000) _____ the mass of the hydrogen atom. Cathode rays have a (*negative, zero*, or *positive*) _____ charge. The modern term for cathode rays is _____ . The technical term for the picture tube of a TV set is _____ .

1-11 X-rays are produced in a cathode ray tube when _____

_____ . X-rays are (*particles* or *waves*) _____ that have a (*zero, small*, or *large*) _____ mass. X-rays are similar in nature to (*cathode rays* or *gamma rays*) _____ . X-rays have a (*negative, zero*, or *positive*) _____ charge. When X-rays pass through matter, some fraction of the X-rays is absorbed. This fraction depends upon the material. When a medical X-ray photograph is obtained, the bone absorbs (*more, equal*, or *less*) _____ X-rays than the surrounding flesh.

1-12 In his experiments with X-rays, Moseley showed that the atomic number Z of an element revealed something about the structure of the atom. What was this revelation?

1-13 Marie Curie found that, in addition to uranium, the element _____ was also radioactive. When she measured the radioactivity of the mineral pitchblende, she found that

_____

_____ .

She therefore concluded that _____

_____ .

Using a chemical separation procedure to analyze the pitchblende material, Curie discovered the elements _____ and _____ .

1–14 Rutherford suggested a scattering experiment, and this experiment was performed by Hans Geiger.

    a. Briefly describe this experiment, indicating what data were obtained.

    b. Ernest Marsden joined Geiger in the scattering experiment. What experiment did Rutherford suggest? What was the most startling observation of Geiger and Marsden's experiment? Could this result be explained in terms of Thomson's model of the atom?

    c. What model of the atom did Rutherford propose? Why?

1–15 According to Bohr's model of the hydrogen atom, the electron moves in the following way: _____

_____ .

However, the electron is restricted to move in

_____ .

Light is emitted by a hydrogen atom when

_____ .

1–16 Describe how Rutherford discovered the proton. Which is more easily absorbed in passing through matter, a proton or an alpha particle?

1–17 Write a nuclear reaction equation showing how boron is changed into carbon when boron is bombarded by alpha particles.

1–18 Describe how Chadwick discovered the neutron.

1–19 Enter the mass and charge of the particles listed in the accompanying tabulation. Take the mass of a hydrogen atom to be 1 amu (rounded off to a whole number), and the charge of an electron to be -1. The mass of an electron is known accurately to be 1/1833 times that of a hydrogen atom.

	Mass (amu)	Charge
Electron	1 / 1833	-1
Hydrogen atom	1	0
Alpha particle	_____	_____
Beta ray	_____	_____
Gamma ray	_____	_____
Proton	_____	_____
Neutron	_____	_____
X-ray	_____	_____

Questions for Calculation

1–1 A cube of aluminum with a 1-cm side length has a mass of _____ grams. (See Fig. 1–4)

1–2 What would be the masses of copper blocks having the same dimensions as those in Fig. 1–4? Gold blocks?

1–3 An atom of calcium has a mass about _____ times that of a hydrogen atom, about _____ times that of a bromine atom, and about _____ times that of a neon atom. Since a hydrogen atom has a mass of 1.67×10^{-27} kg, an atom of calcium has a mass of _____ kg.

1–4 A molecule of carbon dioxide (CO_2) contains 1 atom of carbon and 2 atoms of oxygen. How many kilograms of oxygen will combine with 1 kg of carbon?

†1–5 a. Two grams of hydrogen will combine with _____ g of oxygen to form _____ g of water. In 2 g of hydrogen there are _____ $\times N_A$ atoms and, in this amount of oxygen (_____ g), there are _____ $\times N_A$ atoms. In the amount of water formed, _____ g, there are _____ $\times N_A$ molecules of water.

b. The molecular mass M is defined as the sum of the atomic masses (amu) of all atoms in the molecule. A sample of a chemical compound having a mass of M grams (where M is the molecular mass of a molecule of the compound) contains _____ molecules.

c. Use the results of **a** to show that the statement in **b** is correct.

d. Find the molecular mass of a carbon dioxide molecule. There are N_A molecules in _____ g of carbon dioxide.

†1–6 The density of copper is 8.91 g/cm^3. The atomic mass of copper is 63.55 amu.

a. One cubic centimeter of copper has a mass of _____ g.

b. Show that 1 cm^3 of copper contains 8.44×10^{22} atoms.

c. Show that the diameter of a copper atom is 2.28×10^{-10} m.

d. Find the mass of a copper atom.

1–7 a. Use Balmer's formula to calculate the wavelength of light emitted by hydrogen when n equals 4. What color is this?

b. According to Bohr's model of the hydrogen atom, this light is emitted by the following process: _____ _____ .

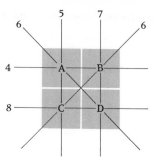

FIG. 1–25   Diagram for calculation question 1–8.

1-8 In Fig. 1-25, the integers A, B, C, and D are less than 10. The numbers beside each straight line indicate the average of the two integers through which that lines passes. For example, $(A + C)/2 = 5$. From these averages, determine the values of A, B, C, and D. What is the connection between this problem and the X-ray CT-scan procedure? (Hint: B = 5.)

1–9 Figure 1–18(c), shows the scattering pattern for balls striking a circular-shaped obstacle. What would the scattering pattern look like if the obstacle had been a triangle or a square, as shown in part (a) of this figure? Sketch the scattering patterns.

1–10 If the radius of the first allowed orbit in the Bohr model of the hydrogen atom is 0.528×10^{-10} m, then the radius of the second allowed orbit is _____ .

1–11 Uranium-235 and uranium-238 are isotopes of the element uranium. An atom of uranium-235 ($Z = 92$, $A = 235$) has _____ electrons, _____ protons, and _____ neutrons. An atom of uranium-238 has _____ electrons, _____ protons, but _____ neutrons.

†A dagger by a question indicates that the answer to the question appears at the back of the book.

Footnotes

[1] Jacob Bronowski, *The Ascent of Man* (Boston: Little, Brown and Co, 1973), p. 330.

[2] Morris H. Shamos (Ed.), *Great Experiments in Physics* (New York: Henry Holt and Company, 1959), p. 202.

[3] Godfrey N. Hounsfield, "Computed Medical Imaging," *Science* 210 (3 October, 1980): 22.

[4] Ibid., p. 22.

[5] John Harris, SCATTER: Unit on Particle Scattering, Chelsea Science Simulation Project (London: Edward Arnold, 1975; also Iowa City, Iowa: Conduit).

[6] Excerpt from *Rutherford and the Nature of the Atom* by E. N. da C. Andrade, p. 111. Copyright © 1964 by Doubleday and Company, Inc. Reprinted by permission of the publisher.

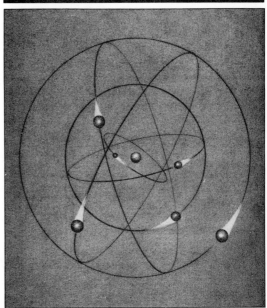
Rutherford–Bohr nuclear model of the atom.

PART II

Classical Physics
with Applications
to Modern Physics

Introduction

At the heart of each of the many exciting discoveries in modern physics resides the concept of **motion:** the *passage* of cathode rays in a cathode ray tube, the *emission* of rays from a radioactive source, the backward *scattering* of alpha particles, and the *circular motion* of an electron around the nucleus of a hydrogen atom. Thus, as we start Part II, which deals with the basic principles of classical physics, the most logical place to begin is with the subject of motion.

We'll first concentrate on a situation that is easily understood: the motion of a car along a highway. From this, we can extract the basic relationship involving distance, speed, and time. With this preparation, we'll then examine another situation, one with which we have no experience and can only imagine: an electron traveling around the nucleus of a hydrogen atom. This will be our theme throughout Chapter 2 (and in the rest of Part II as well)—deriving principles from familiar, concrete examples and extending them to explore the abstract world of the atom. We will, for instance, investigate changes in speed, taking the discussion through cars to Rutherford's scattering experiment with alpha particles (see Chapter 1), to the maneuvering of giant oil tankers. Finally, we'll consider a phenomenon always associated with the name of that great Italian physicist Galileo: the motion of an object falling under the influence of gravity.

Galileo

CHAPTER 2
Motion

Distance, Speed, and Time

ON THE HIGHWAY

At one time or another, all of us have used the basic relationship involving distance, speed, and time. In discussing this relationship, we will pretend that the United States has adopted the metric system of measurement. (Refer to Appendix B for a complete description of this unit system.) Imagine that you have just driven from Chicago to Cincinnati, a distance of 480 kilometers (abbreviation: km), in 6 hours. Without a moment's hesitation, you determine that the speed, or, more accurately, the **average speed** was 80 kilometers per hour (km/hr) by dividing the distance by the time. That is,

$$\text{Average speed} = \frac{\text{distance traveled}}{\text{time interval}},$$

or, using symbols,

$$\bar{v} = \frac{d}{t}. \tag{1}$$

The symbol for average speed, $\bar{v}$, is read as v-bar. (In statistics, a bar over a symbol indicates an average value of that quantity.) The speed during the trip was obviously not constant, and the term *average speed* reflects these variations. On the freeways, you probably drove faster than 80 km/hr, and most likely slower than that along the outskirts of Chicago and Cincinnati. You also undoubtedly stopped for lunch.

If you aren't used to thinking in metric units, and want to know how fast you "really" had been going, you might convert the metric units to more familiar ones. Since 1 mile equals 1.6 km, 480 km corresponds to 300 miles, and 80 km/hr corresponds to 50 miles per hour (mph).

You certainly don't want to drive very fast during a vacation trip through the spectacular Colorado Rockies, and you often might stop completely to view scenery. Here, your average speed may be, say, only 30 km/hr. At that rate, you can only drive 180 km in 6 hours. That is, you obtain the **distance** by multiplying the average speed by time, or

$$d = \bar{v}t. \tag{2}$$

If, alas, there is dense fog and your speed is reduced to 20 km/hr, then a distance of 180 km will require 9 hours. This **time** was obtained by using

$$t = \frac{d}{\bar{v}}. \tag{3}$$

Equations 2 and 3 are, of course, just different versions of Equation 1.

Suppose that, while in the Rockies, you see lightning strike a tree on a nearby mountain and then 10 seconds (sec) later hear the thunder. While the lightning and thunder are produced at the same time, you see the lightning in an instant because the speed of light (3×10^8 m/sec) is so fast. You can find the distance to the mountain (Equation 2) by multiplying the speed of sound, 340 m/sec, by the time delay of 10 seconds, and obtain a distance of 3400 m (Fig. 2–1).

IN THE HYDROGEN ATOM

A single electron revolves around the nucleus of a hydrogen atom in allowed circular orbits (Chapter 1). The first allowed orbit has a radius r_1 of 0.528×10^{-10} m, and the electron travels at a speed v_1 of 2.19×10^6 m/sec. As it happens, the electron's speed in the second allowed orbit is only one-half of that in the first; in the third, only one-third of that in the first. Table 2–1 shows the relationships among the radius, speed, and time for the first three allowed orbits.

Let's compare the times for the electron to make one revolution around the three allowed orbits. We'll use Equation 3, where the distance d is the circumference of the circular orbit. In the first allowed orbit, the time t_1 is given by the circumference ($2\pi r_1$) divided by the speed v_1:

$$
\begin{aligned}
t_1 &= \frac{2\pi r_1}{v_1} \\
&= \frac{2(3.14)(0.528 \times 10^{-10}\text{m})}{2.19 \times 10^6\,\text{m/sec}} \\
&= 1.51 \times 10^{-16}\,\text{sec}.
\end{aligned}
$$

FIG. 2–1 The number of seconds between the time you see lightning flash and the time you hear the thunder can be used to figure out how far away from you the lightning struck. In this instance, the time delay is 10 sec, so the distance to the mountain is 3400 m.

Table 2–1 shows the results of similar calculations for the other orbits and compares the times to t_1, the time for the electron to make one revolution around the first orbit. An electron in the second orbit takes eight times as long to make one revolution as one in the first, because the circumference is four times larger, and the speed is only one-half as much as in the first allowed orbit.

Acceleration

So far, we have eliminated fluctuations in speed by considering average speeds. To examine these fluctuations in detail, we must introduce a new concept called **acceleration.** Acceleration specifies the rate at which the velocity is changing:

$$\text{Acceleration} = \frac{\text{change in velocity}}{\text{time interval}}$$

$$a = \frac{v_f - v_i}{t}, \qquad (4)$$

where v_f is the final velocity at the end of the time interval t, and v_i is the initial velocity at the beginning of the time interval t.

TABLE 2–1 TIME FOR ELECTRON TO TRAVEL AROUND ALLOWED ORBITS IN HYDROGEN ATOM

ALLOWED ORBIT NUMBER	RADIUS	SPEED	TIME FOR ONE REVOLUTION
1	r_1	v_1	t_1
2	$4r_1$	$\frac{1}{2}v_1$	$8t_1$
3	$9r_1$	$\frac{1}{3}v_1$	$27t_1$

Note: For each value of the radius and speed, the time for an electron to make one revolution around an allowed orbit is given by 2π (radius/speed). Values are expressed in terms of r_1, v_1, and t_1, where r_1 equals 0.528×10^{-10} m, v_1 equals 2.19×10^{-6} m/sec, and t_1 equals 1.51×10^{-16} sec.

You notice immediately that the word **velocity** is used rather than *speed*. The layman often uses these terms interchangeably, but the term *velocity* conveys more information. For example, 20 m/sec (44.7 mph) is a speed, but 20 m/sec north is a velocity. To specify the velocity, a speed *and* a direction are needed.

Acceleration occurs when *either* the speed or the direction changes. For example, a car is said to accelerate when it makes a turn at a constant speed because its direction changes. Perhaps you wonder, Why not simply define acceleration as the rate at which the speed changes? To make the definition of acceleration seem plausible, imagine that you are a passenger in a car. As long as this car travels with constant speed in a straight line, it seems no different than sitting in your favorite easy chair at home. However, if the driver speeds up, slows down, or turns, you sense it immediately. Using this physical sensation as a criterion for acceleration, it seems reasonable to class change in direction as well as a change in speed as acceleration.

Since we shall consider only changes in speed in this chapter, we shall use the term *speed* rather

than *velocity* in the following discussions. In Chapter 3, changes in direction will be considered.

Distance Traveled During Acceleration

Suppose that a car accelerates from rest to 20 m/sec (44.7 mph) in 10 seconds. Its acceleration is given by

$$a = \frac{20\,\text{m/sec} - 0}{10\,\text{sec}}$$
$$= 2\,\text{m/sec per sec (or, 2 m/sec}^2).$$

The units of acceleration are handled just like the division of fractions:

$$\frac{\text{m/sec}}{\text{sec}} = \frac{\text{m/sec}}{\text{sec}/1}$$
$$= \frac{\text{m}}{\text{sec}} \times \frac{1}{\text{sec}}$$
$$= \frac{\text{m}}{\text{sec}^2}.$$

The expression m/sec^2 is the most commonly used unit of acceleration.

An acceleration of 2 m/sec^2 means that the speed changes by 2 m/sec during 1 second. Since the car starts from rest, its speed after 1 second is 2 m/sec; after 2 seconds, 4 m/sec; after 3 seconds, 6 m/sec; . . . after 10 seconds, 20 m/sec. Figure 2-2 shows a graph of speed versus time.

To find how far the car travels during these 10 seconds, we use the relationship d equals $\bar{v}t$. The question is, What is the average speed $\bar{v}$?

Since the speed varies in a regular way, let us average the speeds occurring at 2-second intervals: 0, 4, 8, 12, 16, and 20 m/sec. Adding these speeds and dividing by 6, we find the average speed to be 10 m/sec. Figure 2–2 shows that the car has a speed of 10 m/sec at 5 seconds. Before that time, its speed is less than 10 m/sec and after that, greater than 10 m/sec. So, from the idea of what an average means, it seems reasonable that the average speed over the 10-second time interval is 10 m/sec. We also see that the average speed could just as easily be obtained by averaging the initial speed and the final speed:

$$\bar{v} = \frac{0 + 20\,\text{m/sec}}{2}$$

$$= 10\,\text{m/sec}.$$

This can be expressed in a general way as the average of the initial speed v_i and the final speed v_f:

$$\bar{v} = \frac{v_i + v_f}{2}, \qquad (5)$$

where v_i is the speed at the beginning of the time interval, and v_f is the speed at the end of the time interval. This is valid even when the initial speed is not zero.

As we have just seen, the average speed during 10 seconds was 10 m/sec. Multiplying this average speed by 10 seconds (Equation 2), we find that the car travels a distance of 100 m.

After 5 seconds, the car's speed is 10 m/sec. Its average speed is

$$\bar{v} = \frac{0 + 10\,\text{m/sec}}{2}$$

$$= 5\,\text{m/sec}.$$

The distance traveled equals 5 m/sec × 5 sec, or 25 m. Figure 2–3(a) shows the speedometer and odometer for this example. The other odometer readings are calculated in the same way. Figure 2–3(b) shows the same readings for a car traveling at a constant speed of 10 m/sec.

To see that Equation 5 can be used when the initial speed is not zero, let's consider the following example. We see from Fig. 2–3(a) that the car travels a distance of 75 m from 5 to 10 seconds. Now let's see if our definitions of average speed (Equation 5) and distance traveled (Equation 2) also give a distance of 75 m. At 5 seconds, its speed was 10 m/sec and at 10 sec-

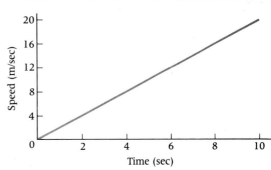

FIG. 2–2 Graph of speed versus time for a car accelerating at 2 m/sec². The speed increases by 2 m/sec after each second in motion.

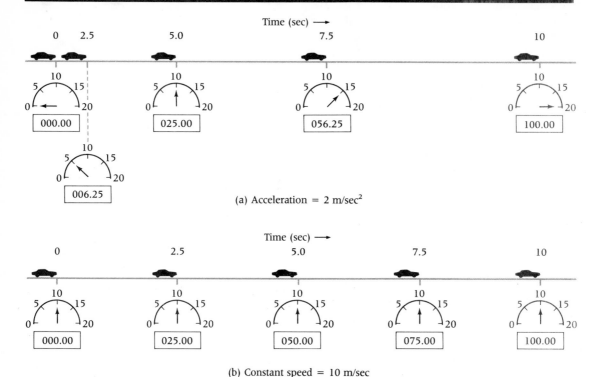

FIG. 2−3 (a) Speedometer and odometer readings for car accelerating from rest at 2 m/sec². (b) Speedometer and odometer readings for car traveling with constant speed of 10 m/sec.

onds, 20 m/sec. The average speed during the 5-second interval is

$$\frac{10\,\text{m/sec} + 20\,\text{m/sec}}{2} = 15\,\text{m/sec.}$$

The distance traveled equals 15 m/sec × 5 sec, or 75 m, as we expected.

In any situation involving constant acceleration, only three relationships are needed: (1) $d = \bar{v}t$, where d is the distance traveled during the time interval t, or the distance from the point where the speed was v_i; (2) the definition of acceleration; and (3) the definition of average speed.

In the following situations, it is usually easier to use a single equation equivalent to these three

steps: (1) initial speed is zero and the speed increases; (2) final speed is zero and the speed *decreases*. In these situations, the distance can be found from the time, using

$$d = \frac{1}{2}at^2. \tag{6}$$

In the first situation, where v_i equals 0, the distance can also be found from the final speed v_f, as

$$d = \frac{v_f^2}{2a}, \tag{7}$$

and, in the second situation, where v_f equals 0, the distance can be found from the initial speed v_i, as

$$d = \frac{v_i^2}{2a}. \tag{8}$$

Let's see that these equations "work" for our car accelerating from rest at 2 m/sec². Using Equation 6, we find that the distance traveled during 10 seconds is $d = \frac{1}{2}(2 \text{ m/sec}^2)(10 \text{ sec})^2$, or 100 m; from 0 to 5 seconds,

$$d = \frac{1}{2}(2 \text{ m/sec}^2)(5 \text{ sec})^2$$

$$= 25 \text{ m}.$$

After 5 seconds, the speed is 10 m/sec; using Equation 7, we find that the distance traveled is

$$d = \frac{(10 \text{ m/sec})^2}{2(2 \text{ m/sec}^2)}$$

$$= 25 \text{ m}.$$

We can derive Equations 6 and 7 quite simply. When the initial speed is zero, the average speed becomes half the final speed, and the distance traveled becomes $\frac{1}{2} v_f t$. Here, Equation 4 can be written as

$$v_f = at,$$

and the distance becomes

$$d = \frac{1}{2} v_f t$$

$$= \frac{1}{2}(at)(t)$$

$$= \frac{1}{2}at^2.$$

Alternatively, Equation 4 can be written as

$$t = \frac{v_f}{a},$$

and the distance becomes

$$d = \frac{1}{2} v_f t$$

$$= \frac{1}{2}(v_f)\left(\frac{v_f}{a}\right)$$

$$= \frac{v_f^2}{2a}.$$

TABLE 2–2 STOPPING DISTANCE OF A CAR AT VARIOUS SPEEDS

SPEED OF CAR (mph)	STOPPING DISTANCE (range in ft)
20	40–44
40	108–124
60	228–268
80	422–506

Note: The range of values of the stopping distance takes into account driver reaction time, the condition of the pavement, and types of brakes.
Source: "Illinois Rules of the Road," published by Illinois Secretary of State.

Deceleration and Stopping Distances

STOPPING DISTANCE OF A CAR

Anyone who drives surely has had the sinking feeling of looking into the rearview mirror and wondering, Will the car behind me be able to stop in time? This is especially true if the pavement is icy and the slowing down of cars is much less than normal. It is absolutely essential for all drivers to realize that the stopping distance is *not proportional* to the speed; that is, when the speed is doubled, the stopping distance is simply *not* doubled. The driver's license examination usually includes a question about this concept. The stopping distance versus speed is shown in Table 2–2.

We see that the stopping distance when traveling at 80 mph is about 4 times that at 40 mph. When the initial speed is doubled, the stopping distance is 4 times as large. We can understand these effects by looking at Equation 8. Recall that the acceleration a in this equation describes situations where the speed decreases. The term **deceleration** specifically indicates this "slowing down." The stopping distance is given by $v_i^2/2a$. Assuming that the deceleration is the same for all speeds shown in Table 2–2, we see that the stopping distance is proportional to the *square* of the initial speed. Therefore, when the initial speed doubles, we expect that the stopping distance should quadruple, or increase by a factor of 4. (The data shown in Table 2−2 include driver reaction time and other effects, so that doubling the speed does not yield precisely a factor of 4 as, for example, when comparing the stopping distances for 20 mph and 40 mph.)

When a car is traveling on an icy pavement and the brakes are applied, the car skids rather than rolling to a stop. Since the friction between the tires and ice is so very small, the deceleration of the car is also quite small. We know what happens: Either the car requires a very long distance to stop or . . . crash! Note that Equation 8 conveys this information also. Since the deceleration is in the denominator, a small deceleration yields a large stopping distance.

STOPPING DISTANCE AND THE RUTHERFORD SCATTERING EXPERIMENT

We can gain a great deal of insight into Rutherford's scattering experiment by noting the similarities between the stopping distance of an alpha particle that hits a gold foil and the stopping distance of a car. Before exploring this, however, let's consider a purely hypothetical situation and ask, What is the deceleration of a car traveling at 40 m/sec (90 mph) and "coming to a stop on a dime"? That is, the deceleration of a car traveling only 1.7 centimeters (cm) before coming to a stop? Without making any calculations, we know that indeed the deceleration is tremendous! But what can we find out by comparing the deceleration of the car with the deceleration of the alpha particle?

First, let's look at the deceleration of a car traveling at 40 m/sec with a stopping distance of 1.7 cm. Rearranging Equation 8, we find

$$a = \frac{v_i^2}{2d}.$$

Substituting the initial speed of 40 m/sec and the distance of 0.017 m, we find the deceleration to be

$$a = \frac{(40\,\text{m/sec})^2}{2(0.017\,\text{m})}$$
$$= 47{,}000\,\text{m/sec}^2.$$

As expected, the deceleration is a whopping number. The speed decreases by 47,000 m/sec after each second with this deceleration. This car comes to a stop in a fraction of a second. "Hypothetical" is surely the correct description for this deceleration. Let's find out if the deceleration of the alpha particle is larger than this.

Recall that when Rutherford and his collaborators Geiger and Marsden measured the scattering of alpha particles by a thin gold foil, they were incredulous to find that one 1 particle in 8000 was scattered in the backward direction. Most of the alpha particles passed through the foil without being deflected at all from a straight line path, and some were scattered in the forward direction by only a few degrees. Figure 2−4 shows the most extreme case of backward-scattering, where the alpha particle completely turns around and retraces its path. This is analogous to a ball being thrown vertically into the air: The ball comes to a stop momentarily at its highest point before falling to the ground. In this example, we shall assume that the alpha particle turns around at point S (although it could turn around at any point inside the foil). We can now calculate the deceleration of the alpha particle in Fig. 2−4. We cannot be sure that the deceleration of the alpha particle is constant, but we shall assume that it is because the calculation is so simple to carry out. If, in fact, the deceleration varies as the alpha particle travels through the foil, then this calculation will represent the average deceleration. Since the speed is so large (2×10^7 m/sec) and the foil so thin (4×10^{-7} m), the deceleration is going to be very large, larger than 47,000 m/sec^2. The deceleration is given by

$$a = \frac{v_i^2}{2d}$$

$$= \frac{(2 \times 10^7 \, \text{m/sec})^2}{2 \, (4 \times 10^{-7} \, \text{m})}$$

$$= 0.5 \times 10^{21} \, \text{m/sec}^2.$$

It's larger, all right—10,000,000,000,000,000 (10^{16}) times larger than the deceleration of the car! Rutherford must have pondered, What is going on inside the foil to produce this deceleration? Why do a few alpha particles experience this tremendous deceleration, when most of them are not affected in the least (an acceleration of zero)? It took Rutherford two years to finally conclude that the positively charged alpha particle is repelled by the nucleus of the gold atom. This repulsion brings the alpha particle to a stop and then causes it to retrace its path. For backward-scattering, the alpha particle must approach the nucleus head-on. Since this occurs so rarely, it was reasonable to assume that the nucleus must be very small compared to the total size of the atom, and that the alpha particle must interact only with the nucleus of *one* atom.

Note that the alpha particle in Fig. 2–4 passes most of the way through the foil before it encounters a nucleus of a gold atom. Alpha particles not coming close to any nucleus would not be deflected. With this model of the atom, Rutherford was able to explain the experimental observations. His crucial first step was to realize the tremendous deceleration of these backward-scattered alpha particles, a deceleration far greater than any he had ever seen before.

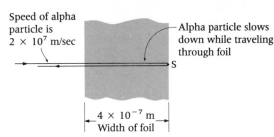

FIG. 2–4 An alpha particle traveling with a speed of 2×10^7 m/sec approaching a gold foil of thickness 4×10^{-7} m. The alpha particle stops momentarily at point S and then retraces its path.

STOPPING DISTANCE OF AN OIL TANKER

In December, 1976, an oil tanker with a mass of 18,500 tons, or 1.675×10^7 kilograms (abbreviation: kg) ran aground off Nantucket Island and spilled 7.5 million gallons of oil into the Atlantic. This particular tanker was small compared to the so-called medium-size supertankers, which have a mass of 250,000 tons (2.27×10^8 kg). Describing the supertankers as difficult to maneuver is an understatement. The deceleration in a medium-size supertanker is extremely small, and one traveling at 16 knots (8.22 m/sec) requires 22 minutes (or 1320 seconds) to stop. This yields a deceleration given by

$$a = \frac{8.22 \text{ m/sec}}{1320 \text{ sec}}$$
$$= 0.006227 \text{ m/sec}^2.$$

Using Equation 8, we find that the stopping distance is given by

$$d = \frac{(8.22 \text{ m/sec})^2}{2(0.006227 \text{ m/sec}^2)}$$
$$= 5425 \text{ m}$$
$$= 3.4 \text{ miles}.$$

The Acceleration of Gravity

The child in Fig. 2–5 seems intent on studying gravity by watching toys fall to the floor. What do we know about gravity from our everyday experience? Probably not a great deal more than we learned as a child. Objects fall to the floor in a fraction of a second. A feather falls to the ground more slowly than a penny because **air resistance** has a greater effect on the feather. The reason for this difference is that the feather encounters many more air molecules (in proportion to its mass) than the penny. An object falling from a second-story window takes longer to reach the ground than one dropped from the height of a table. We know much more from ordinary experience about the motion of a car. We can look at the speedometer and measure its speed and/or see the speed change. Physically, we can even sense the acceleration of a car by the change in pressure on our back, or feel our body sway as the car makes a turn. But what do we know about how an object falls to the ground? Is its speed constant? Is its acceleration constant? Is the motion perhaps even more complicated than constant speed or constant acceleration?

The Greek philosopher Aristotle (384–322 B.C.) proposed that the speed of a falling object was constant and proportional to its weight. For example, suppose that two objects, one five times heavier than the other, are dropped from the same height. According to Aristotle's hypothesis, the heavier one will travel five times faster than the light one, and, hence, the heavier one will take only one-fifth as long as the light one to reach the ground. Aristotle did not carry out such an experiment, although from today's viewpoint that does seem quite absurd.

Galileo (1564–1642) is often called the father of modern science precisely because he initiated the comparison between theory and experiment, or what is usually called the **scientific method.** In its modern form, the scientific method consists of three steps:

1. Perform an experiment and collect reproducible data.
2. Formulate a scientific law that describes the results of the experiment and that can make predictions as well.
3. Formulate a scientific theory or model that explains the results of the experiment and the scientific law in more fundamental terms.

Essentially, the early Greek philosophers, such as Aristotle, usually bypassed the first two steps and went directly to formulating theories. The early Greeks were interested in determining the *causes* of natural occurrences and were not concerned with describing them. They attempted to acquire knowledge of nature by pure thought.

MEASURING THE ACCELERATION OF GRAVITY

According to legend, Galileo disproved Aristotle's theory that the speed of a falling object was proportional to its weight by dropping two unequal weights from the Leaning Tower of Pisa.

At some time and place, he did drop two unequal weights and found that they did not reach the ground with the great difference that Aristotle had pre-

dicted. But it appears from modern scholarship that he never did so, at least publicly, from the Pisan tower.[1]

Actually, the heavier weight does strike the ground *slightly ahead* of the ligher one because air resistance affects the lighter weight more than it does the heavier. Galileo was aware of this effect. These experimental results suggested to Galileo that, excluding air resistance, two unequal weights will fall with the same acceleration. And if they fall from the same height, they'll hit the ground at the same time. From this, he derived the law

$$d = \frac{1}{2} at^2.$$

Galileo then set out to verify this relationship. He could not accurately time the descent of a *freely* falling object because the time is so very short, but he thought of a clever way around this difficulty. He used a plane inclined at a small angle with the horizontal to dilute gravity (Fig. 2–6) and a water clock to measure the much larger time of this descent. A water clock simply measures the amount of water that flows from one container into another. The amount of water, then, is proportional to the time. As the angle of the plane reaches 90°, the motion approaches that of a freely falling object, as shown in Fig. 2–6(b). Galileo measured the time of descent for various distances along the inclined plane and found that the distance d was proportional to t^2. This relationship indicated constant acceleration, as he had anticipated. When the angle of inclination increased, the time of descent (for the same distance) decreased, indicating that the acceleration was larger. Galileo determined that the **acceleration of gravity** was a constant, but he did not determine its value, which is now known to be 9.80 m/sec².

Figure 2–7 shows a modern method for measuring the acceleration of gravity, denoted by the symbol g. Air resistance effects are so small that we neglect them here.

The paper tape showing the location of the falling object at every one-twentieth of a second is shown in Fig. 2–8. The tape shows that the

FIG. 2–5   A child apparently studying effects of gravity.

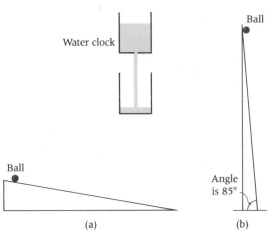

FIG. 2–6   (a) The apparatus used by Galileo to study gravity. The ball rolls down a groove in the inclined plane. (b) The plane is inclined at angle of 85° so that the ball approximates the motion of a freely falling object.

MOTION

47

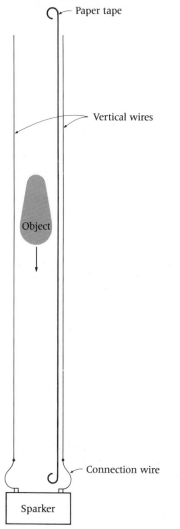

Paper tape

Vertical wires

Object

Connection wire

Sparker

FIG. 2-7 Experimental apparatus for measuring the acceleration of gravity (*g*). A specially designed object falls freely between two vertical wires. A paper tape is placed between the object and one wire. The sparking device is connected to the two vertical wires and every one-twentieth of a second a spark jumps across the two short gaps between the object and the vertical wires. Each spark makes a pinhole burn in the paper tape, and the passage of the falling object is recorded by a series of pinhole burns in the paper tape as shown in Fig. 2-8.

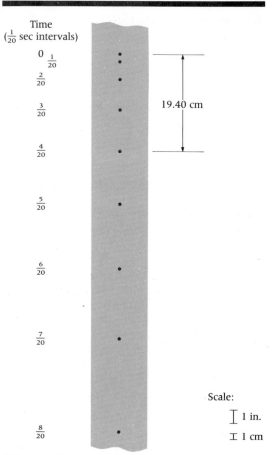

Time
($\frac{1}{20}$ sec intervals)

0 $\frac{1}{20}$

$\frac{2}{20}$

$\frac{3}{20}$

19.40 cm

$\frac{4}{20}$

$\frac{5}{20}$

$\frac{6}{20}$

$\frac{7}{20}$

Scale:

1 in.

$\frac{8}{20}$

1 cm

FIG. 2-8 Paper tape showing the passage of the falling object. The pinhole burns occur every one-twentieth of a second.

TABLE 2-3   CALCULATION OF THE ACCELERATION OF GRAVITY USING DISTANCE AND TIME MEASUREMENTS FROM THE PAPER TAPE IN FIG. 2-8.

TIME (*t*) (sec)	DISTANCE (*d*) (m)	ACCELERATION OF GRAVITY (*g*) (m/sec^2)
0	0	
1/20	0.0120	9.60
2/20	0.0485	9.70
3/20	0.1095	9.73
4/20	0.1940	9.70
5/20	0.3035	9.71
6/20	0.4385	9.73
7/20	0.5965	9.74
8/20	0.7795	9.74

Note: Average value *g* equals 9.71 m/sec^2.

falling object accelerates because the distance between adjacent pinholes increases from the top to the bottom of the tape. Table 2−3 shows the distance of each pinhole from the starting point and the corresponding time. The pinholes are about ½ mm (.0005 m) in size, and this limits the accuracy of the distance d. For this reason, the number in the fourth decimal place of the distance is either 0 or 5. The acceleration of gravity g can be found from the distance d and the time t. Substituting g for the acceleration in Equation 6, we find

$$d = \frac{1}{2}gt^2.$$

Solving this for g, we obtain

$$g = \frac{2d}{t^2}.$$

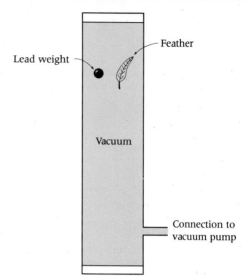

FIG. 2−9 A lead weight and a feather have the same acceleration in a vacuum and will fall side by side.

Substituting the distance and time measurements into this equation, we obtain the values of the acceleration of gravity listed in Table 2−3. The spread of values for g is inherent in any kind of measurement and indicates the precision of the measurement. The average of the values is 9.71 m/sec², and it differs by only 1 percent from the accepted value of 9.80 m/sec².

ACCELERATION IN A VACUUM

Figure 2−9 shows that, in a vacuum, a lead weight and a feather fall side by side (that is, have the same acceleration). The **vacuum** was obtained by linking the vessel to a vacuum pump and evacuating air from the vessel. When air resistance is eliminated as a factor, the acceleration of gravity is the same for all objects and does not depend upon the object's mass or shape. This fact was dramatically illustrated during the *Apollo-15* moon flight when astronaut David R. Scott dropped a hammer and a feather at the same time and saw them reach the moon's surface simultaneously.

MOTION

TIME (2.5 sec intervals)		SPEED (m/sec)	DISTANCE (m)
0	●	0	
2.5	●	24.5	30.625
5.0	●	49.0	122.5
7.5	●	73.5	275.625
10.0	●	98.0	490.0

FIG. 2–10 A ball falling from a position of rest under the influence of gravity.

ACCELERATION OF GRAVITY COMPARED TO THE ACCELERATION OF A CAR

Earlier, we investigated the motion of a car accelerating from rest to a speed of 20 m/sec (44.7 mph) in 10 seconds, or an acceleration of 2 m/sec². Figure 2–3(a) showed the speed of the car and the distance from the starting point. Now let's find the speed and distance traveled by a ball, falling from rest, during the same 10-second time interval. An acceleration of 9.8 m/sec² means that the speed of the ball increases by 9.8 m/sec during each second. After each 2.5-second time interval the ball's speed increases by 9.8 m/sec² × 2.5 sec, which equals 24.5 m/sec. Therefore, after 2.5 sec, its speed is 24.5 m/sec; after 5 sec, its speed is 49.0 m/sec; and so on, as shown in Fig. 2–10.

The distance from the starting point is found using $d = \frac{1}{2}gt^2$. After 2.5 sec,

$$d = \frac{1}{2}(9.8\,\text{m/sec}^2)(2.5\,\text{sec})^2$$

$$= 30.625\,\text{m},$$

and so forth. During the 10-second time interval, the ball travels a distance nearly 5 times that of the car shown in Fig. 2–3. At the end of this time, the ball's speed is 98.0 m/sec (219 mph), compared to the car's speed of 20 m/sec.

To get a "feel" for the size of the acceleration of gravity, let us realize that our car in this example would have to accelerate from rest to a speed of 20 m/sec (44.7 mph) in only 2 seconds to have an acceleration of 10 m/sec², which is the same as the acceleration of gravity.

EFFECTS OF THE MOON'S SMALLER ACCELERATION OF GRAVITY ON THE ASTRONAUTS: AN EXAMPLE

Between 1969 and 1972, we watched in awe the TV coverage of the United States' astronauts on the moon. They moved with such ease and grace, it was almost like watching a film in slow motion. Why were their movements so different on the moon? The reason, pointed out often by the TV narrators, is that the acceleration of gravity on the moon is only one-sixth that on earth. Let's examine how this phenomenon affected the astronauts' movements.

Suppose that, while still on earth, an astronaut jumps vertically into the air. He bends his knees and pushes his feet against the ground as he straightens his legs. As he pushes away from the ground, his speed, which is initially zero, increases so that it is 3 m/sec at the instant his toes leave the ground. Now suppose that astronaut executes the same jump on the moon, so that his speed is 3 m/sec as his toes leave the moon's surface. In each situation, how long does it take the astronaut to reach his maximum height? What is his maximum height?

On the earth, after the astronaut leaves the ground, gravity acts, so that his speed decreases by 9.8 m/sec during each second. The maximum height occurs when his speed is zero. After this, of course, gravity pulls the astronaut back to earth. So, we need to know the time for the speed to change from 3 m/sec to zero. Because a change of 9.8 m/sec occurs in 1 second, a change of 3 m/sec will occur in a time given by

$$t = \frac{3 \, \text{m/sec}}{9.8 \, \text{m/sec}^2}$$
$$= 0.31 \, \text{sec}.$$

The average speed must be found before the maximum height can be determined. Using Equation 5, we obtain

$$\bar{v} = \frac{(0 + 3 \, \text{m/sec})}{2}$$
$$= 1.5 \, \text{m/sec}.$$

The maximum height may be found from Equation 2:

$$d = \bar{v}t$$
$$= 1.5 \, \text{m/sec} \times 0.31 \, \text{sec}$$
$$= 0.465 \, \text{m},$$

which corresponds to a distance of 1.35 ft.

When the astronaut jumps again on the *moon*, the acceleration of gravity is 1.63 m/sec^2 (one-sixth of earth's gravity). This means that his speed changes 1.63 m/sec during each second. So the change in speed from 3 m/sec to zero will occur in a time given by

$$t = \frac{3 \, \text{m/sec}}{1.63 \, \text{m/sec}^2}$$
$$= 1.84 \, \text{sec}.$$

This is 6 times longer than on earth and explains why the astronauts seem to move in "slow motion." Since the initial and final speeds are the same as on earth, the average speed $\bar{v}$ is also the same, 1.5 m/sec. The maximum height is found from:

$$d = 1.5 \, \text{m/sec} \times 1.84 \, \text{sec}$$
$$= 2.76 \, \text{m}.$$

Because the height is also 6 times larger on the moon, the astronaut's motion is not merely equivalent to running a motion picture in slow motion.

It's amusing to let the imagination wander and to picture how a ballet or a sporting event might appear on the moon. A ballet dancer able to jump 1 m on earth would be able to jump 6 m on the moon. In a broad-jump competition of the Moon Olympics, the distances would be 6 times larger than those on earth. But in pole-vaulting, the situation is more complex since the pole will not bend as much because the pole-vaulter's weight is reduced by a factor of 6. The speed, attained as the pole straightens, is likewise reduced, and it is not valid merely to multiply the height reached on earth by 6. What size would a baseball field have to be on the moon? Thinking along these lines, you can come up with many intriguing situations.

From astronauts moving on the moon to auto traffic moving on earth, the situations involving motion seem endless. But, as we've learned, the principles involved are few and may be applied to all these myriad circumstances. In our discussions, however, we have employed certain terms. Talking about gravity, for instance, we have used the term *pull of gravity* and, in describing Rutherford's scattering experiment, we spoke of the *repulsion* between an alpha particle and the nucleus. In both cases, it would seem that "something" is needed to make an object speed up or slow down. Specific terms, such as *push, pull, attraction,* and *repulsion,* can be described more generally as "forces." Our next task, then, is to investigate the relationship between force and acceleration.

Summary

The basic relationship among distance, average speed, and time is expressed by the equation: Distance equals average speed multiplied by time, $d = \bar{v}t$. This relationship may be applied to any motion—from the movement of a vehicle to the movement of electrons in an atom. Velocity requires information about the speed *and* the direction of motion.

Acceleration is defined as the change in velocity divided by the change in time, $a = (v_f - v_i)/t$. Acceleration occurs when the speed alone changes, when the direction alone changes, or when both change. If there is no concern about change in direction, then, for example, a constant acceleration of 4 m/sec² means that the speed changes by 4 m/sec during each second. When an object travels with constant acceleration, the average speed during the time interval t is given by the expression: Average speed equals one-half the initial speed plus the final speed, $\bar{v} = (v_i + v_f)/2$. The distance can be found from $d = \bar{v}t$. When either the initial speed or the final speed is zero, the distance can be determined using the expression: Distance equals one-half the acceleration multiplied by the time squared, $d = \frac{1}{2}at^2$. When the final speed is zero, the stopping distance d can be found from the equation: Distance equals the initial speed squared divided by twice the acceleration, $d = v_1^2/2a$.

When an alpha particle strikes a gold foil, comes to a stop, and then retraces its path, its deceleration is a tremendously large number (about 10^{21} m/sec²). Yet, other alpha particles striking the foil are not scattered at all, and their acceleration is zero. Rutherford's model of the atom and its nucleus accounts for this wildly different behavior of the alpha particles.

According to legend, Galileo dropped two unequal weights from the Leaning Tower of Pisa and found that the heavier one reached the ground slightly ahead of the lighter one. He concluded that, if air resistance could have been eliminated, the weights would have reached the ground at the same time. This suggested to him that the acceleration of gravity was a constant, independent of the weights of the objects. In experiments, he showed that the acceleration of balls rolling down slightly inclined planes was a constant. He demonstrated this fact by measuring the descent times of the balls and by knowing the length of the plane.

The acceleration of gravity is 9.8 m/sec², meaning that the speed of an object influenced by gravity changes by 9.8 m/sec (22 mph) during each second. To mimic gravity, a car would have to accelerate from rest to 22 mph in exactly 1 second.

True or False Questions

Indicate whether the following statements are true or false. Change all of the false statements so that they read correctly.

2–1 When a car travels at a constant speed for 10 seconds, the distance traveled during the first second is equal to that traveled during the tenth second.

2–2 When a car has a constant acceleration for 10 seconds, the distance traveled during the first second is equal to that traveled during the tenth second.

2−3 When a car has a constant acceleration for 10 seconds, the change in speed from the start to 5 seconds is the same as that from 5 to 10 seconds.

2−4 When a car has a constant acceleration for 10 seconds, the average speed from the start to 5 seconds is the same as that from 5 to 10 seconds.

2−5 When a car has a constant acceleration for 10 seconds, the distance traveled from the start to 5 seconds is the same as that from 5 to 10 seconds.

2−6 When a car starts from rest and has a constant acceleration, the distance traveled from the start to 5 seconds is one-fourth the distance traveled from the start to 10 seconds.

2−7 An acceleration of 3 m/sec^2 means that the speed changes by 3 m/sec during each second.

2−8 In describing the motion of an object, the terms *speed* and *velocity* can be used interchangeably.

2−9 For the same reason that the average of the numbers 10, 20, 30, 40, and 50 is equal to the average of 10 and 50, the average speed for an object undergoing constant acceleration is given by the average of the initial speed and final speed.

2−10 When the drivers of identical cars, one traveling at a speed of 25 mph and the other at 50 mph, slam on the brakes, the car traveling at 50 mph travels twice as far as the other in coming to a stop.

Questions for Thought

2−1 Match the name or term with its description by inserting the correct letter from the lettered column into each blank in the descriptive column. Entries in lettered column may be used more than once.

a. Aristotle
b. Newton
c. Galileo
d. velocity
e. acceleration of gravity on earth
f. acceleration of gravity on the moon
g. average speed
h. speed
i. distance
j. acceleration

____ father of the scientific method
____ believed that an object fell with constant speed, which was proportional to its weight
____ carried out experiments with inclined planes to dilute the effect of gravity
____ rate at which distance changes with time
____ rate at which velocity changes with time
____ speed plus a direction
____ distance traveled divided by time
____ 9.8 m/sec^2
____ showed that the acceleration of gravity was constant
____ 1/6 × 9.8 m/sec^2

2−2 Refer to Table 2−1, which deals with Bohr's model of the hydrogen atom. In the time that an electron in the third orbit makes one revolution, an electron in the second orbit would make _____ revolutions, and an electron in the first orbit would make _____ revolutions.

2−3 In Table 2−1, three allowed orbits are shown. Looking at the sequence of numbers for the radii and speeds in Table 2−1, what are the appropriate values for the radius and speed of an electron in the fourth allowed orbit? What is the time for one revolution of an electron in the fourth allowed orbit?

2−4 Describe the scientific method and show
    a. how Rutherford used it in determining the structure of the atom.
    b. how Bohr used it to determine the structure of the hydrogen atom. (Refer to the discussions in Chapter 1.)

2–5 Suppose that a 2-kg object takes 0.5 seconds to reach the ground. According to the theories of Aristotle, a 4-kg object, falling from the same height, would take _____ seconds to reach the ground. However, experiments show that the 4-kg object takes _____ seconds, and the 2-kg object takes _____ seconds.

2–6 Cut a piece of paper the same size as a quarter. Place the paper on top of the quarter and let them fall to the ground. What do you observe? Explain these results. Ask a friend the question, Which reaches the ground first, a quarter or a piece of paper? Then demonstrate.

Questions for Calculation

2–1 The speed of a car is 40 mph. How far does it travel in 30 minutes? In 45 minutes? In 2 hours?

2–2 A car trip between Cincinnati and Chicago, a distance of 480 km, required 8 hours. What was the average speed?

2–3 You have reservations at a motel 572 km, or 358 miles (mi) away, and the speed limit is 88 km/hr (55 mph). What is the shortest time in which you can reach the motel without exceeding the speed limit?

2–4 The distance from the sun to the earth is 1.5×10^{11} m (93 million mi), and the speed of light is 3×10^8 m/sec (186,000 mi/sec). How long does light from the sun take to reach the earth in seconds? In minutes?

2–5 The particles produced at the Fermi National Accelerator travel inside a circular ring 2000 m in diameter (1.25 mi in diameter). The circumference of this ring is 6283 m. The particles have a speed nearly equal to the speed of light, which is 3×10^8 m/sec. How long does it take a particle to go once around the circular ring? How many times does a particle go around the ring in 1 second?

2–6 While watching a summer thunderstorm, you see a bolt of lightning in the sky and 3 seconds later hear the thunder. The speed of sound is 340 m/sec. How far away were you from the lightning?

2–7 A car traveling in a straight line has an acceleration of 2.5 m/sec^2. With this acceleration, the speed of the car changes by _____ during each second.

  a. If the car is initially at rest, then, after 1 second, the speed is _____ . After 2 seconds, the speed is _____ . After 4 seconds, the speed is _____ . After 4.6 seconds, the speed is _____ .

  b. If the car is initially traveling at 15 m/sec (34 mph) and the speed is increasing, then after 1 second, the speed is _____ . After 2 seconds, _____ . After 4 seconds, _____ . After 4.6 seconds, _____ .

  c. If the speed of the car is initially 15 m/sec and the brakes are applied so that the deceleration is 3 m/sec^2, then after 1 second the speed is _____ . After 2 seconds, _____ . After 4.6 seconds, _____ . The car comes to a stop after _____ seconds.

2–8 Suppose that the brakes of a car are applied when the car is traveling at 50 mph. If the deceleration is 15 mph/sec, how long does it take the car to stop?

2–9 You are driving a sports car on a straight road and wish to measure its acceleration. The car has a speedometer and you have a watch with a second hand. What measurements would you make to find the acceleration of the car? Make up some sample measurements and calculate the acceleration of the car.

2–10 A car accelerates from 15 m/sec (34 mph) to 17 m/sec in 6 seconds and a bicycle accelerates from 3 m/sec to 5 m/sec in 6 seconds.

Insert in the blanks: *larger than*, *smaller than*, or *the same as*.

a. The acceleration of the car is _____ _____ the acceleration of the bicycle.

b. The average speed of the car is _____ _____ the average speed of the bicycle.

c. The distance traveled by the car during 6 seconds is _____ that traveled by the bicycle.

2−11 Figure 2−11 shows three of many possible ways that a car can accelerate from complete rest to a speed of 24 m/sec in 12 seconds. The table shows the speeds at 0, 4, 8, and 12 seconds.

a. In case _____ , the acceleration is constant, while in case _____ , the acceleration increases, and, in case _____ , the acceleration decreases. The car travels the greatest distance in case _____ and the shortest distance in case _____ . Explain the reason(s) for your answers to this last question.

b. For case B, determine the distance traveled during 12 seconds, using the relationship d equals $\bar{v}t$.

c. For case B, find the area of the trianglular section between the sloping straight line and the time axis. Show that this area is equivalent to the distance found in **b.**

d. The small, shaded square represents a distance of 2 m. Find the distance traveled by the car in case A by estimating the number of squares beneath the curve.

2−12 A car accelerates from rest at 2.2 m/sec².

a. The speed after 10 seconds is _____ . The average speed during 10 seconds is _____ . The distance traveled during 10 seconds is _____ .

b. Using the relationship $d = \frac{1}{2}at^2$, find the distance traveled after 10 seconds. This is _____ .

c. The speed after 4 seconds is _____ . The average speed from the start to 4 seconds is _____ . The distance traveled during this 4-second interval is _____ .

Time	Speed (m/sec) A	B	C
0	0	0	0
4	3.5	8	11
8	12.0	16	19
12	24	24	24

FIG. 2−11   Graph for calculation question 2-11.

d. Using the distances found in **a** and **c,** find the distance traveled from 4 to 10 seconds. This is _____ .

e. The car's average speed from 4 to 10 seconds is _____ . The distance traveled is _____ .

f. Can the relationship $d = \frac{1}{2}at^2$ be used to find the distance traveled from 4 to 10 seconds? Explain the reason for your answer.

2−13 The speed of a car changes from 14 m/sec (31 mph) to 22 m/sec (49 mph) in 4 seconds. The constant acceleration of the car is _____ . The average speed is _____ . The distance traveled is _____ .

2−14 If an alpha particle striking the gold foil in Fig. 2−4 travels only halfway through the foil before turning around, what is its deceleration? If it travels only one one-thousandth of the way, what is its deceleration?

2–15 A rock is dropped from rest into a mine shaft. The acceleration of gravity is 9.8 m/sec². This means that the rock's speed increases by _____ during each second in motion, or increases by _____ during each half-second in motion. After 0.5 seconds, the rock's speed is _____ . After 1 second, the speed is _____ . After 2 seconds, the speed is _____ . The average speed during the first second is _____ , and the distance traveled is _____ . Using the relationship $d = \frac{1}{2}at^2$, the distance traveled during the first second is _____ .

†2–16 A heavy object is dropped from the top of the John Hancock Building in Chicago. It takes 8.4 seconds to reach the ground. With what speed does the object strike the ground? How high is the John Hancock Building?

2–17 Susan lets her toy truck roll down a slide 3 m long. It takes 1.5 seconds to reach the end. Find the acceleration of the truck and compare it with the acceleration of a freely falling object.

†2–18 A bank robber, standing on a bridge, throws his gun into the murky river below. The gun has a speed of 8 m/sec as it leaves his hand and enters the water after 1 second.

    a. What is the gun's speed when it enters the water?

    b. How high is the bridge above the water?

    c. If the robber simply dropped the gun from rest into the water, how long would it take to enter the water?

†2–19 The driver of a car slams on the brakes because the bridge, only 32 m ahead, has been washed out. If the car is traveling at 21 m/sec (47 mph), and the maximum deceleration of the car is 7 m/sec², will the driver be able to stop in time?

Footnote

[1] Bernard Cohen, "Galileo," *Scientific American* 181(1949):40.

Introduction

We'll now continue discussing motion, this time focusing on *the forces* responsible for it. We'll start with a commonsense notion of force, then define force precisely through Newton's three laws, and review fundamental situations where force is required: when a car speeds up, slows down, or travels in a circle; when an object falls; when an electron travels in a circular orbit around the nucleus in a hydrogen atom.

The latter part of this chapter will turn to the subject of force and the structure of the atom. Since the atom contains charged particles, we'll examine the forces between those charged particles first. The law describing these forces is called Coulomb's law. Rutherford's scattering experiment (introduced in Chapter 1 and used to illustrate acceleration in Chapter 2) will be brought back into play. Here, we'll examine the connection between acceleration and force. There must be a tremendous force of repulsion (described by Coulomb's law) between an alpha particle and a gold nucleus to cause backward-scattering. We shall see how this led Rutherford to propose a nuclear model of the atom. Then, we'll conclude with our ongoing discussion of Bohr's model of the hydrogen atom.

Charles Augustine Coulomb

CHAPTER 3

Forces: Newton's Laws and Coulomb's Law

What Is Force?

One of the first signs of spring is the neighborhood baseball game. Sometimes, the baseball hits a window: The glass shatters, and the baseball is slowed down and finally stops as a result. The baseball exerted a force on the glass and the glass moved—it broke. But the motion of the baseball changed too. Forces have been exerted on both the glass *and* the baseball. When a child pulls (that is, exerts a force) on his or her mother's hair, the mother's head moves. When automobiles collide, forces crumple the fenders. Each one of these examples illustrates our commonsense notion of **force** as a push or a pull, or a kind of influence on an object. So, forces produce motion (the glass shatters, the head moves) or changes in motion (the baseball stops, the fenders crumple).

Newton's Laws of Motion

Newton's three laws of motion describe precisely how an object moves when a force is applied to it. These laws, stated in 1686, are valid for all situations except those in which the speeds being considered approach the speed of light. The three laws of Newton are:

1. An object remains at rest or moves with constant speed in a straight line unless a net force acts on it.

2. The acceleration produced by a net force is directly proportional to the net force and inversely proportional to the mass of the object being accelerated. This law can be written in symbols

$$a = \frac{F_{net}}{m},$$

or

$$F_{net} = ma. \tag{1}$$

3. Whenever one object exerts a force on a second object, the second object exerts a force equal in magnitude and opposite in direction on the first object.

The terms *net force* and *mass* are extremely important to any understanding of Newton's laws. First, then, we'll clarify the meaning of net force, and next we'll establish a precise definition for mass.

NET FORCE: AN EXAMPLE

You are walking two dogs. Not resisting, you allow the dogs to pull you in *opposite directions* (see Fig. 3–1). If the dogs pull with the same force, then you certainly feel these forces, but you will not move. You remain where you are because the forces acting on you are equal and are being exerted in opposite directions. According to Newton's laws, this situation has

the same effect on you as if the dogs were exerting no force at all on your arms. In either case, you do not move, so, in both instances, the net force is zero.

Now, suppose that dog B pulls harder than dog A. You will then, as a result, move (or accelerate) toward the right. The net force is then

$$\text{Net force} = \text{pull of B} - \text{pull of A},$$
$$F_{net} = F_B - F_A. \tag{2}$$

You would move in exactly the same way if a single dog pulled to the right with a force given by F_{net} in Equation 2.

When both dogs pull in the *same direction*, you will accelerate in that direction (see Fig. 3–2). The net force here is given by

$$F_{net} = \text{pull of A} + \text{pull of B}$$
$$= F_A + F_B. \tag{3}$$

Again, you would move in exactly the same way if a single dog pulled you to the right with a force given by F_{net} in Equation 3. Thus, a **net force** is a single force that is equivalent to several forces acting in unison on an object. The net force causes an object to move in the same way it would if several forces were acting on it simultaneously.

(The dogs in our example can, of course, pull at an angle, so that the directions of the forces involved are neither the same nor opposite. This particular situation is more complex and need not concern us at this point.)

THE DEFINITION OF MASS

In our daily language, we commonly interchange the words *mass* and *weight*, but, in physics, both of these terms carry very exact meanings. The implications of Newton's laws would surely be much different if "mass" were replaced by "weight"!

Mass can be defined as the quantity of matter in an object. Thus, the mass of an object is the same anywhere in the universe. Mass also refers

FIG. 3–1   Forces exerted on you by dogs pulling your arms in opposite directions.

FIG. 3–2   Forces exerted on you by dogs pulling your arms in the same direction.

to **inertia,** or the resistance of an object to change in motion.

Consider the Lunar Rover, the vehicle that the astronauts used to explore the moon's surface (Fig. 3–3). It has the same mass on the moon as it does on earth. According to Newton's second law, the net force needed for a given acceleration is the same on the moon as it is on the earth. If the astronauts had made such a comparative determination, they would have found that Newton's second law is correct.

You undoubtedly know that the weight of an object on the moon is only one-sixth of that on earth. How does weight differ from mass? This question will be answered shortly in the subsequent section "The Force of Gravity." Before we get into that, however, and before we actually get down to applying Newton's laws to gravity and practical cases involving motion, let's clarify the first law a bit.

Clarification of Newton's First Law

The first law does seem reasonable where objects at rest are concerned. But it also states that the net force is zero when an object travels with constant speed. You may thus have trouble relating Newton's first law to your everyday experience. You know that anything you start into motion will eventually stop, whether the moving object is a billiard ball on a smooth table or your new Italian 10-speed bicycle on a level road. Unless you give that ball a push or pedal that bike from time to time, they will stop. In other words, moving things sooner or later stop completely unless something acts to keep them going. But Newton seems to be saying the opposite: that moving things stay in motion unless something acts to stop them.

In reality, there is no conflict. The reason you observe moving things slowing down and stopping is that, here on earth, there is *always* a net force acting on them, just as Newton called for—

that is, the force of **friction.** Friction can occur between a moving object and the air, between the object and the ground it travels over, or even between any moving parts of the object itself. The resistive force we call friction always acts to *oppose* motion, so that any moving object on earth slows down and eventually stops.

To find the circumstances under which an object, once set into motion, could keep moving forever, we have to get far away from earth (or any celestial object) and go deep into space. Free of friction, a space capsule would, for example, travel through space with constant speed in a straight line, using further force only to change its speed or its direction. In deep space, Newton's first law makes perfect sense.

For example, when our moonbound capsule is, say, halfway to the moon, the gravitational attraction to the earth and to the moon is extremely small. As a result, the capsule's acceleration is very nearly zero. Here, the capsule travels with essentially constant speed in a straight line, without the use of fuel.

The Force of Gravity

Consider an apple that falls from a tree to the ground. Because the falling apple accelerates, Newton's laws maintain that a force must be exerted steadily on it. We learn the *amount* of this force by multiplying the apple's mass by its acceleration. However, Newton's laws completely ignore *how* the force was exerted, as well as the *nature* of the force producing this acceleration. Newton, and other scientists before him, attributed this force to the attraction of the apple to the earth, a phenomenon difficult to visualize. We see no "wind" blowing the apple to the ground, no "hand" pulling it down. All that we can perceive is the *effect* of the force of gravity: the apple accelerates. The force of gravity is called **weight.** Thus,

Weight = mass × acceleration of gravity.   (4)

On earth, this would be

$$\text{Weight} = mg,$$

where g equals 9.8 m/sec². On the moon, this would be

$$\text{Weight} = mg_{\text{moon}},$$

where g_{moon} equals 1.63 m/sec². The weight of an apple with a mass of 0.2 kg is

$$\begin{aligned}\text{Weight of apple on earth} &= 0.2\,\text{kg} \times 9.8\,\text{m/sec}^2 \\ &= 1.96\,\text{kg-m/sec}^2 \\ &= 1.96\,\text{newtons,}\end{aligned}$$

and its weight on the moon is 0.326 newton. (Note that in the unit kg-m/sec², the hyphen between the "kg" and the "m" denotes multiplication.) The **newton** (abbreviation: N) is defined as the amount of force that will give a 1-kg mass an acceleration of 1 m/sec², or

$$1 \text{ newton} = 1 \text{ kg-m/sec}^2. \qquad (5)$$

Since the acceleration of gravity on the moon is one-sixth of what it is on earth, the weight of any object on the moon, such as the Lunar Rover, is one-sixth of what it would be on earth. If the astronauts had needed to lift the Lunar Rover, the force required to overcome moon gravity would be one-sixth of that needed for this same job on earth. However, if the astronauts tried to push the Rover across the moon's surface, the same net force would, for a given acceleration, be needed on the moon as on the earth. The Rover's weight is different on the moon and on the earth, but its mass is the same everywhere.

Consider again the falling apple illustrated in Fig. 3–4(a). According to Newton's third law, the earth is likewise attracted to the apple with a force of 1.96 newtons. We can paraphrase the earlier statement of Newton's third law, as:

When the earth exerts a force of 1.96 newtons on the apple, the apple exerts a force of 1.96 newtons on the earth, but in the opposite direction.

From Newton's second law, we can see why the earth doesn't move as a result: The acceleration

FIG. 3–3 Astronaut James B. Irwin, lunar module pilot, walking away from the Lunar Rover during the third *Apollo 15* lunar surface extravehicular activity at the Hadley-Apennin landing site. (Courtesy of NASA)

FIG. 3–4 (a) Gravitational forces exerted on the falling apple and on the earth. (b) Forces exerted on the apple at rest on the ground and on the earth.

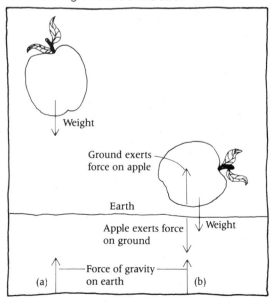

of the earth equals F/m, which equals 1.96 newtons divided by the mass of earth, which approaches zero due to the huge mass of the earth.

When the apple rests on the ground (Fig. 3−4(b), Newton's first law states that the *net* force on it is zero. Gravity still pulls on the apple, but the ground prevents it from falling farther. That is, the ground exerts an upward force on the apple that just balances the apple's weight, and the net force on the apple is zero. Newton's third law implies that forces always come in pairs. Since "the ground exerts an upward force on the apple," then the apple must exert a downward force on the ground. Figure 3−4(b) shows all of these forces, with the apple drawn slightly above the ground so that forces can be clearly indicated. All of the forces are the same size and equal to the weight of the apple, which is 1.96 newtons.

Falling Objects and Air Resistance

In a vacuum, a feather and a lead weight have the same acceleration (see Fig. 2−9, p. 49). We know that this is not the case when these objects fall through the air, so let us now examine **air resistance** and its influence on falling objects.

The size and the speed of a falling object determine the amount of air resistance it will encounter. This "resistance" is actually due to collision with air molecules—the more collisions there are, the more resistance there will be. Imagine a large and a small object falling at the same speed. Because the larger object will encounter more air molecules, the air resistance that acts on it will be larger. Next, picture two objects of the same size, but having different speeds. Because the object traveling at the higher speed will encounter more air molecules per second, the air resistance that acts on it will be

larger. Thus, air resistance depends upon a falling object's size and on its speed.

Suppose that Galileo had dropped an iron ball and a wooden ball of the same diameter from the Leaning Tower of Pisa. There is now a computer program called PISA[1] that reveals the effects of air resistance on falling balls such as these. Figure 3−5(a) displays these two balls, each with a 20-cm diameter, falling from the Tower. The iron ball strikes the ground slightly ahead of the wooden one. The calculations show that the iron ball takes 3.344 seconds to reach the ground, and the wooden ball takes 0.068 seconds longer. If they had fallen that distance in a vacuum, the time for each one would be the same—3.337 seconds. From this data, we see that air resistance influences both, but affects the lighter one—the wooden ball—more. Newton's second law can help us understand these results.

Since their diameters are the same, both balls experience the same air resistance at a given speed. The force of gravity mg and the opposing air resistance F_R act on each ball (Fig. 3−6), giving a net force equal to the force of gravity minus the opposing air resistance $(mg - F_R)$. From Newton's second law, we can find the acceleration of each one:

$$mg - F_R = ma$$

$$a = \frac{mg - F_R}{m}$$

$$= g - \frac{F_R}{m}.$$

Since the air resistance F_R is the same for both balls, the term F_R/m is smaller for the iron ball. As a result, the iron ball has a larger acceleration than the wooden one, and therefore will, as noted, reach the ground first.

This equation also shows that an object with a very small mass and a large surface area, such as a feather, will be greatly affected by air resistance when dropped. PISA can be used to investigate this situation. Instead of a feather, though, let's just use a lightweight ball. Figure

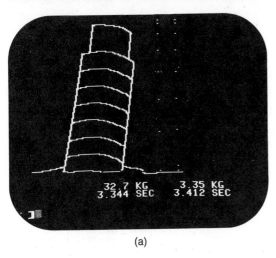

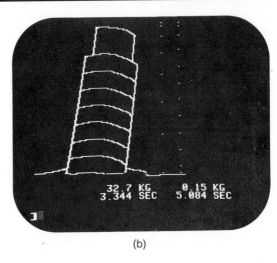

(a)	(b)

FIG. 3−5 (a) In this run of PISA, two balls having a diameter of 20 cm fall from the Leaning Tower of Pisa, a height of 54.6 m. The iron ball on the left has a mass of 32.7 kg, and the wooden ball on the right has a mass of 3.35 kg. The iron ball takes 3.344 seconds to reach the ground, and the wooden one, 3.412 seconds. The positions of the balls are shown at 0.167-second time intervals. (b) This run shows the positions of a 20-cm iron ball on the left and a lightweight ball (diameter = 20 cm, mass = 0.15 kg) on the right. The time interval is the same as in (a).

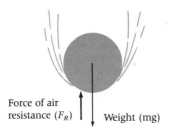

Force of air resistance (F_R)   Weight (mg)

FIG. 3−6   A ball encountering air molecules as it falls to the ground under the influence of gravity. The ball's weight (mg) is a downward force. The force of air resistance (F_R) opposes motion and is an upward force. Since the forces are in the opposite direction, the net force is equal to the difference of the two forces ($mg - F_R$). In the diagram, the ball's weight is larger, so that the ball accelerates. When the force of air resistance equals the weight, the net force is zero, and the ball travels with constant speed.

3−5(b) shows the motion of a 20-cm ball having a mass of only 0.15 kg. It takes 5.084 seconds for this ball to reach the ground. Air resistance greatly increases the descent time. The ball accelerates at first because the distance between adjacent "dots" increases. Quickly, though, the "dots" become equally spaced, indicating that the ball falls with constant speed, which calculations show to be 13 m/sec.

As this ball falls, its speed increases; consequently, its air resistance also increases. Soon the air resistance equals the ball's weight (force of gravity), and the net force on the ball is zero. When this happens, the ball's speed is constant, as described by Newton's first law. This constant speed is called its **terminal speed,** which is the maximum speed that an object can attain falling through air. In this case, the terminal speed of the lightweight ball is 13 m/sec.

FORCES NEWTON'S LAWS AND COULOMB'S LAW

Riding an Elevator

Riding in an elevator, you can sense an increased force on your feet just as the elevator starts to move upward, and a decreased force when the elevator just starts to move downward. A few seconds after the elevator is in motion, the force on your feet returns to normal. Newton's laws explain these effects.

You exert a *downward* force on the floor (F_f), and the floor exerts an *upward* force (F_p) on you, the passenger. From Newton's third law, F_f equals F_p at all times. Your weight (W) is always the same no matter how the elevator moves (see Fig. 3–7).

When the elevator is at rest, or is moving with constant speed, the net force acting on you must be zero. Therefore, the upward force F_p just balances your weight:

$$F_p = W.$$

Since F_f equals F_p, then F_f equals W.

When the elevator accelerates upward, a net force in the upward direction must be exerted on you. Therefore, F_p must be greater than W, and you feel a greater force on your feet. The downward force F_f is also greater than W.

When the elevator accelerates downward, a net force in the downward direction must be exerted on you. Therefore, F_p must be less than W, and you feel a smaller force exerted on your feet. Since F_f equals F_p, the downward force F_f is also less than W.

Now, let's look at an example of these forces at work: an elevator passenger standing on a scale. If we know that the passenger has a mass of 50 kg, and that the elevator accelerates upward or downward at 3 m/sec², we can calculate the scale readings when the elevator is at rest, or is moving with constant speed, accelerating upward and accelerating downward.

First, we find the weight of the passenger:

$$W = mg$$
$$= (50\,\text{kg})(9.8\,\text{m/sec}^2)$$
$$= 490\,\text{newtons}.$$

When the elevator is at rest, or is moving with constant speed, we know from the preceding discussion that F_f equals W, and, therefore, the scale reading (equal to F_f) is 490 newtons.

When the elevator accelerates upward, the net force acting on the passenger is

$$F_{net} = ma$$
$$= (50\,\text{kg})(3\,\text{m/sec}^2)$$
$$= 150\,\text{newtons}.$$

However, the net force is equal to the *difference* of F_p and W_p; that is,

$$F_{net} = F_p - W$$
$$150\,\text{newtons} = F_p - 490\,\text{newtons}$$
$$F_p = 640\,\text{newtons}.$$

The forces on the floor F_f is also 640 newtons, and the scale reads 640 newtons—an increase of 31 percent from 490 newtons.

When the elevator accelerates downward, the net force acting on the passenger is also 150 newtons, but in the downward direction. In this case, the weight W is *larger* than F_p, and the net force is W minus F_p.

$$F_{net} = W - F_p$$
$$150\,\text{newtons} = 490\,\text{newtons} - F_p$$
$$F_p = 340\,\text{newtons}.$$

F_f is also 340 newtons, and the scale reads 340 newtons—a decrease of 31 percent from 490 newtons.

If the scale has been calibrated in pounds (lb) instead, the reading at rest or at constant speed would be 110 lb (50 kg). The scale reading when accelerating upward would have increased by 31 percent to 144 lb. When accelerating downward, the scale reading would have decreased

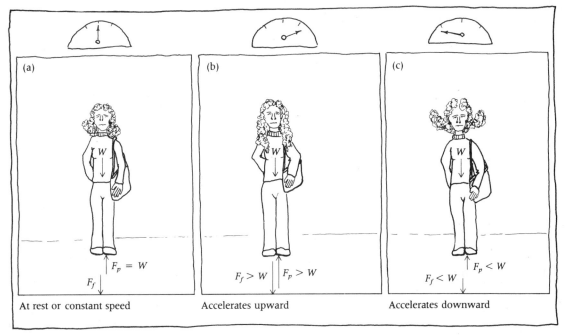

In each panel the elevator contains a standing person labeled W.

(a) $F_p = W$
F_f
At rest or constant speed

(b) $F_f > W$ $F_p > W$
Accelerates upward

(c) $F_p < W$
$F_f < W$
Accelerates downward

FIG 3-7 The forces exerted on a passenger in an elevator when (a) the elevator is at rest or moving at constant speed, (b) the elevator accelerates upward, and (c) the elevator accelerates downward.

by 31 percent or to 76 lb. Try this experiment yourself!

Motion of a Car

CONSTANT SPEED

Since cars probably provide one of our most common experiences with motion, let's talk about a car moving with constant speed in a straight line. Two forces act on the car: One force, F_{gas}, which is due to gasoline consumption, turns the rear wheels of the car. The opposing or resisting forces, together called F_R, are due to road friction and air resistance (both are actually specific types of friction). The net force is then given by

$$F_{net} = F_{gas} - F_R,$$

where F_R is in the opposite direction to F_{gas}. Since F_{net} equals zero for constant speed, F_{gas} and F_R must be equal at any given constant speed. Let's see, then, why gas mileage is more efficient at lower speeds.

Consider two cars: one is traveling at 55 mph and the other, at 70 mph. Because air resistance increases dramatically with increasing speed, F_R is larger at 70 mph than it is at 55 mph. Since F_{gas} equals F_R, and because gasoline consumption is related to F_{gas}, you get better gas mileage at 55 mph than you do at 70 mph. In 1973, the maximum speed limit for automobiles in the United States was lowered from 70 to 55 mph

FORCES NEWTON'S LAWS AND COULOMB'S LAW

in an effort to reduce gasoline consumption. According to the federal Energy Administration, most automobiles get about 21 percent more miles per gallon (mpg) at 55 mph than at 70 mph.

AN EXPERIMENT THAT SHOWS THE EFFECTS OF WIND RESISTANCE

An easy experiment may be performed to demonstrate that the force of wind resistance decreases when the speed of the car decreases. On a little-traveled road, take a car up to 55 mph. Then put it in neutral, so that there will be no further effects from the engine. The force of air resistance (as well as other smaller frictional forces) will decelerate the car. If you measure this deceleration, then you can find the force of air resistance from Newton's second law:

$$F_R = \text{mass of car} \times \text{deceleration of car.}$$

It should take only about 6 seconds for the wind resistance to reduce the car's speed to 50 mph. The car decelerates by 5 mph in 6 seconds. Now, repeat the experiment at 25 mph. You should find that, at this slower speed, it takes the air resistance almost twice as long (11 seconds) to cut the speed by that same 5 mph. Since the deceleration at 25 mph is about one-half what it is at 55 mph, the force of air resistance at 25 mph is about one-half what it is at 55 mph.

ACCELERATION

Suppose a car travels from rest to 24.4 m/sec (54.5 mph) in 10 seconds, an acceleration of 2.44 m/sec². If the car has a mass of 1500 kg or 3300 lb, then the net force acting on the car is

$$F_{net} = (1500 \text{ kg})(2.44 \text{ m/sec}^2)$$
$$= 3600 \text{ newtons.}$$

Since $F_{net} = F_{gas} - F_R$, this becomes
$$F_{gas} - F_R = 3600 \text{ newtons}$$
$$F_{gas} = 3600 \text{ newtons} + F_R.$$

Thus, gasoline is consumed in order to overcome air resistance and other frictional effects, and to provide acceleration.

If you are a passenger in this car and you have a mass of 60 kg (132 lb), then the net force acting on you is

$$F_{net} = 60 \text{ kg} \times 2.44 \text{ m/sec}^2$$
$$= 146.4 \text{ newtons.}$$

This is due to the back of the seat pushing you forward (see Fig. 3–8). Since forces come in pairs, you push backward on the seat, and the upholstery is compressed. Note that the only force acting on you, the passenger, is in the forward direction. To say that "you are pushed backward" is incorrect; this statement really refers to the compressed upholstery that results when you push on it.

DECELERATION

Picture yourself as a passenger in the front seat of a car traveling in a straight line at 50 mph, and the driver suddenly applies the brakes during a "panic stop." You might say that you were thrown against the dashboard. The word *thrown* implies that a force had been exerted on you. Is that what really happened? Well, according to Newton's first law, your body will continue to move in a straight line at 50 mph unless acted upon by another force. This other force could come from friction between your body and the seat of the car, as well as from restraint caused by the seatbelt. If the force of friction were insufficient, and you were not protected by a seatbelt, then you would continue to move at 50 mph while the car stopped. Your head would hit the dashboard, not because a force threw you forward, but because a force had *not* been applied to oppose your forward motion.

Circular Motion

Our attention will now turn to the concept of circular motion and the force behind it. That any object moving in a circle must have a force exerted on it is apparent from Newton's first law. Even if this object moves with constant speed, it isn't moving in a straight line and, hence, somewhere a force must be exerted on it. Newton's second law shows that the force required can be calculated if one knows the mass of the object and also knows the acceleration due to the changing direction. Our goal, then, is to calculate the force required for an object to move with constant speed v in a circle of radius r. The next logical step would be to calculate the acceleration by dividing the change in velocity due to the changing direction, by the time. Remember that velocity is defined by specifying an object's speed *and* its direction. Thus, if speed and/or direction changes, then the velocity changes. Here, we shall use an approach not quite so rigorous, but based upon a familiar example: the forces that a passenger in a car feels as the car travels over a circular path with constant speed.

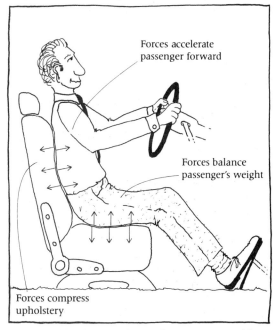

FIG. 3–8 The forces acting on the passenger and the car seat when the car accelerates forward.

A COMMON MISCONCEPTION

Passengers in a car turning sharply to the left often report being thrown against the right-hand door. In explaining what happened, they might say that *centrifugal force* (see the following section) threw them to the right. Let's examine this claim.

Pretend you *are* in a car that turns left too sharply. Will you, in fact, be thrown toward the right-hand door? Has a force actually been exerted on your body? The answer is no. Your body would continue to move in a straight line (a direction that would now seem to be forward and to the right), while the car was turning left (see Fig. 3–9). Eventually, your right shoulder would hit the right-hand door. Your body sways because a force had *not* been applied to your right shoulder to oppose its motion in a straight line.

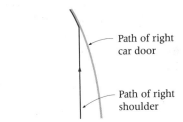

FIG. 3–9 When the car makes a left turn, the passenger's shoulder continues to travel in a straight line. The passenger sways, causing a shoulder to collide with the right-hand door.

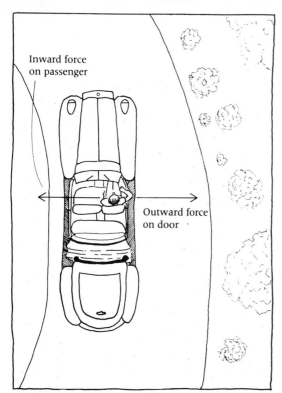

FIG. 3–10 The passenger is seated next to the right-hand door. An inward force causes the passenger to move in a circular path rather than traveling in a straight line.

FIG. 3–11 A car traveling over three circular paths with the same speed.

Okay, you are now seated next to the door as the car makes a sharp left turn (Fig. 3–10). In this case, the door prevents your swaying to the right. That is, the door exerts a force on you directed to the left, and your body travels in a circle, not in a straight line. The question is, How much force is exerted on you and in what direction?

First, let's talk about the *direction* of the force on you, the passenger. In Fig. 3–10, the force on the passenger is directed to the left, or, more generally for any point on the circle, toward the center of the circle. Since the door exerts a force on you to the left (or toward the center of the circle), you also exert a force on the door toward the right (or, away from the center of the circle). If, for example, the car door were not latched properly, it might come open. According to Newton's third law, the force on the car door is equal to the force on the passenger and these forces are oppositely directed, as we have just seen. The force that is directed away from the center of the circle is called the **centrifugal force.** That is, the centrifugal force acts on the car door, not the passenger. Thus, the passenger cannot say that "centrifugal force threw me to the right" because centrifugal force is exerted upon the *door,* not the passenger.

Second, we must take into account *how much* force is exerted on you the passenger as the car travels with constant speed v over the circular path of radius r. Assume you're seated next to the right-hand door in a car that travels at the same speed over the three circular paths shown in Fig. 3–11. On which path do you feel the greatest force? Undoubtedly, on the one with the sharpest curve or the smallest radius, r_1. The important conclusion, then, is as follows:

Conclusion A: When a car travels at the same speed over circular paths, each with a different radius, the force on the passenger is largest when the radius is smallest.

Now, suppose you happen to be a passenger in a car that travels over the same circular path at different speeds. At which speed do you feel the greatest force? Certainly, you find that the force increases when the speed increases. The important conclusion is as follows:

Conclusion B: When a car travels over the same circular path at different speeds, the force on the passenger increases when the speed increases.

Let us now try to construct a formula that relates the mass of the passenger, the speed v, and the radius r to the force. We already know from Newton's second law that the force is proportional to the mass of the passenger. Consider the basic relationship:

$$F = \frac{mr}{v}.$$

Is this relationship plausible? When r increases (and m and v remain constant), then F increases, which is just the opposite of conclusion A. When v increases (and m and r remain constant), then F decreases, which is just the opposite of conclusion B. From this, we obtain:

Conclusion C: In a mathematical formula for circular motion relating the force F to m, v, and r, the radius r must be in the denominator and the speed v in the numerator.

The simplest formula that will satisfy conclusion C is

$$F = \frac{mv}{r}.$$

Is this right? If the mass is given in kilograms, the speed in meters per second, and the radius in meters, then a correct formula for the force must have the units of newtons, or kg-m/sec². If this is correct, then the right-hand side must have units of kg-m/sec². Thus, the formula

$$\frac{mv}{r} = \frac{\text{kg} \times \text{m/sec}}{\text{m}}$$

$$= \text{kg} \times \frac{\text{m}}{\text{sec}} \times \frac{1}{\text{m}}$$

$$= \frac{\text{kg}}{\text{sec}}.$$

The right side does not have the proper units, so it cannot be correct. Since the proper units require the term sec², let us try the following:

$$F = \frac{mv^2}{r}$$

$$\frac{mv^2}{r} = \frac{\text{kg} \times (\text{m/sec})^2}{\text{m}}$$

$$= \text{kg} \times \frac{\text{m}^2}{\text{sec}^2} \times \frac{1}{\text{m}}$$

$$= \text{kg} \times \frac{\text{m}}{\text{sec}^2}.$$

This has the right units and thus constitutes a plausible relationship.

In summary, then:

1. An object of mass m traveling with constant speed in a circle of radius r has an acceleration given by

$$a = \frac{v^2}{r}. \tag{6}$$

2. The net force that must be exerted on an object in order to make it move in a circle is given by

$$F_{\text{net}} = \frac{mv^2}{r}. \tag{7}$$

3. The net force exerted on the object must be directed toward the center of the circle.

CIRCULAR MOTION: AN EXAMPLE

You are driving a car along a curved road at a speed of 15 m/sec (33.6 mph) and you hit a patch of ice. The car starts to go off the road and you pump (rather than slam on) the brakes. To investigate what happens and why you reacted the way you did, let the mass of the car be 1500 kg (3300 lb), and let the curved road have a circular path with a radius of 180 m (590 ft). When the car hits the patch of ice, the frictional force between the tires and the road is reduced by a factor of 5.

First, let's find the frictional force between the tires and the road *before* the car hit the patch of ice. This force, which keeps the car traveling over a circular path, is given by

$$\text{Frictional force} = \frac{mv^2}{r}$$

$$= \frac{(1500\,\text{kg})(15\,\text{m/sec})^2}{180\,\text{m}}$$

$$= 1875\,\text{newtons}.$$

When the car hits the patch of ice, this frictional force is reduced to 1875 newtons/5, or 375 newtons. As a result, the car begins to go off the road. What really happens is that the radius of the circular path becomes much larger. Solving Equation 7 for r, we find

$$r = \frac{mv^2}{F_{net}}$$

$$= \frac{(1500\,\text{kg})(15\,\text{m/sec})^2}{375\,\text{newtons}}$$

$$= 900\,\text{m}.$$

Thus, the car is now traveling over a much wider curve than the road (180 m), and will go off the road unless the speed is reduced.

The next question is, logically, What must be the car's speed if it is to stay on the road as it travels over the ice? Solving Equation 7 for v^2, we find

$$v^2 = \frac{F_{net}r}{m}$$

$$= \frac{(375\,\text{newtons})(180\,\text{m})}{1500\,\text{kg}}$$

$$= 45\,\text{m}^2/\text{sec}^2$$

$$v = 6.7\,\text{m/sec}.$$

When the car hits a patch of ice, its speed must be reduced from 15 m/sec (33.6 mph) to 6.7 m/sec (15 mph) to stay on the road.

Coulomb's Law

Static electricity, which occurs when two different kinds of materials are rubbed together, provides the most familiar example of charged particles. When wet clothes are placed in a dryer, for instance, they acquire a static charge. In common demonstrations of static electricity, a rubber rod rubbed with fur becomes negatively charged; a glass rod rubbed with silk becomes positively charged.

In 1785, the French physicist Charles Coulomb (1736–1806) used the apparatus sketched in Fig. 3–12 to find the force between the charged particles. The dumbbell, which consists of two lightweight balls, is suspended from its center by a fiber that is finer than a human hair. One ball of the dumbbell is negatively charged by static electricity. When another negatively charged ball is brought close to this one, the dumbbell starts to rotate due to the repulsion that occurs between like charges. Because the fiber resists rotation, the rotation stops at a maximum angle (designated by θ_{max}). This **maximum angle of rotation** depends upon the force of repulsion between the two balls. In fact, θ_{max} is proportional to F, the force of repulsion. Figure 3–13 shows how Coulomb obtained the same amount of charge on balls A and B.

The charge on ball A (or B) was cut in half by touching it with a neutral identical ball. One-half of the charge on ball A (or B) flowed onto the ball that came in contact with it. In this way, Coulomb determined the relative amounts of charge on ball A and ball B. In his experiments, Coulomb varied the charge on balls A and B, and he also varied the distance between them. Each time, he measured the angle θ_{max}, which turned out to be proportional to the force of repulsion. From this series of measurements, Coulomb was able to show that the force was proportional to the charge on each ball, and inversely proportional to the square of the distance between them. This relationship is known as **Coulomb's law** and may be expressed mathematically as

$$F \propto \frac{Q_A Q_B}{d^2},\qquad (8)$$

where Q_A represents the amount of charge on ball A, Q_B the amount on ball B, and d is the distance between the centers of the two balls.

As an illustration of Coulomb's law, consider two positively charged particles, one having twice the amount of charge as the other, separated by a distance of 1 m. Even though the charges are not the same, the force on one particle is equal to that on the other (see Fig. 3–14). Let's designate this force F_0. When the distance is 2 m (Fig. 3–14b), the force on each particle is $F_0/4$ due to the fact that the distance is squared in the denominator of Equation 8. That is, if the distance between two charged particles is increased by a factor of 2 (the distance is doubled), then the force on each particle decreases by a factor of 4. Similarly, if the distance between two charged particles increases by a factor of 4, then the force on each particle decreases by a factor of 16 (see Fig. 3–14c).

Note that, in this example, the force was not given in newtons, nor were there units associated with each charge. We simply used Coulomb's law to see how the force changed as the distance changed. In fact, we shall be greatly concerned with such comparisons in the rest of this chapter, and the proportionality in Equation 8 is quite adequate for this purpose. However, should we need the force expressed in newtons, Coulomb's law must be changed into an equation, and we must specify the units of charge. In order to better comprehend the difference between an equation and a proportionality, let's consider an example close to home.

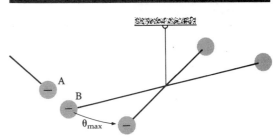

FIG. 3–12 A schematic diagram of the apparatus used by Coulomb to investigate the forces between charges.

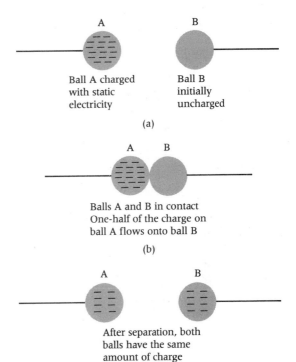

FIG. 3–13 The method used by Coulomb to obtain the same amount of static charge on two balls. The charge on ball A could be cut in half again by touching it with yet another identical uncharged ball.

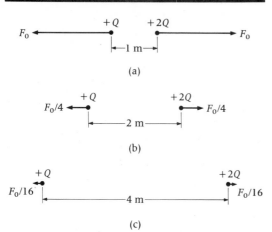

(a)

(b)

(c)

FIG. 3–14   This diagram shows how the forces on two positively charged particles change when the distance between them changes from 1 m to 2 m to 4 m.

FIG. 3–15   Disintegration of a radium-226 nucleus into an alpha particle and a radon-222 nucleus.

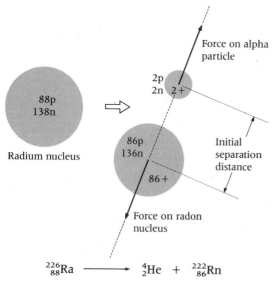

$$^{226}_{88}\text{Ra} \longrightarrow {}^{4}_{2}\text{He} + {}^{222}_{86}\text{Rn}$$

If you have a job, your pay P is proportional to the number of hours H that you worked, thus: P "is proportional to" H; that is, you receive twice as much pay for 8 hours of work as you do for 4 hours. This proportionality does not indicate the amount of your pay in dollars. To obtain this dollar figure, we need the equation P equals SH, where S is your salary in dollars per hour. When both the salary S and the hours H are substituted into P equals SH, the "hours" cancel, and your pay is indicated in dollars. The important point to note here is that the proportionality was changed into an equation by including a **constant**—in this case, S. Similarly, to change Coulomb's law into an equation, we include a constant called K. Then, Coulomb's law becomes

$$F = \frac{KQ_A Q_B}{d^2}. \qquad (9)$$

The charge is specified in terms of a unit called the **coulomb.** One coulomb represents an extremely large amount of charge. The electron, for example, has a negative charge of 1.6×10^{-19} coulombs. Since the force must be expressed in newtons, the constant K must have units of N-m^2/C^2 (where N, remember, is the abbreviation for newtons and C is the abbreviation for coulombs). We can see that Equation 9 has units of newtons by substituting these units into the right-hand side:

$$\frac{(\text{N-m}^2/\text{C}^2)(\text{C})(\text{C})}{\text{m}^2} = \text{N}.$$

The following section shows how the value of K can be determined.

THE CONSTANT K AND THE UNIT OF CHARGE: AN EXAMPLE*

A 3-kg object and a 6-kg object are initially separated by a distance of 1 m. Both objects have the same amount of negative charge. Because of their mutual repulsion, the objects

*An asterisk by an example indicates that the example may be omitted without loss of continuity.

move apart. The 3-kg object has an initial acceleration of 30 m/sec², and the 6-kg object has an initial acceleration of 15 m/sec². We can find the force on each object by using Newton's second law. The force on the 3-kg object is

$$F = ma$$
$$= (3\,kg)(30\,m/sec^2)$$
$$= 90\,newtons,$$

and, similarly, the force on the 6-kg object is 90 newtons. The fact that the forces are equal agrees with Newton's third law. These forces are generated by the mutual repulsion that occurs between the negative charge on each of the two objects. Since all units are defined quantities, let's simply define the charge on each object to be 10^{-4} coulombs. Inserting values into Equation 9, we find

$$90\,N = \frac{K(10^{-4}\,C)(10^{-4}\,C)}{(1\,m)^2}.$$

Solving for K, we obtain

$$K = 9 \times 10^9 \text{ N-m}^2/C^2.$$

This value will be used whenever we want to find the force in newtons.

Alpha Decay and Coulomb's Law

In 1902, Rutherford was studying the nature of radioactivity with a young chemist named Frederick Soddy. These colleagues wondered just what really happened when a radioactive substance emitted alpha rays. From their experiments (which we shall discuss in Chapter 16), they concluded that atoms in a radioactive source break apart—disintegrate—into two pieces: an alpha particle and a smaller atom of a different element. That is, atoms of one element are continually changing into atoms of another element. This was an earthshaking conclusion because, until that time, atoms were thought to have existed *unchanged* since creation. The

complete picture finally came with the discovery of the nucleus in 1913: *the nucleus disintegrates*. For example, when a radium-226 nucleus disintegrates, it breaks up into two parts: an alpha particle and a radon nucleus (Fig. 3–15).

Radium has an atomic number Z equal to 88, which indicates that the radium nucleus contains 88 protons. The atomic mass number A equals 226, which indicates that there are a total of 226 protons and neutrons in the nucleus. Hence, the radium nucleus contains 138 neutrons (226 minus 88). Two protons and two neutrons form the alpha particle, and the remaining 86 protons and 136 neutrons form a nucleus of radon-222. The remaining nucleus has an atomic number Z equal to 86, which the periodic table of the elements (see Fig. 1–3, p. 6) shows to be radon.

Since both the alpha particle and the radon nucleus are positively charged, a force is exerted on each one that tends to drive them apart. The same *amount* of force acts on each one, and this amount is given by Coulomb's law. In Equation 8, one charge is that on the alpha particle, and the other charge is that on the radon nucleus. The symbol d represents the distance between the centers of the two. Note that, in Fig. 3–15, the lengths of the arrows depicting these forces are the same. This provides another good example of Newton's third law. Here, we have two charged entities interacting, and the forces on them are equal in magnitude, but the forces are oppositely directed, as described by the third law.

Although the forces are the same, the alpha particle and the radon nucleus react differently because they have different masses. (Try pushing a heavy box and a lightweight box across the floor with the same force. You'll quickly realize that it's much easier to accelerate the lightweight box.) To find the acceleration, we divide the force on a particle by the particle's mass ($a = F/m$). Since the alpha particle has a

---

FORCES: NEWTON'S LAWS AND COULOMB'S LAW

mass of about 4 amu and the radon nucleus, about 222 amu, the acceleration of the alpha particle is 55.5 times that of the radon nucleus, meaning that, while the radon nucleus travels 1 millimeter (abbreviation: mm), the alpha particle will travel 55.5 mm.

Figure 3–15 indicates how the alpha particle and the radon nucleus move in opposite directions along a straight line. Initially, they are separated by a distance that is approximately equal to the diameter of the radium nucleus, 2×10^{-14} m. As the distance between them widens, the force quickly drops off because of the d^2 term in the denominator of Coulomb's law. For example, a distance of 0.001 m (1 mm) is an infinitely large distance compared to 2×10^{-14} m. And when the distance between the alpha particle and the radon nucleus is infinitely large, the force approaches zero ($1/\infty \rightarrow 0$). At this point, both the alpha particle and the radon nucleus travel with constant speed. Because the acceleration immediately after the disintegration was so large, the alpha particle reaches a speed of 1.52×10^7 m/sec. But, the radon nucleus reaches a speed reduced by a factor of 55.5, or 2.73×10^5 m/sec. Thus, we can view alpha decay as a dramatic example of Coulomb's law.

Coulomb's Law and the Size of the Nucleus

When an alpha particle approaches a gold nucleus, the force of repulsion between them increases as they come closer together. As a result, the alpha particle slows down. In order to explain backward-scattering, Rutherford realized that this force of repulsion, described by Coulomb's law, must be large enough to cause the alpha particle to come to a complete stop. Then, the force of repulsion would drive the alpha par-

ticle backwards. Rutherford concluded that, to explain backward-scattering, the nucleus must be about 10^{-14} m in diameter. But why must the nucleus be so small? Why can't it be 10^{-10} m in diameter?

Consider an alpha particle approaching a gold nucleus of radius R (Fig. 3–16a); for the moment, we shall not specify the value of the radius. The charge on the alpha particle is $+2$, and Rutherford estimated the charge on the gold nucleus to be about $+100$ (today, we know that it is equal to the atomic number Z, which is 79 for gold). The charge on the gold nucleus is spread uniformly throughout the sphere. When the alpha particle is outside the nucleus, the distance d in Coulomb's law is taken to be distance between the alpha particle and the *center* of the nucleus, as shown in Fig. 3–16(a). That is, the charge on the sphere behaves as though it were concentrated at the center of the sphere. (This is similar to the way the mass of an object seems concentrated at its center, called its center of gravity. For example, you can balance a table knife on your index finger by putting your finger directly below the center of gravity of the knife.)

Figure 3–16(b) shows how the force of repulsion increases as the distance d gets smaller. When the distance d goes from $4R$ to $2R$, the distance is cut in half, and the force is 4 times larger. Similarly, when the alpha particle goes from $2R$ to R, the force is again 4 times larger. The force on the alpha particle keeps increasing until it reaches the surface of the nucleus. We must next consider the force acting on the alpha particle when it penetrates the nuclear surface.

Let's take the simplest case first: the alpha particle is at the center of the nucleus ($d = 0$). Here, the alpha particle is surrounded by the *same* amount of charge in all directions. Thus, the force of repulsion in any one direction is exactly balanced by another force in the opposite direction, so that the *net* force acting on the alpha particle is zero.

Figure 3–17 shows the forces acting on the alpha particle when it has penetrated the surface of the nucleus. The nucleus is divided into two parts, A and B. F_A is the force on the alpha particle due to the charge of part A and similarly, with part B. Since the forces are oppositely directed, the net force is F_B minus F_A.

Figure 3–18 shows the force acting on the alpha particle due to parts A and B when the alpha particle is at the surface. Here the net force is F_B plus F_A (neglecting any effects due to the small change in distance). Thus, the net force is larger at the surface than when the alpha particle penetrates the nucleus.

More sophisticated calculations show that the net force inside the nucleus is represented by a straight line (see Fig. 3–16). The most important conclusion is that *the maximum force exerted on the alpha particle occurs at the surface of the nucleus.*

Let us now compare the forces acting on two alpha particles. One approaches a nucleus with a radius of 10^{-10} m and the other, a nucleus with a radius of 10^{-14} m (Fig. 3–19). When the alpha particles are far away from the two nuclei, their speeds are the same. Each alpha particle slows down as it approaches the nucleus. When the alpha particles are the same distance from the nuclei, such as at points A and A', the forces on them are the same. Recall that the charge on a sphere acts as though it were concentrated at the center of the sphere. Since the *charges* on the two nuclei in Fig. 3–19 are the same, only distributed differently, the forces at points, such as A and A', are the same because the distances from the *centers* are the same. Thus, the behaviors of the two alpha particles in region 1 are identical, and they reach points C and P with the same speed.

When the alpha particles travel through region 2, their behaviors are now quite different. As depicted in Fig. 3–19(a), the alpha particle has entered the charged sphere. We know from our previous discussion that the maximum force occurs at the surface. Therefore, the force on the alpha particle decreases as it travels in region 2. Since the alpha particle depicted in Fig. 3–19(b) has not yet reached the nuclear surface, the force on it *increases* until it reaches the surface. As the alpha particle travels from point P to point T, the distance between the alpha particle and the center of the sphere goes from 10^{-10}

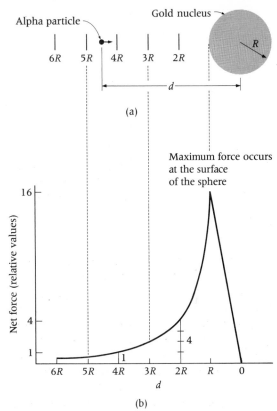

(a)

Maximum force occurs at the surface of the sphere

(b)

FIG. 3–16   Force acting on alpha particle as it approaches a nucleus. The maximum force occurs when the alpha particle is at the surface of the nucleus.

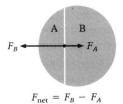

$$F_{net} = F_B - F_A$$

FIG. 3–17   Forces acting on the alpha particle inside the nucleus.

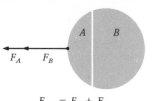

$$F_{net} = F_A + F_B$$

FIG. 3–18 Forces acting on the alpha particle at the surface of the nucleus.

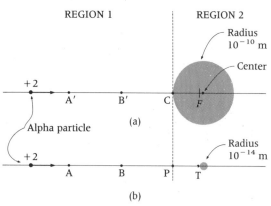

FIG. 3–19 (a) Alpha particle approaching a sphere of positive charge with a diameter of 10^{-10} m. (b) Another alpha particle approaching a nucleus with a diameter of 10^{-14} m. Both spheres have the same amount of positive charge.

m to 10^{-14} m. To see how the force of repulsion changes with distance, let us first calculate the quantity $1/d^2$ for each value. When d is 10^{-10} m (point P and point C), $1/d^2$ is 10^{+20} (neglecting the units for simplicity); when d is 10^{-14} m (point T), $1/d^2$ is 10^{+28}. Therefore, the quantity $1/d^2$ is 10^8 times larger at point T than at point P. Since Coulomb's law depends upon $1/d^2$, the force at T is 10^8 times larger than it is at point P (and at point C). As the alpha particle travels from P to T, the force on the alpha particle increases by a factor of 10^8, or by 100,000,000-fold!

While the puny force exerted on the alpha particle as it travels from C to F is not sufficient to stop the alpha particle, the tremendous force exerted on it as it travels from P to T *is*. It comes to a stop somewhere between P and T (actually quite close to T) and then, because of the tremendous repulsion, retraces its path. Rutherford saw that such a tremendous force could be realized if the atom did in fact have a nucleus of very small size—on the order of 10^{-14} m.

The Hydrogen Atom and Circular Motion

The hydrogen atom, as we know, consists of a negatively charged electron traveling in a certain allowed circular orbit around the comparatively massive, positively charged nucleus (see Chapter 1). Here, we'll take our first step toward analyzing this model and toward finding the speed of the electron in various orbits. (We shall continue this study of Bohr's model in Chapters 4 and 6, as we learn additional basic principles of physics.)

Figure 3–20 displays two orbits in which the electron can travel. Orbit 2 has a radius four times that of orbit 1. We shall call v_1 the speed of the electron in orbit 1, and we want to find the speed in orbit 2. In our discussion of circular motion, we learned that there must be a force directed toward the center of the circle. Here, that force is the attraction of the negatively charged electron to the positively charged nucleus. If we call F_1 the force acting on the electron in orbit 1, then the force on it in orbit

2 must be $F_1/16$. Since the distance between the electron and the nucleus increases by a factor of 4, Coulomb's law says that the force decreases by a factor of 16, as we saw in Fig. 3–14(c). From Newton's second law, we know that the acceleration of the electron is proportional to the force on it: If the force decreases by a factor of 16, so does the acceleration. Therefore, the acceleration of the electron in orbit 2 is related to that in orbit 1 by

$$\frac{\text{Acceleration}}{\text{in orbit 2}} = \frac{1}{16}\left(\frac{\text{acceleration}}{\text{in orbit 1}}\right). \quad (10)$$

Now, remember that the acceleration of any object traveling in a circular path is given by

$$\text{Acceleration} = \frac{(\text{speed})^2}{\text{radius}}.$$

Thus, the acceleration of the electron in orbit 1 is

$$\frac{\text{Acceleration}}{\text{in orbit 1}} = \frac{v_1^2}{r_1}.$$

We can satisfy the requirements in Equation 10 *only if* the speed of the electron in orbit 2 is $v_1/2$: The speed in orbit 2 is half that in orbit 1. Let's check it:

$$\frac{\text{Acceleration}}{\text{in orbit 2}} = \frac{(v_1/2)^2}{4r_1}$$

$$= \frac{1}{16}\left(\frac{v_1^2}{r_1}\right)$$

$$= \frac{1}{16}\left(\frac{\text{acceleration}}{\text{in orbit 1}}\right).$$

The requirements in Equation 10 are satisfied.

The speed of the electron in other orbits can be found by using the same procedure. Table 3–1 compares the force acting on the electron in other orbits to the force F_1, and compares the speed to the speed v_1. You can easily check that the acceleration is proportional to the force in each case.

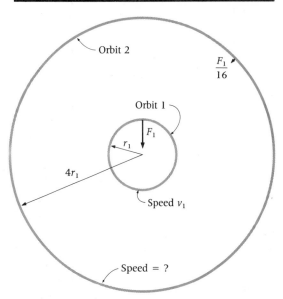

FIG. 3–20 Two orbits in which the electron in the hydrogen atom can travel. In orbit 1, the electron travels in a circular path of radius r_1 with speed v_1 due to the force F_1 acting on the electron. In orbit 2, the force acting on the electron is $F_1/16$, and the electron travels in a circular path of radius $4r_1$. The question is, What is the speed of the electron in orbit 2?

TABLE 3–1 SPEED OF ELECTRON TRAVELING AROUND THE NUCLEUS IN A CIRCULAR ORBIT

RADIUS (r_1)	FORCE (F_1)	SPEED (v_1)
0.50	4.00	1.41
0.75	1.78	1.16
1.00	1.00	1.00
2.00	$1/4$	0.708
3.00	$1/9$	0.576
4.00	$1/16$	0.500
5.00	$1/25$	0.448
7.00	$1/49$	0.379
9.00	$1/81$	0.333
12.00	$1/144$	0.290
16.00	$1/256$	0.250

Note: The radius of the circular path is given in terms of r_1, the radius of the first allowed orbit. The speed is given in terms of v_1, the speed of the electron in the first allowed orbit. r_1 equals 0.528×10^{-10} m, and v_1 equals 2.19×10^6 m/sec. F_1 is the force acting on the electron in the first allowed orbit.

We know that the radius of the smallest allowed orbit is 0.528×10^{-10} m (Chapter 1), and we shall let r_1 have this value. The following section shows that the speed v_1 is 2.19×10^6 m/sec.

Table 3–1 shows that we can find the speed of the electron in *any* circular orbit, but we haven't found a way to select the allowed orbits yet. In Chapter 4, we shall examine how Bohr used another basic principle, called angular momentum, to find the allowed orbits. We shall see that orbits having radii of r_1, $4r_1$, $9r_1$, $16r_1$, and so on, are allowed orbits.

SPEED OF THE ELECTRON IN
THE FIRST ALLOWED ORBIT: AN EXAMPLE*

Consider the electron in the first allowed orbit. We can find the speed of this electron if we

*This example may be omitted without loss of continuity.

substitute into Newton's second law the force due to Coulomb's law and the acceleration due to the electron's circular motion:

$$F = ma$$
$$\frac{KQ_eQ_N}{r_1^2} = \frac{mv_1^2}{r_1}. \tag{11}$$

Solving for v_1^2, we find

$$v_1^2 = \frac{KQ_eQ_N}{mr_1}.$$

The charge on the electron Q_e and the charge on the nucleus Q_N are both 1.6×10^{-19} coulombs, the mass of the electron is 9.11×10^{-31} kg, the radius r_1 is 0.528×10^{-10} m, and K equals 9×10^9 N-m^2/C^2. Recalling that 1 N is equal to 1 kg-m/sec^2, K can also be written as 9×10^9 kg-m^3/sec^2-C^2. Therefore, v_1^2 has units m^2/sec^2. Substituting values into the preceding equation, we find

$$v_1^2 = 4.77 \times 10^{12}\, \text{m}^2/\text{sec}^2$$
$$v_1 = 2.19 \times 10^6\, \text{m/sec}.$$

Summary

Newton's first law describes the behavior of an object when the net force on it is zero: The object remains at rest or moves with constant speed in a straight line. When an object has a net force on it, Newton's second law says that the object accelerates, and the acceleration equals the net force divided by the mass ($a = F_{net}/m$). Newton's third law states that forces occur in pairs: two oppositely directed forces of the same size act on two different objects. A unit of force is the *newton*.

Because a freely falling object accelerates, it must have a force acting on it. This force is called its weight and is due to the attraction of the object to the earth. The weight W is equal to mg, where g is the acceleration of gravity, 9.8 m/sec^2. The mass of an object is the same anywhere in the universe, but its weight depends upon its location.

An object traveling over a circular path with constant speed has an acceleration because its direction changes. The acceleration is given by v^2/r. The force necessary to produce this acceleration is mv^2/r.

The force between charged particles is given by Coulomb's law ($F = KQ_AQ_B/d^2$). The unit of charge is the *coulomb*.

Rutherford showed that the size of the nucleus must be 10^{-14} m. This accounts for the large deceleration that occurs during the backward-scattering of alpha particles by a gold foil.

The concepts of circular motion and the attraction between the nucleus and the electron are applied to Bohr's model of the hydrogen atom to find the speed of the electron in various orbits.

True or False Questions

Indicate whether the following statements are true or false. Change all of the false statements so that they read correctly.

3–1 When an object is traveling with constant speed, some net force must obviously act on it to keep it in motion.

3–2 To lift a 98-newton suitcase from the floor, a force slightly larger than 98 newtons is needed, but to raise it with constant speed only a force of 98 newtons is needed.

3–3 On another planet, where the acceleration of gravity is 5 m/sec^2, an astronaut drops a hammer from rest.

    a. The mass of this hammer is the same as on earth, but its weight is less than on earth.

    b. As the hammer falls, the force acting on the hammer is less than the force acting on the planet, because the planet doesn't move.

3–4 There is a car accident in which a passenger's head collides with a padded dashboard.

    a. The force on the dashboard is larger than the force on the head.

    b. If the dashboard had not been padded, the force on the passenger's head obviously would have been larger.

    c. By padding the dashboard,

        (1) the distance that the head travels in coming to a stop has been increased.

        (2) the average speed in coming to a stop has been increased.

        (3) the time in coming to a stop has been increased.

        (4) the change in speed in coming to a stop has been increased.

        (5) the deceleration in coming to a stop has been increased.

3–5 When the distance between two charged particles is doubled, the force acting on each one is one-half of what it was originally.

3–6 An alpha particle approaches a nucleus of a gold atom head-on.

    a. As the alpha particle gets closer to the nucleus, the force on the alpha particle decreases.

    b. As the alpha particle gets closer to the nucleus, the speed of the alpha particle decreases.

    c. The net force on the alpha particle is zero when it is located at the surface of the nucleus.

    d. The maximum force on the alpha particle occurs when it is located at the surface of the nucleus.

3–7 The force that makes the electron in a hydrogen atom travel in a circular orbit around the nucleus is centrifugal force.

FORCES: NEWTON'S LAWS AND COULOMB'S LAW

Questions for Thought

3-1 Match the name or term with its description by inserting the correct letter from the lettered column into each blank in the description column. Entries in the lettered column may be used more than once.

a.	size of the nucleus	**i.**	Newton's third law
b.	size of the atom	**j.**	air resistance
c.	coulomb	**k.**	static charge
d.	mass	**l.**	inertia
e.	acceleration	**m.**	weight
f.	force	**n.**	$F_{net} = 0$
g.	Newton's first law	**o.**	Coulomb's law
h.	Newton's second law	**p.**	newton

_____ states that forces occur in pairs

_____ object remains at rest or travels with constant speed in straight line

_____ property of object that is the same on the moon as on earth

_____ property of object that is smaller on the moon that on earth

_____ inertia

_____ object's resistance to change in motion

_____ mass multiplied by acceleration of gravity

_____ retarding force on freely falling object that depends upon the object's size and speed

_____ force between two charged particles

_____ obtained by rubbing two surfaces together

_____ unit of force

_____ unit of charge

_____ shorthand for kg-m/sec^2

_____ push or pull

_____ net force divided by mass

_____ 10^{-10} m

_____ 10^{-14} m

_____ v^2/r

3-2 A car at rest has a force of 500 newtons exerted on it due to a strong wind. What is the *net* force on the car? Explain why the car does not move.

3-3 A heavy box of 25 kg (55 lb) rests on the floor. In order to push it across the floor with a constant speed, a horizontal force is applied to it. This horizontal force (choose **a, b,** or **c**)

    a. must equal the weight of the box.

    b. must be sufficient to overcome friction between the box and the floor.

    c. must equal the weight of the box plus the force necessary to overcome friction between the box and the floor.

3-4 Suppose that Newton's second law was F_{net} equals weight multiplied by acceleration instead of mass multiplied by acceleration. How would this affect the horizontal force necessary to push the Lunar Rover on the moon compared with the force required on earth? How would this affect the vertical force necessary to lift the Lunar Rover on the moon compared with that on the earth?

3-5 Figure 3-5(b) shows a 0.15-kg ball falling with a constant speed of 13 m/sec due to the effects of air resistance. Explain why the ball travels with constant speed. What is the force of air resistance acting on the ball?

3-6 The air resistance of a falling object depends only upon that object's size and speed. If the air resistance depended upon the object's mass instead, show that the acceleration *in air* would then be the same for all masses.

3-7 When a charged particle is placed in a magnetic field, the particle travels in a circle. What can be said about the direction of the force on the charged particle?

3-8 A book with a weight of 10 newtons falls to the ground.

    a. As the book falls, the force acting on the earth is _____ newtons.

    b. The acceleration of the earth is _____

        (1) larger than 9.8 m/sec^2.

        (2) equal to 9.8 m/sec^2.

        (3) much smaller than 9.8 m/sec^2.

    c. The reason for the answer in **b** above is as follows: _____

3–9 Insert *larger than, the same as,* or *smaller than* in the blanks: When a radium-226 nucleus (with a charge of $+88$) disintegrates into an alpha particle ($+2$) and a radon-222 nucleus ($+86$), the force of repulsion on the alpha particle is _____ that on the radon-222 nucleus for the following reason: _____

_____

The acceleration of the alpha particle is _____ _____ that of the radon-222 nucleus for the following reason: _____

_____ .

3–10 The radium nucleus (with a charge of $+88$) disintegrates into an alpha particle ($+2$) and a radon nucleus ($+86$). A uranium nucleus ($+92$) fissions into a barium nucleus ($+56$) and a krypton nucleus ($+36$). Compare the force of repulsion between the alpha particle and the radon nucleus with that between the barium nucleus and the krypton nucleus. For simplicity, assume in each case that the particles are initially separated by the same distance. In which case is the force of repulsion larger, and by how much?

3–11 What is the speed of the electron of the hydrogen atom in the fifth allowed orbit?

Questions for Calculation

3–1 A 2000-kg car accelerates from rest to 15 m/sec (33.55 mph) in 5 seconds.
    a. What is the acceleration of the car?
    b. What is the net force acting on the car?
    c. If the car is heading directly into a strong wind that exerts a force of 500 newtons on the car, what is the force on the car due to the consumpion of gasoline?
    d. If the car has a tail wind due to a strong wind that exerts a force of 500 newtons on the car, what is the force on the car due to the consumption of gasoline?

3–2 Two children want to play with the same toy, which has a mass of 2 kg. The 4-year-old child exerts a force of 40 newtons, and the 2-year-old child exerts a force of 30 newtons in the opposite direction.
    a. What is the net force on the toy?
    b. What is the acceleration of the toy?

3–3 A car that normally has a maximum acceleration of 2.5 m/sec^2 is pushing an identical car.
    a. What is the maximum acceleration of the two cars?
    b. What would be the acceleration of the two cars if the car being pushed had only one-half the mass of the other car?

3–4 A golf ball with a mass that equals 0.06 kg is struck by a golf club and reaches a speed of 60 m/sec. The impact lasts for 4×10^{-4} seconds. Find the acceleration of the golf ball and the force exerted on the ball during the impact.

3–5 A 1500-kg car traveling at 20 m/sec hits a parked car and comes to a stop after traveling 2 m.
    a. Find the deceleration of the car, the force exerted on the car, and the force exerted on the parked car.
    b. What force is exerted on a 60-kg driver by the seat belt?

3–6 A woman weighs 110 lb. What is her mass in kilograms? Her weight in newtons? (1 kg = 2.2 lb.)

3–7 Calculate the weight of a 20-kg object on the moon, where the acceleration of gravity is 1.63 m/sec^2.

3–8 A 5-kg mass weighs 200 newtons on another planet. What is the acceleration of gravity on this other planet?

3–9 A 5-lb bag of sugar, having dimensions 15 cm $\times$ 22 cm $\times$ 8 cm, rests on a table. (5 lb = 2.27 kg.)

FORCES: NEWTON'S LAWS AND COULOMB'S LAW

a. What force in newtons does the sugar exert on the table? Does it make any difference which side is lying on the table?
b. Pressure is defined as the force per unit area, or pressure equals force/area. What is the greatest pressure that the bag of sugar can exert on the table?

†**3–10** A passenger with a mass of 50 kg is riding in an elevator that accelerates upward. If the force on the passenger's feet increases by 50 percent, what is the acceleration of the elevator?

†**3–11** A child swings a ball on the end of a string 1 m long in a circle. The ball makes 2 revolutions per second. The ball has a mass of 0.4 kg.
    a. What is the speed v of the ball?
    b. What is the acceleration of the ball?
    c. With what force does the child pull on the string?

3–12 A 20-kg box rests on the floor, and a 10-kg box is placed on top of it.
    a. What is the weight of each box?
    b. What force does the heavier box exert on the lighter one?
    c. What force does the lighter one exert on the heavier one?
    d. What force is exerted on the floor?
    e. What is the net force on each box?

†**3–13** Suppose that the alpha particle striking the gold foil in Fig. 2–4 (p. 45) of Chapter 2 comes to a stop after traveling only one-sixth of the way through the foil and then retraces its path.
    a. Find the *constant* deceleration of the alpha particle, using the methods described in Chapter 2.
    b. Use Newton's second law to find the constant force exerted on the alpha particle as it travels in the foil. (Mass of alpha particle $= 6.65 \times 10^{-27}$ kg.)

c. Assume that the alpha particle comes to a stop due to its repulsion to the nucleus of a gold atom. If the radius of the nucleus is 10^{-10} m, find the maximum force on the alpha particle as it approaches the nucleus. The charge on the gold nucleus is $+100$, and the charge on the alpha particle is $+2$ ($+1 = 1.6 \times 10^{-19}$ coulombs).
d. The force between charged particles varies with the distance between them as given by Coulomb's law. However, an average force F_{AV} can be defined as the constant force (acting over a specified distance) that produces the same effect as the varying force (over the same specified distance). The *required* average force F_{AV} needed to stop the alpha particle after traveling one-sixth of the way through the foil is equal to the force found in part **b** above:

Required $F_{AV} = $ _____ newtons.

e. Compare the maximum force found in part **c** and the required average force F_{AV} found in part **d.** Use the concept of an average value to show why the nucleus cannot have a radius of 10^{-10} m and still account for backward-scattering.

3–14 A 1000-kg car travels at a speed of 20 m/sec over a circular path with a radius of 50 m.
    a. What is the acceleration of the car?
    b. What is the frictional force between the tires and road necessary to keep the car traveling in this circular path?
    c. Suppose that this car now travels over an icy road where the frictional force is only one-sixteenth that found in **b.** The radius of the circular path is 200 m. Find the speed of the car traveling in this larger circular path.
    d. What is the acceleration of the car in part **c**?
    e. The car traveling over these two circular paths has many similarities to the electron traveling in orbit 1 and orbit 2 of Bohr's model of the hydrogen atom

(see Fig. 3–20). Compare the forces acting on the car in the two circular paths with the forces acting on the electron in the two orbits. Make the same comparison for the radii of the paths, the accelerations, and the speeds (see Table 3–1).

3–15 In Bohr's model of the hydrogen atom, the third orbit has a radius 9 times that of the first orbit. In the third orbit, the force of attraction between the electron and nucleus is _____ times that in the first orbit. The acceleration of the electron in the third orbit is _____ times that in the first orbit. To account for these accelerations, show that the speed of the electron in the third orbit must be one-third of that in the first orbit.

3–16 Write an equation similar to Equation 11 for an electron in the second orbit and find its speed. Show that this speed is one-half the speed of the electron in the first orbit.

Footnote

[1] Margaret S. Greenwood. PISA (to be published).

Sir Isaac Newton

CHAPTER 4

Conservation of Momentum

Introduction

What do the following have in common: the collision of cars, the firing of a rifle, the launching of a rocket, and the radioactive decay of a radium nucleus? At first glance, these diverse events (see Fig. 4–1) appear to have little in common. However, one similarity can perhaps be detected in the rifle firing and rocket launching: The bullet moves in one direction while the rifle recoils in the opposite direction; the rocket moves upward because the fuel is ejected downward.

One objective of this chapter is to show that *all* of these events can be interpreted in terms of one basic principle: the conservation of momentum, a phrase meaning that the (total) momentum remains constant. Sometimes we say, "Momentum is conserved."

We will begin our investigation by describing experiments that illustrate conservation of momentum. With the results of these experiments in mind, we shall use our intuition or our commonsense notions to express this conservation principle. Then, we will apply the conservation of momentum concept to all the phenomena depicted in Fig. 4–1. We'll see that there is indeed a great deal of similarity between radioactive decay, say, and the firing of a rifle. Such similarities help us to comprehend the abstract world of the atom. We will look into several illustrations of this. For example, we have learned (Chapter 1) that the scattering of alpha particles by a gold foil provided information about the nucleus of the gold atom, but not about the atom's electrons. Here, we shall see why this is so.

Finally, we'll revisit the Bohr model of the hydrogen atom to look in on an interesting aspect of its development: The criterion used to select the allowed orbits.

Learning about Conservation of Momentum

To "discover" conservation of momentum, we shall see the results of experiments that use an air-track apparatus (see Fig. 4−2). To avoid any friction between the cars and the track, air is blown through the holes in the track, so that the cars literally float on a cushion of air. In our experiments, two cars will collide and then couple together. We are interested in finding out how the velocity of the coupled cars depends upon the mass of each car and the initial velocity of each car.

Experiment one: Car B is at rest in the middle of the track. Car A at the left end is set into motion with a constant speed of 0.8 m/sec. Each car has a mass of 0.4 kg. When the cars collide, they couple together and move as one unit. After the collision, the speed of the coupled cars is measured to be 0.4 m/sec. Certainly, this is a reasonable result, and one you would most likely anticipate. Surely, the speed of both cars after the collision will be less than the speed of car A before the collision. Since both cars have the same mass, and the mass in motion after the collision is doubled, then an educated guess would be that the speed after collision would be reduced by a factor of 2, or that the speed after collision would be 0.4 m/sec. Thus, the results of the experiment agree with what you would expect based on common sense or intuition. Table 4−1 shows the results. Now, let's see how the speed after collision changes when the masses of the two cars differ.

Experiment two: The moving car A has a mass of 0.4 kg, but the resting car B has a mass of 0.8 kg. The first experiment is repeated with the speed of car A remaining at 0.8 m/sec. After the collision, the speed of the coupled cars is measured to be 0.267 m/sec. Again, based upon intuition, you would certainly expect that the speed after collision would be less than 0.4 m/sec. Since car B is more massive than in the first experiment, it will offer more resistance to being set in motion, and, hence, the speed of the coupled cars will be less than it was in the first experiment.

Experiment three: The first experiment is repeated, only this time, the mass of car B is smaller than the mass of car A. In this case, you would no doubt expect that the speed after collision would be greater than 0.4 m/sec and, of

Collision of two cars

Firing a rifle (rifle recoils)

Rifle   Bullet

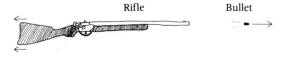

Radioactive decay of radium nucleus
Radium nucleus breaks up into two pieces: radon nucleus and alpha particle

Radium
nucleus

Speed of
radon
nucleus

2 +   Radioactive decay
of radium nucleus

Speed of alpha particle

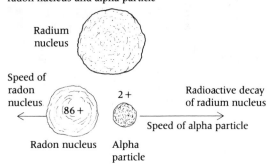

Radon nucleus   Alpha
particle

Launching a rocket

FIG. 4−1 Diverse examples illustrating conservation of momentum.

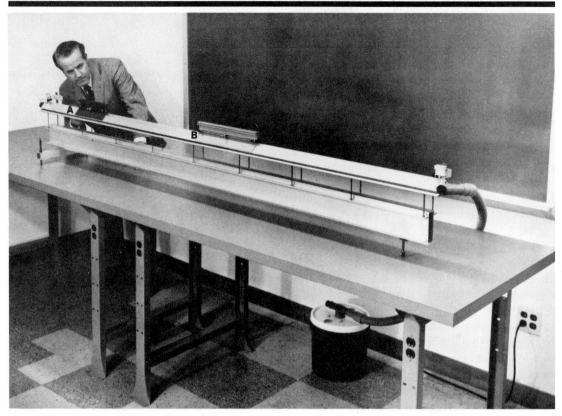

FIG. 4–2 Air-track apparatus demonstrating the principle of conservation of momentum. (Courtesy of the Ealing Corporation)

TABLE 4–1   DATA FROM EXPERIMENTS WITH AIR-TRACK APPARATUS

EXPERIMENT	MASS OF CAR A (kg)	VELOCITY OF CAR A (m/sec)	MASS OF CAR B (kg)	VELOCITY OF CAR B (m/sec)	VELOCITY OF COUPLED CARS (m/sec)
1	0.4	0.8 to right	0.4	0	0.400 to right
2	0.4	0.8 to right	0.8	0	0.267 to right
3	0.4	0.8 to right	0.2	0	0.533 to right
4	0.4	0.8 to right	0.4	0.8 to left	0

course, less than 0.8 m/sec. Results bear out this expectation. When the mass of car A is 0.4 kg, the mass of car B is 0.2 kg, and the speed of car A is 0.8 m/sec, the speed of the coupled cars after collision is measured to be 0.533 m/sec.

Experiment four: Car A is placed at the left end of the track, and car B is placed at the right end. Both are set into motion with a speed of 0.8 m/sec. When they collide, the cars couple together. After collision, the speed of the coupled cars is zero. It appears that the motion of car A to the right is exactly counterbalanced by the motion of car B to the left. This, too, is just what we would expect.

These experiments show that it is quite possible to use one's common sense or intuition to at least estimate the speed of the coupled cars. However, if a general principle could be found that governs the collision of the cars, then it could be used to account for the results of these experiments. More importantly, it could help us to understand other phenomena as well. There is such a principle, and it is called **conservation of momentum.** Let's see how it can aid in comprehending the results of these experiments.

Conservation of Momentum

From our experimentation, we found that the velocity (speed and direction) of the coupled cars depends upon their mass, their speed, and their direction of motion before collision. The term *momentum* specifies the relationship between mass and velocity. The **momentum** of an object is defined as the mass of the object multiplied by its velocity:

$$\text{Momentum} = \text{mass} \times \text{velocity}. \quad (1)$$

Recall that the term *velocity* includes a speed *and* a direction. For example, the velocity of an object moving to the right can be denoted with a positive number, and to the left, with a negative number. The total momentum is found by adding the momentum of each object involved in the action (or in the collision):

$$\text{Total momentum} = \text{momentum of object 1} + \text{momentum of object 2}$$
$$+ \text{momentum of object 3} + \cdots \quad (2)$$

The principle of conservation of momentum states that the total momentum before the action (or collision) must be equal to the total momentum after the action:

$$\text{Total momentum before action} = \text{total momentum after action.} \quad (3)$$

Let's now see if conservation of momentum accounts for the results of our air-track experiments.

In the first experiment, car A, before the collision takes place, has a momentum (Equation 1) given by $(0.4 \text{ kg})(+0.8 \text{ m/sec})$, or $+0.32$ kg-m/sec. This momentum is positive, indicating that car A is moving toward the right. Since car B is initially at rest, its momentum is zero. After the collision occurs, car A and B couple and move as one unit, with a total mass of 0.8 kg. Next, we want to find the velocity v of the coupled cars and compare it with the experiment. Equating the total momentum before the collision to the total momentum after the collision, we find that

$$\text{Momentum of car A} + \text{momentum of car B} = \text{momentum of coupled cars}$$
$$+ 0.32 \text{ kg-m/sec} + 0 = (0.80 \text{ kg})(v).$$

Solving this equation for v, we find that

$$v = \frac{0.32 \text{ kg-m/sec}}{0.80 \text{ kg}}$$
$$= +0.4 \text{ m/sec}.$$

A positive value for v indicates that the coupled cars travel to the right. This is the same as the velocity measured in the experiment.

The only different factor in the second and third experiments is the mass of car B. If we also carried out similar calculations for these cases, we would find that the speed of the coupled cars agrees with those found in the experiments.

However, in the last experiment, the two cars hit head-on. Since car A has a mass of 0.4 kg and moves to the right at 0.8 m/sec, its momentum is $+0.32$ kg-m/sec. The momentum of car B is -0.32 kg-m/sec. Car B has the same mass and speed as Car A, but it moves to the left. The

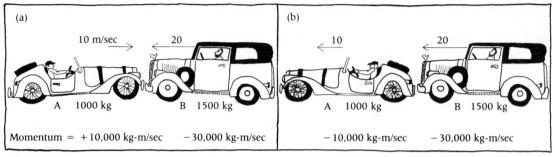

FIG. 4-3 (a) Head-on collision. (b) Tail-end collision. Car A has a mass of 1000 kg (2200 lb) and a speed of 10 m/sec (22.4 mph). Car B has a mass of 1500 kg (3300 lb) and a speed of 20 m/sec (44.7 mph). After the collision, the bumpers of the two cars interlock, and they move together as one unit. The momentum of each car is obtained by multiplying the mass of the car by its velocity, which has a positive value when the car is traveling to the right, and a negative value when the car is traveling to the left.

total momentum before collision is zero. Conservation of momentum requires that the total momentum after the collision is also equal to zero. Therefore, the velocity of the coupled cars must equal zero. This, too, agrees with the results of our experiments, and is exactly what we would anticipate. The direction of motion is quite important. If it had not been considered, then the total momentum before collision would not have been zero, and the calculations would not agree with the experiment.

Collision of Cars

From common observation, we know that a head-on collision between two automobiles usually causes much more damage than a tail-end collision, even if the speeds of the vehicles are the same in both cases before the collision.

We can now use the conservation of momentum principle to describe this observation. Figure 4-3 depicts both a head-on and a tail-end collision between two cars. Let's consider the situation where the bumpers of the two cars interlock, so that after the collision the two cars move together as a unit. The momentum of each car is indicated in the figure. Note that the momentum is positive if the cars travel to the right, and negative if they travel to the left.

For the head-on accident, the total momentum before the collision is $-20,000$ kg-m/sec. After the collision, the total mass of the two cars is 2500 kg, and the momentum is given by (2500 kg)(v), where v is the velocity of the unit. Conservation of momentum gives

$$(2500\,\text{kg})(v) = -20,000\,\text{kg-m/sec}$$
$$v = -8\,\text{m/sec}.$$

(The negative value for the velocity indicates that the coupled cars move to the left after the collision.)

For the tail-end accident, the total momentum before the collision is $-40,000$ kg-m/sec. Conservation of momentum gives

$$(2500\,\text{kg})(v) = -40,000\,\text{kg-m/sec}$$
$$v = -16\,\text{m/sec}.$$

The head-on collision is so disastrous because the cars—and the passengers—undergo a large change in velocity as compared to the tail-end collision. For example, in the tail-ender, car A is traveling at 10 m/sec to the left, and after the impact, travels at 16 m/sec to the left. The change in velocity is 6 m/sec. But, in the head-on collision, car A is initially traveling at 10 m/sec to the *right*, and ends up traveling at 8 m/sec to the *left*. Since the directions are opposite, the change in velocity is 18 m/sec. That is, a change in velocity of 10 m/sec would bring car A to a stop, and an additional change of 8 m/sec would cause it to travel to the left at 8 m/sec. This change in velocity, 18 m/sec, is 3 times that in the tail-end collision.

Similarly, in the tail-end collision, car B's change in velocity is 4 m/sec: from 20 m/sec (left) to 16 m/sec (left). In the head-on collision, car B, initially traveling at 20 m/sec to the left, has a speed of 8 m/sec to the left afterward, yielding a change in velocity of 12 m/sec. Again, this is 3 times that in the tail-end collision. Table 4–2 summarizes these results. So, conservation of momentum bears out what we know from experience and gives us some insight as well. Also, it's interesting to note that, in both collisions, the more massive car B experiences a smaller velocity change than the lightweight car A. Statistics bear this out: You are safer in a heavier car!

CONSERVATION OF MOMENTUM AND NEWTON'S LAWS

When you think about collision of cars, another term besides momentum comes to mind immediately: *force*. Let's reexamine the tail-end collision, focusing on the relationship between momentum and Newton's laws.

We'll use Newton's second law to find the force exerted on each car during the collision. If the impact lasts for 0.1 second, then each car's velocity changes during that time. Newton's second law becomes

$$F = ma = \text{mass} \times \frac{\text{change in velocity}}{t}.$$

TABLE 4–2   CHANGE IN VELOCITY OF CARS INVOLVED IN HEAD-ON AND TAIL-END COLLISIONS (FIG. 4–3)

	INITIAL VELOCITY (m/sec)	FINAL VELOCITY (m/sec)	CHANGE IN VELOCITY (m/sec)
Head-on collision:			
Car A	10 right	8 left	18 left
Car B	20 left	8 left	12 right
Tail-end collision:			
Car A	10 left	16 left	6 left
Car B	20 left	16 left	4 right

The force on each car is given by

$$\text{Force on car A} = \frac{(1000\,\text{kg})(6\,\text{m/sec})}{0.1\,\text{sec}}$$
$$= 60,000\,\text{newtons}$$
$$\text{Force on car B} = \frac{(1500\,\text{kg})(4\,\text{m/sec})}{0.1\,\text{sec}}$$
$$= 60,000\,\text{newtons.}$$

During the impact, the force on car A is equal to the force on car B. This agrees with Newton's third law, which says that, when two objects interact, the forces on each object are equal in size but opposite in direction. When we equate these two forces and cancel the time from both sides, we obtain

$$(1000\,\text{kg})(6\,\text{m/sec}) = (1500\,\text{kg})(4\,\text{m/sec}).$$

CONSERVATION OF MOMENTUM

But this is the same as

$$(1000 \text{ kg})(16 \text{ m/sec} - 10 \text{ m/sec})$$
$$= (1500 \text{ kg})(20 \text{ m/sec} - 16 \text{ m/sec}).$$

Rearranging this equation, we find

$$(1000 \text{ kg})(-10 \text{ m/sec}) + (1500 \text{ kg})(-20 \text{ m/sec})$$
$$= (2500 \text{ kg})(-16 \text{ m/sec})$$
$$+ (1000 \text{ kg})(-16 \text{ m/sec})$$
$$= (2500 \text{ kg})(-16 \text{ m/sec})$$

Identifying each term with momentum, we find

$$\begin{matrix} \text{Momentum} \\ \text{of car A} \end{matrix} + \begin{matrix} \text{momentum} \\ \text{of car B} \end{matrix} = \begin{matrix} \text{momentum} \\ \text{of coupled} \\ \text{cars.} \end{matrix}$$

Thus, conservation of momentum is a direct consequence of Newton's laws of motion.

Rifles and Rockets

The introduction to this chapter points out a similarity between the firing of a rifle and the takeoff of a rocket. How does conservation of momentum apply to these two events?

A 5-kg rifle fires a 10-g bullet with a speed of 800 m/sec. What is the recoil speed of the rifle? Before the rifle is fired, the rifle and the bullet are both motionless. Therefore, the total momentum before the firing is zero.

The momentum of the bullet is 8 kg-m/sec assuming a positive value for the velocity of the bullet. The bullet has a mass of 0.010 kg and a speed of $+ 800$ m/sec. With the recoil velocity of the rifle denoted as v, conservation of momentum becomes

$$\begin{matrix} \text{Total} \\ \text{momentum} \\ \text{before firing} \end{matrix} = \begin{matrix} \text{Total} \\ \text{momentum} \\ \text{after firing} \end{matrix}$$

$$0 = (8 \text{ kg-m/sec}) + (5 \text{kg})(v) \qquad (4)$$
$$v = \frac{-8 \text{ kg-m/sec}}{5 \text{ kg}}$$
$$= -1.6 \text{ m/sec}.$$

The negative sign indicates that the rifle recoils in the opposite direction from which the bullet was fired, which is, of course, what actually happens.

Equation 4 shows that the recoil speed of the rifle will be increased if either the mass of the bullet or its speed is increased. A rocket works on the same basic principle. Because a large amount of gas is ejected at very high speed, the momentum is large. As a result, the rocket takes off, or recoils, in the opposite direction.

Alpha Decay and Momentum

Chapter 3 discussed alpha decay from the standpoint of Coulomb's law. When the radium-226 nucleus, which contains 88 protons (abbreviation: p) and 138 neutrons (abbreviation: n), breaks apart into an alpha particle (2p, 2n) and a radon-222 nucleus (86p, 136n), there is a mutual force of repulsion between them, and they move apart. The alpha particle has a speed of 1.52×10^7 m/sec when it emerges from the radium source. In this section, we'll take a look at a related aspect of alpha decay.

Our discussion concerning the collision of cars showed us that conservation of momentum is a direct consequence of Newton's laws; so, too, in the instance of alpha decay. Momentum conservation results because of the equal, but oppositely directed, forces acting on the alpha particle and the radon nucleus. Given the speed of the alpha particle, we can find the speed of the radon nucleus.

Before the decay, the radium nucleus is motionless; the total momentum before decay is, therefore, zero. If the velocity of the radon nucleus after the decay is denoted as v, then conservation of momentum gives

$$0 = \frac{\text{momentum of}}{\text{alpha particle}} + \frac{\text{momentum of}}{\text{radon nucleus}}$$

$$= (4\,\text{amu})(+1.52 \times 10^7\,\text{m/sec})$$
$$+ (222\,\text{amu})(v) \quad (5)$$

$$v = -2.73 \times 10^5\,\text{m/sec}.$$

The radon nucleus and the alpha particle travel in opposite directions, as indicated by a negative value for the velocity of the radon nucleus and a positive one for the alpha particle. The speed of the radon nucleus is the same as that found in Chapter 3 (see p. 74).

The similarity between alpha decay and the firing of a rifle is also evident. The alpha particle is analogous to the bullet, and the radon nucleus, to the rifle. The formats of the conservation of momentum equations, Equations 4 and 5, are also identical.

Billiards and Momentum

Suppose you're involved in a game of billiards (Fig. 4–4) when one of your friends plays a practical joke on you. Two of the regulation balls are replaced with a solid steel ball and a hollow plastic ball (an oversize Ping-Pong ball). Both balls are the same size as the regular ones and have been painted to look like them. During play, can you pick out the two substitutes? Sure, you can. Just by the way the two phony balls move!

When a fast-moving billiard ball (speed equals v) strikes the oversize Ping-Pong ball, the Ping-Pong ball moves forward (speed greater than v), and the billiard ball continues to move forward (speed only slightly less than v). Because the mass of the Ping-Pong ball is small compared to the regulation ball, the Ping-Pong ball's momentum is also quite small. In order to conserve momentum after the collision, the billiard ball must continue to move in the same direction, with nearly the same speed (Fig. 4–5a).

When a fast-moving billiard ball (speed equals v) strikes another regulation ball head-on, the struck ball moves forward, and the other one comes to a stop (Fig. 4–5b). In order to con-

FIG. 4–4   A game of billiards in progress. There are many similarities between the collision of these billiard balls and the collision of atoms, electrons, and alpha particles. (Rose Skytta, © 1982 Jeroboam, Inc.)

FIG. 4–5   (a) Collision of a regulation billiard ball with the oversize Ping-Pong ball. (b) Collision of two regulation billiard balls. (c) Collision of a regulation billiard ball with the steel ball.

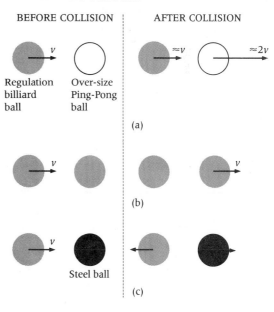

BEFORE COLLISION   AFTER COLLISION

Regulation billiard ball   Over-size Ping-Pong ball

(a)

(b)

Steel ball

(c)

serve momentum, the struck ball has speed v after collision.

When a fast-moving billiard ball strikes the steel ball, the billiard ball rebounds backwards, almost as if had hit a wall, and the steel ball is nudged slightly forward. Because the billiard ball before collision (Fig. 4−5c) has momentum directed to the right, the same amount must be directed to the right after collision. Because the steel ball is so massive compared to the billiard ball (even though the steel ball has a smaller speed), the momentum of the steel ball is much larger than the momentum of the billiard ball. Because the balls travel in opposite directions after collision, the total momentum is found by subtracting the momentum of the billiard ball from that of the steel ball. This results in a total momentum directed toward the right, as required for conservation of momentum.

By observing how the balls move, you can easily pick out the two substitutes. These billiard balls can just as easily be pictured as electrons, atoms, alpha particles, and so on. We now consider two such examples.

Alpha-Scattering Experiment

Rutherford, recall, was able to determine the size of the nucleus by scattering alpha particles from a gold foil (Chapter 1). This experiment, however, provided absolutely no information about the electrons in the gold atom. Why?

An alpha particle is about 8000 times more massive than an electron, and a gold atom is about 50 times more massive than an alpha particle. This information tells us that the interaction between an alpha particle and an electron in a gold atom is analogous to the interaction of billiard balls depicted in Fig. 4−5(a). The motion of the alpha particle (billiard ball) is hardly changed at all. In like manner, the interaction of an alpha particle with a gold

nucleus is analogous to the situation shown in Fig. 4−5(c). In this case, the alpha particle (billiard ball) rebounds backwards. The direction of motion of the alpha particle is *greatly changed*.

Note that the word *interaction* has been used instead of the word *collision*. The reason is that the alpha particle interacts with the charge of the electron and the charge of the gold nucleus, rather than colliding with the mass of the electron and the mass of the gold nucleus. Regardless of how the interaction takes place—whether by collision or by the interaction of charges—conservation of momentum applies.

Momentum and Reactors

In a nuclear reactor, uranium rods are suspended in a tank of water. A fission reaction occurs in the uranium rods when a neutron strikes a nucleus of a uranium-235 (U-235) atom, causing it to break apart into two large fragments and additional neutrons. Such breakups can occur in many different ways. Figure 4−6 shows a typical one.

When a neutron strikes a U-235 nucleus (92p, 143n), it forms a composite nucleus of U-236 (92p, 144n). The U-236 nucleus then breaks apart—fissions—into two neutrons; a barium-144 (Ba-144) nucleus, which has 56 protons and 88 neutrons; and a krypton-90 (Kr-90) nucleus, which has 36 protons and 54 neutrons. (Remember that the number of protons in a nucleus is equal to the atomic number Z, and the number of neutrons is equal to the atomic mass number A minus the atomic number Z.) A chain reaction occurs when one of the newly produced neutrons strikes another U-235 nucleus, causing still another fission reaction. Fission happens much more readily if the neutron is moving very slowly. In this case, the neutron spends more time in the vicinity of the U-235 nucleus and, hence, has a longer time to cause the breakup than would a fast-moving neutron. However, the neutrons produced by fission are extremely fast-moving ones. The question is, How can you slow down fast-moving neutrons?

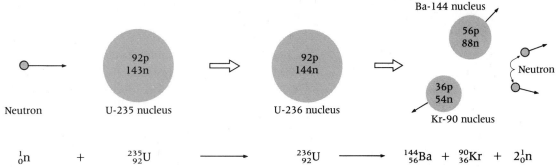

$$\,_0^1 n \quad + \quad \,_{92}^{235}U \qquad\longrightarrow\qquad \,_{92}^{236}U \qquad\longrightarrow\qquad \,_{56}^{144}Ba \;+\; \,_{36}^{90}Kr \;+\; 2\,_0^1 n$$

FIG. 4−6   A neutron causing a fission reaction in a U-235 nucleus. (The letter *p* indicates a proton and *n*, a neutron.) A neutron striking a U-235 nucleus can cause fission, or the neutron may simply scatter from the uranium nucleus, much like the scattering of billiard balls. The chance that fission will occur increases if the neutron speed is extremely small.

Suppose that a fast neutron strikes another uranium nucleus before it emerges from the uranium rod, and that this event does not result in fission. This "billiard ball" collision is analogous to Fig. 4−5(c), and the speed of the neutron is reduced very little.

If, however, the fast neutron strikes the nucleus of a hydrogen atom, then (analogous to Fig. 4−5b) the neutron's speed is reduced to zero! Not all of the collisions will be the head-on type. In any case, we can see that the neutron's speed will be greatly reduced when striking another particle of similar mass.

In a nuclear reactor, a fast neutron emerging from one uranium rod will travel through water and slow down in the process of striking many hydrogen nuclei. When it hits another uranium rod, the chances that this slow neutron will cause fission are greatly increased. The water also serves to remove the heat generated by the fission process.

Angular Momentum

Just as objects that move in a straight line have momentum, objects that move in a circle or rotate about an axis have **angular momentum.** An object of mass m that moves with speed v in a circle of radius r has an angular momentum given by

$$\text{Angular momentum} = mvr. \qquad (6)$$

As with momentum, angular momentum must be conserved.

If you are interested in sports, then no doubt you are aware of the effects of *conservation of angular momentum.* An ice skater executes a spin when the arms, initially outstretched, are brought close to the body. At the beginning of the spin, the hands are at a large distance from the body, which corresponds to a large value of r. Finally, the hands have a small value of r. If the angular momentum expressed in Equation 6 is always the same, then the speed must increase as the value of the radius decreases. This is exactly what you observe in the skater.

The somersaults of a diver are another case involving conservation of angular momentum. The arms and legs, initially outstretched, are brought close to the body. The increased rotational speed allows the diver to make several somersaults before diving into the water.

TABLE 4–3 ANGULAR MOMENTUM OF AN ELECTRON IN A HYDROGEN ATOM TRAVELING IN A CIRCULAR ORBIT AROUND THE NUCLEUS

RADIUS (r_1)	SPEED (v_1)	ANGULAR MOMENTUM $(h/2\pi)$	ORBIT (n)
0.50	1.41	0.707	
0.75	1.16	0.867	
1.00	1.00	1.00	1
2.00	0.708	1.42	
3.00	0.576	1.73	
4.00	0.500	2.00	2
5.00	0.448	2.24	
7.00	0.379	2.65	
9.00	0.333	3.00	3
12.00	0.290	3.48	
16.00	0.250	4.00	4

Note: r_1 equals 0.528×10^{-10} m, v_1 equals 2.19×10^6 m/sec. Values of the angular momentum are expressed in terms of $h/2\pi$, which equals 1.05×10^{-34} kg-m^2/sec.

Any object moving in a circular path has angular momentum. Examples include a car traveling over a circular path, the planets in orbit around the sun, and an electron in orbit around the nucleus. Let's look into this last example.

The Hydrogen Atom and Angular Momentum

In talking about Bohr's model of the hydrogen atom (Chapter 3), we determined the relationship between the speed of an electron traveling in a circular orbit and the radius of the orbit. Table 4–3 lists 11 values of the radii (first column) and their corresponding speeds (second column). We were unable to select the allowed orbits from this list.

To see how Bohr selected the allowed orbits, we turn briefly to the work of the German physicist Max Planck (1858–1947), which was carried out in about 1900. When an object is heated to a high temperature, it emits a continuous spectrum of light, but the intensity varies

with the wavelength of the light. (Our sun, for example, emits a continuous spectrum of radiation, but light of about 500 nm is the most intense.) The continuous spectrum also depends greatly upon the temperature of the object. Planck was trying to explain how light is emitted by a heated object, and was attempting to formulate a theory to compare with the experimental results. He was able to account for the experimental data only when he introduced a constant, now called Planck's constant h, where

$$h = 6.63 \times 10^{-34} \text{ kg-m}^2/\text{sec.}$$

Planck's theory was quite revolutionary, and Bohr felt that somehow Planck's constant must be introduced to explain how atoms emit light. To this time, traditional theories had neither been able to account for the line spectra emitted by elements nor predict the wavelengths.

Noting that Planck's constant h has units of angular momentum, Bohr made the following restriction: The angular momentum of an electron in an allowed orbit must be an integral multiple of $h/2\pi$. Expressed mathematically, this is

$$\text{Angular momentum of electron in } n\text{th allowed orbit} = \frac{nh}{2\pi}, \quad (7)$$

where n equals 1, 2, 3, 4, and so on, and $h/2\pi$ equals 1.05×10^{-34} kg-m^2/sec.

The angular momentum of the electron in the first allowed orbit is given by

$$\text{Angular momentum in first orbit} = mv_1r_1, \quad (8)$$

where m is the mass of the electron, v_1 is the speed of the electron in the first allowed orbit, and r_1 is the radius of the first allowed orbit. They have the following values:

$$m = 9.11 \times 10^{-31} \text{ kg}$$
$$v_1 = 2.19 \times 10^6 \text{ m/sec}$$
$$r_1 = 0.528 \times 10^{-10} \text{ m.}$$

Substituting these values in Equation 8, we find

that the angular momentum of the electron in the first orbit equals $h/2\pi$, as desired.

Table 4–3 indicates that only certain orbits have integral multiples of $h/2\pi$, and these are the allowed orbits. In general, the allowed orbits have a radius given by r_1/n^2 and a speed given by v_1/n.

Summary

The momentum of an object is defined as the object's mass multiplied by its velocity (momentum = mass × velocity). A complete definition of velocity requires a speed and a direction. Since a velocity in one direction is chosen as a positive number, and that in the opposite direction as a negative number, momentum is likewise a positive or negative number. The total momentum is the sum, including signs, of the momentum of each object involved. In the collision of objects, for example, the total momentum is unchanged or conserved. Conservation of momentum follows directly from Newton's three laws of motion. Objects that move in a circle have angular momentum that is defined as the product of its mass, velocity, and radius (angular momentum = mvr).

In Bohr's model of the hydrogen atom, the electron traveling in a circular orbit around the nucleus has angular momentum. Bohr determined the radius of the allowed orbits by restricting the angular momentum to integral multiples of $h/2\pi$. The letter h, Planck's constant, is equal to 6.63×10^{-34} kg-m^2/sec.

True or False Questions

Indicate whether the following statements are true or false. Change all of the false statements so that they read correctly.

4–1 The momentum of an object is given by its mass multiplied by its velocity.

4–2 A car traveling north has the same momentum as an identical car traveling south at the same speed.

4–3 A compact car traveling at 30 mph collides head-on with a van traveling at 30 mph.

   a. The total momentum before the collision is the same as the total momentum after the collision.

   b. The compact car undergoes the same change in momentum as the van.

4–4 It is a fact that an energetic electron, aimed directly at a helium nucleus, can be scattered backwards.

   a. Conservation of momentum says that an energetic alpha particle, aimed at an electron, can likewise be scattered backwards.

4–5 When Rutherford aimed alpha particles at a gold foil and observed the scattering of alpha particles, he could not infer the behavior of the electrons in the gold atom because his measurements were not sufficiently accurate.

4–6 Only objects that move in a circle or rotate about an axis have angular momentum.

4–7 The angular momentum of a rotating wheel changes when the wheel is turned because the axis of rotation changes.

4–8 The ability of a child to ride a bicycle with "no hands" is an example of conservation of angular momentum.

4–9 Bohr used the concept of angular momentum to find the speed of the electron in an allowed orbit.

4–10 In the first allowed orbit, the angular momentum is one-half that in the second allowed orbit.

CONSERVATION OF MOMENTUM

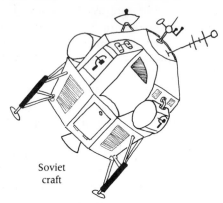

Soviet
craft

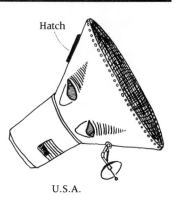

Hatch

U.S.A.

FIG. 4–7   Diagram for thought question 4–1.

Questions for Thought

4–1 In the science fiction film *Marooned*, a Soviet spacecraft tries to rescue an American spacecraft marooned in space and nearly out of oxygen. The spacecrafts are situated as depicted in Fig. 4–7. The drama becomes tense as Houston Control frantically tells the American crew not to blast the hatch off (using explosives located around the hatch), but to open the hatch. The astronauts, barely functioning due to oxygen deprivation, do not comprehend and blast the hatch off anyway. Why was Houston Control so insistent that the hatch not be blased off? What happened when the astronauts did so?

4–2 In the rear-end collision of two cars shown in Fig. 4–3(b), what is the direction of the force exerted on each car? Compare this with the direction of the change in velocity for each car listed in Table 4–2. What is the connection between the direction of force and the direction of the change in velocity?

4–3 Describe the similarities between the radioactive disintegration of a radium nucleus, the fission of a uranium nucleus into two fragments, and the firing of a rifle.

4–4 In Chapter 1 (p. 23) we used, as an analogy for alpha scattering, the following example: a golf ball hitting a bee as the ball sails through the air, and a golf ball hitting a tree. Discuss these examples in terms of basic principles.

4–5 What criterion did Bohr use to choose the allowed orbits of the hydrogen atom?

4–6 How does the speed of the electron in an allowed orbit depend upon n?

4–7 How does the force on an electron in an allowed orbit depend upon n?

Questions for Calculation

4–1 *Insert* larger than, smaller than, *or* equal to *in the blanks.* An empty railroad boxcar A has a mass of 25,000 kg and a velocity of 2 m/sec to the right. It collides and couples with an identical empty boxcar B at rest. After the collision, the coupled boxcars have a velocity of 1 m/sec to the right.

   a. If Boxcar A had been loaded with freight, so that its mass was 50,000 kg and it still had a speed of 2 m/sec, then the speed of the coupled cars after collision would be _____ 1 m/sec.

b. If the speed of the loaded boxcar A increased to 3 m/sec, then the speed of the coupled boxcars after collision would be _____ 1 m/sec.

c. If empty boxcar A, moving at 2 m/sec to the right, collides with boxcar B loaded with freight (mass = 50,000 kg) at rest, then the speed of the coupled cars after collision is _____ 1 m/sec.

d. If empty boxcar A, moving at 2 m/sec to the right, collides and couples with empty boxcar B, moving at 2 m/sec to the left, the speed of the coupled cars after collision is (insert number) _____ m/sec.

e. If empty boxcar A, moving at 2 m/sec to the right, collides and couples with empty boxcar B, moving at 3 m/sec to the left, then after collision the coupled cars move toward the (insert: *left* or *right*) _____ .

4–2 Use conservation of momentum to calculate the speeds after collision in calculation question 4–1. Compare your answers.

4–3 For the head-on collision of two cars shown in Fig. 4–3(a), show that application of Newton's second and third laws yields the principle of conservation of momentum.

†**4–4** A radon-222 nucleus ($Z = 86$) disintegrates by emitting an alpha particle that has a speed of 1.61×10^7 m/sec.

a. What nucleus remains after the disintegration?

b. What is the speed of the resulting nucleus?

†**4–5** When a uranium-235 nucleus absorbs a neutron, a nucleus of uranium-236 is formed. Fission occurs when the uranium-236 nucleus breaks apart into a barium-144 nucleus and a krypton-92 nucleus. The barium-144 nucleus has a speed of 10^7 m/sec.

a. The total momentum before fission is _____ .

b. In order to conserve momentum, the krypton-92 nucleus must move in (insert: *the same* or *in the opposite*) _____ direction as the barium-144 nucleus. The speed of the krypton-92 nucleus must be (insert: *larger than, equal to,* or *smaller than*) _____ 10^7 m/sec.

c. Calculate the speed of the krypton-92 nucleus.

(Note: Chadwick carried out the calculations in calculation questions 4–6 and 4–7 in order to show that the mass of a neutron was equal to the mass of a proton.)

4–6 A neutron traveling at 3.3×10^7 m/sec strikes a proton at rest. As a result of the collision, the neutron comes to a stop and the proton moves forward. Assuming that the mass of the neutron is equal to that of the proton, find the speed of the proton after collision.

4–7 A neutron traveling at 3.3×10^7 m/sec collides with a nitrogen-14 nucleus and knocks it forward with a speed of 0.44×10^7 m/sec. As a result of the collision, the neutron rebounds *backwards* with a speed of 2.86×10^7 m/sec. Use the mass of the neutron as 1 amu and the mass of the nitrogen nucleus as 14 amu, and show that momentum is conserved.

Henry Cavendish

CHAPTER 5

Newton's Law of Gravitation

Introduction

Since earliest time, human beings have looked toward the heavens and attempted to understand the motion of the planets and the sun. Stonehenge, a famous prehistoric example of an early astronomical observatory in England, attests to this interest (Fig. 5–1).

The ancient Greek philosopher Aristotle (384–322 B.C.) described motion in terms of the "natural" motions of objects. In Aristotle's view, an object falls to earth because the earth is its natural resting place, while the natural motions of the planets and the sun are circular, with the earth at the center of the solar system. When the positions of the planets were actually measured, it became apparent that Aristotle's notions were too simplistic. Many complexities (all based upon circular motion) were introduced to force agreement between observations and calculations. In 1530, the Polish astronomer Nicolaus Copernicus (1473–1543) suggested that the motions of the planets could be interpreted more easily if the planets traveled in circular orbits around the sun. Next, a Danish astronomer, Tycho Brahe (1546–1601), working for 20 years with huge protractorlike instruments, measured the position of the planets to an accuracy of within one-sixtieth of a degree. (After hearing about the invention of the telescope, Galileo, in 1609, constructed one. He was the first to explore the heavens with a telescope.) Upon Brahe's death, his assistant, the German astronomer Johannes Kepler (1571–1630), obtained these precise measurements, which Brahe had previously kept to himself. Kepler tried in vain to interpret the data in terms of planets moving in circles around the sun. After 10 years of labor, he concluded that the planets travel in elliptical paths, and he discovered three laws describing their motions.

In 1666, Isaac Newton formulated a single law—the law of universal gravitation—which described the mutual force of attraction between any two masses in the *universe*. This law was the culmination (the zenith, in fact) of more than 20 centuries of thought. Newton was able

FIG. 5–1 Stonehenge, constructed between 1850 B.C. and 1500 B.C. The alignment of the stones indicates the position of the sun and moon on the longest and shortest days of the year. (Courtesy of British Tourist Authority, 680 Fifth Avenue, New York)

to explain the elliptical motions of the planets as well as the acceleration of a freely falling object on earth and, further, was able to demonstrate that these two diverse phenomena had a common base in the law of gravitation. Indeed, this law provided the long-sought unity. Newton completely demolished the Aristotelian notion that the motions of planets and objects on earth were governed by different laws.

Our main concern in this chapter shall be some of the many and far-reaching consequences of Newton's law of universal gravitation. In the course of this pursuit, we will touch on topics mentioned in earlier chapters and raise many interesting questions, such as: Why is the acceleration of gravity the same for heavy objects and for light objects? Why is the acceleration of gravity on the moon only one-sixth that on earth? Before the astronauts landed on the moon, how was this known? (This lays the basis for some notion of the attraction between masses, which will be required when we talk about the formation of stars in Chapters 6 and 19.) How can the mass of the sun and the earth be determined?

We shall see how Newton established a connection between the famous freely falling apple and the motion of the moon around the earth, and was thus able to verify that his law was valid. Finally, we'll examine one of Newton's greatest achievements: showing that the empirical law discovered by Kepler could be derived from the law of gravitation.

Newton's Law of Universal Gravitation

Every mass in the universe attracts every other mass; that is to say, a force of gravitational attraction exists between every two masses in the universe. This mutual force of attraction (Fig. 5–2) is proportional to the product of the two masses and inversely proportional to the square

NEWTON'S LAW OF GRAVITATION

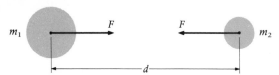

FIG. 5–2 The force of attraction between two masses.

of the distance between their centers. This is expressed mathematically as

$$F \propto \frac{m_1 m_2}{d^2}. \tag{1}$$

This relationship is a proportionality, not an equation. For example, if the distance between the two objects is doubled, the force of attraction decreases by a factor of 4. An equation, with a constant of proportionality, is needed to specify the *size* of the force. This is given by

$$F = \frac{G m_1 m_2}{d^2}, \tag{2}$$

where G is the constant of proportionality. If the masses are specified in kilograms and the distance in meters, the units of G must be N-m^2/kg^2, so that the force is given in newtons. (Recall that N is the abbreviation for the newton unit.) The currently accepted value of G is 6.67×10^{-11} N-m^2/kg^2. Shortly, we shall see how, in 1798, the English scientist, Henry Cavendish (1731–1810) devised an experiment to measure this incredibly small number.

Suppose that two books are placed on a tabletop. According to the law of gravitation, there is a force of attraction between them, yet the books do not move toward each other. Why don't they? They do not because, between the books and the tabletop, there is a force of friction that counteracts the attraction.

THE EXTREME SMALLNESS OF THE FORCE OF ATTRACTION: AN EXAMPLE

You place two 1-kg (2.2 lb) books on a tabletop 0.825 m (32.5 in.) apart. What is the force of attraction acting on each book?

$$\begin{aligned} F &= \frac{G m_1 m_2}{d^2} \\ &= \frac{(6.67 \times 10^{-11}\,\text{N-m}^2/\text{kg}^2)(1\,\text{kg})(1\,\text{kg})}{(0.825\,\text{m})^2} \\ &= 9.8 \times 10^{-11}\,\text{newtons.} \end{aligned}$$

The weight of a 1-kg mass is

$$\begin{aligned} W &= mg \\ &= (1\,\text{kg})(9.8\,\text{m/sec}^2) \\ &= 9.8\,\text{newtons.} \end{aligned}$$

The force of attraction between the two books is one hundred billion (or 10^{11}) times smaller than the weight of a 1-kg book!

Acceleration of Gravity

A freely falling object of mass m has an acceleration g_E equal to 9.8 m/sec^2. (The symbol g_E denotes the acceleration of gravity at the *surface* of the earth.) This means that its speed increases by 9.8 m/sec during each second. If the object is dropped from rest, its speed after 1 second is 9.8 m/sec; after 2 seconds, 19.6 m/sec; and so on. According to Newton's second law ($F_{\text{net}} = ma$), some kind of force must be acting on the object, but the *nature* of that force is not specified. The following illustration will clarify this point.

Suppose that a child's toy car ($m = 0.1$ kg) accelerates across the floor at 1 m/sec^2. This means that the net force acting on the car is

$$\begin{aligned} F_{\text{net}} &= ma \\ &= (0.1\,\text{kg})(1\,\text{m/sec}^2) \\ &= 0.1\,\text{newtons.} \end{aligned}$$

The force could be due to the child pushing the car, a windup mechanism, or batteries in the car, or perhaps the car is iron and a large magnet is attracting it. The fact that the toy car is accelerating provides absolutely no information about the nature of the force causing the acceleration.

Similarly, the force on a freely falling object of mass m must equal mg_E, but the nature of the force is not described. Newton's law of universal gravitation does describe this force: It is due to the attraction of mass m to the mass of the earth m_E. In Equation 2, the two masses are m and m_E, but what is the distance d? This question can be answered by considering the situation described in the section that follows.

CENTER OF GRAVITY

A piece of lumber 2 m in length is carried most easily by holding it at the center; that is, the mass of the wooden plank seems to be concentrated at its center. This point is called the **center of gravity.** A framed picture will not hang properly unless suspended directly over its center of gravity. Note that the law of gravitation expressly specifies the distance between the *centers* (of gravity) of two masses.

The center of gravity of a given sphere is at the center of that sphere. Even though the mass of the earth is distributed throughout the sphere, the mass acts as if it were concentrated at the earth's center. Therefore, for an object located at the surface of the earth (meaning that the distance above ground is extremely small compared to the radius of the earth r_E), the distance between the object and the center of the earth is r_E.

For an object located at the surface of the earth, Newton's second law becomes

$$\frac{Gm_E m}{r_E^2} = mg_E. \tag{3}$$

The mass of the object m is on both sides of the equation and thus cancels out. The expression g_E is then given by

$$g_E = \frac{Gm_E}{r_E^2}. \tag{4}$$

From this, we can see that the acceleration of a freely falling object does not depend upon its mass. The acceleration is the same for all masses, as Galileo was the first to observe from his experiments. The law of gravitation fundamentally explains the acceleration of gravity at the surface of the earth. Equation 4, moreover, can give us additional knowledge.

If G and r_E can be determined, then the mass of the earth m_E can also be found, since it is the only unknown in the equation.

Equation 4 is valid only for objects located at the surface of the earth. As the distance d of the object from the center of the earth increases, the attraction to the earth decreases and, likewise, the acceleration decreases. When d is much larger than r_E, the acceleration g is given by

$$g = \frac{Gm_E}{d^2}. \tag{5}$$

For example, if d is twice the radius of the earth, then g is smaller than g_E by a factor of 2^2, or

$$g = \frac{g_E}{2^2}$$
$$= \frac{(9.8\,\text{m/sec}^2)}{4}$$
$$= 2.45\,\text{m/sec}^2.$$

SLIGHT VARIATIONS IN
THE ACCELERATION OF GRAVITY

The acceleration of gravity at the North Pole is

$$g_E = 9.83 \text{ m/sec}^2,$$

and, at the equator,

$$g_E = 9.78 \text{ m/sec}^2.$$

These slight variations are due to the fact that the earth is not a perfect sphere. Rather, the distance from the North Pole to the center of the earth is slightly *smaller* than the distance from the equator to the center of the earth. Since the distance appears in the denominator, Equation 4 shows that g is *larger* at the North Pole than at the equator. (Or vice versa—measuring the acceleration of gravity at many points on the earth's surface provides a way of determining the shape of the earth.)

Acceleration of Gravity on the Moon

An equation analogous to Equation 4 can be written for the acceleration of gravity on the moon's surface. (This is represented by the symbol g_M.) The value of g_M depends upon the moon's mass and its radius. The mass of the earth is 81.3 times the mass of the moon, and the radius of the earth is 3.67 times the radius of the moon. Because the mass is in the numerator in Equation 4, the smaller mass of the moon tends to make g_M smaller than g_E by a factor of 81.3. Because the acceleration of gravity is inversely proportional to the radius squared, the smaller radius of the moon tends to make g_M larger than g_E by a factor of $(3.67)^2$. Combining the effects of mass and radius, the acceleration of gravity g_M on the moon is one-sixth that on earth, as shown by

$$g_M = \frac{(3.67)^2}{81.3} g_E$$

$$= \frac{1}{6.04} g_E$$

Since g_E equals 9.80 m/sec^2, then g_M equals 1.62 m/sec^2.

Measurement of G and the Mass of the Earth

In 1798, about seven decades after Newton's death, Cavendish demonstrated the validity of Newton's law of gravitation and measured the value of G. (The Cavendish Laboratory of Cambridge University in England is named in Henry Cavendish's honor. We'll visit this famous laboratory in later chapters.) As noted, the force of attraction between masses of ordinary size is extremely small, so the measurement of G was indeed a difficult undertaking.

Figure 5–3 shows a diagram of the apparatus used in the Cavendish experiment. Due to the attraction between each small mass m on the dumbbell and the larger mass M placed nearby, the dumbbell arrangement rotates. (Remember that Coulomb, in 1785, used a similar apparatus to investigate the force between charged particles—see Fig. 3–12.) However, because the very fine suspension fiber resists rotation, the rotation in that direction stops at a *maximum angle*, which is designated by θ_{max}. At angle θ_{max}, the force of attraction is exactly counterbalanced by the force of resistance in the fiber. One step in the experiment is to determine the force needed to rotate the fiber through various angles. Once this relationship has been established, a measurement of θ_{max} determines the force of attraction F between the masses. By varying the masses and the distance between them, Cavendish established the validity of Equation 2. Since F, the masses m and M, and the distance between the masses are known, the gravitation constant G in Equation 2 can be determined.

Because the force of attraction between m and M is exceedingly small, the angle θ_{max} is likewise extremely small. For this measurement, a mirror is mounted on the fiber and a beam of light is reflected by the mirror. The reflected beam is observed on a screen at some distance from the mirror. As the mirror rotates, the light beam moves across the screen, and the maximum angle θ_{max} can be determined. The mirror acts as a "magnifier" and enables the small value of angle θ_{max} to be determined accurately (see Fig. 5–4).

Cavendish also determined the mass of the earth. As previously mentioned, once G has been determined, the mass of the earth m_E is the only unknown in Equation 4. The radius of the earth was known at that time. In fact, the Greek mathematician and astronomer Eratosthenes had determined the radius of the earth in 240 B.C.! This measurement is so interesting that we'll take some time here to consider it.

THE RADIUS OF THE EARTH

The earth's axis of rotation is tilted by 23.5°, and on June 22, at noon, the sun is overhead at the Egyptian city of Syene (Fig. 5–5). Syene, 480 miles due south of Alexandria, is located on the Tropic of Cancer, which means that the noontime sun on June 22 is *directly* overhead. On that day, and that day only, Eratosthenes could see the reflection of the sun from the water of a deep well. At noontime on the same day in Alexandria, the sunlight struck the earth at an angle of 7.2° off the vertical. The angle between Syene, the center of the earth, and Alexandria must also be 7.2°. If a distance of 480 miles corresponds to 7.2°, then the circumference of the earth, corresponding to 360°, must be

$$\frac{480\,\text{mi} \times 360°}{7.2°} = 24{,}000\,\text{mi}.$$

The radius of the earth is 24,000 mi/2π, or 3820 mi. Today, the accepted value for the radius at the equator is 3963 miles, or 6378 kilometers.

The mass of the earth is found from Equation 4 to be

$$m_E = \frac{gr_E^2}{G}$$
$$= \frac{(9.78\,\text{m/sec}^2)(6.378 \times 10^6\,\text{m})^2}{6.67 \times 10^{-11}\,\text{m}^3/\text{kg-sec}^2}$$
$$= 5.96 \times 10^{24}\,\text{kg}.$$

Earlier, the units of G were given as N-m²/kg². Since a newton is equivalent to kg-m/sec², the units of G are given as m³/kg-sec² in the preceding equation.

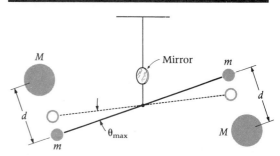

FIG. 5–3　A diagram of the Cavendish experimental apparatus used for measuring the gravitational constant G.

TOP VIEW

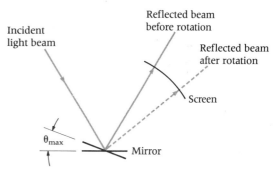

FIG. 5–4　The method used to measure the extremely small angle θ_{max}.

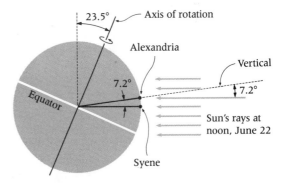

FIG. 5–5　At noon on June 22, the sun is directly overhead at Syene, Egypt, but inclined at 7.2° from the vertical at Alexandria, Egypt.

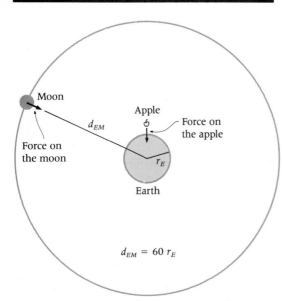

FIG. 5–6   The circular path of the moon around the earth. The distance between the earth and the moon (d_{EM}) is 3.83×10^8 m. The circumference of the orbit is 24.06×10^8 m.

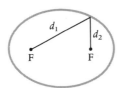

FIG. 5–7   For an ellipse, d_1 plus d_2 equals a constant. The shape of the ellipse depends upon this constant and the distance between the foci F.

The Cavendish experiment was not carried out until a little over seven decades after Newton's death. What evidence did Newton himself have for the validity of his law? The following section discusses this evidence.

The Connection Between the Apple and the Moon

Newton figured that the attraction to the earth was responsible for the apple falling to the ground, *as well as* for the moon traveling in a circular path around the earth (Fig. 5–6). If he were correct, then these two motions could prove his law (Equation 1) correct. Let's rewrite Equation 1 for the moon. The force acting on the moon due to its attraction to the earth is given by

$$F_{\text{moon}} \propto \frac{m_E m_{\text{moon}}}{d_{EM}^2},$$

where m_E is the mass of the earth, m_{moon} is the mass of the moon, and d_{EM} is the distance between the earth and the moon, which was known in Newton's time. To find the acceleration of the moon, we divide the force exerted on it, by the moon's mass. Dividing both sides of the last equation by m_{moon}, we obtain

$$a_{\text{moon}} \propto \frac{m_E}{d_{EM}^2},$$

where a_{moon} is the acceleration of the moon. Similarly, the acceleration of the apple is given by

$$a_{\text{apple}} \propto \frac{m_E}{r_E^2},$$

where r_E is the radius of the earth. This proportionality is very similar to Equation 4, except that here the constant G is missing.

We see that both the acceleration of the moon and that of the apple depend upon the mass of the earth, but the distances in the denominators are different. The distance between the earth and the moon is about 60 times larger than the radius of the earth. This means that the acceleration of the moon is only 1/60², or 1/3600 times as large as the acceleration of the apple. The apple has an acceleration of 9.8 m/sec² and, in order for his law of universal gravitation to be correct, Newton realized that the moon must have an acceleration of (9.8 m/sec²)/3600 or 0.00272 m/sec². Does it?

The moon has an acceleration due to its change in direction as it travels around the earth with constant speed. This acceleration is given by v_{moon}^2/d_{EM}, where v_{moon} is the moon's speed in its circular orbit, and d_{EM} is the radius of this orbit. If we know both of these terms, we can determine the moon's acceleration.

First, let's find the moon's speed, v_{moon}. The moon takes 27.3 days or 2.36×10^6 sec to go around the orbit once. This distance is equal to the circumference of the circle, or 24.06×10^8 m. Since speed is defined as the distance divided by the time,

$$v_{moon} = \frac{24.06 \times 10^8 \, m}{2.36 \times 10^6 \, sec}$$
$$= 1.02 \times 10^3 \, m/sec.$$

The moon's acceleration is then given by

$$a_{moon} = \frac{v_{moon}^2}{d_{EM}}$$
$$= \frac{(1.02 \times 10^3 \, m/sec)^2}{3.83 \times 10^8 \, m}$$
$$= 0.00272 \, m/sec^2.$$

This shows that Newton's law of universal gravitation is correct. Two motions that appear quite different to us are in fact governed by the same law.

Kepler's Laws and Universal Gravitation

Kepler discovered three laws that describe planetary motion:

1. Each planet moves in an elliptical path with the sun at one focus.

2. The line from the sun to any planet sweeps out equal areas in equal time intervals.
3. r^3 is proportional to T^2, where r is the average sun-planet distance and T is the time for one revolution.

Figure 5−7 illustrates some of the properties of an ellipse. Each fixed point is called a **focus.** At any point on the ellipse, d_1 plus d_2 equals a constant. This constant determines the shape of the ellipse. An ellipse can be drawn by taking a length of string and fastening its two ends by thumb tacks to a drawing board. A pencil placed beside the string and pushed in a curve will trace out an ellipse.

Newton was able to show that a planet moved in an elliptical path with the sun at one focus due to the gravitational attraction between the planet and the sun. He developed the mathematics of **calculus** for this purpose. While the planets do travel in elliptical paths, the orbits of most planets (with the exception of Mercury and Pluto) deviate only slightly from a circle. For example, the earth travels in an elliptical path, but the length of the shorter axis is equal to 0.99986 times the length of the longer axis. Thus, the earth's orgit is very nearly circular.

While we cannot discuss elliptical motion in detail here, we can consider one special case— circular motion, which, as we have just seen, is a good approximation for many planets.

Kepler's third law can be derived from Newton's second law ($F = ma$) and the law of universal gravitation. The force F is due to the attraction between the planet and the sun and is given by

$$F = \frac{Gm_sm_P}{r_{SP}^2}$$

where r_{SP} is the distance between the planet and the sun. Since the planet moves in a circle, its acceleration is given by v_P^2/r_{SP}, where v_P refers to the velocity of the planet as shown in Fig.

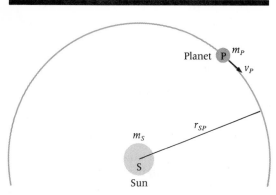

FIG. 5-8 A planet traveling in a circular orbit around the sun.

FIG. 5-9 The relationship between conservation of momentum and Kepler's second law.

5-8. In the time T, the planet makes one complete revolution. Its speed v_P is given by the circumference of the circular path divided by T, or v_P equals $2\pi r_{SP}/T$. The acceleration becomes $4\pi^2 r_{SP}/T^2$. Newton's second law ($F = ma$) becomes

$$\frac{Gm_S m_P}{r_{SP}^2} = m_P \left(\frac{4\pi^2 r_{SP}}{T^2} \right),$$

which simplifies to

$$\frac{Gm_S}{4\pi^2} = \frac{r_{SP}^3}{T^2}. \tag{6}$$

This shows that, for the special case of circular motion, r^3 is proportional to T^2 in agreement with Kepler's third law. Table 5-1 illustrates Kepler's third law. The distance from the sun to the earth is taken to be 1 unit, called an **astronomical unit** (AU). When the time T is expressed in years (yr), we see

$$\frac{r^3}{T^2} = \frac{1.00 \, \text{AU}^3}{\text{yr}^2}.$$

From Equation 6 we see that

$$\frac{Gm_S}{4\pi^2} = \frac{1 \, \text{AU}^3}{\text{yr}^2}.$$

Today, we know that the sun-earth distance (1 AU) is equal to 1.496×10^{11} m (usually remembered as 93 million miles), and the mass of the sun m_s is found to be 2.0×10^{30} kg from the preceding equation.

Kepler's second law is really another version of the law of conservation of momentum (see Fig. 5-9). As the planet travels in an elliptical path, its speed does not remain constant, but is such that it sweeps out equal areas in equal time intervals. In the same length of time, the planet travels from point A to point B and also from point C to point D. Since the path from point A to point B is larger than from point C to point D, the planet obviously has a larger speed while traveling over the path A to B. For the small

TABLE 5–1   VALUES FOR KEPLER'S THIRD LAW

PLANET	r DISTANCE FROM SUN (AU)	TIME T FOR ONE REVOLUTION (yr)	r^3	T^2
Mercury	0.387	0.241	0.0596	0.0581
Venus	0.723	0.615	0.3779	0.3782
Earth	1.000	1.000	1.0000	1.0000
Mars	1.520	1.880	3.5120	3.5340
Jupiter	5.200	11.860	140.6100	140.6600
Saturn	9.540	29.460	868.2500	867.8900

distances shown, the shaded areas are approximately triangles. The area of a triangle is given by ½(base) × (height). Kepler's second law states that the areas are equal, and

$$\frac{1}{2}(v_1 t)(r_1) = \frac{1}{2}(v_2 t)(r_2). \qquad (7)$$

Multiplying Equation 7 by the mass of the planet m_p and simplifying yields

$$m_p v_1 r_1 = m_p v_2 r_2.$$

This is the law of conservation of angular momentum.

Summary

Newton's law of universal gravitation is given by

$$F = \frac{Gm_1 m_2}{d^2},$$

where G is equal to 6.6×10^{-11} N-m^2/kg^2, and the force F is exerted on mass m_1, and another force F is exerted on mass m_2. The Acceleration of mass m_2 is F/m_2, or

$$\text{Acceleration of } m_2 = \frac{Gm_1}{d^2}.$$

Let us summarize results obtained from these equations of force and acceleration, beginning with the acceleration of gravity.

A freely falling object of mass m_2 has an acceleration of 9.8 m/sec^2 at the surface of the earth (when air resistance is negligible). In this case, m_1 equals the mass of the earth m_E, and d is the radius of the earth r_E. Since the equation for acceleration does not contain the mass m_2, the acceleration is the same for all freely falling objects. Furthermore, since all terms, except the mass of the earth m_E, are known in the acceleration equation, the mass of the earth can be determined (as Cavendish did after he determined G). Since the earth is flattened at the poles, the acceleration of gravity is largest there since r_E is smallest. Just as g_E (the acceleration of gravity on earth) depends upon the mass of the earth and the radius of the earth, the acceleration of gravity on the moon (g_M) depends upon the mass of the moon and the radius of the moon. Due to these differences in mass and radius, the acceleration of gravity on the moon is one-sixth that on the earth.

To measure the constant G and show the validity of Newton's law of universal gravitation, Cavendish suspended a dumbbell (with mass m at each end) from a very fine fiber. Adjacent to each mass m, he placed a massive sphere of mass M at a distance d. Due to the attraction between m and M, the dumbbell arrangement rotated, and the maximum angle of rotation yielded the force of attraction. The only unknown in the force equation was the constant G, which Cavendish was thus able to determine.

To test his theory of gravitation, Newton compared the acceleration of an apple falling to the ground with the acceleration of the moon in its circular orbit around the earth. For both the apple and the moon, the mass m_1, in the acceleration equation, equals the mass of the earth. For the apple, the distance d is the radius of the earth r_E; for the moon, the distance d is the distance from the center of the earth to the moon r_{EM}, where r_{EM} equals $60 \times r_E$. If his theory were correct, the acceleration of the apple should be 60^2 times larger than the acceleration of the moon. Newton determined the speed of the moon by dividing the circumference of its orbit by the time to make one revolution. The acceleration of the moon is equal to the speed squared divided by the radius of its orbit. Newton calculated the acceleration of the moon and found it to be 9.8 m/sec^2 divided by 3600—exactly as his theory had predicted.

From the precise astronomical observations of Brahe, Kepler formulated three laws that describe the elliptical orbits of planets around the sun. Newton showed that these laws can be deduced from his laws of gravitation and from his laws of motion. With the exception of Mercury and Pluto, most of the planets travel in nearly circular orbits. With this simplification, we shall see how to obtain Kepler's third law from Newton's laws.

The force of attraction between a planet and the sun causes the planet to travel in orbit around the sun. In the acceleration equation, m_2 is equal to the mass of the planet, m_1 is equal to the mass of the sun, and d is the radius r_{SP} of the (nearly) circular orbit. Calculating the acceleration of the planet (in the same way as we did for the moon), we find it equal to $4\pi^2\, r_{SP}/T^2$, where T is the time for one complete revolution around the sun. Substituting this result into the left side of the acceleration equation and simplifying, we find that r_{SP}^3/T^2 is equal to a constant. That is, r_{SP}^3 is proportional to T^2, which is Kepler's third law. Furthermore, this constant is equal to $Gm_S/4\pi^2$; from this relationship, the mass of the sun can be determined.

True or False Questions

Indicate whether the following statements are true or false. Change all of the false statements so that they read correctly.

5–1 A 2-kg ball and a 4-kg ball, at the same height above the ground, are dropped from rest.
   a. According to Newton's law of gravitation, the force of attraction between the 4-kg ball and the earth is twice that between the 2-kg ball and the earth.
   b. Both balls have the same acceleration because F/m—the force divided by the ball's mass—is the same for both.

5–2 If the distance between two objects is tripled, the force of gravitational attraction between them is one-third of what it was originally.

5–3 On a planet where the mass and radius are twice that of the earth's, the acceleration of gravity would be the same as on earth.

5–4 Henry Cavendish showed that Newton's law of gravitation was correct and measured the value of the gravitational constant G.

5–5 The gravitational constant G is a very large number.

5–6 The distance between the earth and the moon is known to be 60 times the radius of the earth.
   a. The force of attraction on an apple on the earth is 3600 times larger than the force of attraction between the earth and moon.

b. The acceleration of the apple should be 3600 times larger than the acceleration of the moon, according to Newton's law of gravitation.

c. The acceleration of the moon can be determined by knowing that the moon takes 27.3 days to go around the earth, as well as knowing the radius of the moon's path.

d. Newton tested his law of gravitation by comparing the acceleration of an apple on earth with the acceleration of the moon in its circular orbit.

5−7 a. The gravitational force acting on the sun is much larger than that acting on the earth because the sun is much more massive than the earth.

b. The acceleration of the sun is much larger than that of the earth because the sun is much more massive than the earth.

c. As a result of the attraction of the sun and the earth, the sun must also move in a very small (compared to the earth) circular orbit.

5−8 The planet Pluto, traveling in an elliptical orbit around the sun, has its greatest speed when it is the farthest from the sun.

5−9 Kepler's second law shows that the angular momentum of the planet doesn't change.

5−10 For a planet traveling in a circular orbit, Kepler's third law can be obtained from Newton's second law of motion and Newton's law of gravitation.

5−11 The mass of a planet can be obtained by observing the moons of the planet and by applying Newton's law of universal gravitation.

5−12 If the mass of the sun were larger, then the earth (still located 93 million miles from the sun) would take more than 1 year to make one revolution around the sun.

Questions for Thought

5−1 During the moon shots, TV viewers often heard the announcer say, "The astronauts are now out of the earth's gravitational pull and are entering the moon's gravitational pull." Is this statement accurate? Is the gravitational pull of the earth or the moon ever zero? Discuss the analogy between this question and calculation question 5−2.

5−2 Read calculation question 5−2. Is it possible to place object C somewhere between A and B so that the net force is zero? Discuss this question only. Do not make any numerical calculations.

5−3 Describe how Newton showed the validity of the law of universal gravitation.

5−4 Discuss how the observation of a moon circling a planet provides information about the mass of the planet.

Questions for Calculation

†5−1 The electron in the first allowed orbit of the hydrogen atom has a radius of 0.53×10^{-10} m. The mass of the electron is 9.1×10^{-31} kg, and the mass of the hydrogen nucleus is 1.67×10^{-27} kg. The charge on the hydrogen nucleus and the charge on the electron are 1.6×10^{-19} coulombs. ($K = 9 \times 10^{9}$ N-m²/C².) Calculate the force due to gravitational attraction, and due to the attraction between the charges. Which one is larger, and by what factor?

5−2 Object A (mass = 4 units) and object B (mass = 2 units) are separated by 8 distance units. Object C (mass = 2 units) is placed between A and B. For simplicity, assume that the gravitational constant $G = 1$.

---

a. Suppose that object C is 2 distance units from A. What is the force of attraction acting on C due to A? In a diagram, show the direction of this force. What is the force of attraction on C due to B? Show the direction in the diagram. What is the net force acting on C? Indicate its direction.

b. Suppose that object C is 2 distance units from object B. (Answer the same questions as in part **a**).

5–3 The radius of the planet Uranus is about 4 times that of the earth, but the acceleration of gravity on Uranus is (nearly) the same as that on earth. This shows that the mass of Uranus is about (insert: 4, ¼, 16, or ¹⁄₁₆) _____times that of the earth.

†5–4 The mass of Mars is one-ninth times that of the earth and the radius of Mars is one-half that of the earth. What is the acceleration of gravity on the surface of Mars?

†5–5 An artificial satellite is placed in a circular orbit around the sun. Its distance to the sun is 0.86 AU. Find the time for one revolution.

Introduction

Conservation of energy is one of the most fundamental concepts in physics. It is probably expressed most commonly as, "Energy cannot be created or destroyed." This idea means that, while energy may be transformed from one form into another, the total amount of energy involved in the interaction must remain unchanged or constant. We shall be examining many different kinds of energy in this chapter: potential energy, which is the energy due to position; kinetic energy, which is the energy of motion, heat, and light; and, extremely significant to an understanding of the modern world, the energy stored in matter, nuclear energy.

We'll take a phenomenon that is now familiar to us—the motion of a freely falling object—and reexamine it from the new perspective of conservation of energy, thus establishing this principle. With a foundation in place, we can then apply conservation of energy to more complicated motions, such as a moving roller coaster or a pendulum, both of which represent problems that are difficult to solve using Newton's laws as a starting point. These situations involve the transformation of potential energy to kinetic energy. (The potential energy derives from the object's height above the ground, and the kinetic energy derives from the increased speed as the object falls.) The conservation of energy principle will be generalized to include *all* other forms of energy, some applications of which will be explored.

One such example concerns the formation of a star, where potential energy is transformed into kinetic energy and starlight as the star contracts due to gravity. (Does this kind of transformation explain how our sun could emit light for 4.5 billion years, the current estimate of the sun's age? We shall see that the answer is no and, as a result, scientists searched for another process to account for sunlight.)

We'll approach conservation of energy yet again through another situation we have already studied: the head-on and tail-end collision of cars (Chapter 4). This time, we'll use conser-

J. P. Joule

CHAPTER 6

Conservation of Energy

vation of energy to show why a head-on collision is the more disastrous of the two. With this familiarity under our belts, we shall see how Chadwick used essentially the same ideas to discover the neutron. Also, we shall see how the kinetic energy of two colliding particles can be transformed into the mass of yet a third particle via $E = mc^2$ (energy equals mass multiplied by the speed of light squared), which relates the transformation of energy into mass and vice versa. Our understanding of the atom is due largely to the ability to make such extensions.

Next, we'll look at another "episode" in our story of Rutherford's model of the atom and Bohr's model of the hydrogen atom; specifically, at how Rutherford estimated the size of the nucleus to be 10^{-14} m and how Bohr predicted the energy of light rays emitted by the hydrogen atom.

Energy: The Ability to Do Work

We already have some understanding about the concept of **energy.** One can hardly read a newspaper or watch TV these days without being bombarded by information and controversies concerning the energy crisis. Fuel provides energy; that is, the ability to do the work we desire. The phrase "the ability to do work" is the key ingredient in the concept of energy. Something possesses energy if it can do work. Such a notion is indeed quite familiar to us. Undoubtedly, you have said to yourself upon occasion, "I just don't have enough energy to do that task now. I need something to eat first." This implies that your body possesses energy, which is expended as you do work, whether playing golf, cleaning house, reading a textbook, or simply breathing. If the task requires

more energy than you have, then that energy must be replenished. Let us now investigate the various forms of energy using the same idea: Energy is the ability to do work.

Work and Potential Energy

Imagine that you are a camper who wants to drive a tent stake into the ground (Fig. 6–1). You lift the rock with both hands to height h. To balance the rock's weight (designated as mg), each hand exerts an upward force of $mg/2$. Since the *net* force on it is zero, the rock moves upward with *constant* speed. When it reaches a certain height h, you bring it to a stop and then let it fall *freely*. It hits the tent stake, driving it into the ground.

There is no question that you do work in lifting the rock. In fact, the heavier the rock, the more work you do. Likewise, the higher you lift the rock, the more work you do. Therefore, a reasonable definition for **work** in this example is

$$\text{Work done on rock} = mg \times h.$$

A general definition follows from this: The work done on an object is equal to the force exerted on the object times the distance through which the force was exerted. That is,

$$\text{Work} = \text{force} \times \text{distance}. \qquad (1)$$

At height h the rock has energy, called **potential energy** (abbreviated PE) because it has the *potential to do work* when it falls, in this case, to drive the stake into the ground. The work done in lifting the rock has been transformed into potential energy.

$$\text{Work done on rock} = \text{potential energy of rock}.$$

Since the left-hand side of this equation is mgh, the potential energy of the rock is also mgh. The potential energy of any object of mass

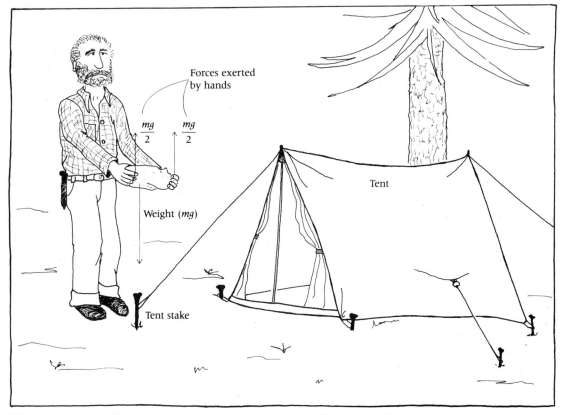

FIG. 6–1 A camper lifting a heavy rock to height h above ground. The rock falls freely to drive the tent stake into the ground.

m a distance h above the ground is, in general, given as

$$\text{Potential energy} = mgh. \qquad (2)$$

We can extend this to include the transformation of work into *any* form of energy:

$$\frac{\text{Work done}}{\text{on object}} = \frac{\text{increase in energy}}{\text{of object}}. \qquad (3)$$

When the mass is expressed in kilograms, the acceleration in m/sec^2, and the height h in meters, the units of potential energy are

$$\begin{aligned}
PE &= mgh \\
&= \text{kg} \times \text{m/sec}^2 \times \text{m} \\
&= \text{kg-m}^2/\text{sec}^2.
\end{aligned}$$

(Remember that the hyphen between the "kg" and the "m" denotes multiplication). For simplicity, this unit of energy is called the **joule** (abbreviation: J), where

$$1 \text{ joule} = 1 \text{ kg-m}^2/\text{sec}^2 \qquad (4)$$

This unit of energy was named after the English scientist James P. Joule (1818–1889), who investigated the relationship between energy and heat. Work and any other kind of energy must also have units of joules.

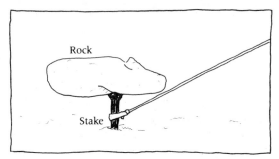

FIG. 6–2 Because it can drive the stake into the ground, this rock possesses energy, which depends upon its speed and its mass.

Power

A powerful weight lifter could probably lift the heavy rock depicted in Fig. 6–1 much more easily than you could. The weight lifter would exert a force equal to mg through the distance h, doing the same amount of work that you did, but lifting the rock with a much larger constant speed and, hence, lifting it in a much shorter time than you did. On this basis, a reasonable definition for **power** is

$$\text{Power} = \frac{\text{work (or energy)}}{\text{time}}. \qquad (5)$$

The unit of power is given in joules/sec, which is defined as a **watt** (abbreviation: W). This term is most familiar in referring to the intensity or wattage of a light bulb: the electrical energy consumed per second.

Kinetic Energy

Let us use the same example of the rock and the tent stake to define **kinetic energy** (abbreviation: KE), the energy of motion. Only, let's

approach the situation from a slightly different viewpoint, the one shown in Fig. 6–2. The instant before the rock hits the stake it possesses energy (because it has the ability to drive the stake into the ground). This kinetic energy must depend upon the speed of the rock and its mass (or weight). A large rock moving with the same speed as a small one will obviously push the stake farther into the ground. We can get a clue about kinetic energy's dependence upon mass and speed by looking at the units for energy in Equation 4 and noting that m^2/sec^2 is the unit for speed squared. Thus, a unit for energy is equivalent to mass (m) multiplied by speed (v) squared. Kinetic energy is defined as

$$\text{Kinetic energy} = \frac{1}{2}mv^2. \qquad (6)$$

Conservation of Energy

The falling rock has two kinds of energy: potential (due to its height above ground) and kinetic (due to its motion). The principle of **conservation of energy** requires that the sum of these two kinds of energy must remain constant. Thus, as the rock falls, its potential energy decreases and its kinetic energy increases, but the total amount of energy involved is unchanged.

As an example, consider a 5-kg rock falling from rest at a height of 1 m above the ground. We want to find its potential energy and kinetic energy at certain heights above the ground and see that their sum is always the same. The distance that the rock travels from its starting point is given by

$$d = \frac{1}{2}gt^2,$$

where g is the acceleration of gravity, 9.8 m/sec². After 0.2 seconds, this distance is ½(9.8 m/sec²)(0.2 sec)², or 0.196 m, indicating that the rock is 0.804 m above the ground. Table 6–1 shows these values. The first column gives entries for other times; the second column gives the corresponding heights, computed in a similar way.

Next, we want to find the speed of the rock

TABLE 6–1   EXAMPLE OF CONSERVATION OF ENERGY

TIME (sec)	HEIGHT ABOVE GROUND (m)	SPEED (m/sec)	POTENTIAL ENERGY (joules)	KINETIC ENERGY (joules)	TOTAL ENERGY (joules)
0	1	0	49.0	0	49.0
0.1	0.951	0.98	46.6	2.4	49.0
0.2	0.804	1.96	39.4	9.6	49.0
0.3	0.559	2.94	27.4	21.6	49.0
0.452	0	4.43	0	49.0	49.0

Note: Values reflect a 5-kg rock falling to the ground from a height of 1 m.

at a given height above ground. An acceleration of 9.8 m/sec^2 means that the rock's speed increases by 9.8 m/sec after each second. After 0.1 second, the speed is 0.98 m/sec; after 0.2 seconds, 1.96 m/sec; and so on. That is, after a time *t* the speed of an object falling from rest is given by multiplying the acceleration 9.8 m/sec^2 by the time: speed equals (9.8 m/sec^2) (time). The third column of Table 6–1 shows the values of the speed.

With these heights and speeds, we can now calculate the potential energy (Equation 2) and kinetic energy (Equation 6), and find their energy sum, or total energy (abbreviation: TE). After 0.2 seconds, the potential energy *mgh* is (5 kg)(9.8 m/sec^2) (0.804 m), or 39.4 joules; the kinetic energy is ½(5 kg) (1.96 m/sec)^2, or 9.6 joules. The total energy is 49.0 joules. The other values for the potential energy, kinetic energy, and total energy are calculated in a similar way. As the rock falls, its potential energy decreases and its kinetic energy increases, but the total energy is always the same. The energy changes from one form into another, but the total energy remains constant.

Let's now apply our basic knowledge about conservation of energy to more complex situations.

APPLICATION OF CONSERVATION OF ENERGY TO COMPLEX MOTIONS

Figure 6–3 depicts several complicated problems that can be solved using the principle of conservation of energy. For simplicity, we shall let the objects in Fig. 6–3 all have the same mass.

At point A, all of the objects are motionless and have only potential energy, but the same amount, since they are the same height above ground and have the same mass. When the objects are released from rest at point A, the potential energy decreases and the kinetic energy increases. At point B, for example, all of the objects have the same potential energy. They must also have the same kinetic energy and the same speed. Thus, we can "solve" the problem: We can find the speed at any height. Such ideas are vital in the design of roller coasters (see Fig. 6–4).

Perhaps you are amazed that conservation of energy can be applied to so many diverse situations. Advanced physics students can show the same results using Newton's laws as a starting point, but doing so is a quite complicated matter. We must recognize that there are differences in these examples. If all of the objects are released from point A at the same time, they will reach point B with the same speed but *not* at the same time. Since the pendulum bob travels a greater distance than a freely falling object, it requires a longer time to reach point B. Another difference can be seen in acceleration. The freely falling object and the ball rolling down the inclined plane have constant acceleration, but the acceleration of the pendulum bob and the car traveling over the loop-the-loop varies over their paths.

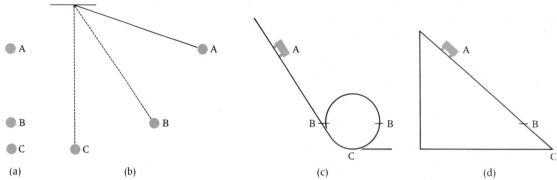

FIG. 6–3   Conservation of energy applies equally well to (a) a freely falling object, (b) a pendulum bob, (c) a car traveling over the loop-the-loop, and (d) a car rolling down an inclined plane. All of the objects are released from rest at point A. As they fall, their potential energy decreases and their kinetic energy increases. At point B, all objects have the same potential energy, since they are the same height above ground and have the same mass. Since energy is conserved, all objects have the same kinetic energy at point B. The same arguments apply to point C.

FIG. 6–4   The speed of this roller coaster at any point can be found using conservation of energy if we know its speed and its height at any one point. (Courtesy of Mariott's Great America Theme Park, Gurnee, Illinois)

A complete law of energy conservation must cover *all* forms of energy, and not merely the transformation of potential energy to kinetic (or vice versa). These represent one type of energy transformation. You are surely aware of others. The burning of coal, for instance, provides heat to change water into steam, which in turn drives electrical generators. Many contemporary homes utilize solar energy, which is light from the sun converted into heat. When the explosive TNT blows up, stored chemical energy transforms into heat and blast energy due to the large kinetic energy of the air molecules. In a nuclear reactor, the energy stored in matter—nuclear energy—is transformed into heat and, finally, into electricity. (Nuclear energy is released when a very small part of the nucleus is converted into other forms of energy. Einstein's famous equation $E = mc^2$, which holds that mass and energy are interchangeable, governs this process.) The total energy (TE) must include all forms of energy:

$$\text{TE} = \frac{\text{kinetic}}{\text{energy}} + \frac{\text{potential energy}}{\text{of various kinds}}$$

$$+ \text{heat} + \frac{\text{nuclear}}{\text{energy}}$$

$$+ \text{sound} + \text{light} + \frac{\text{chemical}}{\text{energy}} \qquad (7)$$

$$+ \frac{\text{electrical}}{\text{energy}} + \frac{\text{any other}}{\text{forms of}}$$
$$\text{energy.}$$

Shortly, we'll examine many examples of the transformation of one form of energy into another. First, though, we shall see that heat itself is a form of energy.

Heat as a Form of Energy

You must have heard the commonplace expression, "Heat flows from a hot object to a cooler one." This phrase is a throwback to a widely held theory of heat that prevailed during the 1800s. According to this theory, a substance called *caloric* (origin of the familiar term *calorie*) flows into a heated object and surrounds each atom. Supposedly, the presence of caloric made the temperature of the object increase. Friction, however, posed difficulties for the caloric theory. When two objects are rubbed together, their temperatures increase. For example, if you rub your hands together on a cold blustery day, they become warm even though you are not holding them over a heat source. No heat has been applied, only work has been done in rubbing your hands together. No "caloric" has been added. Why, then, should the temperature increase?

The caloric theory had been on shaky ground for about 40 years when, in 1847, Joule carried out the decisive experiment that showed the relationship between heat and work. Figure 6–5 illustrates Joule's experiment. He let a weight fall to the ground with constant speed, causing a paddle wheel to stir the water. At height h above the ground, the weight has potential energy given by mgh. Since it falls with constant speed, its kinetic energy does not increase. Instead, work is done by the falling weight in stirring the water. Joule found that the temperature of the water increased, and he recorded this increase. From the number of times the weight had fallen, he determined the total work done in stirring the water.

Joule could have achieved the same temperature rise by heating the same amount of water, as shown in Fig. 6–5(b). A **calorie** is defined as the amount of heat required to raise 1 gram of water by 1°C. Thus, Joule could determine the amount of heat in calories needed to raise the temperature of the water in Fig. 6–5(b).

From the work done in Fig. 6–5(a), and the calories added in Fig. 6–5(b), Joule found that 1 calorie of heat was equivalent to 4.19 joules of work.

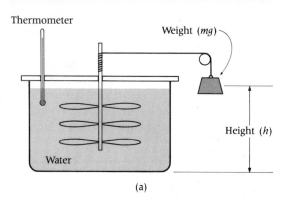

Thermometer

Weight (mg)

Height (h)

Water

(a)

Thermometer

Water

Flames

(b)

FIG. 6–5 Joule's experiment showing the relationship between (a) work done in stirring the water and (b) the heat added to the water.

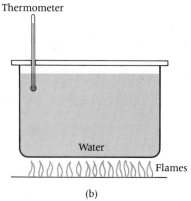

This relationship between work (or energy) and heat suggested that heat is really a form of kinetic energy, the kinetic energy of atoms. Further, the motion of atoms would be governed by Newton's laws. The basis of this so-called *kinetic theory* is that the temperature of an object is related to the kinetic energy of the atoms (or molecules) of the object. Not all of the atoms have the same kinetic energy, however, but when the temperature increases, the *average* kinetic energy of the atoms increases. In a solid, the atoms vibrate about an equilibrium position, much like a mass vibrates on the end of a vertical spring. When the temperature increases, the atoms in a solid vibrate more rapidly and with greater excursions from the equilibrium position.

When two objects are rubbed together, the energy or work done is simply transferred to the atoms, causing an increase in their kinetic energy, which we perceive as a rise in temperature.

When we say that "heat flows," then, what we really mean is that kinetic energy is transferred from the atoms of the hot object to the atoms of the cooler one. Eventually, the atoms of both objects have the same average kinetic energy and, thus, the same temperature.

When atoms collide, both momentum and energy are conserved. To illustrate this point, let's consider the collision of two balls (see Fig. 6–6). Ball A, traveling at a speed of 2 m/sec, collides with ball B, at rest. After the collision, some of the momentum and energy of ball A has been transferred to ball B. Also after the collision, ball A rebounds with a speed of 0.67 m/sec, and ball B has a speed of 1.33 m/sec. Figure 6–6 shows the momentum and kinetic energy of each ball before and after the collision, as well as the total momentum and total energy. We can see that momentum and energy are conserved. It is important to note that the kinetic energy of ball A decreases, while the kinetic energy of ball B increases. Now suppose these balls are atoms. After many collisions between atoms A and B, then, their average kinetic energies will be the same, and they will

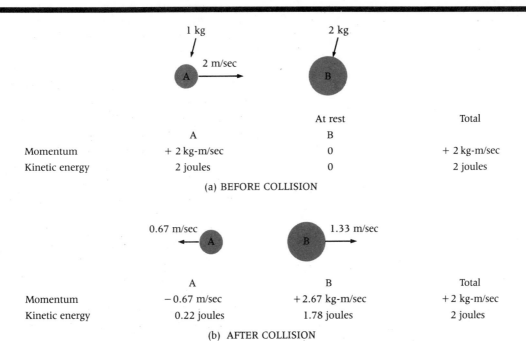

FIG. 6-6   The collision of two balls illustrates conservation of momentum and energy. After the collision, the kinetic energy of ball B increases, but that of ball A decreases.

have the same temperature. In general, conservation of momentum and energy show that kinetic energy will be transferred from the more energetic object to the less energetic, never the reverse.

EXPANSION

When heated, most solids expand. As the temperature rises, the atoms gain a larger kinetic energy and vibrate through larger distances. Because the atoms are now separated by larger average distances, they expand. If sufficient heat is added to a solid, the vibrations of the atoms become so violent that the attractive bonds between atoms are greatly reduced (due to the larger separation). As a result, the atoms can move about more freely: The solid melts and becomes liquid.

For similar reasons, a liquid left out in the air evaporates. The more energetic molecules can overcome the force of attraction and escape from the surface of the liquid. Adding heat quickens this process considerably by raising the average kinetic energy.

GAS PRESSURE AND TEMPERATURE

Imagine a gas trapped in a cubic container. The kinetic theory also predicts that there is a relationship between the **pressure** exerted by the gas on the walls and the temperature of the gas. (*Pressure* is defined as the force per unit area.)

The pressure is due to atoms rebounding from the walls. The walls exert a force on the atoms to make them change direction. According to Newton's third law, the atoms will exert the same amount of force on the wall. When the temperature of the gas increases, this force on the wall likewise increases (because the atoms are moving faster and thus require a greater force to change direction). Their increased speed will allow the atoms to travel between two opposite walls in a shorter time. Therefore, when the gas temperature increases, the atoms exert a larger force when they strike the wall, and they also strike the wall more frequently. The pressure in an automobile tire, for instance, will increase after a long car trip because the temperature of the air inside that tire has increased.

KINETIC ENERGY TRANSFORMS INTO HEAT

Pretend, in the situation where you are dropping a rock on a tent stake, that you do not aim carefully and the rock misses the stake. Just before hitting the ground, the rock has kinetic energy. This energy cannot be lost, but must be converted to other forms. For example, you can hear the rock hit the ground and possibly notice its imprint if the ground is soft enough. Or perhaps the rock breaks into several pieces. There can also be another effect: The temperature of the rock and the ground both increase, which means that some of the kinetic energy has changed into heat. A complete conservation of energy equation is

$$PE_{top} = KE_{bottom}$$

$$= heat + sound + \begin{matrix} energy \ to \ deform \\ rock \ and \ ground. \end{matrix}$$

Heat and Radioactivity

The Curies found that radium had a remarkable property: it remained 1.5°C above room temperature. Puzzled, they raised the question, What was the source of this heat?

Radium emits energetic alpha particles. Rutherford, who experimented extensively with these alpha particles, saw a connection between kinetic energy and heat, and was able to solve the puzzle of the heat source. The alpha particles produced on the surface of the radium source escape into air, but those produced inside the radium collide with many radium atoms and transfer kinetic energy to them. After many such collisions (perhaps billions), the alpha particle has no kinetic energy left, and stops. Thus, the heat generated by the radium source reflects the transfer of kinetic energy from the alpha particles to the radium atoms. But this solution raised yet another question: What is the source of energy for the alpha particles? The answer is the transformation of mass into kinetic energy, or *nuclear energy,* which we shall discuss shortly.

X-Ray Production

In a cathode ray tube, energetic electrons strike the anode, slow down, and finally stop. Because the kinetic energy of charged particles changes into heat, the anode, certainly, gets hot. This, however, is not the only energy transformation process that occurs. Sometimes the kinetic energy of the electrons changes instead to X-rays, as illustrated in Fig. 6–7. X-rays are similar in nature to visible light, but have about 5000 times more energy. As the electron slows down, one or perhaps several X-rays may be produced. If only one X-ray results, then its energy must equal the kinetic energy of the electron. If several X-rays are produced, then the kinetic energy is shared among them, but their total energy must equal the kinetic energy of the electron. Therefore, X-rays produced in a cathode ray tube have a continuous range of energies: from nearly zero to a maximum energy equal to the kinetic energy of the electron.

Formation of Stars and Starlight

Especially intriguing questions whose answers involve gravity and energy transformation processes are: How are stars formed? How long will the sun (only one among billions of stars) continue to bathe the earth with its sunlight?

FORMATION OF STARS

Initially, the universe appears to have consisted of a cloud of gas (primarily, a mixture of hydrogen and helium) that pervaded all space. This gas was almost uniformly distributed, but in some regions the density was slightly greater than in others (Fig. 6–8a). The density became still higher as more gas atoms were attracted to the newly forming stars through the force of gravitational attraction (Fig. 6–8b). This process continued until the gas became quite compressed. Figure 6–8(c) shows the formation of two stars. Even today, stars form in the same way (see Fig. 6–9).

ENERGY TRANSFORMATION IN STARS

The light emitted by any star, our sun included, comes from the conversion of one type of energy into light. How does this occur?

Let's first consider the transformation of potential energy into kinetic energy as the star contracts. Figure 6–10 shows the similarity between a ball falling to the ground, a phenomenon we have already studied (see p. 115), and the formation of a star. The ball falls due to its attraction to the earth: Its potential energy decreases as its kinetic energy increases. A similar process occurs in the formation of stars. The gas atoms, attracted to regions of higher density, gain kinetic energy as they "fall" toward those regions, but must lose potential energy as a result. Therefore, the potential energy of a star decreases as its size (or the radius of the sphere) decreases.

The gas atoms have more potential energy in Fig. 6–10(a), but more kinetic energy in Fig. 6–10(b). Earlier, we learned that there is a relationship between kinetic energy and tempera-

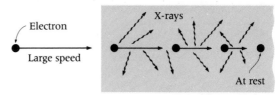

FIG. 6–7   The production of X-rays in a cathode ray tube. The kinetic energy of an electron striking the anode transforms into X-rays when the electron comes to rest.

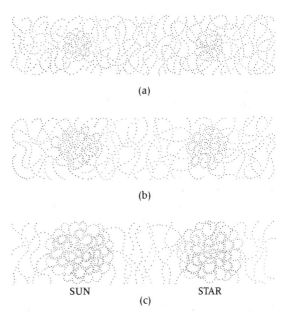

FIG. 6–8   The formation of two stars from a cloud of gas. The gas molecules are attracted to regions of slightly higher density.

FIG. 6-9   Star formation taking place in the Orion nebula. (Yerkes Observatory photograph)

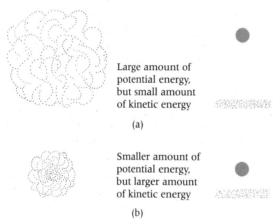

Large amount of potential energy, but small amount of kinetic energy

(a)

Smaller amount of potential energy, but larger amount of kinetic energy

(b)

FIG. 6-10   Comparison between the formation of a star due to gravitational contraction and a ball falling to the ground due to gravity.

ture. Thus, as the star contracts, and the kinetic energy of its atoms increases, the star gets hotter. By estimating the kinetic energy of atoms at the center of a star, scientists can determine the temperature there. The increasing temperature as the star contracts leads, in turn, to another effect: starlight.

TEMPERATURE AND STARLIGHT

It's a common observation that extremely hot objects emit light. For example, a heated piece of iron first glows red and then white-hot as its temperature increases. The heating coil of an electric stove glows red, and the filament of a light bulb, white. (Objects at room temperature also emit radiation, but our eyes are not sensitive to it.) Thus, the temperature of an object determines the intensity and color of the light emitted by its surface. In fact, scientists determine the surface temperature of a star by analyzing its starlight. The important point to remember is that, as a star contracts, some of its potential energy transforms into kinetic energy (or heat) and starlight.

CONSERVATION OF ENERGY IN STARS

We can now use conservation of energy to probe the question, How much light is emitted by a star as it contracts?

Initially (Fig. 6-10a), the gas atoms have little kinetic energy; the gas is very cold. (We shall assume that its kinetic energy is zero.) The gas does have potential energy, denoted by the symbol PE_i, which can be calculated. (Such calculation goes beyond our present discussion and will thus not be done here.)

Finally (Fig. 6-10b), the gas atoms have both kinetic energy KE_f and potential energy PE_f, which can likewise be determined. In addition, the contracted star emits starlight. The question is, How much? From conservation of energy:

$$TE_i = TE_f$$

$$PE_i = PE_f + KE_f + \text{light energy.}$$

This equation becomes

$$\text{Light energy} = PE_i - PE_f - KE_f, \quad (8)$$

where $PE_i > PE_f$. Since all terms on the right-hand side can be calculated, the total amount of energy transformed into light can be determined. Several crucial questions must now be raised: Does this conversion of potential energy into light explain the production of starlight? Furthermore, is this process the *only one* that produces light? In order to answer such questions, we must contrast theory with the results of experiment. Much experimental data are available about one star: our sun.

LIFETIME OF THE SUN

The age of the solar system is about 4.5 billion years, and Darwin's theory of evolution indicates that the evolutionary process requires about 2 billion years. Is the light energy obtained in Equation 8 sufficient to last that long?

The light energy per second reaching the earth's surface has been measured. Assuming that the sun radiates the same in all directions, the total amount of energy per second emitted by the sun is 3.8×10^{26} joules/sec. The lifetime of the sun, or the total time during which the sun emits light, can be found as follows:

$$\text{Lifetime of sun} = \frac{\text{light energy from Eq. 8}}{3.8 \times 10^{26} \text{ joules/sec}}.$$

In the 1920s, scientists calculated the sun's lifetime to be only 30 *million* years: 150 times too small! *While the contraction of a cloud of gas does explain a star's formation and incredibly high temperature, such contraction does not produce nearly enough light to account for the longevity of the sun. Thus, another transformation process must also take place.* The most likely choice was the transformation of mass into energy, but this process, called **fusion,** was not well understood until the late 1930s. Basically in the fusion process, 4 hydrogen atoms combine (fuse together) to form 1 helium atom, which has a smaller mass than 4 hydrogen atoms. According to Einstein's mass-energy relationship ($E = mc^2$), this difference in mass is converted into energy, much of which we receive as sunlight. In the sun, 700 million tons of hydrogen are converted into helium every second!

Collision of Cars

Returning to activities on earth, let's revisit the situation first discussed in Chapter 4: the head-on and tail-end collisions of two cars (see Fig. 6–11). Using conservation of momentum and energy, we shall see why the head-on version is so much more disastrous than the tail-ender. Our objective is to comprehend how these conservation laws apply to a familiar example so that we can extend them more easily to abstract examples, such as the discovery of the neutron and the creation of elementary particles.

Figure 6–11 shows the momentum (mv) and kinetic energy of each car. In the head-on collision, the total momentum before the collision is −20,000 kg-m/sec. After the collision, the momentum of the coupled cars is (2500 kg)(v), which must also equal −20,000 kg-m/sec. Therefore, the velocity v after the collision is −8 m/sec, which shows that the unit moves to the left. Before the collision, the kinetic energies total 350,000 joules. After the collision, the kinetic energy of the unit is only 80,000 joules. The rest of the energy, 270,000 joules, is used to crumple the fenders: The energy of the head-on crash is equal to 270,000 joules.

For the tail-end collision, conservation of momentum shows that the velocity of the cars after collision is 16 m/sec directed to the left. Energy conservation shows that the energy used to crumple the fenders is only 30,000 joules: The energy of the tail-end crash is equal to 30,000 joules. Thus, the crash energy in a head-on

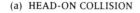

(a) HEAD-ON COLLISION

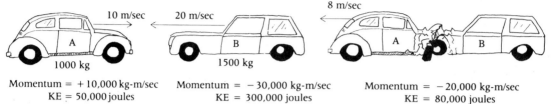

Momentum = +10,000 kg-m/sec
KE = 50,000 joules

Momentum = −30,000 kg-m/sec
KE = 300,000 joules

Momentum = −20,000 kg-m/sec
KE = 80,000 joules

(b) TAIL-END COLLISION

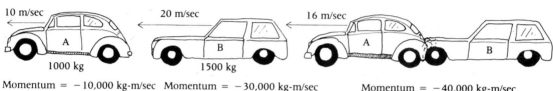

Momentum = −10,000 kg-m/sec
KE = 50,000 joules

Momentum = −30,000 kg-m/sec
KE = 300,000 joules

Momentum = −40,000 kg-m/sec
KE = 320,000 joules

FIG. 6–11   (a) Head-on collision of two cars. (b) Tail-end collision. Car A has a mass of 1000 kg (2200 lb) and a speed of 10 m/sec (22.4 mph). Car B has a mass of 1500 kg (3300 lb) and a speed of 20 m/sec (44.7 mph). After the collision, the bumpers of the two cars interlock, and they move together as one unit. The momentum and kinetic energy of each car are displayed.

collision is 9 times that in a tail-end collision, which seems a reasonable result.

With the confidence we have now gained in this area, let's tackle a problem far removed from our everyday experience: Chadwick's discovery of the neutron in 1932. Many elementary particles have been discovered since that time, and their discovery, just like Chadwick's did, relies upon conservation of energy and momentum.

The Mass of the Neutron

Recall (Chapter 1) how James Chadwick discovered the neutron: When alpha particles bombard a beryllium target, neutrons are produced (see Fig. 6–12). To find the mass of the neutron, Chadwick carried out two experiments and then analyzed his results, using conservation of momentum and energy.

In the first experiment, neutrons emerging from the beryllium target are directed onto a paraffin target. Since a molecule of paraffin contains many hydrogen atoms, neutrons strike many protons (1 amu) and knock them completely out of the paraffin target. Chadwick measured their speed as 3.3×10^7 m/sec.

In the second experiment, neutrons bombarded a container of nitrogen gas, and Chadwick measured the speed of the struck nitrogen nuclei to be 0.44×10^7 m/sec. Because of its larger mass, the nitrogen nucleus (14 amu) was not set into motion as easily as the protons in the paraffin target. Hence, the lower speed of the struck nitrogen nucleus.

Chadwick showed that momentum and energy are conserved *only if* the mass of the neutron is nearly equal to the mass of a proton: The mass of a neutron is about 1 amu. For simplicity, let's use the reverse procedure: Assume that the mass of the neutron is 1 amu and see whether momentum and energy are conserved. If so, then our assumption about the mass was correct.

Figure 6–13 shows the head-on collision of a neutron with a proton. We know that the speed of the proton after collision is 3.3×10^7 m/sec, but we don't know the speed of the neutron before collision. Assuming that the masses are equal, conservation of momentum and energy show that the neutron's speed before collision must have been 3.3×10^7 m/sec, and that the neutron is at rest after collision. (Figure 4–5b p. 91 shows an analogy to this situation.)

In Fig. 6–14, a neutron traveling at 3.3×10^7 m/sec collides with a nitrogen nucleus and knocks it forward with a speed of 0.44×10^7 m/sec. We don't know the neutron's velocity after collision, but we can use conservation of momentum to find it:

$$\frac{\text{Momentum before}}{\text{collision}} = \frac{\text{Momentum after}}{\text{collision}}$$

$$(1 \text{ amu})(3.3 \times 10^7 \text{ m/sec})$$
$$= (1 \text{ amu})(v_n) + (14 \text{ amu})(0.44 \times 10^7 \text{ m/sec})$$
$$v_n = -2.86 \times 10^7 \text{ m/sec}.$$

The negative sign shows that the neutron moves toward the left after collision, as shown in Fig. 6–14. Is energy conserved? That's the crucial last step. Only one form of energy is involved, so that the total kinetic energy before the collision must equal that after the collision. The total (kinetic) energy *before* the collision is equal to the kinetic energy of the neutron:

$$\underset{\text{before}}{\text{TE}} = \frac{1}{2}(1 \text{ amu})(3.3 \times 10^7 \text{ m/sec})^2$$
$$= 5.445 \times 10^{14} \text{ amu-m}^2 / \text{sec.}^2$$

Similarly, *after* the collision, the kinetic energy of the neutron is

$$\underset{\text{after}}{\text{KE}_n} = 4.090 \times 10^{14} \text{ amu-m}^2/\text{sec}^2;$$

the kinetic energy of the nitrogen nucleus is

$$\underset{\text{after}}{\text{KE}_N} = 1.355 \times 10^{14} \text{ amu-m}^2/\text{sec}^2;$$

and the total energy is

$$\underset{\text{after}}{\text{TE}} = 5.445 \times 10^{14} \text{ amu-m}^2/\text{sec}^2.$$

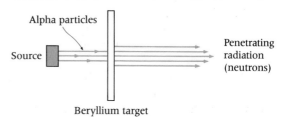

Alpha particles

Source

Penetrating radiation (neutrons)

Beryllium target

FIG. 6–12   Discovery of the neutron. Alpha particles from a radioactive source strike a beryllium target.

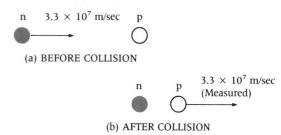

n 3.3×10^7 m/sec   p

(a) BEFORE COLLISION

n   p 3.3×10^7 m/sec (Measured)

(b) AFTER COLLISION

FIG. 6–13   Head-on collision of a neutron with a proton. After the collision, the proton has a speed of 3.3×10^7 m/sec, and the speed of the neutron is reduced to zero. The speed v of the neutron before collision must also be 3.3×10^7 m/sec if momentum and energy are conserved, and if the mass of the neutron is equal to the mass of the proton.

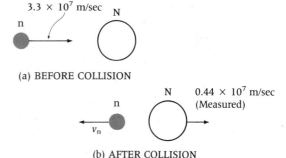

3.3×10^7 m/sec   N

n

(a) BEFORE COLLISION

N 0.44×10^7 m/sec (Measured)

n

v_n

(b) AFTER COLLISION

FIG. 6–14   Head-on collision of a neutron (n) with a nitrogen nucleus (N). The speed of the neutron before collision was calculated to be 3.3×10^7 m/sec. Chadwick measured the speed of the nitrogen nucleus to be 0.44×10^7 m/sec.

CONSERVATION OF ENERGY

Yes, energy is conserved! We must realize that the neutron's speed before collision depends upon the assumption that the neutron mass is equal to the proton mass. Thus, conservation of energy verifies the correctness of this assumption. (Perhaps the most convincing argument is to see that momentum and energy cannot be conserved for another choice of neutron mass; for example, when the neutron mass is chosen as 14 amu. See calculation question 6–16.)

Alpha Decay and Nuclear Energy

Earlier (p. 120), we learned that the heat generated by a radium sample is due to the absorption of energetic alpha particles by the radium itself. We then raised the question, What is the source of energy for the kinetic energy of the alpha particles? Let's look more deeply into the answer we gave—**nuclear energy.**

Many scientists, the Curies included, could not understand how any of the usual forms of energy could transform into the kinetic energy of the alpha particle. Scientists speculated that the energy from radioactivity must somehow be due to the energy stored in the atom. They called this *atomic energy,* but today it is more accurately called *nuclear energy,* for the energy is stored in the nucleus. The following summarizes the energy transformation steps:

Nuclear energy
$\Downarrow$
kinetic energy
of alpha particle
$\Downarrow$
heat generated
by radium.

In Chapter 3, we discussed how alpha particles are emitted by radium. The radium nucleus breaks up into two parts. The alpha particle ($^4_2$He) is ejected, and a smaller radon nucleus remains (see Fig. 6–15):

$$^{226}_{88}\text{Ra} \rightarrow {}^4_2\text{He} + {}^{222}_{86}\text{Rn}.$$

Offhand, you might say that the mass of the radium nucleus is equal to the mass of the radon nucleus plus the mass of the alpha particle. This is a rather basic idea: the total mass remains the same. Tear a piece of paper into two pieces—the total mass is unchanged. However, in 1905, Einstein proposed a theory that violated this commonsense idea. He said that mass and energy are interchangeable and obey the now-familiar equation

$$E = mc^2,$$

where c is the speed of light, 3×10^8 m/sec. Thus, the total mass does not have to remain constant; some of it can change into energy, which is exactly what happens here. Note that, in Fig. 6–15, the radium nucleus is much larger than the radon nucleus and the alpha particle. This representation is intentional. It illustrates that the mass of the radium nucleus is larger than the total mass of the two parts in Fig. 6–15.

The difference in mass is converted into kinetic energy: The kinetic energy of the alpha particle, and the much smaller recoil kinetic energy of the radon nucleus. About 0.002 percent of the mass of the radium nucleus transforms into kinetic energy. The alpha particle has 98 percent of this kinetic energy, and the radon nucleus has the rest. In general, the term *nuclear energy* refers to the transformation of mass into kinetic energy.

Elementary Particles

Recall that Chadwick discovered the neutron by bombarding a beryllium target with alpha particles from a radioactive source. These sources, however, had their limitations. The maximum speed of an alpha particle from a radioactive source, for instance, is only about 5 percent of the speed of light. Obviously, the target could be bombarded only by alpha, beta, or gamma rays. Physicists quickly realized that they could learn much more by using particles of higher energy and by using other kinds of particles, such as protons. For these reasons, they designed machines, called **particle accelerators,** which accelerated the desired particles to a very high energy. In 1932, the same year that Chadwick discovered the neutron, the first particle accelerator also began operation. In Chadwick's experiment, the kinetic energy of the alpha particle was used to knock loose a neutron (from either the beryllium nucleus or from the alpha particle). However, when the beam energy of the particles from the accelerator became sufficiently large, it was possible to do much more: In the collision between the energetic particle and the target nucleus, some of the kinetic energy was transformed into the mass of a third particle, called an **elementary particle.** To date, more than 200 elementary particles have been discovered!

As an example of elementary particle production, suppose that a proton from a particle accelerator strikes a carbon target. When a proton collides with a carbon nucleus, the carbon nucleus moves in the forward direction, but the proton and a new particle called a **meson** emerge at an angle, as shown in Fig. 6–16. As with the collision of cars, momentum and energy must be conserved. Before the collision, only the proton had momentum, directed toward the right. After the collision, the momentums of the participants are as follows: the carbon nucleus, directed toward the right; the proton, directed toward the right and upward; the meson, directed toward the right and downward. Because the proton's upward momentum counterbalances the downward momentum of the meson, the total momentum of all three particles is likewise directed toward the right, just as the conservation of momentum principle requires.

To understand how the meson can be created, let's consider the collision of our two cars. In that case, some of the kinetic energy of the cars changes form: It changes into energy to crumple the fenders. Similarly, in the collision of the proton with the carbon nucleus, some of the proton's kinetic energy changes form: It

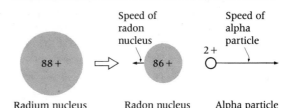

Radium nucleus   Radon nucleus   Alpha particle

FIG. 6–15   A radium nucleus decays (breaks apart) into an alpha particle and a radon nucleus. Some of the mass of the radium nucleus transforms into the kinetic energy of the alpha particle and the much smaller kinetic energy of the radon nucleus. The exaggerated sizes of the particle, before and after the decay, are intended to show this decrease in mass.

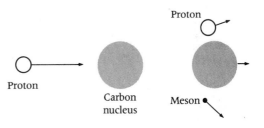

(a) BEFORE COLLISION   (b) AFTER COLLISION

FIG. 6–16   (a) The collision between an accelerated energetic proton and a carbon nucleus. (b) A meson is produced when some of the kinetic energy of the proton transforms into the mass of the meson.

CONSERVATION OF ENERGY

changes into the mass of the meson. Conservation of energy requires that the total kinetic energy of all three particles after collision must be *less* than the kinetic energy of the proton before collision. From conservation of energy, then, we have

Initial KE
of proton

    total KE
= of proton, carbon   + (mass of meson)c^2,
   nucleus, and meson

where the difference in kinetic energy is converted into the mass of the meson. If all kinetic energies are measured in an experiment, the mass of the meson can be determined. Scientists, of course, are interested in obtaining even more energetic particle accelerators. When the kinetic energy of the particle beam increases, more massive elementary particles can be created.

Potential Energy of Charged Particles

Suppose there are two charged particles, Q_1 and Q_2, initially at rest and separated by a distance d. If they are free to move, the forces of attraction (or repulsion) will cause them to move toward each other (or away from each other). In either case, the charged particles *gain* kinetic energy and, hence, they initially had potential energy that transformed into kinetic energy:

PE of charged
particles separated
by distance d

    ↓   transforms into

KE of charged particles,
as particles begin to move
due to attraction (or
repulsion).

The potential energy between two charged particles is given by

$$PE = \frac{KQ_1Q_2}{d}, \qquad (9)$$

where d is the distance between the charges Q_1 and Q_2, and K is Coulomb's constant. In Equation 9, the charges are substituted with their positive or negative signs. If both charges have the same sign, the potential energy is a positive number. These charges will repel each other, and the distance between them will increase. As the distance d in Equation 9 increases, the potential energy decreases. Thus, the potential energy decreases, while the kinetic energy increases, in agreement with conservation of energy.

If the charges Q_1 and Q_2 have opposite signs, the potential energy is a negative number. These charges attract each other, and their separation distance decreases. As the denominator in Equation 9 gets smaller, the potential energy decreases because it is *negative*. (Remember, for example, that $-\frac{1}{4}$ is *larger* than $-\frac{1}{2}$.) Thus, as the particles move toward each other, the potential energy decreases, while the kinetic energy increases.

Equation 9, which defines potential energy, does have units of energy. Remember that K has units of newton-m^2/coulomb^2, or, more simply, N-m^2/C^2. The units on the right-hand side of Equation 9 then become

$$\left(\frac{\text{N-m}^2}{\text{C}^2}\right)(\text{C})(\text{C})\left(\frac{1}{\text{m}}\right) = \text{N-m}$$

$$= \left(\frac{\text{kg-m}}{\text{sec}^2}\right)(\text{m}) = \text{joule}.$$

Recall that a newton is equivalent to kg-m/sec^2.

From the viewpoint of units and conservation of energy, Equation 9 seems to be a plausible definition.

We have just seen that the potential energy of two unlike charges is a negative number. But what is the meaning of *negative energy?* To answer, let's take another look at the example computed in Table 6−1: a 5-kg object, initially

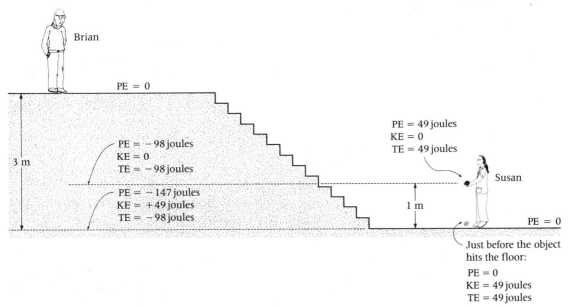

FIG. 6-17   A 5-kg object falls from a height of 1 m and is observed both by Brian and by Susan.

at rest, falls from a height of 1 m. Susan and Brian (in Fig. 6–17) observe such an object fall to the floor. Both of them measure the distances and the potential energy relative to the floor on which they stand. Susan claims that the object has initially a potential energy, *mgh*, of (5 kg)(9.8 m/sec²)(1 m), or 49 joules; when it reaches the floor, its potential energy is zero. Brian says that the object is initially 2 m *below* his floor and finally, 3 m *below.* The initial potential energy, found by letting the height h equal −2m, is −98 joules. Finally, when h equals −3m, the potential energy is −147 joules. According to conservation of energy, the difference in potential energy is converted to kinetic energy. For Susan, the difference in potential energy is 49 joules minus 0 joules, or +49 joules. The quantity "0 joules" is shown for emphasis. For Brian, the difference is (−98 joules) minus (−147 joules), or +49 joules. Thus, both Brian and Susan agree that the object has a kinetic energy of 49 joules just before it hits the floor.

Even though Brian and Susan do not agree on what they call the potential energy at each point, they do agree that initially the potential energy is 49 joules larger than when it hits the floor. So the important information for understanding the physics involved is *the difference between two energies,* and it doesn't matter whether they are positive or negative numbers. The significant questions are: Which energy is larger? and, How much larger is it?

The Size of the Nucleus

The size of the nucleus must (as we saw in Chapter 3) be much smaller than 10^{-10} m, the size proposed by Thomson. Let's learn, now, how Rutherford estimated the size of the gold nucleus.

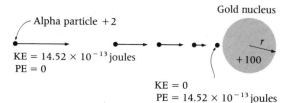

Alpha particle +2

KE = 14.52×10^{-13} joules
PE = 0

Gold nucleus

r

+100

KE = 0
PE = 14.52×10^{-13} joules

FIG. 6–18 An alpha particle slows down as it approaches the gold nucleus due to the mutual repulsion between the positive charge on the alpha particle and that on the gold nucleus. The length of the arrow indicates the speed of the alpha particle.

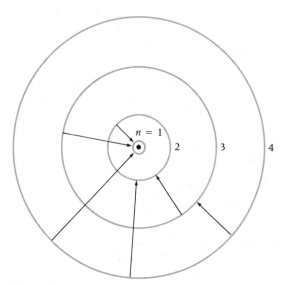

$n = 1$ 2 3 4

FIG. 6–19 A diagram of the hydrogen atom that shows four allowed orbits. Light is emitted when the electron jumps from a larger orbit to a smaller one. All possible jumps between these four orbits are shown.

When alpha particles bombard a gold foil, only a few will be directed exactly toward the center of the gold nucleus (see Fig. 6–18). In these experiments, a certain radioactive source was used. This source emitted alpha particles with a kinetic energy of 14.52×10^{-13} joules.

When Rutherford and his collaborators, Geiger and Marsden, performed the scattering experiments, the charge on the nucleus was not known. Rutherford estimated that this charge was approximately equal to one-half the atomic mass, which for gold was 197 divided by 2, or about 100+ charges. (Today, we know that the number of positive charges on the nucleus is equal to the atomic number, which for gold is 79.)

As the alpha particle approaches the gold nucleus, it slows down due to the repulsion of the positive charge on the alpha particle and the positive charge on the nucleus. Thus, the kinetic energy of the alpha particle decreases, but the potential energy of the charged particle increases. Rutherford predicted that the alpha particle would stop, due to the huge coulomb repulsion, before it penetrated the surface of the nucleus and, in fact, he assumed that it stopped just at the nuclear surface.

At a large distance from the nucleus, the potential energy of the alpha particle is zero, and its kinetic energy is 14.52×10^{-13} joules. At the nuclear surface, the kinetic energy is zero, and the alpha particle has a potential energy of 14.52×10^{-13} joules. In order to calculate the potential energy, we shall assume that the charge of the nucleus acts as if it were concentrated at its center. When the alpha particle has reached the surface of the nucleus, the distance between the alpha particle and the center of the nucleus is, of course, just the radius r. The potential energy is given by

$$PE = \frac{KQ_A Q_N}{r},$$

where Q_A is the charge on the alpha particle and Q_N is the charge on the gold nucleus. Since values are known for all terms except the radius r, the radius r can be determined. This calcu-

INTEGER (n)	RADIUS (r_1)	SPEED (v_1)	KINETIC ENERGY (eV)	POTENTIAL ENERGY (eV)	ENERGY E (eV)
1	1	1	13.6	-27.2	-13.6
2	4	$\frac{1}{2}$	3.40	-6.80	-3.40
3	9	$\frac{1}{3}$	1.51	-3.02	-1.51
4	16	$\frac{1}{4}$	0.85	-1.70	-0.85

Note: The energy E is the sum of the kinetic energy and the potential energy of the electron. The radius r_1 is 0.528×10^{-10} m, and v_1 is 2.19×10^6 m/sec.

lation is carried out in the following section, and the radius of the gold nucleus is found to be 3.17×10^{-14}m.

THE RADIUS OF THE GOLD NUCLEUS: AN EXAMPLE*

Solving the above equation for r, we find

$$r = \frac{KQ_AQ_N}{(\text{PE})}.$$

Since the alpha particle has two positive charges, Q_A equals 2 (1.6×10^{-19} C), or 3.2×10^{-19} C. (Remember that the charge on the electron is 1.6×10^{-19} C.) According to Rutherford's estimate, the charge on the gold nucleus was 100 positive charges, or 1.6×10^{-17} C. The constant K is 9×10^9 N-m^2 / C^2. The potential energy is 14.52×10^{-13} joules or, equivalently, 14.52×10^{-13} N-m. Substituting these values into the preceding equation, we find the radius to be 3.17×10^{-14}m.

The Hydrogen Atom and Energy

We now come to the end of the story of Bohr's model of the hydrogen atom. Chapter 1 informed us that light is emitted when the electron jumps from a larger orbit to a smaller one. Here, we'll examine that process in more depth. Light is a form of energy; so let's look for the energy that transforms into light.

*This example may be omitted without loss of continuity.

The electron traveling in an allowed orbit has two kinds of energy: kinetic and potential. The sum of these two kinds we denote as energy E. We shall find the energy E for the first four allowed orbits (Fig. 6-19) and see that it is smallest for the first orbit and largest for the fourth orbit. The emission of light illustrates conservation of energy. For example, when the electron jumps from the third orbit to the second orbit, the hydrogen atom emits red light (see Table 1-3, p. 24). The energy E is larger for the third orbit. When the electron jumps from the third orbit to the second orbit, the difference in energy ($E_3 - E_2$) transforms into light.

First, we must find the energy E for the four allowed orbits. We already know a great deal about the hydrogen atom (see Chapters 3 and 4), and this information is displayed in the first three columns of Table 6-2.

We begin by finding the kinetic energy of the electron in the first allowed orbit, where its speed is 2.19×10^6 m/sec. Since the mass of the electron is 9.11×10^{-31} kg, the kinetic energy $\frac{1}{2} mv^2$ is 2.18×10^{-18} joules. This is such a small number that we introduce a new unit called the **electron volt** (abbreviation: eV). Chapter 7 discusses the origin of this new energy unit. The electron-volt is related to the joule by

$$1\,\text{eV} = 1.6 \times 10^{-19}\,\text{joules.} \qquad (10)$$

CONSERVATION OF ENERGY

With this definition, the kinetic energy becomes

$$\frac{2.18 \times 10^{-18}\,\text{joules}}{(1.6 \times 10^{-19}\,\text{joules})/\text{eV}} = 13.6\,\text{eV}.$$

In the second orbit, the speed is one-half that in the first orbit. Since the kinetic energy depends upon the speed *squared,* the kinetic energy is one-fourth that in the first orbit, or 3.4 eV. Similarly, in the third orbit, the kinetic energy is one-ninth that in the first orbit, or 1.51 eV. Table 6−2 lists the values for the kinetic energy for four orbits.

Next, we find the potential energy of the electron in the allowed orbits. First of all, since the electron and the nucleus are oppositely charged, the potential energy is a negative number. To find the potential energy of the electron in the first allowed orbit, we use Equation 9, substituting values for the charge on the electron and the nucleus. We let the distance d equal the radius of the first allowed orbit. This calculation, when carried out, shows that the potential energy of the electron in the first allowed orbit is −27.2 eV.

In the second allowed orbit, the distance between the electron and nucleus is 4 times that in the first orbit and, hence, from Equation 9, the potential energy is one-fourth that in the first orbit, or −6.8 eV. In the third orbit, the distance is 9 times that in the first and the potential energy is −27.2 eV/9, or −3.02 eV. Table 6−2 lists the potential energy for the first four allowed orbits. Since these values are negative numbers, the potential energy is largest in the fourth allowed orbit and smallest in the first allowed orbit.

We know that the electron will always jump to a smaller orbit (see Chapter 1). The values

for potential energy in Table 6−2 help us account for this behavior. Although we have not discussed it, any object moves in such a way that its potential energy becomes smaller. For example, as an object falls to the ground, its potential energy decreases. Similarly, an electron in the fourth orbit will jump to a smaller orbit, where the potential energy is smaller. When the electron reaches the first orbit, it will remain there (unless disturbed by an outside influence) because there is no possible smaller potential energy.

THE POTENTIAL ENERGY OF AN ELECTRON IN THE FIRST ALLOWED ORBIT: AN EXAMPLE*

We find the potential energy using Equation 9. In this equation, K is 9×10^9 N-m^2/C^2. The charge on the electron is -1.6×10^{-19} C. The charge on the nucleus is $+1.6 \times 10^{-19}$ C. The distance d is equal to the radius of the first allowed orbit, which is 0.528×10^{-10} m. Using these values in Equation 9, we find the potential energy to be -4.36×10^{-18} joules. Using the relationship between joules and electronvolts in Equation 10, we find that the potential energy becomes

$$\text{PE} = \frac{-4.36 \times 10^{-18}\,\text{joules}}{1.6 \times 10^{-19}\,\text{joules/eV}}$$
$$= -27.2\,\text{eV}.$$

Referring to Table 6−2, we see that the radius of an allowed orbit is given by

$$r = r_1 n^2, \tag{11}$$

where r_1 equals 0.528×10^{-10} m, and n is the number specifying the orbit. The energy E is given by

$$E = \frac{-13.6\,\text{eV}}{n^2}. \tag{12}$$

Light is emitted when the electron jumps to a smaller orbit. Figure 6−19 shows all possible

*This example may be omitted without loss of continuity.

jumps between the first four allowed orbits. According to the principle of conservation of energy, the total energy before and after the electron jumps must be the same:

$$\begin{matrix}\text{Energy } E \text{ of} \\ \text{larger orbit}\end{matrix} = \begin{matrix}\text{energy } E \text{ of} \\ \text{smaller orbit}\end{matrix} + \text{energy of light.}$$

The energy of the light is therefore given by

$$\begin{matrix}\text{Energy of} \\ \text{light}\end{matrix} = \begin{matrix}\text{energy } E \text{ of} \\ \text{larger orbit}\end{matrix} - \begin{matrix}\text{energy } E \\ \text{of smaller orbit.}\end{matrix} \quad (13)$$

In Fig. 6−20, the length of the arrow is equal to the difference in energy of the two orbits, which is the right-hand side of Equation 13. Thus, the length of the arrow represents the energy of the light that is emitted by that transition between orbits. These "arrows" in Fig. 6−20 have only *certain* lengths. Thus, only *certain amounts* of light energy can be emitted by the hydrogen atom. There is a correlation between the energy of light and its wavelength. Since only certain energies of light are possible, only certain wavelengths are found in the spectrum of light from hydrogen: a line spectrum. For example, ultraviolet light has a larger energy than red light. The hydrogen atom emits ultraviolet light when the electron jumps from the second orbit to the first; it emits red light when the electron jumps from the third orbit to the second. From Fig. 6−20, we see that the arrows representing these two transitions differ greatly in length, the longer one being the transition from the second orbit to the first.

Let's take a moment here to review our progress with the hydrogen atom and then see where we are headed. In Chapter 3, we saw that the attraction between the electron and the nucleus caused the electron to travel in a circular orbit. We determined the speed of the electron in 11 orbits. In Chapter 4, we learned how Bohr restricted the angular momentum of the electron to certain values (integral multiples of $h/2\pi$). As a result, we could chose 4 allowed orbits from our original 11 values. In this chapter, we calculated the energy E for these 4 allowed orbits and found the energy of light emitted by

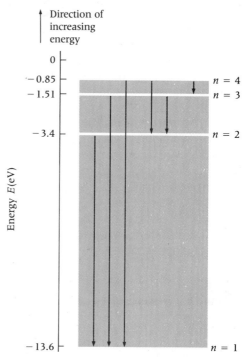

FIG. 6−20 An energy level diagram for the four allowed orbits of the hydrogen atom. The vertical arrow connecting two energy levels represents the difference in energy between them and, thus, the energy of the emitted light.

the hydrogen atom when the electron jumps to a smaller orbit. Further, the hydrogen spectrum was a *line* spectrum: a characteristic of the spectra of light from *all* elements. This was a major breakthrough because, up to then, no theory had adequately described the process by which an atom emits light.

So far, however, we have not determined the wavelengths of light in the hydrogen spectrum. That is, the "correlation between the energy of light and its wavelength" has not been specified. That is the work of later chapters. First, we must investigate the nature of light in Chapters 8 and 10. In Chapter 13, we shall take up our discussion of the hydrogen atom, determine the wavelengths of its line spectrum, and consider the spectra of other elements as well.

Summary

The work done in lifting an object to height h is defined mathematically as: work is equal to force times distance (work $= mg \times h$). Since change in energy equals work done, the energy of the object at height h is potential energy, which equals mgh (PE $= mgh$). When an object falls, its potential energy changes into kinetic energy, or energy of motion (defined mathematically as KE $= \frac{1}{2}mv^2$). The unit of energy and work is the joule (1 joule $= 1$ kg-m^2/sec^2). Power is defined as work or energy divided by time. The watt is the unit of power.

There are many forms of energy, such as heat, sound, light, nuclear energy, kinetic energy, potential energy. One form can transform into another, but the total amount of energy in such a transaction must remain the same, according to the law of conservation of energy. For example, when an object falls to the ground, some of its kinetic energy transforms into heat. Heat itself is a form of kinetic energy, since heat refers to the kinetic energy of the vibrating atoms.

When the temperature increases, the kinetic energy of the atoms increases.

The heat generated by a radium sample is due to the alpha particles emitted by radium coming to a stop and transferring their kinetic energy into heat.

A star is formed by gas-cloud contraction due to gravity. In this phenomenon, the PE decreases, and the KE (and thus the temperature of the star) increases. Light is also emitted in this process, but not enough to account for a star's long existence. Nuclear energy released during fusion of 4 hydrogen atoms into a helium atom accounts for a star's long lifetime.

Conservation of momentum and conservation of energy explain why a head-on collision of cars is more disastrous than a tail-end collision. Chadwick used the same principles to determine the mass of the neutron. When a proton collides with a carbon atom and a meson is produced, the mass of the meson is analogous to the energy used to crumple the fenders of the cars. In each case, the change in kinetic energy is converted into another form of energy.

Nuclear energy results from the conversion of mass into kinetic energy. When a radium nucleus disintegrates into an alpha particle and a radon nucleus, about 0.002 percent of the mass of the radium nucleus transforms into the kinetic energies of the alpha particle and the radon nucleus.

The potential energy between charged particles may be represented mathematically as PE equals $(KQ_1Q_2)/d^2$, where K equals 9×10^9 N-m^2/C^2.

Using conservation of energy, Rutherford determined the size of the nucleus. As the alpha particle approaches the nucleus, its kinetic energy transforms into potential energy. Rutherford assumed that the alpha particle comes to a stop very close to the nuclear surface, where its potential energy is equal to the initial kinetic energy.

We determined the kinetic energy, potential energy, and total energy of the electron in allowed orbits of the hydrogen atom. When the electron jumps to a smaller orbit, the difference in energy is converted into light.

True or False Questions

Indicate whether the following statements are true or false. Change all of the false statements so that they read correctly.

6–1 Brian and Susan lift identical boxes to the same height above the ground, but Brian lifts his in a shorter time.
    a. Brian is more powerful than Susan.
    b. Brian does more work than Susan.

6–2 When a ball is thrown vertically into the air, then, according to conservation of energy, the ball returns to the hand with the same speed as it left the hand.

6–3 A pendulum bob is pulled aside so that it rises a vertical height of 20 cm above its equilibrium position.
    a. When the pendulum bob, released from rest, passes through its equilibrium position, it has a smaller speed than an object falling freely from a height of 20 cm.
    b. The time for the pendulum bob to fall through a vertical height of 20 cm is the same as that for an object falling freely through a height of 20 cm.

6–4 James P. Joule showed that heat and energy are equivalent because water that is vigorously stirred rises in temperature, just as if it were heated by a flame.

6–5 Both the heat generated in a radium source and in the anode of an X-ray tube can be explained by the transformation of kinetic energy into heat.

6–6 As a star contracts during its formation, some of its kinetic energy is transformed into potential energy and light.

6–7 When two cars collide head-on,
    a. the total momentum is the same before and after the collision.
    b. the total kinetic energy before and after the collision is the same.

6–8 When a radium nucleus disintegrates into an alpha particle and a radon nucleus, the mass of the radium nucleus is equal to the total mass after the disintegration.

6–9 Rutherford estimated the size of the gold nucleus by considering the transformation of mass into kinetic energy.

6–10 An electron in a hydrogen atom jumps from the second orbit to the first orbit.
    a. The kinetic energy of the electron is larger in the second orbit.
    b. The potential energy of the electron is larger in the second orbit.
    c. The energy E (where E is the sum of the kinetic energy and the potential energy) is larger in the second orbit.
    d. The energy of light is equal to the energy E in the first orbit minus that in the second orbit.
    e. The electron jumps to the first orbit because its energy E is smaller there.

Questions for Thought

6–1 Match the name or term with its description by inserting the correct letter from the lettered column into each blank in the descriptive column. Entries in the lettered column may be used more than once.

a. potential energy
b. kinetic energy
c. watt
d. joule
e. newton
f. energy
g. power
h. work done on object
i. conservation of energy
j. total energy
k. heat

____ ability to do work
____ unit of work
____ is equal to increase in energy of object
____ unit of energy
____ shorthand for kg-m/sec^2

_____ force × distance

_____ shorthand for joule/sec

_____ shorthand for newton-meter (N-m)

_____ energy per unit time

_____ energy of motion

_____ energy of position

_____ energy of two charged particles at rest

_____ total energy always remains the same

_____ sum of all forms of energy

_____ due to kinetic energy of atoms

6–2 The beautiful fountains at Tivoli Gardens near Rome operate by having the water from a river at a high level flow to the fountains at a lower level. No pumps at all are used to operate the fountains, the tourist guide is quick to point out. The fountains are shown schematically in Fig. 6–21.

    a. If no energy is lost due to heat or friction, what is the maximum height the water from each fountain should reach?

    b. Which speed—v_A, v_B, v_C, or v_D—is largest?

    c. The water from fountain D rises to a maximum height of three-fourths h. How much of the potential energy that it had at the top is converted into heat and lost due to friction?

6–3 Using everyday experiences, describe three situations involving the transformation of one form of energy into another form.

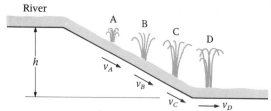

FIG. 6–21 Diagram for thought question 6–2.

6–4 A fragile china cup is dropped from a height of 1 m onto a tile floor and breaks, while an identical cup falls from the same height onto a plush, shag carpet and does not break.

    a. Do they have the same kinetic energy the instant before they strike the floor?

    b. Using conservation of energy, explain why one breaks and the other does not. Are there different energy transformation processes in the two cases?

    c. Which one has the greater deceleration? Explain.

    d. Which one has the greater force exerted on it in stopping? Explain.

6–5 Imagine that you are a participant in an egg-throwing contest for couples at a picnic in the park. You and your partner toss the egg back and forth, always increasing the distance between you. The contest goes on until there is only one unbroken egg left to one of the couples.

    a. Explain how you try to catch the egg so that it doesn't break.

    b. The egg has kinetic energy. The amount of work that you do in stopping the egg is (*less than*, *equal to*, or *greater than*) _____ the kinetic energy of the egg.

    c. Work is defined as force multiplied by distance. In catching the egg so that it doesn't break, you do one of the following:

        (1) maximize the distance in stopping, while also maximizing the force on the egg.

        (2) maximize the distance in stopping, while minimizing the force on the egg.

        (3) minimize the distance in stopping, while also minimizing the force on the egg.

        (4) minimize the distance in stopping, while maximizing the force on the egg.

6–6 Describe the relationship between heat and kinetic energy.

6–7 The Curies found that a sample of radium remained 1.5°C above room temperature. Could they have cited this as a violation of the law of conservation of energy? Explain.

6–8 Suppose that 100 calories are added to each of 100-gram samples of copper and aluminum. Will both have the same temperature rise? Make a distinction between heat and temperature.

6–9 A person often stirs a hot cup of coffee with a spoon so that it cools down more quickly. Shouldn't the stirring cause an increase in temperature? What other effects come into play here?

6–10 Describe the formation of a star and the energy transformation processes involved. What energy transformation process results in the high temperature of a star? What energy transformation process results in starlight?

6–11 Define the term *nuclear energy* and relate it to the disintegration of a radium nucleus into an alpha particle and a radon nucleus.

6–12 Recall the analogy we made between the firing of a rifle and the disintegration of the radium nucleus (Chapter 4). What energy transformation process supplies the kinetic energy of the bullet and the rifle? The kinetic energy of the alpha particle and the radon nucleus?

6–13 Draw an analogy between a ball being thrown vertically into the air and an alpha particle aimed at a gold nucleus. What are the similarities in finding the height reached by the ball and the way Rutherford determined the radius of the gold nucleus?

6–14 This question refers to the Bohr model of the hydrogen atom. Insert: *increases, decreases,* or *remains the same.*

    a. When n increases, the radius of the electron orbit _____ .
    b. When n increases, the angular momentum of the electron _____ .
    c. When n increases, the force on the electron _____ .
    d. When n increases, the kinetic energy of the electron _____ .
    e. When n increases, the potential energy of the electron _____ .
    f. When n increases, the total energy of the electron _____ .

6–15 Imagine that your checkbook has a balance of $500. You write checks for the following amounts: $200, $125, $300, and $50. Make a cash-balance diagram similar to the energy-balance (or energy-level) diagram in Fig. 6–20. Explain the relationship between these two diagrams. What are the checks analogous to?

Questions for Calculation

6–1 Suppose that you lift a heavy box of 10 kg (22 lb) from the floor onto a table 1 m high. This takes you 3 seconds.

    a. How much work did you do in lifting the box?
    b. What is the potential energy of the box?
    c. If the box is inadvertently pushed off the table, what is its kinetic energy when it strikes the floor?
    d. Suppose that you lift another identical box onto the same table, but this time you take only 2 seconds. Did you do more work than in part **a**? Compare the power used in lifting the two boxes.

6–2 A child throws a 0.25-kg ball vertically into the air, and it leaves the child's hand with a speed of 5 m/sec.

    a. Use conservation of energy to find how high the ball reaches. Assume that the potential energy is zero at the point where the ball is released. Measure the height of the ball from this point.
    b. Solve this problem by using the methods of Chapter 2 to find the time it takes the ball to reach its maximum height, its average speed, and then how high the ball reaches.

CONSERVATION OF ENERGY

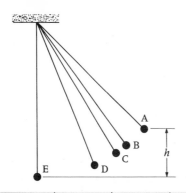

	h	PE	KE	TE
A	0.12 m			
B	0.08			
C	0.06			
D	0.03			
E	0			

FIG. 6–22 Diagram for calculation question 6–5.

6–3 A child lets a 1-kg toy truck roll down a slide. The highest end of the slide is 2 m above the ground and the lowest end is 0.2 m above the ground. Assume that the potential energy is zero at ground level; measure distances relative to the ground.

    a. Use conservation of energy to find the kinetic energy of the truck when it reaches the lower end of the slide.

    b. Find the kinetic energy when it reaches the ground.

6–4 Refer to calculation question 6–3, but here define the point of zero potential energy to be at the lower end of the slide. (Hint: See Fig. 6–17.)

    a. What is the potential energy of the truck at the top of the slide?

    b. What is the truck's kinetic energy at the lower end of the slide? Compare your answer with calculation question 6–3a.

What is more important in solving a problem: the value of the potential energy or the difference in potential energy?

    c. What is the potential energy of the truck on the ground?

    d. What is the truck's kinetic energy when it reaches the ground? Compare your answer with calculation question 6–3b.

6–5 The pendulum, raised a distance h above the lowest point of the swing, has potential energy that can be converted into kinetic energy. The pendulum moves due to gravity and has the same kind of potential energy as an object in free fall. (PE = mgh.)

    a. The pendulum is released from rest at point A. (See Fig. 6–22.) The mass of the pendulum bob is 1 kg, and h equals 0.12 m. At point C, where the height is 0.06 m, the speed of the pendulum bob is 1.08 m/sec. Find the kinetic energy at point C.

    b. Enter values for the kinetic energy, potential energy, and total energy in a table.

†**6–6** A 20,000-kg bus is traveling at 20 m/sec (44.7 mph) and the brakes apply a force of 12,000 newtons for a distance of 200 m.

    a. What is the initial kinetic energy of the bus?

    b. How much work is done by the brakes?

    c. What is the kinetic energy of the bus after traveling a distance of 200 m?

    d. Through what distance must the brakes be applied to bring the bus to a stop?

    e. What happens to the bus's initial kinetic energy?

†**6–7** A roller coaster with a mass of 100 kg travels over a loop (see Fig. 6–23). At the top of the loop, the roller coaster must have sufficient speed, otherwise gravity will cause it to fall. Because the roller coaster is traveling in a circle, it has an acceleration (Chapter 3) given by the formula _____ . In order that the roller coaster does not fall, this acceleration must *at least* be equal to the acceleration of gravity (g = 9.8 m/sec^2).

    a. Find the minimum safe speed of the roller coaster at the top of the loop.

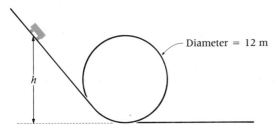

FIG. 6–23   Diagram for calculation question 6–7.

b. Find the minimum safe kinetic energy of the roller coaster at the top of the loop.

c. Find the minimum safe total energy of the roller coaster at the top of the loop.

d. If the roller coaster is released from rest, where *h* equals 14 m, would you ride such a roller coaster? Explain.

6–8 Compare the amount of gasoline needed to accelerate a car from 40 mph to 50 mph, and from 20 mph to 30 mph.

†6–9 A lever is used to lift a rock of 450 newtons (101 lb) to a height of 0.1 m above ground. The rock is located a distance of 2 m from the pivot point of the lever. The wooden plank used as the lever is 8 m long. To lift the rock, you push down at the end of the plank with a force of 150 newtons (33.6 lb) at a distance of 6 m from the pivot point.

a. How much work must be done to lift the rock 0.1 m above ground?

b. Do you do the amount of work found in **a,** even though you exert only a force of 150 newtons? Through what distance do you exert a force of 150 newtons?

c. A lever is one of a class of devices called *machines*. From **a** and **b,** discuss how machines are used to do work, and state what the purposes of machines are.

6–10 A waterfall is 20 m high.

a. What is the potential energy of 1 kg of water at the top of the waterfall?

b. How much additional heat energy does the 1 kg of water have at the bottom of the waterfall?

c. The most familiar unit of heat energy is the calorie, which is defined in terms of joules as 1 calorie equals 4.19 joules. Answer part **b** in terms of calories.

d. A quantity of 1000 calories is necessary to increase the temperature of 1 kg of water by 1°C. What is the increase in temperature of the water at the bottom of the waterfall?

6–11 One gram of radium gives off 100 calories of heat each hour.

a. What kind of energy has been transformed into heat?

b. If this heat is used to heat 10 grams of water, what is the temperature rise of the water?

c. How much energy, in joules, is transferred into heat during each hour?

d. What is the power of this radium source in watts?

6–12 An alpha particle produced in the interior of a radium sample travels only a short distance in the radium before coming to rest. The alpha particle loses about 4.8×10^{-18} joules each time it ionizes a radium atom. That is, the alpha interacts with an electron of the radium atom, and that electron is kicked out of the atom. As a result, the kinetic energy of the alpha particle is decreased by the same amount. If the alpha particle has an initial kinetic energy of 7.7×10^{-13} joules, how many radium atoms does the alpha particle ionize before it comes to a stop?

6–13 In a cathode ray tube, electrons having a kinetic energy of 1.6×10^{-14} joules strike the anode. (This kinetic energy is produced when a voltage of 100,000 volts connects the two electrodes.)

a. If two X-rays of the same energy are produced when one electron comes to rest, what is the energy of the X-rays?

b. If one X-ray has an energy of 0.5×10^{-14} joules, and another X-ray has an energy of 0.8×10^{-14} joules, what is the energy of the third X-ray produced when one electron comes to rest in the target?

c. The maximum X-ray energy is _____ joules, and the minimum X-ray energy is _____ joules.

6-14 A 1000-kg car traveling at a speed of 30 m/sec (67.1 mph) crashes head-on into a 1500-kg car traveling at a speed of 20 m/sec. How much energy is used to crumple the fenders? (The cars couple together after collision.)

6-15 A 1000-kg car traveling at a speed of 30 m/sec crashes into the rear end of a 1500-kg car traveling at a speed of 20 m/sec. How much energy is used to crumple the fenders? (The cars couple together after collision.)

6-16 Suppose that Chadwick had assumed that the mass of the neutron was equal to the mass of the nitrogen nucleus.

a. The following is known from an experiment: After the collision of a neutron with a nitrogen nucleus, the nitrogen nucleus has a speed of 0.44×10^7 m/sec. Assume that the mass of the neutron equals the mass of the nitrogen nucleus, which is known to be 14 amu. Use conservation of momentum to find the speed of the neutron before collision.

b. The following is known from an experiment: After the collision of a neutron with a proton, the proton has a speed of 3.3×10^7 m/sec. The speed of the neutron before collision was found in part **a.** The mass of the proton equals 1 amu; assume that the mass of the neutron equals 14 amu. Consider the collision of the neutron with a proton. Use conservation of momentum to find the speed of the neutron after collision.

c. Consider the neutron colliding with the proton and compare the total energy before collision with the total energy after collision. Is energy conserved? Can the mass of the neutron be equal to the mass of the nitrogen nucleus?

6-17 A nucleus of radium-226 disintegrates into an alpha particle and a radon-222 nucleus. The alpha particle has a speed of 1.52×10^7 m/sec and, according to conservation of momentum, the radon nucleus has a speed of 2.73×10^5 m/sec (see Chapter 4, p. 90). The radium nucleus has a mass of 376×10^{-27} kg; the alpha particle has a mass of 6.65×10^{-27} kg; and the radon nucleus has a mass of 369×10^{-27} kg.

a. Find the kinetic energy of the alpha particle, the kinetic energy of the radon nucleus, and the total kinetic energy of both.

b. Show that the alpha particle has 98 percent of the total kinetic energy.

c. According to Einstein's theory, the total mass after the disintegration is less than the mass before. The difference in mass is transformed into kinetic energy, which is given by: Total kinetic energy equals mass difference $\times c^2$, where c is 3×10^8 m/sec. Using the results of part **a,** find the mass difference. Find the percentage of the mass of the radium nucleus that is transformed into kinetic energy.

6-18 The radon-222 nucleus ($Z = 86$) disintegrates by emitting an alpha particle.

a. What nucleus remains after the disintegration?

b. The alpha particle has a speed of 1.61×10^7 m/sec. Find the kinetic energy of the alpha particle. Using the mass of the alpha particle as 4 amu, the kinetic energy will have units of amu-m^2/sec^2.

c. Neglect the very small kinetic energy of the recoiling nucleus and determine the amount of mass in amu converted into kinetic energy. Compare this mass with the mass of the radon nucleus.

†6–19 A beam of protons from a particle accelerator has a kinetic energy of 9.6×10^{-11} joules. The proton beam strikes a carbon target. As a result of the collision of a proton with a carbon atom, a particle called a *meson* is produced. (See Fig. 6–16, p. 127.) After the collision, the proton, the carbon atom, and the meson have a total kinetic energy of 7.44×10^{-11} joules. Determine the mass of the meson. Compare its mass to the mass of an electron, 9.11×10^{-31} kg.

6–20 A beam of protons from a particle accelerator strikes a carbon target. The kinetic energy of a proton is equal to mass of meson $\times c^2$. Will mesons be produced in this case? To answer this question, first determine the kinetic energy of the particles after collision. Then consider whether this result violates conservation of energy and/or conservation of momentum. (Recall that $c = 3 \times 10^8$ m/sec.)

6–21 Compare the solution of calculation question 6–2a with Rutherford's method for determining the radius of the gold nucleus described in the section "The Size of the Nucleus," (p. 129). What similarities are evident?

6–22 a. Find the energy of the light rays (in eV) emitted by the hydrogen atom when the electron jumps from n_i equals 3 to n_f equals 1, from 4 to 1, from 4 to 3, from 3 to 2.

b. Show that E_{4-1} equals E_{4-3} plus E_{3-1}, where E_{4-1} means the energy of light emitted when the electron jumps from n equals 4 to n equals 1, etc.

6–23 Make entries in Table 6–2 for the fifth allowed orbit. Find the energies of light emitted when an electron in the fifth allowed orbit jumps to two smaller orbits.

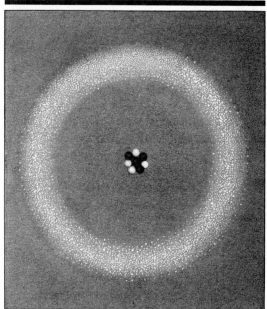

Nucleus containing protons and neutrons surrounded by electron probability cloud.

PART III
Development of Modern Physics

Introduction

Cathode rays and X-rays, apple-green fluorescence and glowing gases: Late in the nineteenth century these various (and then-puzzling) effects captivated the attention of many scientists who, as we have studied, had been experimenting with cathode ray tubes.

In this chapter, we'll concentrate on J. J. Thomson's experiments with cathode rays. Thomson, in 1897, cleverly showed that cathode rays were negatively charged particles. While he could not determine the mass of the electron (modern term for cathode ray, recall) or its charge in coulombs individually, he was able to find the ratio of its mass to its charge (m/e). He compared this ratio with that of the hydrogen ion (obtained from electrolysis experiments), and came to a startling conclusion: The atom *is not* the smallest unit of matter; the electron is part of the atom. The so-called cathode rays result when electrons are detached from the gas atoms remaining in the cathode ray tube.

Since the mass-to-charge ratio of the hydrogen ion was essential for Thomson's interpretation, we shall first discuss *electrolysis,* the decomposition of chemical compounds into elements by an electric current. This topic is also important because electrolysis clearly demonstrates the electrical nature of matter as well as the granular nature of charge. Charge comes in *integral* amounts of a smallest unit, the charge on the electron e: 1e, 2e, 3e; never 1.5e, 2.7e, and so forth.

Although many scientists, including Thomson, attempted to measure the charge on the electron, success in this area did not occur until 1911. Then, the American physicist Robert A. Millikan conducted an ingenious experiment that showed this charge could be measured. We'll see why Millikan was successful where others were not. (A by-product of this measurement was the mass of the electron, determined by multiplying the mass-to-charge ratio by the charge.)

So, the first portion of this chapter deals with such basic concepts as electric current, the

J. J. Thomson

CHAPTER 7

Electricity, Magnetism, and the Electron

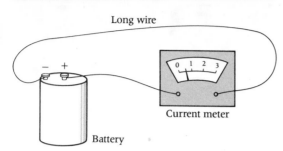

FIG. 7–1   A simple electric circuit.

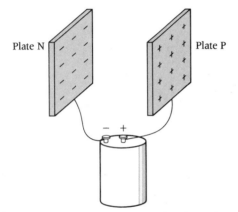

FIG. 7–2   When a battery connects the plates of a capacitor, electrons leave plate P and flow onto plate N.

capacitor, and electric and magnetic fields. These are important to discussions that will occur in later chapters and are required to understand the classic experiments of Thomson and Millikan, discussion of which concludes Chapter 7.

Voltage and Electrical Current

We have talked about static charges (Chapters 1 and 3); now, we'll examine the flow of charges: electric current. Electric current flows through a complete circuit that contains a battery or voltage source (see Fig. 7–1). **Current** is defined as the amount of charge per unit time that flows past any given point in the circuit (through the current meter, for example). Thus,

$$\text{Current} = \frac{\text{charge}}{\text{time}}. \tag{1}$$

The unit of current is the **ampere** (abbreviation: A), where

$$\text{ampere} = \frac{\text{coulomb}}{\text{second}}.$$

The battery in the system converts chemical energy into the energy needed to move charge through the circuit. The amount of current depends upon the voltage of the battery and the length and thickness of the wire. For example, if you double the voltage of the battery, the current in the wire also doubles. For a given wire, the current is directly proportional to the voltage. The wire resists the flow of electrons. If you double the length of the wire while still using the same voltage, the current is cut in half.

The Capacitor:
The Capacity to Store Charge

A current flows for a short time when a battery connects the two parallel metal plates (see Fig. 7–2). Since the charge cannot cross the gap between the plates, charges pile up on the plates. The charge spreads out uniformly over each plate due to the repulsion between like charges. The plates have the capacity to store charge and, hence, are called **capacitors**.

The battery supplies the energy to move negative charges from plate P, through the wire, and onto plate N. Plate N is negatively charged due to an *excess* of negative charge, while plate P is positively charged due to a *deficit* of negative charge.

Negative charges leaving plate P are attracted to the positive charge on it. The battery must be sufficiently powerful to overcome this attraction. The flow of charge from plate P to plate N continues until the battery is not powerful enough. At this point, the current stops, and there is a certain amount of negative charge on plate N and the same amount of positive charge on plate P. We have just learned that the energy to move charge depends upon the voltage of the battery. If you want a greater charge on each plate, then you must use a larger voltage battery. In fact, *the amount of charge on each plate is directly proportional to the voltage.* If you double the voltage, the amount of charge doubles, and so forth.

Consider a charged particle Q in the gap between the two plates. A negative charge Q will move toward plate P, and a positive charge Q will move toward plate N. If you could observe this motion, you could determine the sign of charge Q. (Thomson observed the motion of cathode rays and found them to be negatively charged.)

Next, let us consider the force exerted on this charge Q between the plates. Upon what factors does it depend? If you increase the amount of charge on each plate by increasing the voltage, the force on Q increases because it is more strongly attracted to one plate and more strongly repelled by the other plate. If you push the plates closer together, the force on charge Q increases because it is now closer to all of the charges on plates N and P. The force on charge Q also depends upon the amount of its charge. For example, if you replace charge Q with a particle having twice the amount of charge, the force on this charge

is twice as large. It turns out that the force on charge Q does *not* depend upon its location between the plates, although this may be unexpected. The force on charge Q is the same everywhere between the plates, and the direction of the force is perpendicular to the plates. The force on charge Q is given by

$$F = \frac{QV}{S}, \qquad (2)$$

where V is the voltage of the battery connecting the plates and S is the distance between the plates. This relationship is plausible because F increases when V increases, F increases when S decreases, and F increases when Q increases, as expected.

ELECTRON-VOLT OF ENERGY

Consider a hypothetical electron near the surface of plate N. It will experience a force given by Equation 2, where Q is the charge on the electron, usually designated by the symbol e. Due to this force, the electron will accelerate (its speed will increase) toward plate P. Remember (Equation 1, Chapter 6) that the work done on an object is found by multiplying the force by the distance through which the force acts; work equals force times distance. Also recall (Equation 3, Chapter 6) that work always produces an increase in energy. Here, that energy is kinetic energy. If the electron were at rest at plate N, the kinetic energy at plate P would be equal to the work done on the electron:

$$\begin{aligned} \text{KE at plate P} &= (\text{force})(\text{distance}) \\ &= \left(\frac{eV}{S}\right)(S) \qquad (3) \\ &= eV, \end{aligned}$$

since the force has been exerted on the electron as it traveled a distance S. When e is expressed in coulombs and V in volts, the kinetic energy

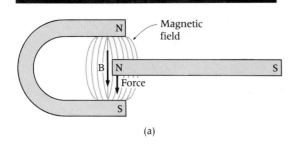

(a)

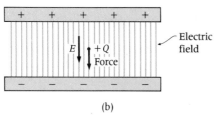

(b)

FIG. 7–3   The horseshoe magnet sets up a magnetic field that acts on the bar magnet; similarly, the charged capacitor plates set up an electric field that acts on the charged particle. (a) The force exerted on the north pole of the bar magnet indicates the direction of the magnetic field of the horseshoe magnet. (b) The force exerted on a positively charged particle indicates the direction of the electric field.

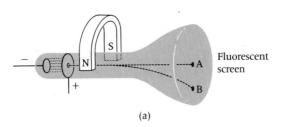

(a)

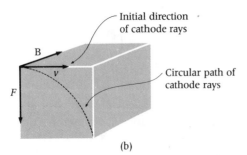

(b)

FIG. 7–4   (a) A magnetic field causes the cathode rays to deflect. (b) The force on the cathode rays is perpendicular to both the velocity and the magnetic field.

is expressed in joules. The charge on the electron e equals 1.6×10^{-19} coulombs, as we shall learn. Suppose that the voltage of the battery is 1 volt. The kinetic energy at plate P is given by

$$\text{KE at plate P} = (1.6 \times 10^{-19}\,\text{coulombs})(1\,\text{volt})$$

$$= 1.6 \times 10^{-19}\,\text{joules}.$$

It's convenient to express the kinetic energy of the electron in terms of the voltage. For example, if the voltage through which the electron (or any other singly charged particle) travels is 1 volt, then the change in kinetic energy, either an increase or a decrease, is called 1 **electron-volt,** or 1 eV. If the voltage is 2 volts, the change in kinetic energy is 2 eV, and so on. Therefore,

$$1\,\text{eV} = 1.6 \times 10^{-19}\,\text{joules}. \qquad (4)$$

For a particle having a charge of $2e$, its change in kinetic energy in traveling through one volt is *two* electron-volts. The electron-volt of energy will be used extensively in later chapters.

Electric and Magnetic Fields

MAGNETIC FIELDS

If you hold the north pole* of a bar magnet between the poles of a horseshoe magnet, you feel a tug pulling it toward the south pole (Fig. 7–3a). We say that the horseshoe magnet sets up a **magnetic field,** and this field exerts a force on the bar magnet. We use the symbol B to denote the strength of the magnetic field, and we indicate its direction with an arrow pointing in the same direction as the force on a north pole at that point.

---

*The end of a freely suspended magnet that points to the north we call the *north pole* of the magnet; the pole opposite, which points to the south, we call the *south pole* of the magnet.

ELECTRIC FIELD

Similarly, the charges on the plates of a capacitor set up, in the gap between the plates, an **electric field** that exerts a force on a charged particle between the plates (see Fig. 7–3b). It seems reasonable that this force will depend upon both the amount of the charge Q on a particle and the strength of the electric field, which is designated by the symbol E. Therefore, the force on a charged particle is given by

$$F = QE. \qquad (5)$$

The direction of the electric field is the same as the direction of a force on a positively charged particle at that point.

Chapter 8 reveals that a light wave consists of electric and magnetic fields propagating through space. For that discussion, we must use the field concept.

MAGNETIC FIELD AND CHARGED PARTICLES

Figure 7–4 illustrates the effect of a magnetic field on cathode rays. Some of the cathode rays pass through the hole in the anode and produce a bright spot when they strike the fluorescent screen. When the north pole faces the reader, the cathode rays are deflected downward because the magnetic field exerts a downward force on cathode rays. In this example, note that cathode rays are moving to the right with velocity v, the force F is downward, and the magnetic field B is pointing from the north to the south pole. These directions are indicated by arrows at the corner of the box in Fig. 7–4(b). The force is given by

$$F = QvB, \qquad (6)$$

where the magnetic field B is perpendicular to the velocity v, and the force F is perpendicular to both v and B. Equation 6 applies equally well to any charged particle moving in a perpendicular magnetic field. By inserting a positive or negative number for the charge in Equation 6, we show that the force on negatively charged particles is opposite to that for positively charged particles. For example, if cathode rays had instead been positively charged particles, the deflection in Fig. 7–4 would have been upward. Thus, the orientation of the magnet and the direction of the deflection indicate that cathode rays are negatively charged particles.

CIRCULAR MOTION

An object traveling in a circle must have a force directed toward the center of the circle (Chapter 3); that is, the velocity and force are always perpendicular. Figure 7–4(b) clearly shows that the force F and velocity v are perpendicular. Thus, a charged particle moving in a perpendicular magnetic field travels in a circle, as shown. Newton's second law, F equals ma, becomes

$$QvB = \frac{mv^2}{r}, \qquad (7)$$

where a equals v^2/r. We can obtain the ratio m/Q by multiplying both sides of Equation 7 by r/Qv^2, yielding

$$\left(\frac{mv^2}{r}\right)\left(\frac{r}{Qv^2}\right) = (QvB)\left(\frac{r}{Qv^2}\right)$$

or,

$$\frac{m}{Q} = \frac{Br}{v}. \qquad (8)$$

This shows how the ratio of the mass of the charged particle to its charge depends upon the magnetic field, the radius of its path in a magnetic field, and the speed of the particle. Multiplying both sides of Equation 7 by r/QvB, we find another useful form:

$$r = \frac{mv}{QB}. \qquad (9)$$

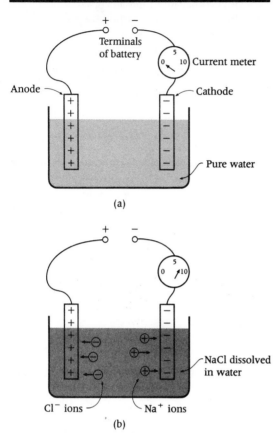

(a)

(b)

FIG. 7–5   (a) An unusual type of capacitor.
(b) Electrolysis of sodium chloride (NaCl).

Electrolysis and the Electrical Nature of Matter

Figure 7-5(a) shows an unusual type of capacitor. Two metal rods are suspended in a container of water that has no impurities: *pure* water. The anode connects to the positive terminal of the battery, and the cathode connects to the negative terminal. A current meter is also included in the circuit. When the circuit is completed, a current flows momentarily as the battery acts to transfer negative charge from the anode onto the cathode. As a result, the anode is positively charged and the cathode, negatively charged. The current then drops to zero, and no more charges are transferred. While tap water contains many impurities and does conduct a current, pure water is an extremely poor conductor of electricity.

If we add sodium chloride (NaCl) to the water, we find that a current now flows in the circuit (see Fig. 7–5b). A molecule of sodium chloride contains 1 atom of sodium and 1 atom of chlorine, and is electrically neutral. However, when sodium chloride is dissolved in water, a molecule breaks up into a positively charged sodium ion (Na^+) and a negatively charged chlorine ion (Cl^-). This process is called **ionization.** These charged fragments of the molecule are called **ions** to distinguish them from atoms, which are neutral. The sodium ion Na^+ is a sodium atom with one electron missing, and the chlorine ion Cl^- is a chlorine atom with an extra electron. The sodium ion travels to the cathode, where it neutralizes one negative charge, and deposits on the cathode. Similarly, the chlorine ion travels to the anode, where it neutralizes one positive charge. Chlorine gas bubbles from the solution.

The breakup of a sodium chloride molecule causes the number of charges on the cathode and anode to be reduced by one unit of charge. The battery acts to restore the same number of charges as there were initially, and one unit of negative charge, called *e*, travels from the anode to the cathode. The **electrolysis,** as this breakup process is called, of sodium chloride continues, and a current flows through the circuit.

Chapter 1 taught us that a sample of an element with a mass in grams equal to its atomic mass contains N_A atoms, where N_A is Avogadro's number. Thus, 23 g of sodium contains N_A atoms, as do 35.5 g of chlorine. These samples combine to form 58.5 g of sodium chloride containing N_A molecules.

Suppose that we pour 58.5 g of sodium chloride into the water in Fig. 7–5(b) and monitor the current. We would find that a current of 6.7 amps flows for 4 hours and, at the end of 4 hours, 23 g of sodium are deposited on the cathode and 35.5 g of chlorine are released at the anode.

We can rearrange Equation 1 to find the amount of charge transferred from the anode to the cathode:

$$\text{Charge} = (\text{current})(\text{time}).$$

Since 4 hours is 14,400 seconds, the charge is (6.7 amps)(14,400 sec), or 96,500 coulombs. The breakup of one sodium chloride molecule causes one sodium ion to deposit on the cathode and one chlorine ion on the anode. This causes one unit of electrical charge e to pass through the circuit and be detected by the current meter. The breakup of N_A molecules causes 23 g of sodium to be deposited on the cathode and the release of 35.5 g of chlorine at the anode. The amount of charge that passes through the current meter is $N_A e$. From this, we find

$$23 \text{ g of sodium} = (N_A)(\text{mass of sodium ion})$$

$$96,500 \text{ coulombs} = N_A e$$

$$35.5 \text{ g of chlorine} = (N_A)(\text{mass of chlorine ion}).$$

Dividing the first equation by the second, we find the mass-to-charge ratio of the sodium ion (because N_A cancels out):

$$\frac{\text{Mass of sodium ion}}{e} = \frac{23 \text{ grams}}{96,500 \text{ coulombs}}$$
$$= 2.4 \times 10^{-4} \text{ g/C},$$
$$= 2.4 \times 10^{-7} \text{ kg/C}. \quad (10)$$

Dividing the third equation by the second, we find that the mass-to-charge ratio of the chlorine ion is 3.6×10^{-7} kg/C.

To observe the granular nature of electric charge, we repeat the electrolysis experiment using other chlorine compounds. When 47.7 g of magnesium chloride is dissolved in water, 35.5 g of chlorine is released and 12.2 g of magnesium is deposited. From Fig. 1–3 (p. 6), we see that the atomic mass of magnesium is 24.3 amu and that of chlorine, 35.5 amu; that is, 12.2 g of magnesium contains $N_A/2$ atoms, and 35.5 g of chlorine contains N_A atoms. Thus, 2 chlorine ions are deposited for each magnesium ion, which shows that the chemical formula for magnesium chloride is $MgCl_2$. Since the molecule is electrically neutral before its decomposition, the charge of the magnesium ion is $+2e$, and the charge on each chlorine ion is $-1e$.

Similarly, the passage of 96,500 coulombs through a solution of aluminum chloride ($AlCl_3$) releases 35.5 g of chlorine, but only 9 g of aluminum (atomic mass equals 27 amu). The aluminum ion has a charge of $+3e$.

From electrolysis, we see the granular nature of charge: Charge comes in integral multiples of e and cannot be subdivided further.

MASS-TO-CHARGE RATIO OF THE HYDROGEN ION

The mass-to-charge ratio of the hydrogen ion may be found from the electrolysis of water. As mentioned, pure water is a very poor conductor of electricity. To overcome this difficulty, we add a small amount of sulfuric acid to the water. Some chemical reactions take place, and the end result is that hydrogen is released at the cathode and oxygen at the anode. After 96,500 coulombs travel through the circuit, 1 g of hydrogen and 8 g of oxygen are released. This corresponds to N_A atoms of hydrogen and $N_A/2$ atoms of oxygen. (The atomic masses of hydrogen and oxygen are, respectively, 1 amu and 16 amu.) Thus, for every 1 oxygen atom released there are 2 atoms of hydrogen; the chemical formula is thus H_2O. The mass-to-charge ratio of the hydrogen ion is found in much the same way as for the sodium ion in Equation 10:

$$\frac{\text{Mass of hydrogen ion}}{e} = \frac{1 \text{ gram}}{96,500 \text{ coulombs}}$$
$$= 10^{-5} \text{g/C,}$$
$$= 10^{-8} \text{kg/C.}$$

We stated in the introduction that the value of the mass-to-charge ratio of the hydrogen ion was an extremely significant piece of information. Thomson obtained the mass-to-charge ratio for cathode rays (electrons) and found it was surprisingly 1000 times smaller. Assuming that the hydrogen ion and the electron have the same *amount* of charge (although opposite signs), Thomson concluded that the mass of the electron is about 1000 times smaller than the mass of the hydrogen ion, which demonstrated that the hydrogen atom was not the smallest unit of matter. In the next major section, we'll look at how Thomson obtained the mass-to-charge ratio of the electron.

AVOGADRO'S NUMBER

There are many methods for determining the value of Avogadro's number N_A, but electrolysis is an exceptionally accurate one. From the electrolysis of sodium chloride, we learned that

$$96,500 \text{ coulombs} = N_A e.$$

If you know the value of e, then you can determine N_A, and vice versa. Later (p. 156), we shall see how Millikan accurately determined the charge e, thus leading to a precise value for N_A. (Recall that in Chapter 1 Avogadro's number was used to determine the masses of atoms and their diameters as well.)

J. J. Thomson Discovers the Electron

There was a great deal of controversy about the nature of cathode rays when they were first discovered. German physicists generally held that cathode rays were a special kind of vibration of the ether. (The ether theory had been introduced to explain the propagation of light. The ether was, supposedly, a massless and transparent medium that permeated all space, even the vacuum of outer space. Light waves allegedly traveled through this medium much like water waves travel on the surface of water. Chapter 8 goes into the ether theory; Chapter 9 tells how Einstein demolished it.) In 1897, Thomson set out to determine the nature of cathode rays. In several experiments, he found strong evidence that cathode rays were charged particles.

The charge of the cathode rays may be detected in two ways: (1) A beam of cathode rays is deflected when it enters the region of a magnetic field, and (2) also when it enters the region between charged capacitor plates. Researchers working with cathode ray tubes easily observed the deflection when a cathode ray beam entered a magnetic field. However, when the cathode ray beam passed between capacitor plates, a deflection could not be observed unless the voltage connecting the capacitor plates was very large. This effect could not be understood at all. Even a small voltage should cause a slight deflection of the cathode rays. Thomson was the first to demonstrate the deflection of cathode rays passing between capacitor plates connected to a voltage as small as 2 volts. He was successful only because he had obtained a high-quality vacuum. During the preceding 10 years or so, great strides had been made in improving vacuum pumps due to the demands of the electric lamp industry.[1] Box 7–1 illustrates the basic principle for producing such a vacuum. The rea-

BOX 7–1   GEISSLER VACUUM PUMP

The air is evacuated from the cathode ray tube in a series of steps:

1. Stopcock C is closed and stopcock D is opened. The mercury reservoir R is raised until the mercury level reaches D, forcing air molecules out the top.
2. Stopcock D is now closed and R lowered. Since there is no way for air to enter, chamber B is evacuated.
3. Stopcock C is opened, and the air in the cathode ray tube rushes into the evacuated chamber B. The pressure in the cathode ray tube and chamber B is less than atmospheric pressure.

This process is repeated many times to produce a good vacuum. This tedious and time-consuming method was used to evacuate the cathode

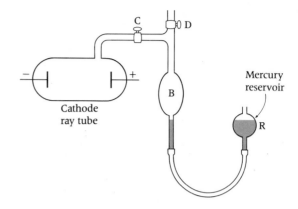

ray tube and the X-ray tube until 1913, when the diffusion pump was invented.

Source: Lloyd W. Taylor, *Physics: The Pioneer Science* (Boston: Houghton Mifflin, 1941), p. 770.

---

son why a good vacuum was so essential is outlined in thought question 7–15.

Thomson used a magnetic field and charged capacitor plates and he arrived at the following conclusion:

As the cathode rays carry a charge of negative electricity, are deflected by an electrostatic force (charged capacitor plates) as if they were negatively electrified, and are acted on by a magnetic force in just the way in which this force would act on a negatively electrified body moving along the path of these rays, I see no escape from the conclusion that they are charges of negative electricity carried by particles of matter. The question next arises, What are these particles? Are they atoms, or molecules, or matter in a still finer state of subdivision? To throw some light on this point, I have made a series of measurements of the ratio of the mass of these particles to the charge carried by it.[2]

Figure 7–6 shows Thomson working with cathode ray tubes in his laboratory. Figure 7–7 shows the tube he used to measure the mass-to-charge ratio of the cathode rays. First, he reduced the pressure to about one ten-thou-

sandth of atmospheric pressure. Here, it is important to realize that even at this low pressure, about 10^{15} molecules of air per cubic centimeter remain! This realization was crucial for Thomson's interpretation of the nature of cathode rays.

The cathode rays, produced by a high voltage connecting the cathode C and the anode A, travel from the cathode toward the anode, where some of them pass through a slit in the anode. Slit B forms a narrow beam. The cathode rays strike the fluorescent screen, producing a bright spot. Thomson read the amount of deflection, if any, from a scale on the fluorescent screen. Plate N connects to the negative terminal of a battery, and plate P connects to the positive terminal. This causes a downward force on the cathode rays when they enter the region between the plates. A magnetic field is set up in the region surrounding plates N and P, perpendicular to

---

ELECTRICITY, MAGNETISM, AND THE ELECTRON

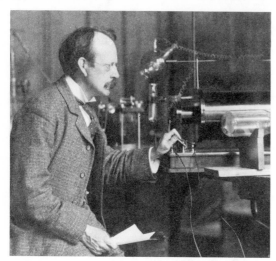

FIG. 7–6   J. J. Thomson at the Cavendish Laboratory. (Courtesy of Cavendish Laboratory)

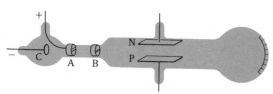

FIG. 7–7   Thomson's cathode ray tube used to determine mass-to-charge ratio of cathode rays, or electrons. [*Source:* J. J. Thomson, *Philosophical Magazine* 44 (1897):293]

the plane of this paper. This results in an upward force on the cathode rays. Thomson's experiment consisted of two steps. Thomson actually observed the deflection of the cathode ray beam as it passed between charged capacitor plates with the magnetic field turned off. However, in step 1 below we shall imagine that Thomson did the reverse: He observed the deflection in the magnetic field only. The experiment will be described as though Thomson had performed it this way. As a result, Thomson's experiment is far simpler to analyze.

Step 1: Deflection by a magnetic field only

In this part of the experiment, Thomson disconnected the battery, and the magnetic field caused the cathode rays to deflect upward. The cathode rays travel in a circle while in the magnetic field, and in a straight line otherwise. Figure 7–8 shows how the radius of the circle can be obtained. At the boundary of the magnetic field, straight lines are drawn tangent to the circle. This shows that circles of different radii yield different deflections on the screen.

Thus, by observing the deflection on the fluorescent screen, Thomson was able to "work backwards" and find the radius of the circle. Using the symbol e for the charge on the cathode ray and m for its mass, Equation 8 becomes

$$\frac{m}{e} = \frac{Br}{v}. \tag{11}$$

Since Thomson knew the strength of the magnetic field B, and he deduced the radius r from the deflection on the fluorescent screen, the only unknown on the right-hand side of Equation 11 was the speed v. Thomson determined this next.

Step 2: Magnetic field and charged capacitor plates used simultaneously

In this case, two forces act on the cathode rays (Fig. 7–9). The charged capacitor plates exert a downward force, which is given by

$$F_{\text{down}} = \frac{(e)(V)}{S},$$

where V is the voltage connecting the plates and S is the distance between the plates. The magnetic field exerts an upward force given by

$$F_{upward} = evB.$$

The cathode rays are deflected in the direction of the larger force. However, when Thomson adjusted the magnetic field B (or the voltage V), so that the upward force exactly balanced the downward force, the cathode rays were not deflected at all and struck the center of the fluorescent screen. In this case

$$F_{upward} = F_{down}$$

$$evB = \frac{eV}{S}$$

$$v = \frac{V}{BS}. \qquad (12)$$

Since Thomson had measured all of the terms on the right-hand side of Equation 12, he determined the speed of the cathode rays to be about 3×10^7 m/sec, 10 percent of the speed of light. Substituting this speed into Equation 11, he found the mass-to-charge ratio m/e to be on the order of 10^{-11} kg/C. (The presently accepted value is 0.569×10^{-11} kg/C.)

In his first trial, the cathode was made of aluminum. Next, he repeated the experiment, but used a platinum cathode instead. He found essentially the same value for m/e. He also repeated the experiment using various gases, such as hydrogen and carbonic acid, in the cathode ray tube (at very low pressure) rather than air. He still found the same value for m/e. Whatever the nature of the cathode rays, it did not depend upon the cathode material or the gas in the tube.

SURPRISING CONCLUSION: THE ATOM
IS NOT THE SMALLEST UNIT OF MATTER

Thomson was quite surprised by the small value of m/e. For example, the mass-to-charge ratio for the hydrogen ion found in electrolysis was 10^{-8} kg/C. The mass-to-charge ratio for cath-

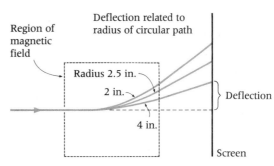

FIG. 7–8   The deflection of the cathode rays can be used to determine the radius of the circular path of the cathode rays in a perpendicular magnetic field.

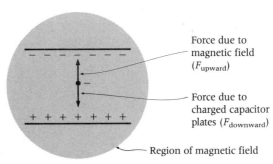

FIG. 7–9   The forces acting on an electron due to a magnetic field and charged capacitor plates. The plates, separated by a distance S, are connected to a voltage V. The electron is traveling to the right with speed v.

ode rays was 1000 times smaller. Thomson concluded:

Thus for the carriers of the electricity in the cathode rays m/e is very small compared with its value in electrolysis. The smallness of m/e may be due to the smallness of m or the largeness of e, or to a combination of these two. That the carriers of the charges in the cathode rays are small compared with ordinary molecules is shown, I think, by Lenard's results.[3]

The German physicist Philipp Lenard (1862–1947) carried out still another experiment with cathode rays. He put a thin foil window at the end of a cathode ray tube so that the cathode rays could pass through it and then travel in air. The cathode rays traveled much farther in air than one would expect for molecules. Lenard realized that the mass of the cathode rays must be quite small compared with the air molecules.

Thomson assumed that the charge on the cathode rays was the same amount as on the hydrogen ion. Since the mass-to-charge ratio for the cathode rays was 10^{-11}, while that for the hydrogen ion was 10^{-8}, the mass of the cathode rays must be 1000 times smaller than the mass of the hydrogen ion.

Let us remember now that about 10^{15} gas molecules per cubic centimeter remained in the cathode ray tube. Thomson concluded that all atoms contained electrons and that the so-called cathode rays resulted from the splitting or ionization of these gas atoms (see Fig. 1–14, p. 15). The electrons of gas atoms (or molecules) near the cathode experience a tremendous force of repulsion and split apart from the atom. The negatively charged electron then travels toward the anode, and the positively charged ion toward the cathode. This explained to Thomson's satisfaction why the ratio *m/e* was the same for all gases in the tube and for all cathode materials. Thomson explained:

Thus we have in the cathode rays matter in a new state, a state in which the subdivision of matter is carried very much further than in the ordinary gaseous state: a state in which all matter—that is, matter derived from different sources such as hydrogen, oxygen, etc.—is of one and the same kind; this matter being the substance from which all chemical elements are built up.[4]

No longer could the atom be considered an indivisible entity. Thomson had discovered the first subatomic particle: the electron.

The Charge on the Electron: Millikan's Oil-Drop Experiment

After the discovery of the electron, several scientists, including Thomson himself, attempted to measure the charge on a drop of water. These measurements indicated that the charge on the electron was about 10^{-19} coulombs, but there were large uncertainties in the measurements. Since *m/e* equals 10^{-11} kg/coulombs the mass of the electron was about 10^{-19} coulombs $\times$ 10^{-11} kg/coulombs, or 10^{-30} kg. This quantity was about 1000 times smaller than the mass of the hydrogen atom, and it agreed with Thomson's conclusions.

In 1911, Robert A. Millikan (1868–1953) determined the charge on the electron by using oil drops instead of water. Oil has a big advantage: It doesn't evaporate appreciably, and the mass of an oil drop remains constant.

Millikan sprayed oil drops from an atomizer, and the drops fell through a hole in the top plate of a capacitor (Fig. 7–11). The drops acquired a static charge due to friction as they passed through the nozzle. Millikan viewed the drops through a telescope. The voltage across the two capacitor plates could easily be changed and switched so that either the top plate was positive and the bottom plate negative, or vice versa.

Millikan focused his attention on a single oil drop. The oil drop, which we shall assume to be negatively charged, had two forces exerted on it: (1) Its weight acting downward; and (2) the force due to the charged capacitor plates, which is upward when the top plate is positively charged. By adjusting the voltage and the charge on the top plate, Millikan could make the drop move upward or downward so that it would not strike either plate. When he adjusted the voltage properly, the drop came to a stop. This occurred when the drop's weight exactly balanced the upward force due to the capacitor plates:

$$F_{\text{up}} = F_{\text{down}}$$

$$\frac{Q_D V_R}{S} = m_D g$$

$$Q_D = \frac{m_D g S}{V_R}, \qquad (13)$$

where V_R is the *voltage* required for the drop to remain *at rest*; S is the distance between the plates; and m_D is the *mass of the drop*. Millikan determined V_R by adjusting the voltage appropriately. Therefore, the only unknown on the right-hand side of Equation 13 was the mass of the drop. Millikan determined this in the second step of the experiment, and then he determined the charge Q_D on the oil drop from Equation 13.

MASS OF THE OIL DROP

In this part of the experiment, the capacitor plates are not charged (the voltage equals zero), and the drop falls under the influence of gravity. However, the force of air resistance opposing motion acts upward. [Air resistance increases with the increasing size of the object and also with increasing speed because, in each case, the object encounters more air molecules (see Chapter 3).] For a special object, such as a spherical oil drop, the force of air resistance is given by

Force of air resistance

= (constant)(radius of drop)(speed of drop).

As the drop falls and its speed increases, the force of air resistance increases. Eventually, the

FIG. 7–10 Robert A. Millikan. (AIP Niels Bohr Library)

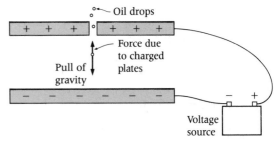

FIG. 7–11 Diagram of Millikan's oil-drop experiment.

TABLE 7–1   DATA FROM MILLIKAN'S
OIL-DROP EXPERIMENT.

TRIAL NUMBER	Q_D (coulombs)	n	$e = Q/n$ (coulombs)
1	11.50×10^{-19}	7	1.642×10^{-19}
2	19.72	12	1.643
3	22.90	14	1.636
4	8.206	5	1.641
5	6.558	4	1.640
6	14.82	9	1.647

Source: Great Experiments in Physics edited by Morris H. Shamos. Copyright © 1959 by Holt, Rinehart and Winston, Inc. Reprinted by permission of Holt, Rinehart and Winston.

force of air resistance exactly equals the drop's weight, and since the net force is zero, the drop moves with a constant speed called its **terminal speed.** *The terminal speed depends upon the radius of the drop:*

Increased radius of drop
↓
Increased weight of drop
↓
Increased force of air resistance when drop travels with *constant* speed, since $F_{\text{air resistance}}$ equals weight of drop
↓
Drop will reach larger terminal speed, producing increased force of air resistance.

The larger the radius of the drop, the larger is the terminal speed.

Millikan observed the drop through a telescope and measured its terminal speed by timing its passage between two points on a vertical scale visible through the telescope. This terminal speed yielded the radius of the drop and, since Millikan knew the density of the oil (mass per unit volume), he determined the mass of the drop.

EXPERIMENTAL DATA

Table 7–1 shows some of Millikan's data.[5] The charge on each drop may be due to one or more electrons on it, so that Q_D equals ne, where n is the number of electrons and e is the charge on the electrons. The problem is to determine the number of electrons on each drop.

Looking at trials 2, 3, and 5 in Table 7–1, it might seem reasonable that n equals 6, 7, and 2, respectively, yielding e equal to 3.2×10^{-19} coulombs. However, during the experiment, the charge on a drop would spontaneously change due to collision with air molecules. It seems reasonable that the charge on the drop would change by one electron, as in trials 4 and 5, where the difference in the two charges is 1.648×10^{-19} coulombs.

From the results of hundreds of measurements, Millikan determined the charge on the electron to be 1.603×10^{-19} coulombs. Today the generally accepted value is 1.60206×10^{-19} coulombs. Indeed, Millikan's experiment was extremely accurate!

MASS OF THE ELECTRON

The mass of the electron can be found by multiplying the mass-to-charge ratio by the charge on the electron.

$$\text{Mass of electron} = \left(\frac{m}{e}\right)(e)$$
$$= (0.5685 \times 10^{-11} \text{kg/C})(1.602 \times 10^{-19}\text{C})$$
$$= 9.11 \times 10^{-31}\text{kg}.$$

Summary

The current in an electrical circuit is defined as the amount of charge per unit time flowing past any point in the circuit. (1 ampere = 1 coulomb/sec.)

If you connect a battery to two parallel metal plates (not in contact), a current flows momentarily as electrons travel from one plate to the other. The one with excess electrons is negatively charged, and the other, with a deficit of electrons, is positively charged. The amount of charge on each plate is proportional to the voltage V connecting the plates. If a charged particle Q is placed anywhere between the two plates, the force on the particle is given by Equation 2 ($F = QV/S$), where S is the distance between the plates. The kinetic energy of this charged particle will change. If the voltage through which a singly charged particle travels is 1 volt, then the change in kinetic energy is 1 electron-volt (1 eV); if doubly charged, 2 eV.

A magnetic field exerts a force on a north or south magnetic pole, and an electric field exerts a force on a stationary charged particle. A magnetic field also exerts a force on a moving charged particle. This force is given by F equals QvB, where the magnetic field B is perpendicular to the velocity v, and the force F is perpendicular to both B and v. The charged particle in a magnetic field travels in a circular path because the force and velocity are perpendicular.

To set up the electrolysis process, connect two metal rods, suspended in a sodium chloride solution, to the terminals of a battery. An electrical current will flow through the circuit. A sodium chloride molecule breaks up into a Na^+ ion (sodium atom with one electron missing) and a Cl^- ion (chlorine atom with one extra electron). These ions travel to oppositely charged rods. From the breakup (or ionization) of each molecule, one ion deposits on each rod and one electron travels through the circuit. The mass-to-charge ratio of each ion can be found by dividing the mass deposited on the rod by the total charge passing through the circuit. The electrolysis of water yields the mass-to-charge ratio of the hydrogen ion.

J. J. Thomson measured the mass-to-charge ratio of the electron (or cathode rays). In the first part of his experiment, he subjected the cathode rays to a magnetic field and essentially measured the radius of their circular path. To determine the speed of the cathode rays, which was needed to determine the mass-to-charge ratio (see Equation 11, p. 154), he subjected the cathode rays to two forces in opposite directions: one due to charged capacitor plates and the other, to a magnetic field. He adjusted the voltage on the plates until the beam was not deflected, which indicated that the two forces were equal. This led to the speed of the cathode rays (Equation 12, p. 155) and, hence, to the mass-to-charge ratio, which was, however, 1000 times smaller than the ratio for the hydrogen ion. Thomson obtained the same value for the ratio no matter what material was used for the cathode or what (rarefied) gas was in the cathode ray tube. He concluded that electrons were part of the atom. Cathode rays were produced when an electron was torn from a gas atom due to the repulsion to the cathode.

Robert Millikan conducted an experiment to determine the charge on an electron. A charged oil drop was passed through a hole in the top plate of a capacitor. Two forces in opposite direction were exerted upon the drop: the pull of gravity downward, and an upward force due to the positive charge on the top plate. By adjusting the voltage connecting the plates, Millikan made the two forces equal ($F_{net} = 0$) and the drop came to a stop. Millikan could determine the charge on the drop if he could determine the mass of the drop (Equation 13, p. 157). To do this, he turned off the voltage and let the drop fall under the influence of gravity. When the drop moved downward, the force of air resistance was upward. Eventually, the force of air resistance was equal to the weight, and the drop traveled with constant speed. This speed, called the terminal speed, depends upon the radius of the drop. Millikan measured this terminal speed and, hence, was able to determine the radius of the drop and its mass. From these measurements, Millikan was able to determine the charge on the drop, which was an integral multiple of e, ($Q_D = ne$). The object was to determine the basic unit of charge e, the charge on the electron. To do this, he had to determine n. Millikan obtained the charge Q_D on hundreds of drops and noted the *smallest difference* between the charges on two drops. He concluded that this difference was e. Then, he determined the integer n for all of his measurements and obtained an extremely accurate value for the charge on the electron e. Since Thomson had determined the mass-to-charge ratio m/e, and Millikan had determined e, the mass of the electron could also be found.

True or False Questions

Indicate whether the following statements are true or false. Change all of the false statements so that they read correctly.

7–1 The two plates of a capacitor are connected to the terminals of a battery.
    a. A current flows continuously in the wires connecting the plates to the battery terminals.
    b. The plate connected to the positive terminal of the battery becomes negatively charged and the other plate, positively charged.
    c. To double the charges on the plates, double the voltage of the battery.

7–2 A charged particle is placed between two parallel metal plates connected to a battery. The force on this charged particle increases when
    a. the charge on the metal plates increases.
    b. the voltage of the battery is increased.
    c. the distance between the metal plates increases.

7–3 The kinetic energy acquired by an electron in accelerating through 1 volt is called 1 electron-volt (eV).

7–4 A charged particle travels in a circular path in a magnetic field.
    a. The velocity and magnetic field point in the same direction.
    b. The path is circular because the force on the particle is perpendicular to its velocity.
    c. The force on the particle does not depend upon its mass.
    d. The acceleration of the particle does not depend upon its mass.

    e. If the particle had a larger speed, it would travel in a circle of proportionately larger radius.

7–5 In the electrolysis of potassium chloride,
    a. a molecule of potassium chloride breaks up into a K^+ ion and a Cl^- ion.
    b. the potassium ions deposit on the cathode and the chlorine ions, at the anode.
    c. The amount of charge deposited on the cathode is equal to that deposited on the anode.
    d. The mass of potassium deposited on the cathode is equal to the mass of chlorine released at the anode.
    e. The mass-to-charge ratio of the potassium ion can be found by dividing the mass deposited on the cathode by the total charge passing through the wire connecting the anode and cathode.

7–6 a. J. J. Thomson showed that cathode rays can be deflected when traveling in a magnetic field and also when passing between the parallel plates of a charged capacitor.
    b. In both cases, the direction of the deflection indicated that cathode rays were negatively charged.

7–7 To measure the charge of cathode rays, Thomson set up his experiment so that the cathode rays traveled in a straight line through a magnetic field and charged capacitor plates simultaneously.

7–8 Suppose that it were possible to repeat Thomson's experiment with a negatively charged particle having twice the charge and twice the mass of an electron, but the same speed as an electron.
    a. In a magnetic field, the force on this particle would be twice that on an electron.
    b. In a magnetic field, the acceleration of

this particle would be twice that of an electron.

c. The deflection of this particle in a magnetic field would be twice that of an electron.

d. The deflection of this particle when passing between charged capacitor plates would be twice that of an electron.

e. The mass-to-charge ratio of this particle would be twice that of an electron.

f. In this repetition of Thomson's experiment, it would be possible to see differences between this particle and an electron.

7–9 In Millikan's oil-drop experiment, a statically charged oil drop remains at rest between two charged capacitor plates because its attraction to the top plate is equal to the weight of an electron.

7–10 Suppose that it were possible to repeat Millikan's oil-drop experiment by replacing all electrons on a given drop with particles having twice the negative charge and twice the mass of an electron.

a. In order for the particles to remain at rest between the charged capacitor plates, the voltage connecting the plates must be doubled.

b. When the capacitor plates are not charged (voltage = 0), the particles will fall with a terminal speed that is twice as large as that for electrons.

c. In this repetition of Millikan's experiment, it would be possible to see differences between this particle and an electron.

Questions for Thought

7–1 Match the name or term with its description by inserting the correct letter from the lettered column into each blank in the descriptive column. Entries from the lettered column may be used more than once.

a.	cathode rays	**i.**	ampere
b.	ether	**j.**	straight line
c.	ionization	**k.**	circle
d.	R. A. Millikan	**l.**	kg/coulomb
e.	J. J. Thomson	**m.**	coulomb
f.	electrolysis	**n.**	electron-volt (eV)
g.	negative ion	**o.**	electric field
h.	positive ion	**p.**	magnetic field

_____ scientist who determined the mass-to-charge ratio of the electron

_____ scientist who determined the charge on the electron

_____ another name for electrons

_____ an atom or molecule with one electron detached

_____ an atom or molecule with one additional electron

_____ decomposition of an atom into an ion and an electron, or decomposition of a molecule into positive and negative ions

_____ decomposition of chemical compounds into elements by the passage of an electric current

_____ medium supposedly responsible for the propagation of light

_____ exists in a region of space where a force is exerted on a stationary charged particle

_____ exists in a region of space where a force is exerted on a magnetic north or south pole

_____ path of a moving charged particle in a perpendicular magnetic field

_____ path of a cathode ray subjected to a force due to a magnetic field and an equal force in the opposite direction due to charged capacitor plates

_____ unit of electric current

_____ unit of electric charge

_____ unit defined as coulomb/sec

_____ unit of energy

_____ units of mass-to-charge ratio

ELECTRICITY, MAGNETISM, AND THE ELECTRON

TABLE 7–2   TABLE FOR THOUGHT QUESTION 7–2

VOLTAGE V IN VOLTS	CHARGE ON PLATE B	IN TRAVELING BETWEEN PLATES, KE OF ELECTRONS (CIRCLE CORRECT ONE)		DO ELECTRONS REACH PLATE B? (CIRCLE CORRECT ONE)		KE OF ELECTRONS REACHING PLATE B
2	+	Increases	Decreases	Yes	No	_____
1	+	Increases	Decreases	Yes	No	_____
0	+	Increases	Decreases	Yes	No	_____
1	–	Increases	Decreases	Yes	No	_____
2	–	Increases	Decreases	Yes	No	_____
2.95	–	Increases	Decreases	Yes	No	_____
3.05	–	Increases	Decreases	Yes	No	_____
4	–	Increases	Decreases	Yes	No	_____
5	–	Increases	Decreases	Yes	No	_____
6	–	Increases	Decreases	Yes	No	_____

7–2 The beam of electrons approaching plate A has a kinetic energy of 3 eV (Fig. 7–12). There is no force on the electrons until they enter the region between the plates by passing through a hole in plate A. If the electrons reach plate B, there is a reading on the current meter. If the electrons stop before they reach plate B, they return to plate A because of their attraction to plate A. The battery can be connected to the two plates so that plate B is positively charged and plate A is negatively charged, or vice versa. The voltage of the battery and the way that it is connected are shown in Table 7–2. The object is to see if the electron beam reaches plate B

and, if so, to determine the kinetic energy as it strikes plate B.

a. Complete Table 7–2.

b. The charged capacitor plates can be considered as a small particle accelerator for electrons when plate B is (insert: *positively* or *negatively*) _____ charged.

c. Suppose that the initial kinetic energy of the electron beam had not been known. Explain how the kinetic energy could be determined experimentally by adjusting the voltage V and observing the current meter. (Note: A similar procedure is used to determine the kinetic energy of electrons ejected from a metal when light strikes the metal. This effect, called the *photoelectric effect*, will be discussed in Chapter 10.)

Thought questions 7–3 to 7–15 refer to Thomson's experiment with cathode rays.

7–3 Briefly describe Thomson's experimental apparatus.

7–4 Since cathode rays are invisible, how was the behavior of cathode rays observed?

7–5 What forces could be exerted on the cathode rays?

7–6 Thomson's experiment consisted of two parts. For each part, describe the force or forces exerted on the cathode rays.

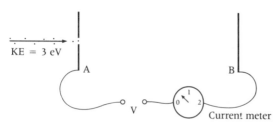

KE = 3 eV

FIG. 7–12   Diagram for thought question 7–2.

7-7 What observations were made? What quantity was determined?

7-8 In order to determine the mass-to-charge ratio of the cathode rays, what three quantities must be determined or observed in Thomson's experiments?

7-9 The value of the mass-to-charge ratio of the cathode rays was about _____ times smaller than the mass-to-charge ratio of the hydrogen ion.

7-10 What result did Thomson find when the cathode material was changed to a different element, and the gas in the cathode ray tube was changed?

7-11 What did Lenard's experiment with cathode rays show about the mass of cathode rays compared to the mass of air molecules?

7-12 If the charge on cathode rays is the same as that on a hydrogen ion, then the mass of a cathode ray is _____ times smaller than the mass of a hydrogen ion.

7-13 What is the nature of cathode rays and how are they produced in a cathode ray tube?

7-14 Suppose that it were possible to achieve a *perfect* vacuum by evacuating *all* air molecules from a cathode ray tube. Would cathode rays be produced?

7-15 Thomson describes the necessity of a good vacuum in order to observe the deflection of cathode rays as follows:

Hertz [Heinrich Hertz, German physicist, 1857–1894—ed.] *made the rays travel between two parallel plates of metal placed inside the discharge-tube (cathode ray tube) but found that they were not deflected when the plates were connected with a battery of storage cells; on repeating this experiment I at first got the same result, but subsequent experiments showed that the absence of deflection is due to the conductivity conferred on a rarefied gas by the cathode rays. On measuring this conductivity it was found that it diminished very rapidly as the exhaustion increased; it seemed that on trying Hertz's experiment at very high exhaustions (good vacuum) there might be a chance of detecting the deflection of the cathode rays by an electrostatic force (force due to charged capacitor plates).*[6]

The phrase "conductivity conferred on the rarefied gas by the cathode rays" refers to the fact that air molecules located in the region between charged capacitor plates ionize or break up into positive and negative ions when struck by a cathode ray.

    a. In what direction do these charged ions travel?

    b. What happens to the total charge on the plates of the capacitor?

    c. Explain Hertz's result: Cathode rays were not deflected when "the plates were connected with a battery."

    d. Explain why Thomson obtained deflection of the cathode rays by charged capacitor plates when he obtained a good vacuum in the cathode ray tube.

Thought questions 7–16 to 7–20 refer to Millikan's oil-drop experiment.

7-16 Briefly describe Millikan's experimental apparatus.

7-17 Millikan's experiment consisted of two parts. For each part, describe the forces exerted on the oil drop. What observations were made? What quantity was determined?

7-18 Suppose that a negatively charged oil droplet is at rest between two capacitor plates connected to a voltage V. What happens to the oil droplet if the voltage is increased? What happens to the oil droplet if the voltage is decreased?

7-19 Describe how Millikan determined the charge on the electron by determining the static charge on oil drops.

†7-20 Consider two spherical oil drops, one having a radius two times that of the other. The oil drops fall under the influence of gravity and encounter air resistance. The weight of the larger oil drop is _____ times that of the smaller one. When both drops reach their terminal speed,

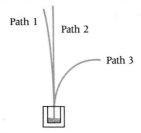

FIG. 7–13 Diagram for thought question 7–21.

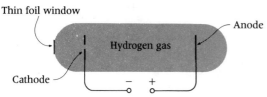

FIG. 7–14 Diagram of an ion source for thought question 7–22.

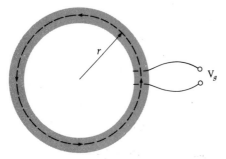

FIG. 7–15 Diagram of a synchrotron for thought question 7–23.

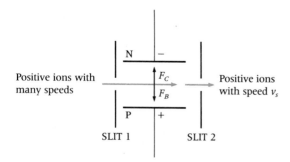

FIG. 7–16 Diagram of a velocity selector for thought question 7–24.

the force of air resistance on the larger drop is _____ times that on the smaller drop. In order for this to occur, the terminal speed of the larger drop must be _____ times that of the smaller one.

7–21 Figure 7–13 shows the paths of alpha, beta, and gamma rays emerging from the radioactive source and subjected to a perpendicular magnetic field. The charge on the alpha ray is _____ , the charge on the beta ray is _____ , and the charge on the gamma ray is _____ . The mass of the alpha ray is (insert: 8, 80, 800, or 8000) _____ times more massive than the beta ray (see Chapter 1). Since the force on the gamma ray due to the magnetic field is _____ , the gamma ray travels over path number _____ in Fig. 7–13. The force on the alpha ray is in the (insert: *same* or *opposite*) _____ direction to the force on the beta ray. Because the alpha ray has a larger mass than the beta ray, the alpha ray is (insert: *easier* or *harder*) _____ to bend than the beta ray, and the beta ray therefore shows a (insert: *greater* or *smaller*) _____ deflection than the alpha ray. The beta ray travels over path number _____ , and the alpha ray over path number _____ . The same conclusions can be reached using Equation 9. Since the alpha ray has the larger mass, the radius of the alpha ray's path is (insert: *larger* or *smaller*) _____ than the beta ray's, and this means that the path of the alpha ray is (insert: *a sharper curve*, or *a smaller curve*) _____ compared to the path of the beta ray.

7–22 The cathode ray tube shown in Fig. 7–14 is called an *ion source*. For example, suppose that this cathode ray tube contains hydrogen gas at low pressure. Ions are produced when a cathode ray (electron) strikes a hydrogen molecule (consisting of two atoms of hydrogen). The positively charged hydrogen ion travels toward the (insert: *cathode* or *anode*) _____ . Similarly, the negatively charged hydrogen ion travels toward the _____ . In particle accelerators ion sources produce the charged particles that are accelerated. An ion source containing hydrogen produces hydrogen ions called protons, which pass through the thin foil window.

7–23 In a type of particle accelerator called a *synchrotron*, protons (or other charged particles as well) travel in a circular path of *constant* radius due to a perpendicular magnetic field. (Fig. 7–15).

    a. Show how the radius of the proton's path depends upon the strength of the magnetic field B, the charge on the proton, the mass of the proton, and the speed of the proton by defining the symbols in Equation 9 (p. 149).

    b. With each revolution, the kinetic energy of the protons increases because the protons pass between charged capacitor plates (or a similar device that increases the proton's kinetic energy). In order to maintain the *same radius* path of the protons, the magnetic field must (insert: *increase, decrease,* or *remain the same*) _____ with each revolution.

    c. The fragment *synchro* in the word *synchrotron* refers to another word, *synchronized.* Use part **b** to explain what is synchronized.

7–24 The device in Fig. 7–16 is called a *velocity selector.* A perpendicular magnetic field exists in the region surrounding the charged capacitor plates. Positively charged ions with a range of speeds pass through slit 1, but only those with one value of the speed, called v_s, pass through slit 2. F_C shows the force on a positively charged ion due to the charged capacitor plates, and F_B, due to the magnetic field.

    a. For ions having speed v_s, F_B is (insert: *greater than, equal to,* or *less than*) _____ F_C.

    b. The force F_C on an ion (insert: *does depend* or *does not depend*) _____ upon the speed of the ion.

    c. The force F_B on an ion (insert: *does depend* or *does not depend*) _____ upon the speed of the ion.

    d. For ions having a speed greater than

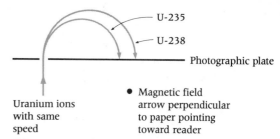

FIG. 7–17   Diagram of a mass spectrograph or mass separator for thought question 7–25.

v_s, F_B is (insert: *greater than, equal to,* or *less than*) _____ F_C. These ions will be deflected (insert: *upward* or *downward*) _____ and strike (insert: *below slit 2* or *above slit 2*) _____ .

    e. For ions having a speed less than v_s, F_B is (insert: *greater than, equal to,* or *less than*) _____ F_C. These ions will be deflected (insert: *upward* or *downward*) _____ and strike (insert: *below slit 2* or *above slit 2*) _____ .

    f. Show how the speed v_s depends upon the voltage connecting the plates N and P, the distance between plates N and P, and the magnetic field B.

    g. How can the desired velocity be selected?

7–25 Uranium ions all having the same speed (obtained by using a velocity selector described in thought question 7–24) and +1 charge pass through a slit and enter the region of a perpendicular magnetic field (Fig. 7–17). One path corresponds to the ions of the isotope uranium-238 (U-238) and the other to ions of uranium-235 (U-235). (Remember that 235 and 238 indicate the relative masses of the atoms or ions.)

    a. Show how the radius of the path depends upon the mass of the ion, the speed of the ion, the charge of the ion, and the magnetic field by defining the symbols in Equation 9 (p. 149).

b. The U-238 ions travel over the path of (insert: *larger* or *smaller*) _____ radius, because part **a** shows that the radius is (insert: *proportional* or *inversely proportional*) _____ to the mass of the ion, while all other terms are the same for U-235 and U-238 ions. (Note: The device in Fig. 7–17 is called a *mass spectrograph* or *mass separator,* depending upon its use. It demonstrates the existence of isotopes. By measuring the radius of the path, the masses of isotopes can be accurately determined. During World War II, mass separators were constructed at Oak Ridge National Laboratory to separate the fissionable U-235 from U-238. In natural uranium, only 1 atom in 140 uranium atoms is U-235; the rest are U-238.)

Questions for Calculation

7–1 An electric current of 5 amperes flows out of a car battery for 2 minutes. How much electric charge flows from one terminal and into the other terminal of the battery?

†7–2 a. Equation 2 (p. 147) describes the force on a charged particle located between two capacitor plates. From the definition of electric field (Equation 5, p. 149), find the electric field between the capacitor plates.

b. Charge Q_1 and charge Q_2 are separated by a distance d. What is the force exerted on Q_2 by charge Q_1? What is the electric field set up by Q_1 at the location of charge Q_2? What is the electric field set up by Q_2 at the location of charge Q_1?

7–3 Suppose that the two parallel metal plates of a capacitor, separated by a distance of 2 cm (0.02 m), are connected to a 6-volt battery, and each plate acquires a certain amount of charge, called Q_p. If the voltage increases to 12 volts, the amount of charge on each plate is _____ Q_p. If the voltage is decreased to 2 volts, the amount of charge on each plate is _____ Q_p.

†7–4 An electron (charge $= 1.6 \times 10^{-19}$ coulombs) is between the capacitor plates, separated by a distance of 2 cm and connected to a 6-volt battery. The force on the electron is _____ newtons. The electron is accelerated toward the positively charged plate. The amount of work done by the force in moving the electron a distance of 0.02 m is _____ joules, or _____ eV. If the initial kinetic energy of the electron is zero, then the kinetic energy when it reaches the positive charged plate is _____ joules, or _____ eV. If the electron had an initial kinetic energy of 2 eV, then the kinetic energy when it reaches the positively charged plate is _____ eV.

†7–5 Repeat calculation question 7–4 for a particle having a positive charge twice the charge on an electron (such as an alpha particle) and that accelerates toward the negatively charged capacitor plate.

7–6 In the electrolysis of sodium chloride, 48,250 coulombs deposit _____ g of sodium on the cathode and _____ g of chlorine on the anode, and 96,500 coulombs deposit _____ g of sodium and _____ g of chlorine.

a. The atomic mass of aluminum is 27 amu and the atomic mass of chlorine is 35.5 amu. In 27 g of aluminum, there are _____ atoms of aluminum. Express answer in terms of Avogadro's number N_A.

b. The molecular mass of the compound aluminum chloride ($AlCl_3$) is _____ amu. Consider a sample of aluminum chloride in which the mass in grams is equal to the molecular mass. In this sample, there are _____ molecules of aluminum chloride. In this sample of aluminum chloride, there are _____ atoms of aluminum and _____ atoms of chlorine. Express answers in terms of N_A.

c. When aluminum chloride is dissolved in water, the charge on each aluminum ion is _____ and the charge on each chlorine ion is _____ . During the electrolysis of aluminum chloride, _____ ions of chlorine deposit on the anode while _____ ion(s) of aluminum deposit on the cathode. Therefore, a charge of 96,500 coulombs passing through the circuit deposits N_A atoms of chlorine or 35.5 g of chlorine and _____ N_A atoms of aluminum, or _____ g of aluminum.

Footnotes

[1] George Thomson, *J. J. Thomson: Discovery of the Electron,* (Garden City, N.Y.: Anchor-Doubleday, 1966), p. 57.

[2] J. J. Thomson, *Philosophical Magazine* 44 (1897): 293.

[3] Ibid; p. 293.

[4] Ibid; p. 293.

[5] R. A. Millikan, *Physical Review* 32 (1911):349.

[6] Thomson, p. 293.

Albert A. Michelson

CHAPTER 8

Waves, Sound, and Light

Introduction

"Touchdown!" the TV blares. You dash from the hall where you are standing into the living room to catch the instant replay. While you couldn't see the TV from the hall, you could certainly hear it. When the sound waves struck the doorway, their direction changed, but the direction of the light waves did not (see Fig. 8–1). This situation illustrates one property of waves: diffraction, or the ability to bend around obstacles under certain conditions. Both light waves and sound waves diffract. In our example, however, the sound waves diffract to a much greater extent than the light waves around the wall, which is a large obstacle.

This chapter and Chapter 10 both deal with the nature of light. Here, we shall see *why* light has a wave nature. In 1801, the English physician Thomas Young showed unquestionably that light did indeed have a wave nature. We shall examine how he discovered this.

There are many kinds of waves—light, sound, water, waves made on violin strings, waves made on the coiled spring toy called a Slinky. Amazingly, all of these diverse waves have one feature in common: interference, or the additive effect caused by two waves where they overlap. Light waves, sound waves, and water waves also diffract. These two properties—interference and diffraction—characterize wave motion of *any kind*. We'll first investigate these properties in familiar water waves, and then apply what we learn to the waves of sound and light. As Part III of the text unfolds, and the discussion focuses on various kinds of rays, a primary question we shall ask is, Are the rays particles or are they waves? If they are waves, we must detect the characteristics of interference and diffraction.

We will learn here how to measure the *wavelength* of light, which is the distance between two consecutive peaks of the wave. The wavelength of light determines the color of the light. Remember that each chemical element emits a characteristic set or spectrum of colors (Chapter 1). Why does this happen? In what way, besides mass, are the atoms of elements different? How and why does an atom emit light? To answer such questions, the wavelengths of spectral colors must first be measured. Any theory concerning the structure of the atom (that is, what an atom looks like) must account for these wavelengths.

We'll turn our attention to how light waves travel through space. It seems natural to imagine that light waves, like all other types of waves, travel through a medium of some sort. For light, this medium was at one time considered to be the "ether," a massless and transparent substance allegedly pervading all of space. We'll discuss the classic experiment of two American scientists, Albert A. Michelson and Edward W. Morley, who were unsuccessful in detecting any ether. Finally, we'll determine how electromagnetic waves travel through space.

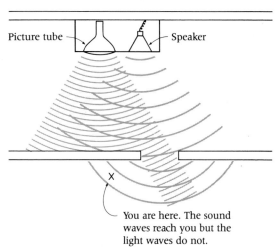

Picture tube

Speaker

X

You are here. The sound waves reach you but the light waves do not.

FIG. 8–1   The sound waves from the TV speaker bend around the doorway, but the light waves from the picture tube do not.

Waves

For studying water waves, a structure called a *ripple tank* is used. This container is constructed with a glass bottom so that ripples occurring on the surface of the water within can be photographed. Photographs taken in this manner are examined to determine the effects of diffraction and interference. Straight or plane waves can be created by *regularly* dipping a narrow bar (shaped like a ruler) into the surface of the water in the tank. Figure 8–2 shows a side view of such waves. The distance between two peaks or between two troughs is called the **wavelength,** which, remember, is denoted by the symbol λ (the lowercase Greek letter lambda). The **amplitude** (denoted by the symbol A) is the maximum height or the maximum depth of the wave.

Suppose we place a cork in a ripple tank. As the waves move past it, the cork will bob up and down *only:* The waves will not push the cork horizontally (see Fig. 8–3). Figure 8–4(a) depicts water waves with a wavelength of 0.06 m just reaching a cork floating on the surface of the water. In Fig. 8–4(b), we see that 10 waves have passed by the cork in 2 seconds. We define wave **frequency** (denoted by the sym-

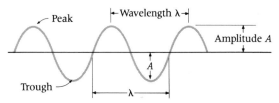

Peak

Wavelength λ

Amplitude A

A

Trough

λ

FIG. 8–2   A side view of plane waves.

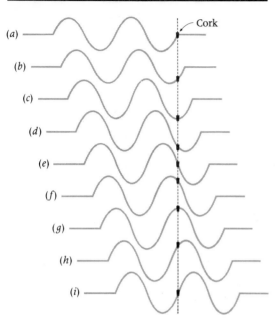

FIG. 8–3   The effect of plane waves on a cork in a ripple tank.

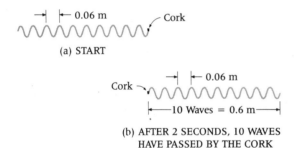

(a) START

(b) AFTER 2 SECONDS, 10 WAVES HAVE PASSED BY THE CORK

FIG. 8–4   A ripple-tank experiment for finding the relationship involving the wavelength λ, the speed v of the water waves, and the frequency f.

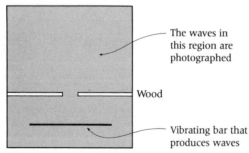

FIG. 8–5   Apparatus set up in a ripple tank so that the phenomenon of diffraction may be observed.

bol f) as the number of waves per unit time that pass any fixed point. The frequency of the water waves in this case is 10 waves per 2 seconds, or 5 waves per second. In 2 seconds, the front of the waves has traveled a distance of 0.6 m, and the speed of the water waves is thus 0.6 m/2 sec, or 0.3 m/sec. From this, we see a relationship involving speed, frequency, and wavelength:

$$0.3 \text{ m/sec} = (5/\text{sec})(0.06 \text{ m})$$
$$v = f\lambda. \tag{1}$$

The frequency is 5 waves per second, but, as the entity "waves" represents a number and does not have any units, the frequency is written as 5/sec.

Diffraction

Now, let's see what happens when we place two pieces of wood in the ripple tank, leaving an opening between them (see Fig. 8–5). We might expect that the plane waves will pass through the opening and disturb only the water directly behind the opening, but this expectation is not borne out. Instead, we find circular waves *and* disturbances well to each side of the opening (see Fig. 8–6a). This ability of waves to bend around corners is called **diffraction.** The water waves bend when they encounter the walls on each side of the opening.

Diffraction, however, depends very much on the size of the opening. In Fig. 8–6, three photographs of plane waves are shown striking the same opening. The diffraction is most pronounced in Fig. 8–6(a), where the wavelength is six-tenths of the width of the opening; it is less pronounced in Fig. 8–6(b), where the wavelength is three-tenths of the width. In Fig. 8–6(c), where the wavelength is only one-tenth of the width, we find nearly a shadow effect—waves occurring only directly behind the opening—as we had anticipated earlier. These experiments demonstrate that diffraction is pronounced when the slit opening and the wavelength are about

170

the same size, and is nearly nonexistent when the opening is much *larger* than the wavelength.

In Fig. 8–7, plane waves strike a peg with a diameter *smaller* than the wavelength. The waves diffract (or bend) in all directions, and a circular wave appears. (Additional examples of this type of diffraction will come up in later chapters. For instance, a diffracted circular wave also appears when X-rays strike an atom.)

Interference of Waves

The regular vibration of an immersed dipper (that is, a sphere about one-half of a centimeter in diameter mounted on the end of a rod) makes circular waves in a ripple tank. Drips of water from a leaky faucet make similar circular waves in a bathtub.

Interference refers to the net effect produced by two or more waves. Figure 8–8 shows the interference pattern that is produced by two dippers placed close together in a ripple tank. In this case, the dippers move up and down together, or, as the jargon of physics puts it: the dippers are *in phase*. The cork at point P, for instance, does not bob up and down at all. It's just as if the dippers weren't operating. The cork at point Q, on the other hand, bobs up and down with an amplitude twice that of each wave separately at that point. We thus find two effects: the amplitudes of each wave add, and they subtract. Let's examine this in more detail.

Figure 8–9 is a diagram of two circular waves coming from the dippers. The circles represent the peaks of the waves.

Anywhere on the solid, straight lines, the two waves are in phase: Whenever there is a peak from one wave, there is a peak from the other; when there is a trough from one, there is a trough from the other. In this case, the cork bobs up and down with amplitude 2A. This additive effect is called **constructive interference** (see Fig. 8–10a).

Anywhere on the dashed lines, the two waves are out of phase: Whenever there is a peak from one wave, there is a trough from the other. Here,

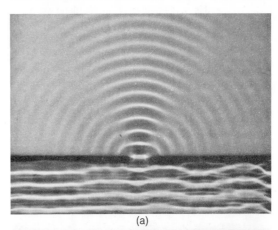

(a)

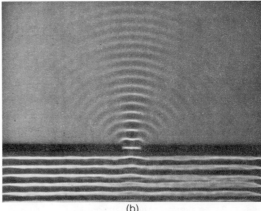

(b)

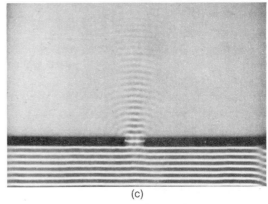

(c)

FIG. 8–6   Plane waves of different wavelengths striking the same opening. Diffraction very much depends on the comparative size of the wavelength and the size of the opening. (*PSSC Physics*, 2nd ed., D. C. Heath & Company and Education Development Center: Newton, Ma., 1965. Photo courtesy of Education Development Center, Newton, Ma.)

171

FIG. 8–7 In this ripple-tank experiment, plane waves traveling in the direction shown strike a small peg. The waves have just passed by the peg and, in the water behind the waves, a diffracted circular wave can easily be seen. It is evident in the upper region as well. (Courtesy of Education Development Center, Newton, Ma.)

Waves travel in this direction

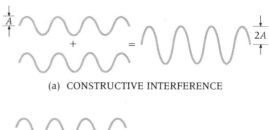

λ

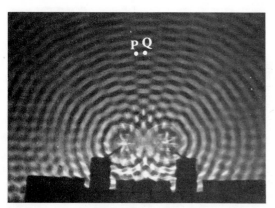

FIG. 8–8 An interference pattern produced by two dippers in a ripple tank. (*PSSC Physics,* 2nd ed., D. C. Heath & Company and Education Development Center: Newton, Ma., 1965. Photo courtesy of Education Development Center, Newton, Ma.)

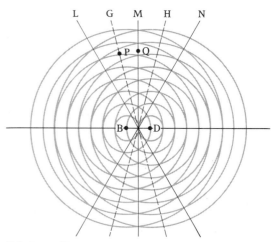

FIG. 8–9 Circular waves from each dipper produce regions of constructive interference (solid lines) and destructive interference (dashed lines).

FIG. 8–10 (a) Constructive interference: two waves in phase. (b) Destructive interference: two waves out of phase.

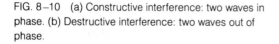

$\frac{A}{\uparrow}$ $+$ $=$ $2A$

(a) CONSTRUCTIVE INTERFERENCE

$+$ $=$

(b) DESTRUCTIVE INTERFERENCE

the cork remains motionless. This canceling effect is called **destructive interference** (see Fig. 8−10b). (The dashed lines in Fig. 8−9 correspond to the lines radiating from the dippers in Fig. 8−8.)

In Fig. 8−11, the dippers are separated by 8 wavelengths rather than 2 wavelengths, as in Fig. 8−9. We see that the straight lines are much closer together than in Fig. 8−9, and it is thus much more difficult to separate regions of constructive and destructive intereference. Therefore, experimenters can observe interference effects most easily if the distance between the dippers (or sources of the waves) is nearly equal to the wavelength.

Finally, let's consider the ripple-tank experiment from the slightly different viewpoint shown in Fig. 8−12. Do the waves have a constructive or a destructive effect on the three corks? (Although Fig. 8−12 is actually a top view, the waves are drawn to represent a side view for convenience.) The waves arriving at cork 1 are in phase, and the cork oscillates with amplitude 2A. The same thing happens for cork 3. The distance from each dipper to cork 1 is 4 wavelengths (4λ). The distances from each dipper to cork 3 are 4 wavelengths (4λ) and 5 wavelengths (5λ). The distances from each dipper to cork 2 are 4 wavelengths (4λ) and 4.5 wavelengths (4.5λ).

We define the **path length** as the distance from a dipper to a cork. In general, constructive interference occurs whenever the *difference in the two path lengths* is an integral number of wavelengths:

$$\text{Difference in path lengths} = n\lambda, \quad (2)$$

where n equals 0, 1, 2, 3, etc. For cork 1, n equals 0; for cork 3, n equals 1. Destructive interference occurs whenever the difference in the two path lengths is a half number of wavelengths:

$$\text{Difference in path lengths} = \frac{n}{2}\lambda, \quad (3)$$

where n equals 1, 3, 5, etc. For cork 2, n equals 1.

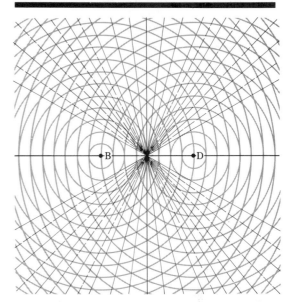

FIG. 8−11 The regions of constructive and destructive interference for two dippers separated by eight wavelengths.

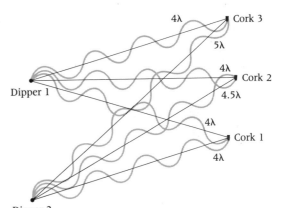

FIG. 8−12 The effect of the two circular waves on each cork.

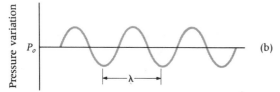

FIG. 8–13 Compressions and rarefactions produced by a vibrating speaker.

Sound

THE NATURE OF SOUND WAVES

The loudspeaker in a TV set produces sound when the cone-shaped part of the speaker, called the *diaphragm*, vibrates back and forth. If the speaker vibrates 440 times per second, then the sound we hear (corresponding to middle A on the piano) has a frequency of 440 oscillations per second. When the diaphragm moves forward, it compresses the air in front of it, causing an increased air pressure; when it moves backward, the air pressure in front of the speaker is reduced. The continual vibration of the speaker produces the compressions and rarefactions to which our ears respond (see Fig. 8–13a). The graph in Fig. 8–13(b) shows the pressure variation, either above or below the normal pressure P_0. These compressions and rarefactions travel away from the speaker at 343 m/sec (1125 ft/sec), which is called the **speed of sound.** The vibration of the speaker causes the air molecules to vibrate at the same rate, producing compressions and rarefactions (see Fig. 8–14).

Suppose that the speaker takes a time T to make one complete oscillation: The time between successive compressions (or rarefactions) is T. If the frequency is 440 oscillations per second (abbreviation: osc/sec), then T equals 1 second per 440 oscillations, or 0.00227 seconds per oscillation.

Line (a) in Fig. 8–14 shows 17 equally spaced air molecules where the pressure is the same everywhere. We shall call the positions of these molecules *equilibrium positions.* Line (b) shows the positions of the molecules producing compressions at the equilibrium positions of molecules 5 and 13; and rarefactions at 1, 9, and 17. Lines (c) through (j) show the positions of the molecules after time intervals of $T/8$.

The compression at position 5 moves to 6 in line (c), 7 in line (d), and so forth to position 13 in line (j). Similarly, the rarefaction at position 1 in line (b) moves to 9 in line (j); the rarefaction at position 9 in line (b) moves to 17 in line (j). The compressions and rarefactions travel to the right with constant speed. This is

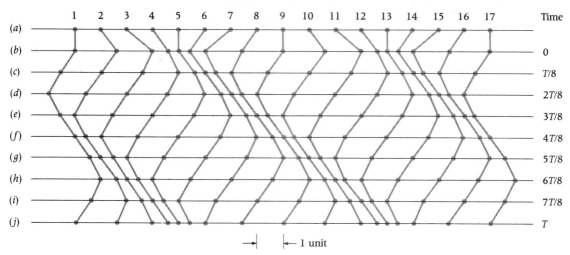

FIG. 8-14 The vibration of air molecules produces compressions and rarefactions.

analogous to the peaks and troughs of water waves in Fig. 8–3 moving to the right.

Looking at position 5, we see that initially a compression is found there. After a time $T/2$ in line (f), a rarefaction has moved into that location. After a time T in line (j), another compression is found there. The speaker makes one complete oscillation in time T and, hence, at a given location the time between successive compressions is also time T. (This is analogous to the cork bobbing up and down in Fig. 8–3.)

The vibration of the air molecules is actually responsible for the compressions and rarefactions. Comparing lines (a) and (b), we see that the molecules 1, 5, 9, 13, and 17 have not moved from their equilibrium positions. However, molecules 2, 3, and 4 have moved to the right; 6, 7, and 8 to the left; and 10, 11, and 12 to the right. This creates a compression at position 5 and a rarefaction at position 9.

Each molecule makes one complete oscillation in time T. Looking at molecule 1, we see that it moves to the left and comes to a stop in line (d), traveling 1 unit left. Then it moves to the right, passing through the equilibrium position in line (f); it comes to a stop in line (h), traveling 1 unit right. It comes back to the equi-

librium in line (j) after time T. Molecules 9 and 17 have identical motions. The motions of 5 and 13 are similar, but they travel to the right first. Molecule 2, for example, moves quite differently from 1. However, molecule 2 makes one complete oscillation in time T. In line (j), its position is the same as in line (b). Molecule 2 travels one unit to the left of its equilibrium position and one unit to the right.

If the volume of the speaker were turned louder, but the frequency remained the same, the speaker would make the same number of oscillations in 1 second. However, it would travel a greater distance both forward and backward: Its amplitude would be greater. This would produce a greater pressure in the compression and a lesser pressure in the rarefaction. Correspondingly, the amplitude in Fig. 8–13(b) would be larger. In Fig. 8–14, the molecules would travel a maximum distance greater than 1 unit to the right and to the left.

WAVES, SOUND, AND LIGHT

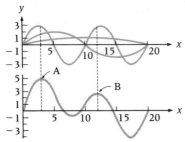

FIG. 8–15.   Interference of three waves of different wavelength (or frequency) and of different amplitude. A equals 4.91, which is the sum of 2.85, 1.61, and 0.45; B equals 2.63, which is the sum of 2.85, −1.17, and 0.95.

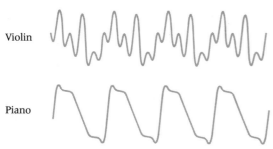

Violin

Piano

FIG. 8–16   Complex waveforms produced by a violin and a piano, each sounding middle A.

Several times, we have noted the similarities of sound and water waves, but there is one essential difference: In Fig. 8–3, the cork bobs up and down as the waves move to the right. The cork's motion is perpendicular to the direction in which the waves travel. This phenomenon is called **transverse wave motion.** But as a sound wave travels to the right, the molecules vibrate in a direction *parallel* to the direction of motion. This phenomenon is called **longitudinal wave motion.**

DIFFRACTION OF SOUND WAVES

To see why the diffraction of sound waves in Fig. 8–1 was so pronounced, we must calculate the wavelength, using Equation 1. Our experience with water waves leads us to believe that the wavelength of the sound waves must be comparable to the width of the doorway for sizeable diffraction effects to occur. If the frequency is 440 oscillations per second, then the wavelength is given by

$$\lambda = \frac{v}{f}$$
$$= \frac{343\,\text{m/sec}}{440/\text{sec}}$$
$$= 0.780\,\text{m}.$$

This wavelength, as we expected, is comparable to the width of the doorway, which is about 1 m. This situation is analogous to Fig. 8–6(a) where the diffraction of water waves is pronounced.

INTERFERENCE OF SOUND WAVES

We have learned that two waves of the same wavelength and amplitude could be added to produce constructive interference, if the waves were in phase, or destructive interference, if the waves were completely out of phase. Figure 8–15 shows the result of adding three waves of different wavelength and amplitude. A rather complex waveform is obtained. Let's use this concept to understand another interesting effect.

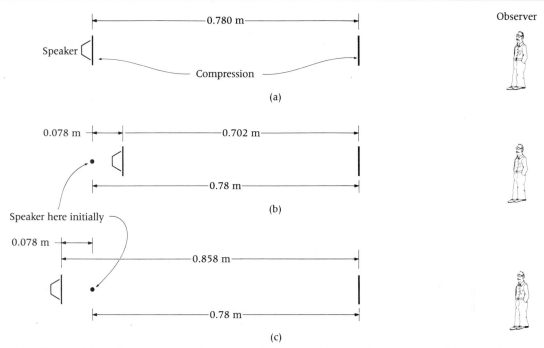

FIG. 8–17 The distance between successive compressions when (a) the speaker is stationary, (b) the speaker is moving toward the observer, and (c) the speaker is moving away from the observer.

When both a piano and a violin sound middle A (frequency = 440 osc/sec), we can certainly tell the instruments apart. Actually, this sound is not composed entirely of the so-called fundamental frequency of 440 osc/sec, but also contains frequencies, called **overtones,** of 880 osc/sec, 1320 osc/sec, and so on. A piano and violin sound differently because they produce different amounts of overtones. The total waveform found by adding the fundamental frequency and all of the overtones (much like in Figure 8–15) is quite different for a piano and violin (see Fig. 8–16).

THE DOPPLER SHIFT

As previously mentioned, a speaker sounding a frequency of 440 osc/sec produces compressions that are separated by 0.78 m (λ = 0.78 m). (See Fig. 8–17a.) However, if the speaker is moving, then the distance between these

compressions changes, and the pitch or frequency heard by a stationary observer differs from 440 osc/sec.

Suppose, for example, that the speaker is traveling at 10 percent of the speed of sound, or 34.3 m/sec (76.8 mph). If the speaker initially produces a compression, then another one is produced after a time T, where T is defined as the time for one oscillation. During time T, the first compression will have traveled a distance of 0.78 m from the speaker (see Fig. 8–17, parts a, b, and c). During time T, the speaker has traveled a distance of 0.078 m. When the speaker is moving toward the observer (Fig. 8–17b), then the compressions are separated by a distance of 0.702 m. From Equation 1, the frequency heard by the observer is given by (speed)/(wavelength), which is (343 m/sec)/(0.702 m), or 489 osc/sec. Similarly, where the speaker is moving away from the observer (Fig. 8–17c), the wavelength is 0.858 m, and the observer hears a frequency of 400 osc/sec. In general, the frequency perceived by the observer is given by

$$f_{obs} = \frac{f_{source}}{1 \pm v_{source}/v_{sound}}, \qquad (4)$$

where v_{source} is the speed of the source of the sound and v_{sound} is the speed of sound.

The minus sign is used when the source is moving toward the observer; the plus sign is used when the source is moving away from the observer. For example, in Fig. 8–17(b), the observed frequency is given by

$$f_{obs} = \frac{440 \text{ osc/sec}}{1 - 0.1}$$
$$= 489 \text{ osc/sec},$$

where v_{source} equals 34.3 m/sec and v_{sound} equals 343 m/sec.

Thus, the pitch of a moving sound source is higher than normal when it approaches a stationary listener, and lower than normal when moving away. This phenomenon is called the **Doppler effect.**

Young's Double-Slit Experiment

For centuries, scientists puzzled over the question, Is light composed of waves or of particles? If light has a wave nature, then experiments must show that light exhibits both diffraction and interference effects. In 1801, the brilliant physician Thomas Young (1773–1829) showed conclusively that light had a wave nature by observing interference effects. Young had a truly amazing mind that enveloped many interests. He carried out medical research on the eye; conducted investigations on sound, light, and optics; studied the elastic properties of materials; and developed a theory of tides.

In fact, Young made so many contributions to both science and literature that he began making some of them anonymously, so he wouldn't be accused of slighting his professional tasks! He was even involved in archeology and philology. He left an active practice in 1814 to work full time on his scientific projects and continued to do so until his death. His contemporary and colleague Sir Humphrey Davy (1778–1829), of the Royal Institution, said of Young:

Had he limited himself to any one department of knowledge, he must have been the first in that department. But as a mathematician, a scholar, a hieroglyphist, he was eminent, and he knew so much that it was difficult to say what he did not know.[1]

(Any comments about the value of a liberal education combined with a curious disposition would be superfluous.)

Figure 8–18 shows the apparatus Young used in his double-slit experiment. Sunlight fell on a single pinhole (slit S), and the diffracted wave then struck two pinholes about 1 mm apart. Because the light striking the two pinholes had the same phase (as shown in Fig. 8–18), the diffracted waves emerging from the two pinholes were in phase. This is exactly the same situation that we witnessed for water waves. Young concluded that the light and dark regions on the screen were due to interference effects. The light regions corresponded to constructive interference; the dark regions, to destructive interference.

The single pinhole S was crucial, for without it the experiment would not have worked. Sunlight falling on the two pinholes *directly* would not be in phase (their phases would not be correlated), and no light and dark regions would be observed on the screen.

Young not only demonstrated the wave nature of light, but he measured the wavelengths of

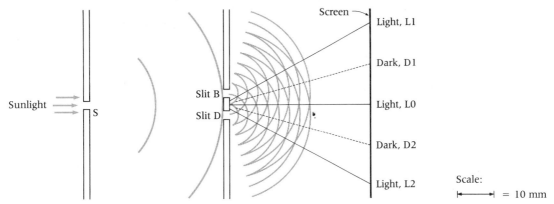

FIG. 8–18 Schematic diagram of Young's experiment for observing the interference of two diffracted circular light waves. The wavelength is grossly exaggerated. Only one wavelength is shown, although sunlight contains light of many wavelengths.

various colors as well. To see how he did this, look at the photographs in Fig. 8–19. The bottom photograph (b) shows red light illuminating two slits, and the top photograph (a) shows blue light illuminating the same two slits. The regions of destructive interference, corresponding to the black regions, are separated by different distances. These distances reflect the different wavelengths for red and for blue light. When sunlight struck the two slits, the interference patterns for each color were superimposed on the screen, and a spectrum of colors appeared. By noting the interference pattern produced by each color, Young was able to determine the wavelength. He found that visible light ranges from 423 nm (violet) to 706 nm (red), where 1 nm equals 10^{-9} m. He wrote:

From a comparison of various experiments, it appears that the breadth of the undulations constituting the extreme red light must be supposed to be, in air, about one 36 thousandth of an inch (or 706 nm), and those of the extreme violet about one 60 thousandth of an inch (or 423 nm); the mean of the whole spectrum, with respect to the intensity of light, being about one 45 thousandth (or 564 nm). From these dimensions it follows, calculating upon the known velocity of light, that almost 500 millions of millions of the slowest of such undulations must enter the eye in a single second.[2]

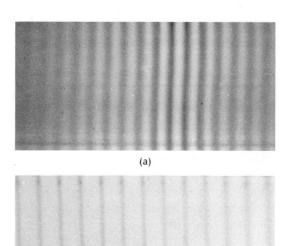

FIG. 8–19 Interference pattern produced by (a) blue light illuminating two slits and (b) red light illuminating the same two slits. (*PSSC Physics*, 2nd ed., D. C. Heath & Company and Education Development Center: Newton, Ma., 1965. Photo courtesy of Education Development Center, Newton, Ma.)

WAVE, SOUND, AND LIGHT

Today, we know that visible light ranges from 400 nm to 700 nm. By comparing these values with Young's, we see how remarkably accurate his measurements were.

Since the wavelength of light is so small, diffraction of light is not a phenomenon that we encounter daily. As we learned when we discussed experiments with water waves, diffraction occurs only when the obstacle (or opening) is on the order of the wavelength. This idea also explains the situation depicted in Fig. 8–1. The TV light, striking the doorway, did not diffract because that obstacle was simply too large compared to the wavelength of light. The sound waves from the TV, on the other hand, with a wavelength of about 0.8 m, did diffract around the doorway.

MEASURING THE WAVELENGTH
OF LIGHT: AN EXAMPLE

Use Fig. 8–18 to determine how Young measured the wavelength of light. A light spot occurs at L0 because the difference in *path lengths* (that is, the distance D to L0 minus the distance B to L0) is zero. A light spot occurs at L1 because the difference in path lengths (that is, the distance D to L1 minus distance B to L1) is 1 wavelength. A dark region occurs at D1 because the difference in path lengths is ½ wavelength.

Use the scale shown in Fig. 8–18 to measure the path lengths directly from the diagram. The path length from the center of slit D to L1 is 46.0 mm; the path length from the center of slit B to L1 is 43.0 mm. The difference is 3.0 mm, which is equal to 1 wavelength.

Now, compare this with the wavelength measured from the circles that represent the waves in Fig. 8–18. Since the error is less in measuring longer distances, we measure the distance between six circles (which equals 5 wavelengths) and then divide by 5 to find a wavelength of

2.8 mm. If you consider the errors involved in measuring the distances, these two values for the wavelength—2.8 mm and 3.0 mm—compare favorably. Thus, it is possible to determine the wavelength by measuring the distance between two consecutive regions of constructive (or destructive) interference.

Prism and Spectra

As we pointed out, a natural first step toward unlocking the structure of the atom is to break up the emitted light into a spectrum of colors and then to measure the wavelengths of these colors. Both the *prism* and a device called a *diffraction grating* can do this, each operating on completely different principles. Let's discuss the **prism** first, since it is the older method.

The prism breaks up a beam of white light into a spectrum of colors, as shown in Fig. 8–20. This happens because each color of light has a different speed in glass, but the same speed in air. Figure 8–20(b) shows the boundaries of the beam greatly enlarged. The top part of the beam reaches point D at the same time as the bottom part reaches point B. The bottom part then enters the glass, where its speed is less than it is in air. Therefore, during the same time that the top part travels to point E, the bottom part travels only to point H. The top part takes the same amount of time to travel from point E to point I as the bottom part takes to travel from point H to point J. This effect of changing the direction of the light beam at the boundary of any two substances is known as **refraction.**

By itself, the prism cannot be used to determine the wavelength: It must be calibrated. First, we must measure the wavelengths of various colors by performing a double-slit experiment (or by using a diffraction grating, as we shall see in the next section). Then, we can send this light through a prism and calibrate the apparatus; that is, discern the relationship between wavelength and position on the screen. With this calibration, we can then determine the previously unknown wavelengths of spectral lines by noting their positions on the screen.

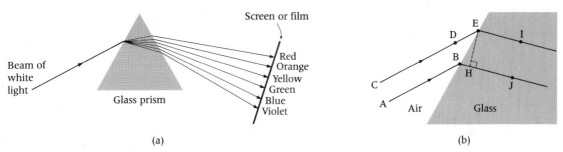

FIG. 8–20 (a) A prism breaks up a beam of white light into a spectrum of colors. (b) An enlarged representation of a light beam striking the side of a prism.

Table 8–1 shows the wavelength for various colors of the visible spectrum, as well as for other types of radiation.

The Diffraction Grating

Diffraction gratings are akin to the double-slit experiment. In the construction of these devices, a diamond point is used to rule parallel and equidistant grooves (usually, thousands of grooves per inch) on glass or metal plates (see Fig. 8–21). Light cannot pass through the grooves scratched on the glass, but can of course pass through the untouched slits. When metal is used, the light is reflected by the untouched metal between the grooves. In either case, a grating consists of thousands of slits rather than only two, as in Young's experiment.

Joseph von Fraunhofer (1787–1826) made the first diffraction grating, which had 4000 grooves covering one-half inch of glass. During the late nineteenth century, Henry A. Rowland (1848–1901) at Johns Hopkins University designed a ruling engine that produced 20,000 grooves per inch, with quite uniform spacing.

Albert A. Michelson (Fig. 8–22), who devoted a lifetime to investigating light, was truly a genius with light measurements (his famous experiments with Morley are discussed later in this chapter). In 1892, even after he became the chairman of the physics department at the Uni-

FIG. 8–21 Schematic diagram of grooves on a diffraction grating.

TABLE 8–1 WAVELENGTHS OF VARIOUS RADIATIONS

TYPE	WAVELENGTH (nm)
Gamma rays	5×10^{-4} to 0.14
X-rays	0.01 to 10
Ultraviolet light	5 to 400
Visible light:	400 to 700
violet	400 to 450
blue	450 to 500
green	500 to 570
yellow	570 to 590
orange	590 to 610
red	610 to 700
Infrared	700 to 2×10^5
Short radio	10^5 to 5×10^{10}
Broadcast	5×10^{10} to 8×10^{11}
Long radio	8×10^{11} to 10^{16}

Note: 1 nm equals 10^{-9} m.

WAVES, SOUND, AND LIGHT

181

FIG. 8–22 Albert A. Michelson in his laboratory. (Photo courtesy of Argonne National Laboratory)

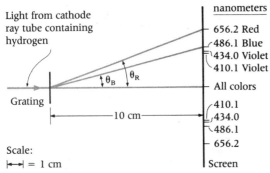

FIG. 8–23 This diffraction grating (5000 grooves per centimeter) breaks up light from a hydrogen discharge tube into four spectral lines.

versity of Chicago, Michelson continued his measurements of the speed of light. In 1899, he began the design of a ruling engine to produce diffraction gratings. Seeking funds, he wrote the president of the university:

The science of spectroscopy [study of the spectral lines of elements—ed.] *has accomplished a number of remarkable feats within the past decade and there appears every reason to expect that an improvement in the essential element of the spectroscope, that is, the "grating," will be followed by a corresponding unfolding of hitherto hidden secrets of Nature's laboratory. It is with the manufacture and improvement of such diffraction gratings that I am chiefly occupied. The process consists in the ruling of exceedingly fine lines (with a diamond point) upon an optically true metal surface—the difficulty to be overcome being the extraordinary accuracy required of a precision screw which I am having constructed in my workshop.*[3]

The screw was the essential element of Michelson's ruling engine, for the screw moves the diamond point the incredibly small distance from one groove to the next. Why do the grooves have to be so closely spaced *and* cover such a large region?

First, let's see what happens when light from a hydrogen discharge tube (a cathode ray tube containing hydrogen) strikes a diffraction grating containing 5000 grooves per centimeter. Figure 8–23 shows the results: The diffraction grating breaks up the light into a spectrum of colors, much like a prism does with one exception. Here, the red light bends more than the violet light, while the reverse is true for a prism. The position of the screen is a measure of the wavelength. Alternatively, the angles, such as θ_B and θ_R, also depend upon the wavelength.

MEASURING THE WAVELENGTH

Because the slits are so close together, the narrow beam of light illuminates about 500 slits of this grating (see Fig. 8–23). A circular diffracted wave emerges from each slit, and, therefore, 500 waves strike *every point* on the screen. Just as in the double-slit experiment, these 500 waves may interfere constructively and destructively. The bright regions in Fig. 8−23 correspond to constructive interference and all other regions correspond to destructive interference.

Consider the red light emerging from four slits of a diffraction grating (see Fig. 8−24a). Because the four slits are so close together and the screen relatively far away, the lines from the center of each slit to the red spot on the screen are parallel. These lines make angle θ_R with the horizontal. Angle θ_R is shown in Fig. 8−23. As before, the length of the line from the center of the slit to a point on the screen is called the **path length.** Constructive interference occurs at angle θ_R because the difference in path lengths between two adjacent slits is equal to 1 wavelength. For example, the difference in path lengths for ray 2 and ray 3 is the distance KL minus the distance AB, which is 1 wavelength.

Figure 8−24(b) shows that destructive interference occurs for red light at angle θ_D. Because the difference in path lengths between two adjacent slits is *not* equal to 1 wavelength, the 500 waves arriving at the screen will have all possible phases. For example, 4 of the 500 waves may have the following height when they strike the screen: $0.8A$, $0.3A$, $-0.3A$, and $-0.8A$. This produces destructive interference. Similar effects will occur for the other 496 waves.

Violet light has a smaller wavelength than red light. Figure 8−24(c) shows that constructive interference for blue light occurs at a smaller angle than for red light: angle θ_B is less than angle θ_R.

Thus, there is a definite correlation between the angle and the wavelength; that is, if you measure the angle, then you can determine the wavelength.

DETERMINING WAVELENGTH FROM
ANGLE MEASUREMENT: AN EXAMPLE

The angle θ_R in Fig. 8−24(a), as well as in Fig. 8−23, is 19.15°, and the grating has 5000 grooves per centimeter. This means that the distance OA

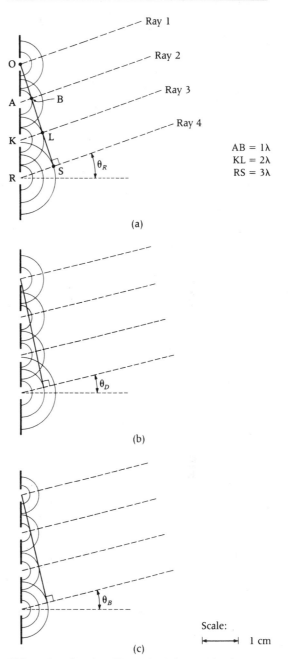

$AB = 1\lambda$
$KL = 2\lambda$
$RS = 3\lambda$

(a)

(b)

(c)

Scale:
|———————| 1 cm

FIG. 8−24 Circular diffracted waves emerging from four slits of a diffraction grating illuminated by light from a hydrogen discharge tube. (a) Constructive interference occurs for red light at angle θ_R. (b) Destructive interference occurs for red light at angle θ_D. (c) Constructive interference occurs for blue light at angle θ_B.

183

is given by

$$OA = \frac{1\,cm}{5000}$$

$$= 2 \times 10^{-4}\,cm$$

$$= 2 \times 10^{-6}\,m \times \frac{1\,nm}{10^{-9}\,m}$$

$$= 2000\,nm$$

From the scale shown in Fig. 8−24(a), the distance OA is 1 cm and represents 2000 nm. The distance AB is about 0.35 cm and represents a distance given by

$$AB = 0.35\,cm \times \frac{2000\,nm}{1\,cm}$$

$$= 700\,nm.$$

Since the distance AB is 1 wavelength, the wavelength of the red light is 700 nm. Considering the errors involved in measuring the small distances, this value compares favorably with the well-known value of 656.2 nm.

From this construction, you can see that it is possible to determine the wavelength if you know the angle for constructive interference and the distance between two adjacent grooves in the grating. It is possible to find a formula for the wavelength in terms of the angle and the distance between two adjacent grooves, and to calculate the wavelength. This example gives the gist of such a calculation.

DETAILS OF THE DIFFRACTION GRATING

There are some important details about the diffraction grating that should be discussed. Why, for instance, must the grooves of a diffraction grating be so closely spaced? Why must they cover such a large area? Michelson, who was awarded a Nobel Prize in 1907 for his precision optical instruments, devoted about 16 years of his life to perfecting diffraction gratings. In light of this degree of invested effort, our questions merit serious consideration.

Why must the grooves be so closely spaced? If we repeat the experiment depicted in Fig. 8−23 with a grating having 10,000 grooves per centimeter (twice the number indicated), and do not change the distance between the screen and the grating, we would find the hydrogen spectrum spread over a larger region on the screen. For example, the separation distance between the two violet lines would be tripled. Our question is thus answered: Reduce the distance between the grooves of the diffraction grating, and the spectral lines become more widely separated. Scientists can therefore obtain and distinguish the wavelengths of two closely spaced spectral lines, which is extremely significant for comparing theory and experiment.

Why must the grooves cover such a large area? Very often, the spectrum is recorded on photographic film. For example, in Fig. 8−23, we could replace the screen with a photographic plate. Studying such a photograph, we would find that a spectral line has a definite width, let's say, 1 mm wide. If another spectral line has a width of 1.5 mm, it must consist of two, or possibly more, closely spaced spectral lines; it is impossible to resolve or separate them because the widths of the lines overlap. We say that the "resolution of the photograph is too poor." The resolution, however, depends upon the *number of grooves* of the grating *illuminated* by the light source. Illuminate a larger number of grooves, and the width of a single spectral line on the photograph *decreases*. In fact, the resolution is directly proportional to the number of grooves illuminated. Illuminate 5 times as many grooves on the grating, and the width of a single spectral line goes from 1 to 0.2 mm. Figure 8−25 shows these effects. Our second question is thus answered: Michelson made diffraction gratings up to 9.4 inches wide to improve the resolution of photographs, so that extremely close lines could be separated. His motivation: again, the comparison between experiment and atomic theories.

The Speed of Light and the Ether

Light certainly has several puzzling features. For example, light from the sun travels to the earth through a vacuum, which has a pressure so low it cannot yet be duplicated in the laboratory. How can light travel through a vacuum—through nothingness? We say that the wavelength of light is the distance between two peaks. But peaks of what? What is forming waves?

Our minds always grope for a model in order to visualize an abstract idea. When scientists considered propagation of light through space, they constructed a model based upon phenomena they already understood. For instance, sound waves travel through air—the vibration of air molecules transmits sound; sound cannot travel through a vacuum. Water waves travel on the surface of the water. Sound from a piano is produced by the vibration of the strings. In short, every type of wave motion that scientists encountered before had a medium in which the wave traveled. So, they concluded, must light. They constructed the following model: Light travels through a medium called the *ether* just as water waves travel on the surface of water. The wave motion is due to the vibration of the ether. The ether had to have unusual properties indeed: It had to pervade all of space, be transparent, and have no mass. Scientists pondered the question, How can we test to see if the ether actually exists?

Albert A. Michelson (1852–1931) accepted this challenge. He came to this area quite naturally since, in 1878, he had started experiments to measure the speed of light. The value he obtained in 1882 was accepted until he made an even more accurate measurement in 1927. To do this, he placed a mirror on Mt. San Antonio in California and a strong light source on Mt. Wilson, 22 miles away. Essentially, he measured the transmission time for the light beam to strike the mirror on Mt. San Antonio and return. Today, the accepted value for the speed of light is 2.9979×10^8 m/sec. We shall use the symbol c for the speed of light, and, in most calculations, we shall use 3×10^8 m/sec.

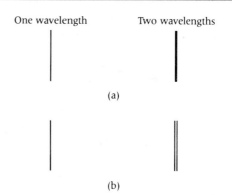

FIG. 8–25   (a) Two spectral lines cannot be resolved, and appear as one broad line. (b) Two spectral lines can be resolved because a larger area of diffraction grating was illuminated.

According to the ether theory, light travels through the ether with speed c. If, however, the ether is traveling toward an observer with speed v, then the observer should measure the speed of light to be c plus v; if the ether travels away from the observer, then c minus v. Because the speed of light is so great, such small variations would be extremely difficult to measure. In 1881, while studying in Europe, Michelson began experiments to detect the ether by attempting to look for these small variations, but his experiments were not successful.

Design of the Michelson-Morley Experiment

In 1885, Michelson, then a professor in the physics department at Case School of Applied Science in Cleveland, Ohio (now Case-Western Reserve University), and Edward W. Morley (1838–1923), then a professor in the chemistry department, joined forces. They made some improvements in Michelson's original design and once again attempted to find the small variations in the speed of light predicted by the ether theory.

They searched for a change in the speed of light as the earth travels around the sun (see Fig. 8–26a). Assume for the moment that the ether is motionless and that the earth moves to the right with speed v. According to observers on earth, the ether flows past them to the *left* with speed v (see Fig. 8–26b). In the Michelson-Morley apparatus, one light beam traveled from point A to C and then returned to A; another beam traveled from A to B and then returned to A. For a moment, let's focus our attention on the beam traveling to and from point B. While traveling from A to B, the light beam would have a speed, relative to the earth, of c minus v. Like a swimmer in a stream, the

light beam has to buck the ether current. But the light beam traveling from B to A would have a speed of c plus v.

The speed v, in this case, is the earth's orbital speed around the sun. To understand what a monumental task Michelson and Morley set for themselves, let's compare v and c. In one year, the earth makes one complete revolution around the sun. The orbital speed, obtained by dividing the circumference (9.39×10^{11} m) by the time (365.25 days, or 3.16×10^7 sec), equals 2.97×10^4 m/sec. Therefore, the ratio v/c is about 0.0001.

Michelson and Morley's experimental apparatus is shown in Fig. 8–27. Mirror 1 is partially silvered, so that part of the light striking it is reflected and part is transmitted. One beam strikes mirror 2 where it is reflected, passes through mirror 1, and enters the telescope. The other beam strikes mirror 3 where it is reflected and then travels back to mirror 1. Here, it is reflected and enters the telescope. These two light beams interfere with each other, producing a pattern, which Michelson and Morley could observe through the telescope. They then rotated the entire apparatus, still looking through the telescope. *They expected to see a changing interference pattern because the light beams should be affected differently by the ether current as the apparatus rotates.* Such a changing interference pattern would demonstrate the existence of the ether. That was their goal: to detect the ether by observing small changes in the speed of light.

We, of course, need to understand how light beams would be affected by the "ether current." To do this, let's first consider the more mundane case of how two swimmers in a stream are affected by the stream's current.

STREAM ANALOGY

Suppose two swimmers, Larry and Brian, can swim only with a constant speed c in still water. They now swim in a stream that has a current v. Larry swims across the stream, while Brian swims parallel to the shoreline (see Fig. 8–28a). Both swim the same distance W. If Larry does not make a correction for the stream current,

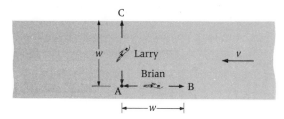

(a) Larry returns to A before Brian does

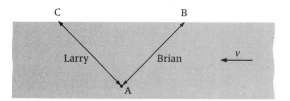

(b) Larry and Brian return to A simultaneously

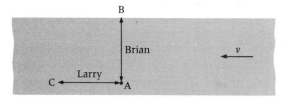

(c) Brian returns to A before Larry does

FIG. 8–28   Larry and Brian start from point A at the same time and swim to point C and point B, respectively. They then return to point A. The speed of the current is v.

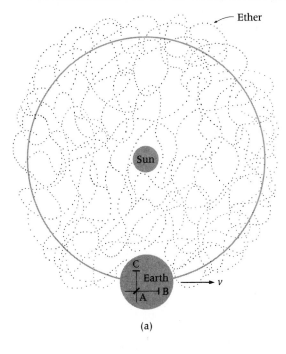

(a)

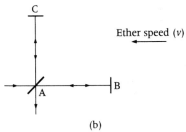

Ether speed (v)

(b)

FIG. 8–26   (a) The earth travels through the motionless ether with speed v. The distance between the sun and earth is 93 million miles, or 1.50×10^{11} m. (b) To observers on earth, the ether travels to the left with speed v.

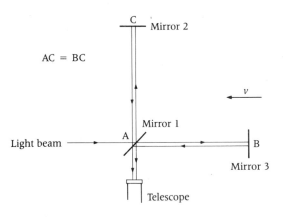

AC = BC

FIG. 8–27   The Michelson-Morley experiment.

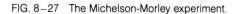

187

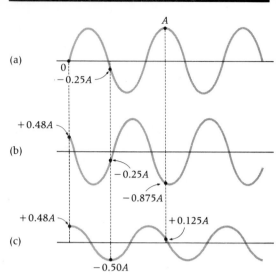

(a)

0
−0.25A

(b)

+0.48A

−0.25A
−0.875A

(c)

+0.48A
+0.125A

−0.50A

FIG. 8–29 Interference pattern (c) produced by the addition of two waves, (a) and (b). The waves in (a) and (b) have the same amplitude A, but are not in phase. The result of adding these two waves produces a wave of amplitude 0.5A. The dashed lines show three examples of this addition. For example, when one wave contributes A and the other −0.875A, the resultant wave has a height of +0.125A.

he will not arrive at point C, but will be pushed downstream. Therefore, Larry has to aim upstream, as shown in Fig. 8−28(a), so that he will arrive at point C.

If Larry and Brian start out at the same time, they obviously will not return to their starting point at A simultaneously, because each is affected differently by the current. Only when the stream current drops to zero will they return together. No matter what the (nonzero) stream current is, the swimmer (here Larry in Fig. 8−28a) who swims across the stream will return first. The time interval between their return to point A increases when the stream current speed increases.

In Fig. 8−28(b), Brian has to buck the current when he swims from point A to point B; Larry has to buck the current in exactly the same way when he swims from point C to point A. Each is affected in the same way for the other half of the trip. Therefore, Larry and Brian return to point A simultaneously.

In Fig. 8−28(c), the situation is identical to that in Fig. 8−28(a), except that the roles of Larry and Brian are interchanged. Here, Brian will return to point A before Larry.

ANTICIPATED RESULTS

From the stream analogy, we know that two swimmers arrive back at the starting point at different times because the current affects them differently. Similarly, the light beams, analogous to the swimmers, will enter the telescope at different times. While Michelson and Morley could not measure such a short time interval directly, they could observe the interference of the two beams. Given the ether model, they expected that the interference pattern would change as they rotated their apparatus through 90°. Let's see why.

When the two light beams start out from point A in Fig. 8−27, they are identical because they originated from a single beam. Now, if the two beams return to point A simultaneously, then the beams will arrive at the telescope in phase (that is, peak on peak, and trough on trough).

There will be constructive interference. In this case, the light observed in the telescope will be of maximum brightness.

If, however, the two light beams do *not* arrive at the telescope at the same time, then the light beams will not be in phase, and they will not interfere constructively. Most likely, the interference will be somewhat between the extremes of constructive and destructive interference shown in Fig. 8−10 (p. 172). We see an example of this in Fig. 8−29. Here, two light beams, each one having amplitude A, interfere so that the resultant beam has amplitude $0.5A$. Since the brightness (or intensity) of the light is related to the amplitude of the (resultant) wave, the brightness will be less than maximum.

Suppose that the apparatus is initially oriented as shown in Fig. 8−27, and is then rotated in a counterclockwise direction about point A by 90°. The initial position corresponds to the stream analogy in Fig. 8−28(a); the 45° rotation corresponds to Fig. 8−28(b); and the 90° rotation corresponds to Fig. 8−28(c). In Fig. 8−28(b), the swimmers return at the same time. Since the light beams are supposedly affected by the ether current in the same way as the swimmers are affected by the stream current, the light beams return at the same time. As we discussed, this means that the light will be of maximum brightness at 45°, and the light will not be as bright at 0° and 90°.

If the ether model were correct, then Michelson and Morley expected that the brightness of the light, viewed through the telescope, would vary as they rotated the apparatus through 90°. The light should be brightest at 45°. Such observations would then confirm the ether theory.

Einstein's Postulates and the Results of the Michelson-Morley Experiment

When Michelson and Morley carried out their experiment, they were astonished to find that *the brightness did not vary at all* as they rotated the apparatus through 90°. They repeated the experiment many times—during the day and night, and during all seasons of the year—in order to discount any peculiar motion of the ether. (Recall from our disccision of Fig. 8−26 that we assumed the ether to be motionless.) Their results were always the same: no variation in the brightness during the 90° rotation.

Michelson concluded that both of his "etherdrift experiments had been miserable failures. The ether was out there, but, at present, his instruments were unable to detect its presence."[4]

Several scientists attempted to explain these null results, but never achieved a basic understanding. In 1905, Albert Einstein published his postulates of relativity and discarded the familiar and comfortable concept of the ether: The speed of light is c, and is the same for *all* observers.

Einstein's postulates permit an understanding of the Michelson-Morley experiment: Either there is no ether or the speed of the ether current is zero ($v = 0$). Hence, each light beam always has a constant speed c. Since each beam travels the same distance with the same speed, both beams take the same amount of time for the round trip; thus, they arrive at the telescope in phase. Since the speed of light *does not* vary as the rotation takes place, there is no corresponding variation in the brightness during rotation.

Chapter 9 examines the far-reaching consequences of such a seemingly simple statement: The speed of light has the same value, 3×10^8 m/sec, for *all* observers. However, answers were still required for some important questions, such as, How *does* light travel through space since "ether" is a useless trapping? or, What is the height of a light wave?

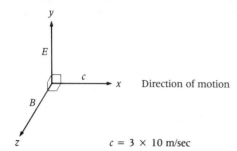

FIG. 8–30   The directions of the electric field and magnetic field, and the direction of motion of an electromagnetic wave.

$c = 3 \times 10$ m/sec

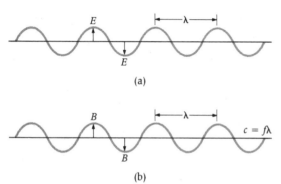

(a)

$c = f\lambda$

(b)

FIG. 8–31   (a) Oscillation of the electric field of an electromagnetic wave. (b) Oscillation of the magnetic field of an electromagnetic wave.

Electromagnetic Waves

Today, we know that light waves are **electromagnetic waves** that possess energy, but no mass. An electromagnetic wave consists of an electric field and a magnetic field that travel through space with speed c. The direction of motion, the electric field, and the magnetic field are mutually perpendicular, as shown in Fig. 8–30. The strength and direction of the electric and magnetic fields oscillate, as shown in Fig. 8–31. The height of the wave at any point indicates the strength of the electric (or magnetic) field. The two arrows in Fig. 8–31(a) show the maximum electric field and, similarly, the maximum magnetic field for Fig. 8–31(b).

What effect does an electromagnetic wave have upon an electron? Electromagnetic waves are to water waves as an electron is to a floating cork (see Fig. 8–3, p. 170). The electron "bobs" up and down much like the cork. As the wave moves to the right, the whole waveform moves to the right just as in the case of water waves. We have learned that the force on a charged particle Q due to an electric field E is F equals QE (Chapter 7). Since the charge on the electron is negative, the direction of the force is opposite to the direction of the electric field. When the electric field passing by the electron is directed upward, the force on the electron is downward. Similarly, the north or south pole of a magnet would be affected by the magnetic field of the electromagnetic wave and would vibrate back and forth as the wave passes by.

Radio waves are electromagnetic waves. The electrons in a radio antenna oscillate as the radio waves are received. This produces a current, which is amplified, and ultimately produces sound from the radio. The various kinds of radiation listed in Table 8–1 (p. 181) are all electromagnetic waves. We simply give names to the various ranges of the wavelength.

Summary

A wave is described by its wavelength (the distance between two peaks) and its amplitude (its maximum height or depth). The number of waves per second passing a given point is called its frequency. The speed of propagation of a wave is given by frequency multiplied by the wavelength ($v = f\lambda$).

Wave motions of all kinds exhibit diffraction and interference effects. Diffraction is the ability of a wave to bend around obstacles or to spread out when passing through openings. Diffraction is most pronounced when the wavelength and obstacle are of comparable size. Two waves interfere constructively when the waves are in phase: the peaks from each wave arrive at the same point. Two waves interfere destructively when waves are out of phase: a peak from one wave and a trough from the other wave arrive at the same point.

Thomas Young demonstrated the wave nature of light. He observed the interference of two circular waves on a screen. The light and dark regions were due to the constructive and destructive interference of the light waves.

Light waves consist of electric and magnetic fields propagating through space. The amplitude (height of peaks) refers to the maximum strength of the electric and magnetic fields. The wavelength is the distance between two successive crests or two successive troughs.

A prism breaks up light into a spectrum of colors. This phenomenon is due to the difference in the speed of light in air and in glass.

A diffraction grating is much more effective than a prism in spreading the spectrum over a large region on a screen. With this instrument, the wavelength of light can very accurately be determined from the angle through which the beam is diffracted and the distance between two grooves in the diffraction grating.

According to the ether model for light propagation, light travels with speed c relative to the ether. If the ether moves with speed v relative to the earth, then an observer on earth would measure the speed of light to be slightly different: no more than c plus v, and no less than c minus v.

In the Michelson-Morley experiment two perpendicular light beams travel the same distance and return to their starting point. In their initial position (0°), the light beams do not return at the same time because they are affected differently by the ether drift. These light beams interfere, but the resulting brightness is not maximum. As the apparatus is rotated, the interference changes. At 45°, Michelson and Morley expected maximum brightness because both beams are affected in the same way by the ether drift and they return at the same time. At 90°, the brightness should be the same as at 0°.

Michelson and Morley found no change in brightness whatsoever, even after constantly repeating their experiment at different times and in different seasons. In 1905, Einstein postulated that all observers will measure the same value for the speed of light—the ether concept is useless trapping. If the light beams always have the same speed, then they will always arrive back at the starting point at the same time, and there will be no change in the brightness as the apparatus is rotated. The Michelson-Morley experiment was not a "miserable failure," but a confirmation of Einstein's postulate.

True or False Questions

Indicate whether the following statements are true or false. Change all of the false statements so that they read correctly.

8–1 Two water waves travel at the same speed, but wave 1 has a wavelength twice that of wave 2.
    a. The frequency of wave 1 is twice that of wave 2.
    b. A cork set into motion by wave 1 will make the same number of oscillations in 1 second as another cork set into motion by wave 2 because the waves travel with the same speed.

8–2 Plane water waves of 1-cm wavelength move toward a barrier with a hole in it.

WAVES, SOUND, AND LIGHT

a. If the hole is 1 cm wide, it stands to reason that plane waves of 1-cm wavelength will appear on the other side of the barrier.

b. Similarly, if the hole is 15 cm wide, plane waves of 1-cm wavelength will appear on the other side of the barrier.

8–3 In a ripple tank, two vibrating dippers, placed 2 cm apart, produce a pattern on the water: lines appear to radiate in all directions from the dippers.

a. These lines are due to the constructive interference of the circular waves produced by the dippers.

b. Constructive interference occurs at a point where the peak of one wave coincides with the peak of the other wave.

c. When a cork is placed equidistant from each dipper, the cork does not vibrate, showing that destructive interference of the waves occurs at this point.

8–4 Sound and water waves are two examples of transverse wave motion.

8–5 A vibrating loudspeaker sets air molecules into motion. These struck air molecules then travel to our ears, creating a pressure, which in turn produces the sensation of sound.

8–6 The driver of an automobile sounds the horn while approaching a pedestrian. The pitch of the horn sounds higher to the pedestrian than when the car is at rest.

8–7 a. In 1801, Thomas Young illuminated two slits with sunlight and observed light and dark regions on a screen behind the slits.

b. Young concluded that light has a wave nature by realizing that the light and dark regions were due to diffraction of the light waves.

8–8 Light illuminates two slits, producing a pattern on the film placed behind the slits.

a. When red light is used, the distance between dark lines on the film is larger than that obtained when blue light is used.

b. By determining the distance between two adjacent dark lines, one can determine the wavelength of the light.

8–9 A beam of light strikes a diffraction grating and its spectrum appears on a screen behind the grating.

a. When the diffraction grating is replaced with a prism, the spectrum covers a larger region on the screen.

8–10 The Michelson-Morley experiment was designed to detect

a. the electric and magnetic fields of light waves traveling through space.

b. exceedingly small variations in the speed of light by light beams traveling in different directions.

c. an interference pattern that did not change when the apparatus was rotated through 90°.

Questions for Thought

8–1 Define the following terms: *wavelength, frequency,* and *amplitude.* What are the units of each one?

8–2 Think of some events that occur frequently in your everyday life. What is the frequency of these events, and what are the units? Compare the units with the frequency of waves.

8–3 What is the relationship between wavelength, frequency, and the speed of waves? Show that the units are consistent.

8–4 The water waves move toward the right, as shown in Fig. 8–32. The cork is

a. moving upward.

b. moving upward and to the right.

c. moving downward.

d. moving downward and to the left.

8–5 Define the following terms: *diffraction, constructive interference, destructive interference.*

8−6 In order to study the diffraction of water waves, we place two pieces of wood separated by 5 cm in a ripple tank. Plane water waves strike the opening. Make a sketch of the water waves emerging from the opening if the wavelength is 2 cm; if the wavelength is 0.5 cm.

8−7 Light, striking a thin film of foil, is reflected from the top and bottom surfaces. Explain how a spectrum of colors results from interference effects.

8−8 Bring your thumb and forefinger together and look through this small opening toward a bright light. Describe what you see before they touch. Explain what you see by considering two beams of light: one diffracted by your thumb and one diffracted by your forefinger.

8−9 Suppose that our eyes were sensitive to light of wavelength 1 mm. How would this affect our ability to see around corners? What would happen if you tried to thread a needle?

8−10 Two children pull on a Slinky. Each child jerks on an end, producing the pulses shown in Fig. 8−33. What happens when these pulses meet?

8−11 You do not find the same effects as in calculation question 8−6 when you listen to your stereo. Why not?

8−12 Define the terms *transverse wave motion* and *longitudinal wave motion*. Are light waves longitudinal or transverse? Are sound waves longitudinal or transverse?

8−13 Briefly describe Young's double-slit experiment.

8−14 Describe how the wavelength of light can be found from a double-slit experiment.

8−15 Suppose that the speed of light in glass were the same as that in air for all colors of light. Could a prism be used to obtain a spectrum of colors?

8−16 Why was the development of the diffraction grating so essential for investigations of the atom?

8−17 a. Explain why the ether, a medium responsible for the propagation of light, was introduced.

FIG. 8−32   Diagram for thought question 8−4.

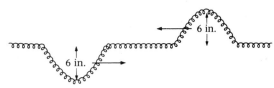

FIG. 8−33   Diagram for thought questions 8−10.

b. According to the ether theory, should the speed of light be the same for all observers? What did the Michelson-Morley experiment show?
c. What is the nature of a light wave? What does the amplitude of a light wave refer to?

8−18 A swimmer swims with a speed *c* relative to the water, and the water in the stream travels with speed *v* relative to the bank. Show that the swimmer's maximum speed relative to the bank is *c* plus *v* and the minimum speed is *c* minus *v*. When does the swimmer have these values?

8−19 a. Michelson and Morley looked for a (insert *10 percent, 1 percent, 0.1 percent,* or *0.01 percent*) _____ change in the speed of light.
b. In essence, they observed the transit times for two light beams. Explain how they made this observation.
c. What results did they expect to find? Did the experimental results agree with their expectations?
d. Explain how Einstein's postulates of relativity can be used to interpret the null results of the Michelson-Morley experiment.

WAVES, SOUND, AND LIGHT

Questions for Calculation

†8–1 **a.** In a ripple-tank experiment, a dipper vibrates 8 times per second, and the wavelength of the circular waves is 4 cm (0.04 m). What is the speed of propagation of the waves?

b. If the dipper vibrates 4 times per second, what is the wavelength of the waves?

c. Place a cork in the ripple tank. With what frequency does it bob up and down in **a?** in **b?**

8–2 A visible light ray has a wavelength of 500 nm; an X-ray, 0.1 nm; and a gamma ray, 2×10^{-3} nm (1 nm = 10^{-9} m). The speed of light is 3×10^{8} m/sec. Find the frequency of each ray.

8–3 Middle C on the piano has a frequency of 262 osc/sec. The speed of sound is 330 m/sec. Find the wavelength of the sound waves.

8–4 In a ripple-tank experiment, two dippers produce waves having a wavelength of 2 cm. The dippers are separated by 4 cm.

a. How do the waves affect a cork 10 cm from one dipper and 12 cm from the other dipper?

b. How do the waves affect a cork 10 cm from one dipper and 10 cm from the other?

c. How do the waves affect a cork 10 cm from one dipper and 11 cm from the other?

d. How do the waves affect a cork 10 cm from one dipper and 10.5 cm from the other?

8–5 Repeat calculation question 8–4, except that the water waves have a wavelength of 1 cm.

†8–6 A sound generator is connected to two loudspeakers separated by 3 m. The speakers produce sound waves having a wavelength of 1 m.

a. If you stand 4 m from one speaker and 5 m from the other, do the waves interfere

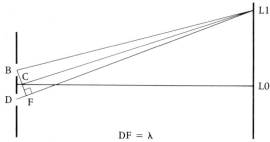

FIG. 8–34 Diagram for calculation question 8–7.

constructively (an intense sound) or destructively (no sound)?

b. If you stand halfway between the speakers, what do you hear?

c. If you stand 4 m from one speaker and 4.5 m from the other, what do you hear?

8–7 **a.** Measure the lengths in Fig. 8–34 and show that

$$\frac{DF}{BD} = \frac{\text{distance between L0 and L1}}{\text{distance between C and L1}}.$$

Usually, the distance between C and L1 is nearly the same as the distance between the slits and the screen W. Using $d = BD$ and $\lambda = DF$, show that the wavelength is given by

$$\lambda = \frac{d}{W} \times \text{distance between two bright spots.}$$

b. Red light from a helium-neon laser illuminates two parallel slits 0.1 mm apart when the screen is 1 m away from the double slit. The first red spot is 6.3 mm away from the central red spot. What is the wavelength of the red light? Compare this value with the known value of 632.8 nm.

8–8 Figure 8–24(a) shows red light emitted by hydrogen passing through a diffraction grating where adjacent grooves are separated by 2000 nm. The wavelength was determined (p. 183) by measuring the distance AB to scale. Draw a larger diagram to scale and find a more accurate value for the wavelength. Draw the distance OA as 10 cm. Use a protractor to draw angle θ_R as 19.2°. Measure the distance AB to scale, determine the wavelength, and compare it with the accepted value of 656.2 nm.

8–9 Repeat calculation question 8–8 for blue light from hydrogen. The angle θ_B in Figs. 8–23 and 8–24(c) is 14.1°. Compare the wavelength with the accepted value of 486.1 nm.

8–10 Red light emitted by hydrogen passes through a diffraction grating where adjacent grooves are separated by 1000 nm. Using the known wavelength as 656.2 nm, draw a diagram to scale and use a protractor to find the angle through which the red light is diffracted. Let the distance between grooves be represented by a distance of 10 cm in the diagram. Compare the angle obtained in the diagram with an experimental value of 41.0°.

†8–11 Suppose that the speed of light were 1000 mph and you saw a car approaching you traveling at 100 mph. If you measured the speed of light from the headlights, what value might you expect to find? What would be the value obtained in the experiment?

Footnotes

[1] Morris H. Shamos, *Great Experiments in Physics*, (New York: Henry Holt and Company, Inc., 1959), p. 95. Quotation reference: John Davy, *Life of Sir Humphrey Davy* (London, 1839). (From *Great Experiments in Physics* edited by Morris H. Shamos. Copyright © 1959 by Holt, Rinehart and Winston, Inc. Reprinted by permission of Holt, Rinehart and Winston.)

[2] Ibid; p. 103. From *Great Experiments in Physics* edited by Morris H. Shamos. Copyright © 1959 by Holt, Rinehart and Winston, Inc. Reprinted by permission of Holt, Rinehart and Winston.

[3] Bernard Jaffe, *Michelson and the Speed of Light*, (Garden City, N.Y.: Anchor Books, 1960), p. 131. (Excerpt from *Michelson and the Speed of Light* by Bernard Jaffe. Copyright © 1958 by Doubleday & Company, Inc. Reprinted by permission of the publisher.)

[4] Dorothy Michelson Livingston, *The Master of Light*, (New York: Charles Scribner's Sons, 1973), p. 132.

Introduction

Albert Einstein's postulates of relativity have completely revolutionized our concepts of time, length, and mass. We were introduced to his second postulate in Chapter 8, where we learned that Einstein discarded the concept of the ether, the medium through which light supposedly traveled, and proposed that the speed of light is the same for *all* observers. In this chapter, we shall look in detail at the startling consequences of Einstein's postulates. To do this, we shall consider a familiar example to see what happens at low speeds and then, at speeds near that of light. We shall see that Einstein's second postulate alters our concept of time and length, while the first postulate shows that mass and energy are interchangeable.

CHAPTER 9
Relativity

Einstein's Postulates of Relativity

One might expect that Einstein introduced his theory of relativity to explain the null results of the Michelson-Morley experiment. Such was not the case. Einstein, in fact, proposed his theory to solve a completely different problem.

To put Einstein in perspective, we must first examine Newton's laws comparably applied in two different locations: (1) at rest relative to the earth and (2) on a moving train. If an experimenter in each location throws a ball vertically into the air with an initial speed of 4 m/sec, the ball will reach a height of 0.82 m and then return to the experimenter's hand after 0.41 sec, as long as the train moves with constant speed in a straight line. The experimenter on the train records exactly the same results as the stationary counterpart. Only when the train accelerates will the train experimenter note any unusual effects. If the train speeds up while the ball is in the air, the ball will not return to the experimenter's hand, but instead will appear to move slightly to the rear of the car.

We can easily imagine what happens to a cup of hot coffee served in the dining car when the train rounds a curve or speeds up quickly. There is no difficulty with the cup of coffee, however, when the train travels with constant speed in a straight line (that is, with constant velocity).

From these examples (and you can think of many others), it appears that Newton's laws are valid for both the stationary and the moving observer, as long as the train travels with constant velocity.

Similarly, Einstein realized that *all* of the laws of physics should be valid for two observers moving relative to each other with constant velocity. Thus, his **first postulate** states:

The laws of physics are the same for any two observers moving relative to each other with a constant velocity.

Einstein looked at the laws governing electricity. He found that these laws could be valid for two observers moving relative to each other with constant velocity *only if* the speed of light is the same for both. This led to his **second postulate,** which states:

The velocity of light (in a vacuum) is the same for all observers and does not depend upon the motion of the light source or upon the motion of the observer.

Let's look at the consequences of the second postulate now, beginning with the notion of relative speed. We'll examine this concept with a low-speed example and then see how a similar example at high speeds is affected by the second postulate.

FIG. 9–1 Albert Einstein. (Courtesy of National Archives, USA)

Relative Speed

EXAMPLE A: LOW SPEED

Refer to Fig. 9–2. Marie, on the station platform, watches Frank on the train as it travels east at 1 m/sec.

We can easily figure out the relative speed between Frank and Marie by using "common-sense" notions. For example, if Frank walks east on the train at 3 m/sec, and Marie is at rest in the station, then Frank moves away from Marie

RELATIVITY

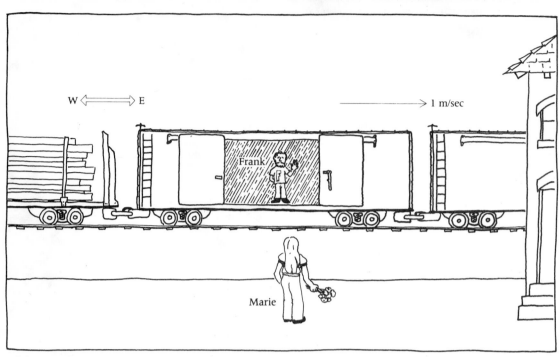

FIG. 9–2   The relative speed between Frank and Marie depends upon the speed of the train, Frank's speed relative to the train, Marie's speed along the station platform, and the directions in which Frank and Marie travel.

TABLE 9–1   RELATIVE SPEED BETWEEN FRANK AND MARIE

CASE	MARIE'S MOTION IN STATION (m/sec)	FRANK'S MOTION RELATIVE TO TRAIN (m/sec)	RELATIVE SPEED BETWEEN FRANK AND MARIE (m/sec)
1	0	3 east	$1 + 3 = 4$
2	0	3 west	$3 - 1 = 2$
3	2 east	3 east	$1 + 3 - 2 = 2$
4	2 west	3 east	$1 + 3 + 2 = 6$
5	2 east	3 west	$3 - 1 + 2 = 4$

at the rate of 4 m/sec. Frank's motion (relative to Marie) is enhanced by the motion of the train in the same direction. If, however, Frank walks at 3 m/sec to the west, then Frank moves toward Marie at the rate of 2 m/sec. Frank's motion is reduced by the motion of the train in the opposite direction. Table 9–1 presents these two cases, plus several more.

EXAMPLE B: THE SPEED OF LIGHT

A light source on the train sends out beams in two directions with speed c. The train moves at the rate of $c/3$ to the east, and Marie has the unique ability to travel at the rate of $c/2$ (see Fig.

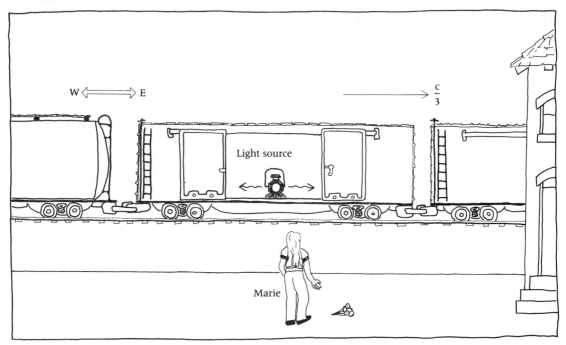

FIG. 9−3 We might expect that the speed of light measured by Marie would depend upon the speed of the train and Marie's speed. However, Marie always measures the same speed c.

TABLE 9−2 SPEED OF LIGHT MEASURED BY MARIE

CASE	MARIE'S MOTION IN STATION	DIRECTION OF LIGHT	INCORRECT RELATIVE SPEED PREDICTIONS	SPEED OF LIGHT MEASURED BY MARIE
1	0	c east	$0.33c + c = 1.33c$	c
2	0	c west	$c - 0.33c = 0.67c$	c
3	0.5c east	c east	$0.33c + c - 0.5c = 0.83c$	c
4	0.5c west	c east	$0.33c + c + 0.5c = 1.83c$	c
5	0.5c east	c west	$c - 0.33c + 0.5c = 1.17c$	c

9−3). (Einstein frequently used thought "experiments" such as this to illustrate his ideas.)

If Marie measures the speed of the two light beams, what value or values does she obtain? Perhaps you are inclined to say that these measurements can be predicted using the concept of relative speed, which does seem to be a reasonable approach. The light source is on the train and the train moves. Therefore, you might expect that Marie will observe different speeds

for the two beams. However, Einstein's second postulate states that the speed of light is the *same* for *all* observers, and does not depend upon the motion of the source or of the observer. Thus, for both beams, Marie always measures the speed

RELATIVITY

TABLE 9–3 A COMPARISON OF RELATIVE SPEEDS WITH THOSE PREDICTED BY THE R POSTULATE		
	ACTUAL RELATIVE SPEED (m/sec)	RELATIVE SPEED PREDICTED BY THE R POSTULATE (m/sec)
L–B	3–2 = 1	3
L–S	3 + 2 = 5	3
M–B	3 – 2 = 1	3
M–S	3 + 2 = 5	3

c. Therefore, what you expect based upon commonsense notions does not agree with the experiment, as Table 9–2 shows. The column referring to the relative speed predictions has been crossed out to emphasize its incorrectness. The last column shows the speed of light actually measured by Marie.

In Example A, the relative speed between Marie and Frank does depend upon the motion of the train and also upon Marie's motion in the station. Contrast this with Example B. Here, the speed of light relative to Marie does *not* depend upon the motion of the train, or upon Marie's motion in the station. Thus, what we expect based upon concepts of relative speed does not happen in Example B.

Our next step will be to demonstrate that the disparity between Example A and Example B results in a reformulation in the concepts of time, distance, and mass. Again, these new effects will be illustrated by contrasting a familiar situation with an unfamiliar one.

Observing Two Events

EXAMPLE C: LOW SPEED

Brian and Susan are on the train and move as shown in Fig. 9–4, while Larry and Margaret are in the station, watching them. Since Brian and Susan are 12 m from the ends of the car, it

will take them 4 seconds to reach the opposite ends. During this time, the train will travel 8 m. Therefore, after 4 seconds, Larry will see Brian directly in front of him, and Margaret will see Susan directly in front of her. Brian and Susan, noting the times on their watches, agree that it took them 4 seconds to arrive at the ends of the car. Any other observer standing on the train will observe the same time.

Do Larry and Margaret agree that it took Brian and Susan 4 seconds? You undoubtedly will answer, "Yes, the time is the same for everyone, for all observers." But let's investigate this a little more.

Table 9–3 shows the relative speed between two observers. Initially, Brian is 4 m from Larry and moves toward him at 1 m/sec. Therefore, Larry finds that Brian takes 4 seconds to reach him, since

$$t = \frac{d}{v}$$

$$= \frac{4\,\text{m}}{1\,\text{m/sec}} \tag{1}$$

$$= 4\,\text{sec}.$$

Initially, Brian is 20 m from Margaret and moves away from her at 1 m/sec. When Brian reaches the end of the car, he is 24 m from her. Therefore, Brian has traveled a distance of 4 m. Margaret finds that it takes Brian 4 seconds to reach the end of the car, since

$$t = \frac{d}{v}$$

$$= \frac{4\,\text{m}}{1\,\text{m/sec}} \tag{2}$$

$$= 4\,\text{sec}.$$

Initially, Susan is 20 m from Margaret and moves toward her at the rate of 5 m/sec. Therefore, Margaret sees that it takes Susan 4 seconds to reach her, since

$$t = \frac{d}{v}$$

$$= \frac{20\,\text{m}}{5\,\text{m/sec}} \tag{3}$$

$$= 4\,\text{sec}.$$

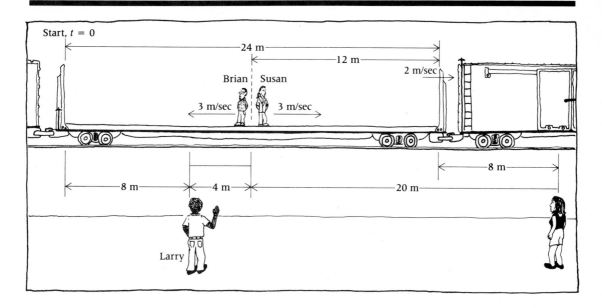

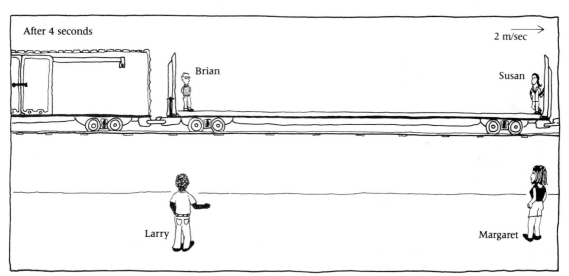

FIG. 9-4 Brian and Susan find that it takes them 4 seconds to reach the ends of the car. According to commonsense notions, the relative speed between Brian and Larry (or Margaret) is 1 m/sec, and between Susan and Margaret (or Larry) the relative speed is 5 m/sec. Then, Larry and Margaret also find that Brian and Susan take 4 seconds to reach the ends. If, however, the speeds of Brian and Susan are always 3 m/sec (R postulate, see Example D, p. 202), Brian will reach his end first.

Initially, Susan is 4 m from Larry and moves away from him at the rate of 5 m/sec. When Susan reaches the end of the car, she is 24 m from Larry and has traveled 20 m. Larry sees that Susan takes 4 seconds to reach the end of the car, since

$$t = \frac{d}{v}$$

$$= \frac{20\,\text{m}}{5\,\text{m/sec}} \qquad (4)$$

$$= 4\,\text{sec}.$$

These results show that all observers—observers in the station as well as observers on the train—agree: Brian and Susan reached opposite ends of the train car *simultaneously* after 4 seconds. These results agree with what you expect, and you are not greatly surprised.

EXAMPLE D: A MODIFICATION

Now, suppose we consider a ridiculous postulate (hereafter called the R postulate), which states: Brian's and Susan's speeds are 3 m/sec for all observers.

You will certainly argue: That sure is a ridiculous postulate! Someone standing on the train in Fig. 9−4 will indeed see Brian and Susan walk at 3 m/sec. That is obvious. But, observers (Larry and Margaret) in the station, watching Brian and Susan move on the train (which has a speed of 2 ft/sec), will *not* observe their speed as 3 m/sec. All of these objections are certainly valid. The R postulate is ridiculous and incorrect. Even so, let us *accept* the R postulate and see what consequences it does have. (You note, of course, that the wording of the R postulate is very similar to Einstein's second postulate.) Table 9−3 compares the actual relative speeds with those predicted by the R postulate.

Brian and Susan have a speed of 3 m/sec on the train and travel a distance of 12 m. Therefore, according to their watches, they still require 4 seconds to travel to the ends of the train car. Brian and Susan, as well as any other observers on the train, agree that they reach the ends *simultaneously,* and that it requires 4 seconds.

Do Larry and Margaret agree that it took Brian and Susan 4 seconds? In order to answer this question, review Equations 1 through 4. According to the R postulate and Table 9−3, the speed v in the denominators of these equations should always be 3 m/sec. Therefore, Larry and Margaret will find that Brian took 1.33 seconds to reach Larry (and the end of the car). Likewise, Larry and Margaret will find that Susan took 6.67 seconds to reach Margaret (and the end of the car). *They do not arrive simultaneously!* There are some strong disagreements here. Brian and Susan agree that it took them 4 seconds and they arrive simultaneously. The important conclusion of this example is:

As a result of the R postulate, times are recorded differently by observers in the station and by those on the train. Events simultaneous to observers on the train are *not* simultaneous to observers in the station.

We have observed the similarity between the R postulate and Einstein's second postulate. The motion of Brian and Susan (speed = 3 m/sec) is analogous to the motion of a light beam (speed = c). We have seen the astounding results of the R postulate. Let us now extend these ideas to the behavior of light by applying Einstein's second postulate.

Suppose that a light source is located at the center of a train car moving at the speed $0.67c$ (analogous to 2 m/sec). The light source is turned rapidly on and off, sending two pulses of light toward opposite ends of the train car. Observers on the train will agree that both pulses of light reach the ends simultaneously. The central question is, Do observers in the station agree that the pulses arrive simultaneously? We have just discussed the R postulate. Therefore, extending the analogy, we would have to conclude that observers in the station do not see

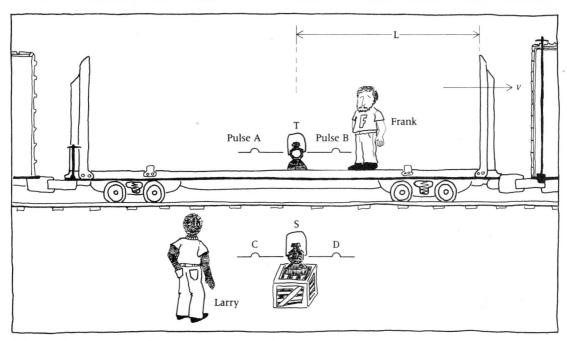

FIG. 9−5   Pulses of light, emitted by adjacent sources S and T, travel toward the ends of the train car.

the pulses arrive simultaneously, and that the pulse traveling west arrived first. Let's consider the following example to further illustrate this result.

EXAMPLE E: THE BEHAVIOR OF LIGHT

Consider two light sources: one located at the center of the train car at point T, and the other located in the station at point S (Fig. 9−5). At the instant that the two light sources are adjacent, pulses of light are emitted from each source. Frank is standing at rest on the train, and Larry is standing in the station. How do the two observers, Larry and Frank, view the four light pulses?

According to Frank, all of the pulses have speed c. After a time t, measured by Frank, all of the pulses are the same distance *from point T.* Therefore, pulse A is adjacent to pulse C, and pulse B is adjacent to pulse D. Pulses A and B

travel the same distance and have the same speed. Therefore, pulses A and B arrive at opposite ends simultaneously.

The same type of arguments can be made for Larry. Larry, however, observes the distances traveled from point S, since point S is fixed for him. Larry will also conclude that pulse A is adjacent to pulse C, and pulse B is adjacent to pulse D.

Thus, one important conclusion is that both Frank and Larry agree that pulse A is adjacent to pulse C, and pulse B is adjacent to pulse D. Frank's observation of the pulses is shown in Fig. 9−6 and Larry's observation is shown in Fig. 9−7.

RELATIVITY

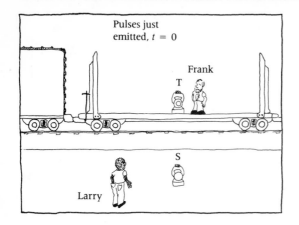

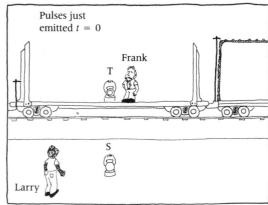

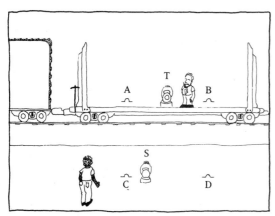

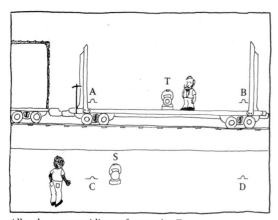

All pulses are equidistant from point T

FIG. 9-6 Frank's observation.

All pulses are equidistant from point S

FIG. 9-7 Larry's observation. In the bottom diagram, pulses A and C (located to the left of source S) are the same distance from S as pulses B and D. Since this distance exceeds the boundary of the figure, pulses A and C are not shown.

According to Larry, the left end of the car approaches pulse C, while the right end of the car moves away from pulse D. Therefore, pulse C reaches one end of the car before pulse D reaches the other end. Since pulse C is adjacent to pulse A, and pulse D is adjacent to pulse B, Larry observes that pulse A reaches the end of the car before pulse B reaches the other end of the car.

You may indeed be tempted to ask, Do the pulses A and B arrive simultaneously or not? Which is correct? We have now arrived at the central point of this discussion: *Both are correct.* Frank makes one observation about the two events and Larry makes another. They do not agree about their observations. Both, however, are correct. Frank's observation is correct and Larry's observation is also correct, even though they do not agree with each other.

If you were standing next to Larry, you would agree with his observation. If you were standing next to Frank, you would agree with Frank's observation. However, you cannot be both places at the same time! Or, you cannot make both sets of observations at the same time. Therefore, there is no contradiction between the two observations.

The conclusions are inescapable. Frank observes two events (arrival of the light beams) as simultaneous, while Larry does not observe simultaneity. The speed v of the train is the relative speed between Larry and Frank. The conclusions can be stated in a general way:

Accepting Einstein's second postulate makes it necessary to alter our concept of time. Two events that occurred simultaneously to one observer will not be simultaneous to another observer, if these two observers move relative to each other with a speed v in a straight line. Time is measured differently by the two observers.

To demonstrate further how our usual concept of time is altered by Einstein's second postulate, let's consider a couple of additional cases. Example F will illustrate the low-speed case, where familiar results are obtained. Example G will illustrate the effects that occur when speeds are comparable to the speed of light.

Time Dilation

EXAMPLE F: LOW SPEED

Susan walks directly *across* the train car from point A to point B, and back to point A, at the rate of 3 m/sec. Figure 9−8 shows a *top* view. Margaret, standing on the train, clocks Susan's round-trip time as 3 seconds. The total distance is 9 m and the speed is 3 m/sec:

$$t = \frac{d}{v}$$

$$= \frac{9\,\text{m}}{3\,\text{m/sec}} \tag{5}$$

$$= 3\,\text{sec.}$$

Lois is standing in the station and also observes the time. What time does she find? You undoubtedly answer, "At low speeds the time is the same for everyone and, so, Lois will also find a time of 3 seconds." Your answer is certainly correct. Lois sees that Susan moves along the path shown in Fig. 9−9, where the distance is greater than 9 m and Susan's speed is greater than 3 m/sec. However, when the distance is divided by the speed, the time is still 3 seconds:

$$\frac{\text{Time observed}}{\text{by Lois}} = \frac{\text{distance observed by Lois}}{\text{speed observed by Lois}}$$

$$= \frac{\text{greater than 9 m}}{\text{greater than 3 m/sec}}$$

$$= 3\,\text{sec.}$$

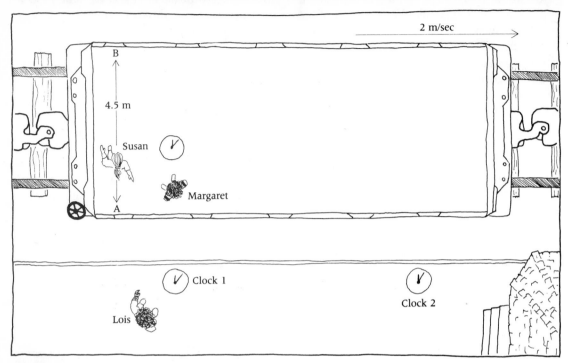

FIG. 9–8   A top view showing Susan walking across the train car from A to B, and back to A. Margaret observes the time on her clock. Lois also observes the time interval on two clocks located along the station platform.

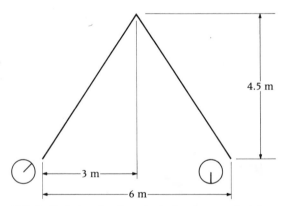

FIG. 9–9   Lois notes Susan's starting time on clock 1 and her return time on clock 2; she then determines the time interval. Because the train moves while Susan is walking across the train, Lois sees an inverted V-shaped path.

EXAMPLE G: THE SPEED OF LIGHT

A pulse of light leaves point A, travels across the car to point B, where it is reflected by the mirror, and returns to point A. The train moves with a speed v, which is *not* small compared to the speed of light. Margaret is on the train and Lois is in the station (Fig. 9–8).

Margaret says that the pulse travels a total distance of $2W$ with speed c, where W is the width of the train car. Hence, the time for the round trip observed by Margaret is

$$t_M = \frac{2W}{c}. \qquad (6)$$

According to Lois, the mirror has moved during the time it takes the pulse to travel from point A to point B. Lois observes that the light pulse travels a distance greater than $2W$. According to Einstein's second postulate, the speed of light is always c, and this does not depend upon whether the source or observer is moving. Hence, the speed of light *actually* observed by Lois is c. Therefore, the time for the round trip of the light pulse observed by Lois is

$$t_L = \frac{\text{distance greater than } 2W}{c}, \qquad (7)$$

and

$$t_L > t_M. \qquad (8)$$

This result can be directly attributed to the fact that, in Equations 6 and 7, the denominators are the same. In Example F, the same times were obtained because the denominators were different.

The exact relationship between t_L and t_M is given by

$$t_L = \frac{t_M}{\sqrt{1 - v^2/c^2}}. \qquad (9)$$

If v is less than c, the denominator is less than 1, and t_L is greater than t_M, as required in Equation 8.

Note also that, unless v is less than c, the denominator is an imaginary number. Hence, because time must be a real number, the speed v must always be less than c. If v is much smaller than c, then v/c is an extremely small number, and the denominator is very nearly equal to 1, yielding t_L equal to t_M. This is indeed comforting because it is what you expect at low speeds. All phenomena with which you are acquainted deal with speeds extraordinarily small compared to the speed of light, resulting in times that are the same for all observers.

A GENERALIZATION

In Example G, there are two events: the pulse of light leaving point A, and the pulse of light returning to point A. The time observed by Lois is t_L.

Relative to Lois, these two events occurred at *different* places. That is, when the first event occurred (light pulse leaving point A), Lois was adjacent to point A. When the second event occurred (light pulse returning to point A), Lois and point A were separated by a distance because the train had moved. Therefore, t_L can be replaced by t_{diff}.

According to Margaret, the two events occurred at the *same* place. When the first event occurred (light pulse leaving point A), Margaret was adjacent to point A. When the second event occurred (light pulse returning to point A), she was still adjacent to point A. Thus, relative to Margaret, these two events happened at the same place. Therefore, t_M can be replaced by t_{same}. The relative speed between Margaret and Lois is v.

Equation 9 can be generalized by making these replacements:

$$t_{\text{diff}} = \frac{t_{\text{same}}}{\sqrt{1 - v^2/c^2}}. \qquad (10)$$

RELATIVITY

207

The term t_{diff} indicates the time between two events that occurred at different places according to one observer. The term t_{same} is the time between the two events that occurred at the same place according to another observer, moving with speed v relative to the first observer. (Equation 10 is derived in calculation question 9–5.) When t_{diff} is larger than t_{same}, there is a **time dilation.**

Length Contraction

Lois measures the length of the train car at rest and calls it L_0. Trying to see what happens to the length when the train car is moving, she measures the time for it to pass by her.

Figure 9–10 shows the two events. Lois records the time t_{same} because both events occur in front of her, while Margaret records the time t_{diff}. According to Lois, the train travels past her at a speed v, and the length of the car is vt_{same}.

According to Margaret, Lois moves along the length of the car with speed v, and the length of the car is vt_{diff}. Since the time t_{same} is shorter than the time t_{diff}, Lois measures a shorter length than Margaret does. In fact, Margaret is at rest on the train and must measure the rest length L_0. Since t_{same} and t_{diff} differ only by the factor $\sqrt{1 - v^2/c^2}$, the length of the car can differ only by the same factor. That is, the distance d_L measured by Lois is

$$\text{Length of train car measured by Lois} = \sqrt{1 - v^2/c^2}\, L_0 .$$

Another way to see this is to multiply both sides of Equation 10 by v, identifying vt_{diff} as L_0 and the distance vt_{same} as d_L.

L_0 is the length of an object at rest. If this object travels with speed v relative to an observer, this observer will find that its length is given by

$$\text{Contracted length} = L_0 \sqrt{1 - v^2/c^2}. \quad (11)$$

New Way to View Relative Speed

Repeatedly, this chapter has assailed the familiar concept of relative speed. At speeds small compared to the speed of light, the familiar concept of relative speed holds true, but near the speed of light, the familiar concept is not correct. Is there, perhaps, *one* concept that can be used to understand *both* low-speed and high-speed behavior? Let's see.

Bernice, standing motionless, watches Lois move away from her with speed U, while object A moves towards Bernice with speed V, as shown in Fig. 9–11.

If the speeds U and V are small compared to light, then Lois sees object A move toward her with speed U plus V. The relative speed between Lois and object A is U plus V, and this is a familiar result.

However, if the speeds U and V are comparable with light, then Lois will not observe U plus V, but will observe the relative speed given by

$$\text{Relative speed} = v_{\text{rel}} \quad (12)$$
$$= \frac{U + V}{1 + UV/c^2}$$

Equation 12 is stated without proof,[1] but we can see that it is plausible. One requirement is that the relative speed must be equal to U plus V at speeds small compared to light. When this is so, UV/c^2 becomes an *extremely* small number because c is so enormously large. Therefore, the denominator of Equation 12 is quite close to 1, resulting in a relative speed U plus V.

Another requirement of Equation 12 is that *all* observers must measure the same value for the speed of light. In Fig. 9–12, Bernice, standing motionless, observes the light beam move toward her with speed c. Einstein's second postulate requires that Lois also observe the same speed c, even though Lois is moving toward the beam. Does this occur? The relative speed between Lois and the light beam is given by Equation 12, where V is equal to c. Equation 12 becomes

First event

Margaret

Lois

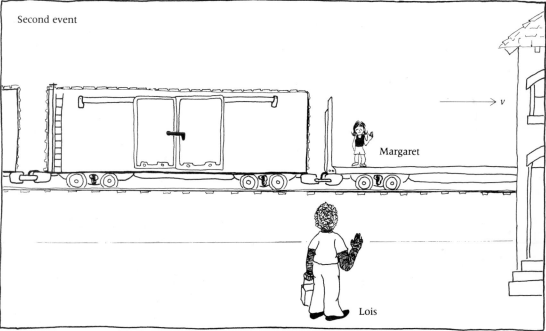

Second event

Margaret

Lois

FIG. 9–10 Lois and Margaret record the times between two events. In the first event, the front end of the car is in front of Lois; in the second event, the rear end of the car is in front of Lois. Lois records a time of t_{same} between these events, while Margaret records a time t_{diff}.

$$v_{\text{rel}} = \frac{U + c}{1 + Uc/c^2}$$

$$= \frac{U + c}{c/c + U/c}$$

$$= \frac{U + c}{(1/c)(c + U)}$$

$$= \frac{1}{1/c}$$

$$= c.$$

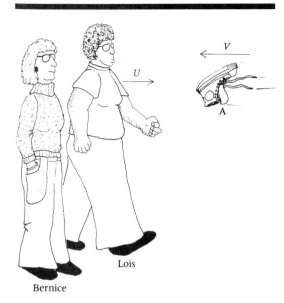

FIG. 9–11   Bernice and Lois observe object A move toward them. The relative speed between Bernice and Lois is *U*. To Bernice, the object approaches with speed *V*.

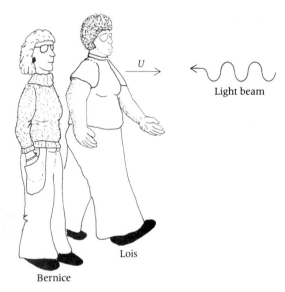

FIG. 9–12   The relative speed between Bernice and Lois is *U*. A light beam approaches. Both Bernice and Lois find that it approaches with speed *c*.

This formulation then gives exactly what is predicted by Einstein's second postulate. Both Bernice and Lois observe the light beam move with speed *c*. Thus, Equation 12 is successful in predicting both low-speed and high-speed behavior.

If object A moves in the opposite direction to that in Fig. 9–11, then Lois will observe the relative speed

$$v_{\text{rel}} = \frac{U - V}{1 - UV/c^2}. \qquad (13)$$

In both Equations 12 and 13, the numerator is the relative speed at low speeds, and the denominator is different from 1 only at speeds comparable with light.

We shall soon see how this formulation of the relative speed in Equations 12 and 13 results in a drastic change in our concept of mass, but first let's look at some applications of time dilation and length contraction.

Is Length Contraction Real or an Optical Illusion?

Suppose a truck, 8 m long when at rest, travels at a speed of 2.4×10^8 m/sec relative to two observers, Larry and Earl. Both observers have a can of red spray paint, which they fire simultaneously when the respective ends of the truck are directly in front of them. When the truck stops, Larry, Earl, and the truck driver look at the red paint spots. How far apart are these spots?

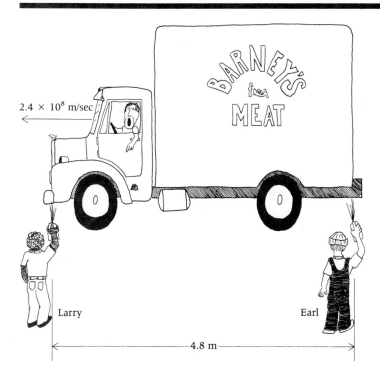

2.4×10^8 m/sec

Larry

Earl

← 4.8 m →

FIG. 9–13 When the ends of the truck are in front of them, Larry and Earl simultaneously fire spray cans of red paint.

To Larry and Earl, the truck does not measure 8 m long when it is in motion. It is smaller:

$$\frac{v}{c} = \frac{2.4 \times 10^8 \, \text{m/sec}}{3 \times 10^8 \, \text{m/sec}}$$

$$= 0.8$$

Length of truck in motion $= L_0 \sqrt{1 - v^2/c^2}$

$$= (8\,\text{m}) \sqrt{1 - 0.8^2}$$

$$= 4.8\,\text{m}.$$

To Larry and Earl, the truck in motion has a length of 4.8 m, and they must stand this distance apart if they are to fire the spray cans simultaneously and hit the ends of the truck (see Fig. 9–13).

Suppose, now, that the truck is brought to a stop, and all three look at the red paint spots. They find them exactly at the ends of the truck. How can the truck driver understand these results? To him, the truck is 8 m long, but Larry and Earl were not separated by 8 m. Obviously,

to the truck driver, Larry and Earl *did not fire the spray cans simultaneously,* which is the only way that paint spots could be found at each end.

Let's look at this situation from the truck driver's point of view (Fig. 9–14). Relative to the truck driver, Larry and Earl are in motion, and this distance between them is *shorter* than 4.8 m. This distance is given by

Distance between Larry and Earl as observed by truck driver $= (4.8\,\text{m}) \sqrt{1 - 0.8^2}$

$$= 2.88\,\text{m}.$$

According to the truck driver, Larry and Earl are a distance of 2.88 m apart. When Larry passes the front end of the truck, he fires, but at that instant Earl is near the truck cab, a distance

RELATIVITY

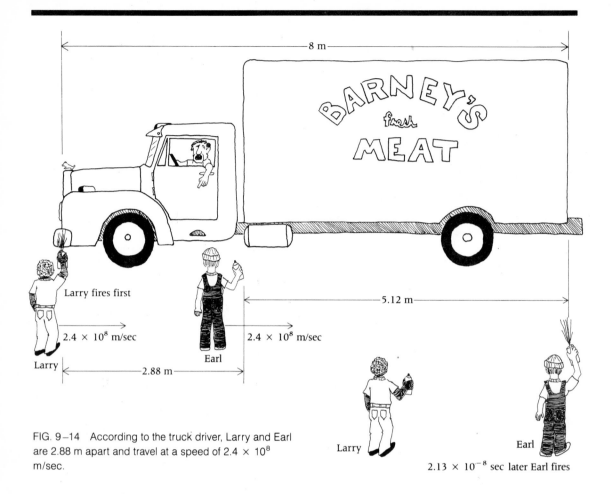

FIG. 9–14 According to the truck driver, Larry and Earl are 2.88 m apart and travel at a speed of 2.4×10^8 m/sec.

2.13×10^{-8} sec later Earl fires

of 5.12 m from the back end of the truck. It takes Earl (5.12 m)/(2.4 × 10^8 m/sec), or 2.13×10^{-8} seconds to reach the back end when he fires.

In this application, we see that events that are simultaneous to Larry and Earl are *not* simultaneous to the truck driver. Larry and Earl spray paint on the ends of the truck, and everyone will see that when the truck stops. Since the truck driver sees that Larry and Earl are separated by a distance shorter than the truck's length, Larry fired his spray can first. Larry and Earl and the truck driver do not agree on the order of the events, or the length of the truck, or the distance between Larry and Earl. However, all three readily agree that the paint spots are on the ends of the truck and *understand why* this is so.

The Twin Paradox

Identical twins, Susan and Marian, agree to test Einstein's fascinating theory. Susan climbs aboard a spaceship and travels in a straight line for 3 years at a speed of 0.6c, then turns around and comes back to the earthbound Marian. After Susan's return, the twins will compare their watches to see if they have aged differently. They also decide to carry out the following experiment: Susan says that she will send out a pulse of light after each year, according to her watch. Marian will receive each pulse and record its arrival time on her watch. In this way, they can keep track of the passage of time. The times required to speed up and slow down and come to a stop will be so short that they will neglect them.[2]

According to Susan, the time between the 2 pulses is 1 year. Relative to Susan, the pulses are emitted at the same place. Susan finds that t_{same} is 1 year. Using Equation 10, we find that t_{diff} is

$$t_{diff} = \frac{t_{same}}{\sqrt{1 - v^2/c^2}}$$

$$= \frac{1 \text{ year}}{\sqrt{1 - (0.6c/c)^2}}$$

$$= 1.25 \text{ years.}$$

Thus, t_{diff} is the time recorded by clocks stationary with respect to Marian. These clocks are located adjacent to the spaceship just as each pulse is emitted (see Fig. 9–15). According to Susan, the first pulse is emitted after 1 year; the second, after 2 years; etc. However, the stationary clocks show a time of 1.25 years when the first pulse is emitted and 2.50 years, for the second. Since the spaceship travels at $0.6c$, Marian concludes that the distance it has traveled in 1.25 years is $(0.6c)(1.25 \text{ years})$, which equals $0.75c$, or 0.75 light-years. Since the first pulse was emitted after 1.25 years and has to travel a distance of 0.75 light-years to reach Marian, the first pulse reaches Marian after 2 years. Similarly, the second pulse reaches Marian after 4 years; the third, after 6 years. The fourth pulse is emitted on Susan's return trip. According to the stationary clock, it is emitted after 5 years, but then has to travel a distance of 1.50 light-years, so that it reaches Marian after 6.50 years.

Likewise, the fifth pulse reaches Marian after 7 years and the sixth, after 7.50 years.

In summary, Susan says that she sent out 6 pulses, 1 every year. Marian says that she certainly did receive 6 pulses (3, spaced 2 years apart, and 3, spaced half a year apart) during *7.5 years*. Their watches do not agree when placed side by side, and Marian has aged more than Susan. We could have also obtained these results by using Equation 10 and substituting t_{same} equals 6 years. The time recorded by Marian, t_{diff}, would then be 7.5 years.

More articles about the twin paradox have appeared in physics journals than any other single topic in the discipline, so compelling and revolutionary is this consequence of Einstein's theory. Still, we have not yet discussed how the "paradox" arises.

In the situation we have just examined, Susan is in motion and Marian is considered to be at rest. The result is that Marian has aged more than Susan. However, from Susan's viewpoint, Marian is in motion relative to her. In that case, one ends up with the conclusion that Susan has aged more than Marian, which is just the opposite of our previous conclusion. How can that be? Obviously, both can't be true. This is the paradox.

We ask the question, Are the two situations *exactly* equivalent? If we look at our twins, we shall see that their experiences during the journey are not identical. Susan feels the acceleration as the spaceship speeds up and slows down, while Marian does not. We must conclude that the two situations are not symmetrical. In this case, we really can decide, without a doubt, which one is in motion—Susan. Therefore, only one conclusion is possible: Marian has aged more than Susan. The stay-at-home twin ages more than the astronaut twin.

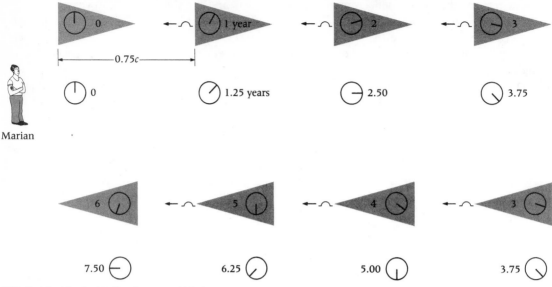

FIG. 9–15 Identical twins, Susan and Marian, compare watches at the end of Susan's voyage into outer space. According to Susan, each leg of the trip lasts three years. The diagram shows the passage of time according to Susan's clock, and according to clocks at rest in relation to earthbound Marian.

Experimental Evidence for Time Dilation and Length Contraction

Muons, which are subatomic particles produced several thousand meters above sea level by cosmic ray collisions, travel at a speed of $0.99c$ and reach sea level. Muons are radioactive and decay into several other particles, one of which is the electron. Muons have an average lifetime of only 2×10^{-6} sec before they decay. A typical muon would then travel only a distance given by

$$d = vt$$
$$= (3 \times 10^8 \, \text{m/sec})(2 \times 10^{-6} \, \text{sec})$$
$$= 600 \, \text{m}.$$

To what distance does the 600 m refer? If it is the distance relative to the earth, then few muons should reach sea level, in contradiction to experimental observations.

Let's consider a clock traveling with the muon. This clock records the time interval between the birth and death of the muon, which is t_{same}. However, clocks at rest relative to the earth record the birth and death on two different clocks, and the time interval is t_{diff}. Using Equation 10, we find t_{diff}:

$$t_{diff} = \frac{t_{same}}{\sqrt{1 - v^2/c^2}}$$
$$= \frac{2 \times 10^{-6} \, \text{sec}}{\sqrt{1 - 0.99^2}}$$
$$= 14.2 \times 10^{-6} \, \text{sec}.$$

Observers on earth find that the high-speed muon lives for 14.2×10^{-6} sec and travels a distance given by

$$d = (3 \times 10^8 \, \text{m/sec})(14.2 \times 10^{-6} \, \text{sec})$$
$$= 4260 \, \text{m}.$$

The muon travels 4260 m before it decays. This explains why scientists observe muons at sea level, although they are produced several thousand meters above.

We can also solve this problem using length contraction. To the muon, the distance 4260 m appears to be only 601 m, as the following calculations show:

$$\text{Distance traveled by muon} = L_0 \sqrt{1 - v^2/c2}$$

$$= (4260\,\text{m})\sqrt{1 - 0.99^2}$$

$$= 601\,\text{m}.$$

The muon can travel this distance in 2×10^{-6} sec. Thus, the fact that muons are found at sea level, in spite of their short lifetime, confirms Einstein's theory of time dilation and length contraction.

Particle accelerators produce particles traveling near the speed of light. Electrons in atoms have speeds about 1 percent of the speed of light. Effects of relativity can surely be seen here. Scientists have carried out so many experiments that the evidence supporting Einstein's theory of relativity is now overwhelming.

So far, we have seen how Einstein's second postulate has caused drastic changes in our concept of relative speed, time, and length. Now, we'll examine the consequences of the first postulate and see how it affects the concept of mass and leads to the idea that mass and energy are interchangeable. To gain familiarity with the first postulate, we'll begin with a low-speed example and look at the well-known laws of conservation of momentum and energy. Then, we shall see what happens at speeds approximating those of light.

The First Postulate and Collisions at Low Speeds

In Fig. 9–16, two identical 1000-kg cars, traveling at 5 m/sec, crash head-on. After the collision, the bumpers lock together. An (indestructible) thermometer placed on car A shows the rise in temperature after the collision.

Let's see how Frank views this event. After the collision, Frank sees that the cars are at rest, as dictated by conservation of momentum. (Recall that the momentum of an object is given by mv.) Because the cars have the same mass and speed, but travel in opposite directions, the total momentum before collision is zero, and must be zero afterwards, too. For this reason, the coupled cars must, therefore, be at rest.

Before the collision, each car has a kinetic energy ($\frac{1}{2}mv^2$) of 12,500 joules; the total energy is 25,000 joules. After the collision, there is, of course, no kinetic energy. However, there are other forms of energy: heat and the energy used to demolish the cars. In what follows, the sum of these energies will be referred to as "heat" for the sake of simplicity. According to Frank 25,000 joules of kinetic energy have been converted to heat, and Frank finds that this corresponds to, let's say, a 5°C rise in temperature. As Larry passes by, he can see the thermometer and observes the 5°C rise, too. Can Larry reconcile this 5°C rise with *his* view of the conservation laws? For example, if Larry's calculation shows a temperature rise other than 5°C, then Newton's laws would not be valid for Larry.

Larry views the collision as shown in Fig. 9–17. Before the collision, car A appears to be at rest, while car B moves toward him and car A at 10 m/sec, using familiar notions for the relative speed. After the collision, the coupled cars A and B have a speed of 5 m/sec to the left.

Let's see if Larry finds that momentum and energy are conserved. Figure 9–17 shows that the total momentum before and after collision are the same and, hence, momentum is conserved. We also see that the total kinetic energy before the collision is 50,000 joules, but only 25,000 joules afterwards, which means that 25,000 joules of kinetic energy must have been

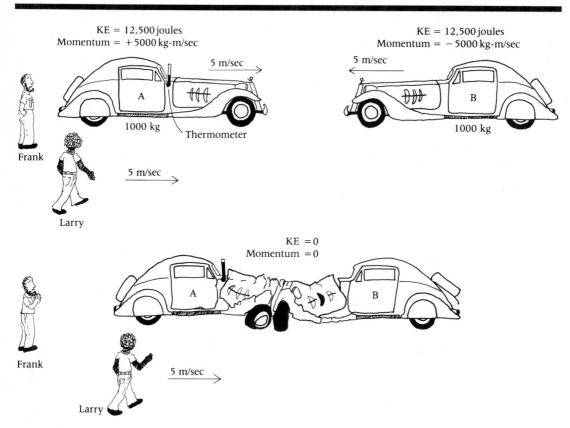

KE = 12,500 joules
Momentum = + 5000 kg-m/sec

5 m/sec

A

1000 kg

Thermometer

Frank

Larry

5 m/sec

KE = 12,500 joules
Momentum = − 5000 kg-m/sec

5 m/sec

B

1000 kg

KE = 0
Momentum = 0

A

B

Frank

Larry

5 m/sec

FIG. 9–16 Frank views the head-on collision of two identical cars. The momentum of each car is given by *mv*, and the kinetic energy by ½ *mv*². The momentum is positive if the car is traveling to the right, and negative if the car is traveling to the left.

converted to heat. Both Frank and Larry find that 25,000 joules of kinetic energy have been transformed to heat. If Frank finds a 5°C temperature rise, then so does Larry. It works out well. Frank and Larry both find that momentum and energy are conserved.

Before going on to a high-speed example, let's take one more look at conservation of momentum and grasp the *essential* point. In Fig. 9–17, momentum is conserved *only because* car B has a speed of 10 m/sec. This relative speed, obtained by using the values in Fig. 9–16, is 5 m/sec plus 5 m/sec.

The First Postulate and Collisions at High Speeds

Let's now see what happens to conservation of momentum when we imagine that the cars travel at six-tenths the speed of light (0.6c), and Larry, likewise, travels at 0.6c, as shown in Fig. 9–18. As before, Frank finds that the coupled cars will be at rest after the collision. This agrees with conservation of momentum.

The question is, Does Larry find that momentum is conserved? As we have just seen in the low-speed example, Larry will find that momentum is conserved *only if* car B has a speed of 0.6c plus 0.6c, or 1.2c. However, we know that this is impossible! Nothing can travel faster than the speed of light. We have learned that Einstein's second postulate forced a drastic

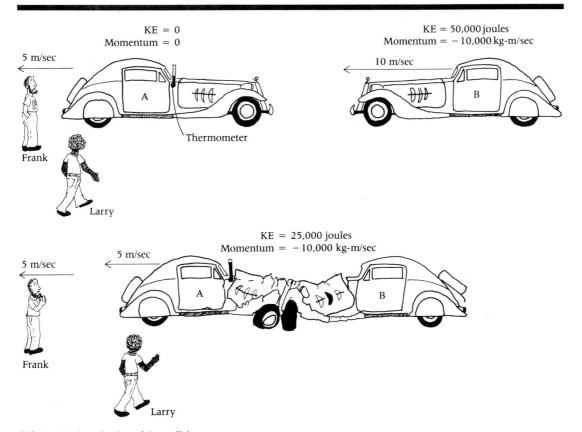

KE = 0
Momentum = 0

5 m/sec ←

Frank

A

Thermometer

Larry

KE = 50,000 joules
Momentum = −10,000 kg-m/sec

10 m/sec ←

B

KE = 25,000 joules
Momentum = −10,000 kg-m/sec

5 m/sec ←
5 m/sec ←

Frank

A

B

Larry

FIG. 9–17   Larry's view of the collision.

change in the concept of relative speed. The relative speed between two objects approaching each other is given by Equation 12:

$$v_{rel} = \frac{U + V}{1 + UV/c^2}$$

Substituting $0.6c$ for U and $0.6c$ for V into this equation, we find that v_{rel} equals $0.8824c$. Larry will see car B move toward him with a speed of $0.8824c$ (see Fig. 9–19).

Conservation of momentum doesn't seem to work. This comes as no surprise, perhaps. Einstein's second postulate forced us to accept a new formulation for relative speed and this, in turn, now forces us to reevaluate the concepts of momentum, energy, *and mass*.

Relativistic Formulation for Momentum and Energy

According to Einstein's theory, the momentum (symbolized by p) of an object is defined as

$$p = \frac{m_0 v}{\sqrt{1 - v^2/c^2}}, \qquad (14)$$

and the energy E is defined as:

$$E = \frac{m_0 c^2}{\sqrt{1 - v^2/c^2}}, \qquad (15)$$

RELATIVITY

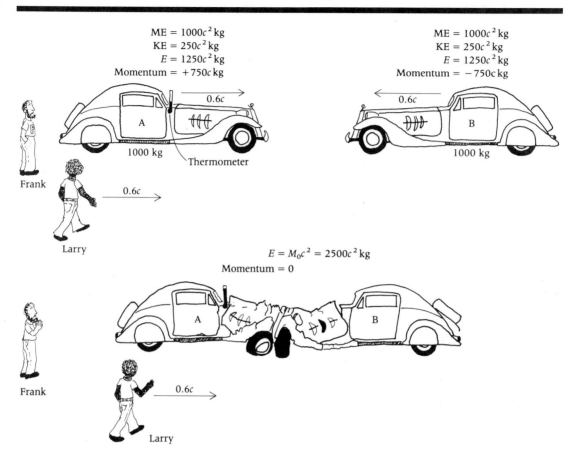

FIG. 9–18 Frank views the head-on collision of two cars traveling at six-tenths of the speed of light.

where m_0 is the mass of an object at rest, v is the speed of the object as *seen by the observer*, and p is the momentum seen by this observer. In these formulas, v will have different values for two observers moving relative to each other.

Note that, when v is much less than c, the denominator in Equation 14 is 1; the momentum p reduces to the familiar formula. It must do so in order to be a valid formula, for we know that p equals m_0v is correct at low speeds.

The energy E is really the sum of two kinds of energy, kinetic energy and mass energy:

$$E = KE + ME. \qquad (16)$$

When an object is at rest, it obviously has no kinetic energy. The only kind of energy it has is its mass energy. Therefore, its mass energy (abbreviation: ME), obtained by letting v equal 0 in Equation 15, is m_0c^2. If we identify m as $m_0/\sqrt{1 - v^2/c^2}$, then Equation 15 becomes the famous equation relating mass and energy ($E = mc^2$).

Einstein obtained these formulas by requiring that two observers, moving relative to each other with constant velocity, find that momentum and energy are conserved. Here, we'll take

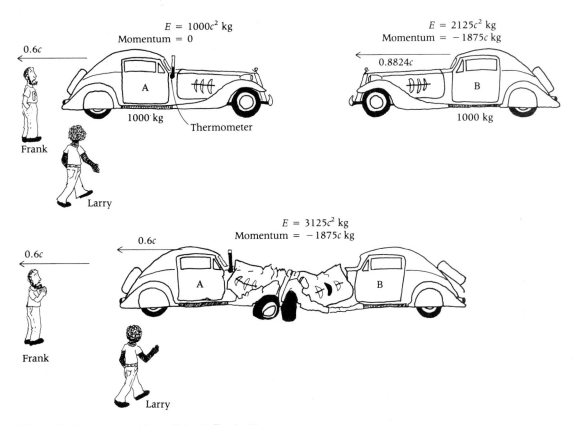

FIG. 9–19   Larry's view of the collision in Fig. 9–18.

the opposite approach, using these formulas to see that the conservation laws are valid for our two observers.

According to Frank, both cars have a speed of 0.6c and this value is used in Equations 14 and 15 to find the momentum and energy of both cars (see Fig. 9–18). The denominator in both equations is $\sqrt{1 - 0.6^2}$, or 0.8:

$$\begin{aligned} \text{Momentum } p \text{ of car A} \atop \text{before collision} &= \frac{(1000\,\text{kg})(0.6c)}{0.8} \\ &= 750c\,\text{kg}. \end{aligned}$$

The momentum of car B has the same value, but it is negative because it is traveling to the left.

$$\begin{aligned} \text{Energy } E \text{ of cars A and B} \atop \text{before collision} &= \frac{(1000\,\text{kg})(c^2)}{0.8} \\ &= 1250c^2\,\text{kg}. \end{aligned}$$

As before, the total momentum before collision is zero. Therefore, after the collision, the coupled cars are at rest, as we still expect.

Figure 9–18 shows that the total energy before collision is $2500c^2$ kg. If energy is conserved, the total energy after collision must also be

$2500c^2$ kg. Since the cars are motionless after the collision, $2500c^2$ kg represents their combined mass energy M_0c^2. Before the collision, the combined mass energy of the two cars was $2000c^2$ kg. That is, after the collision, the combined mass energy increases by $500c^2$ kg. This occurs because the total kinetic energy of the cars before collision ($500c^2$ kg) has been transformed into mass energy afterwards. The combined mass M_0 of the two cars after collision is 2500 kg. In our low-speed example, we found that kinetic energy is transformed into heat; here, we find that kinetic energy is transformed into mass. Mass is a form of energy. Kinetic energy can be transformed into mass energy, and vice versa.

Now, let's see if Larry can use the same formulas for momentum and energy, Equations 14 and 15, and find them valid. As Fig. 9−19 shows, Frank and Larry observe different speeds for the cars. Here, Larry's values are used. Since the speed of car B before the collision is $0.8824c$ and its mass m_0 is 1000 kg, the momentum (Equation 14) of car B is $-1875c$ kg and its energy E (Equation 15) is $2125c^2$ kg. After collision, the combined mass M_0 is 2500 kg and the speed is $0.6c$; the momentum is $-1875c$ kg and the energy is $3125c^2$ kg. Figure 9−19 shows that the total momentum before and after the collision is $-1875c$ kg, and the total energy before and after is $3125c^2$ kg. Momentum and energy are conserved! The laws of physics are the same for Frank and Larry, as Einstein's first postulate requires.

The Antiproton

In 1932, the American physicist Carl D. Anderson (b. 1905) discovered the **positron,** a particle having exactly the same mass as an electron but positively charged. Scientists wondered, is there perhaps a whole family of so-called *antiparticles?*

A proton is the nucleus of a hydrogen atom and is positively charged. The antiproton would have the same mass, but would be negatively charged. The scientists' search for the first artificially produced antiparticle centered on the antiproton. They theorized that it could be produced by having energetic protons from an accelerator strike a hydrogen target. Some of the kinetic energy would be transformed into newly formed particles, according to the following:

$$p + p \rightarrow p + p + p + \bar{p},$$

where p is the symbol for the proton and $\bar{p}$, for the antiproton. (Unfortunately, the symbol for proton and momentum is the same, though we italicize the latter.)

Note that the masses of two *additional* particles must be produced. The kinetic energy of the incident proton would have to be extremely large to accomplish this. In 1954, a high-energy particle accelerator, the Berkeley Bevatron, was constructed with sufficient energy for this purpose. The question here is, How large is "sufficient"?

We have already laid the groundwork in our discussion of Figs. 9−18 and 9−19. Here, we transfer kinetic energy into the mass of a proton and an antiproton.

Figure 9−20(a) shows a proton from the Bevatron traveling to the left with a speed v_a. After the collision, the 3 protons and 1 antiproton must travel to the left (with a speed v_b) in order to conserve momentum. This situation is analogous to that depicted in Fig. 9−19.

Figure 9−20(b), showing how Frank views the collision, is analogous to Fig. 9−18, because the two particles before the collision have the same speed, v_b. Afterwards, the four particles are at rest. In Fig. 9−18, we knew the speeds before collision and found the new rest mass afterwards. Here, we do the reverse: We know the rest mass afterwards, and we want to find the speed v_b.

In Fig. 9−20(b) the total energy after the collision is equal to the rest mass energy of 3 protons and 1 antiproton, or $4Mc^2$, where M is the rest mass of the protons and antiproton. Since the total energy before the collision is equal to that afterwards, we find

$$4Mc^2 = \frac{2Mc^2}{\sqrt{1 - v_b^2/c^2}} \cdot$$

The Mc^2 expressions cancel out on each side of the equation and we obtain

$$\sqrt{1 - v_b^2/c^2} = 0.5,$$

or, squaring,

$$1 - v_b^2/c^2 = 0.25$$
$$v_b/c = \sqrt{0.75}$$
$$= 0.866.$$

Our goal is to find the speed v_a of a proton produced by the Bevatron, where v_a is the relative speed between the two protons:

$$v_{\text{rel}} = \frac{U + V}{1 + UV/c^2} \cdot$$

Since U and V both equal the speed v_b (and v_b is equal to $0.866c$), the relative speed v_a becomes

$$v_a = \frac{0.866c + 0.866c}{1 + 0.866^2}$$
$$= 0.98974c.$$

The physicists designed the Bevatron to produce protons having this speed, and they certainly did find the antiprotons they were searching for! Nature has symmetry.

Experimental Evidence Showing that Mass and Energy Are Interchangeable

In 1932, the English physicist John D. Cockcroft (1897–1967) and the Irish physicist Ernest T. S. Walton (b. 1903) designed an accelerator that produced protons and showed, finally, that mass and energy are interchangeable. They aimed the

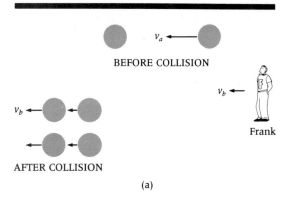

BEFORE COLLISION

AFTER COLLISION

(a)

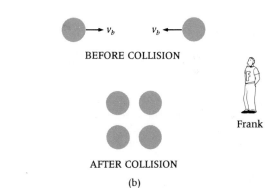

BEFORE COLLISION

AFTER COLLISION

(b)

FIG. 9–20 (a) A proton produced by the Bevatron collides with a stationary proton in the target. (b) Frank's view of the collision.

proton beam at a lithium target. The lithium nucleus absorbed a proton and then disintegrated into two alpha particles:

$$^1_1H + ^7_3Li \rightarrow ^4_2He + ^4_2He.$$

They knew the masses of the hydrogen, helium, and lithium atoms quite accurately. Since the mass of two alpha particles was *less than* that of a hydrogen atom plus a lithium atom, the difference must be converted to the kinetic energy of the alpha particle:

KE of alpha particles = (difference in mass)c^2.

When Cockcroft and Walton measured the kinetic energy of the alpha particles, they found what they had expected from Einstein's theory.

No theory has been tested more thoroughly than Einstein's theory that mass and energy are interchangeable. We have just seen, for example, that mass can be transformed into the kinetic energy of two alpha particles. And remember that, in Chapter 6, we used this theory to understand: (1) why radium emits an alpha particle—a small amount of mass is converted to kinetic energy of an alpha particle, and (2) why light emitted by the sun must be due to conversion of mass into kinetic energy and light.

Summary

Einstein's theory of relativity toppled many cherished concepts. Our "normal" notion of relative speed fell quickly. If a light source is on a train moving with speed v, it seems reasonable that an observer on the train would see speed c, while another observer in the station would see a maximum speed of c plus v and a minimum speed of c minus v, depending upon the orientation of sources and observers. However, according to Einstein's second postulate *all* observers, those on the train as well as those in the station, will find the same speed c. Finally, the null results of the Michelson-Morley experiment (Chapter 8) make sense.

Our usual concept of time also succumbed. Previously, the time between two events was thought to be the same for all observers, but that is no longer true. Two light beams emitted at the center of a train car (directed toward opposite ends) will reach the ends simultaneously, according to a passenger in the car. But to observers in the station, the beams *still* travel with speed c. Since one end moves away from the light beam, while the other end approaches the light beam, the light beams do not reach the ends simultaneously. Observers in relative motion view time differently.

This leads directly to time dilation. Consider a light beam sent across a train car (perpendicular to the direction of motion). Since an observer in the station sees the light beam travel a longer distance than the observer on the train (while the speed of light is the same for both), the station observer measures a longer time for the light beam to travel across the car and return ($t_{diff} = t_{same}/\sqrt{1 - v^2/c^2}$). The expression t_{same} is the time between two events located in the same place according to one observer; t_{diff} is the time between the two events located in different places according to another observer; v is the relative speed between the two observers.

Our normal concept of length also fell quickly. Consider a train car moving past a stationary observer, who records the time (t_{same}) between two events: front end passing and rear end passing in front of her. The train length is vt_{same}. A passenger on the train observes the same two events, but she records t_{diff}. She also measures the length of the car to be L_0, and L_0 equals vt_{diff}. Since t_{same} is always less than t_{diff}, the observer in the station finds that the length of the train car is less than L_0. The contracted length equals $L_0 \sqrt{1 - v^2/c^2}$.

Even our traditional concept of mass changed. The collision of two cars at low speeds showed that momentum and energy are conserved. At high speeds, we saw how momentum and energy must be modified to force conservation of momentum and energy to occur.

Two observers in relative motion viewing the head-on collision of two identical cars will agree that momentum and energy are conserved and will agree on the amount of kinetic energy transformed into heat. They observe the same temperature rise after collision, which illustrates Einstein's first postulate at low speeds: The laws of physics are the same for two observers moving relative to each other with constant velocity. At low speeds, the relative speeds were calculated in the usual way. For example, two cars traveling in opposite directions with speed v have a relative speed of $2v$.

At high speeds, Einstein's second postulate showed that the concept of relative speed must be altered. Then, however, momentum is not conserved for both observers: a violation of Einstein's first postulate. The way out of this dilemma is a reformulation of momentum and energy to find momentum and energy conservation for the two observers in relative motion. The mass after the collision is not the same as before. Mass and energy are interchangeable.

The energy E is the sum of the mass energy and the kinetic energy. The mass energy is m_0c^2.

At low speeds, the relativistic formulas for momentum and energy reduce to the familiar ones.

True or False Questions

Indicate whether the following statements are true or false. Change all of the false statements so that they read correctly.

9–1 A physicist measures the speed of the light coming from an automobile headlight.
   a. When the car approaches the apparatus with speed v, the speed of light is found to be $c + v$.
   b. When the car backs up with speed v, and moves away from the apparatus, the speed of light is found to be c minus v.
   c. The speed of light is always measured to be c when the car goes forward, backs up, and remains at rest.

9–2 A lamp in the center of a train car sends out two pulses of light, one each toward each end of the train car.
   a. Observers on the train say that the pulses reach the ends simultaneously.
   b. Observers in the station say that the pulses reach the ends simultaneously.

9–3 A woman standing in a train station claps her hands twice, while watching a train pass by her with speed v.
   a. The woman in the station denotes the time between the claps as t_{diff}.

   b. The passengers on the train denote the time between claps as t_{same}.
   c. t_{diff} is larger than t_{same}.
   d. When the woman in the station measures the time between claps (with sufficient accuracy), she finds a longer time than the passengers making the same measurement on the train.
   e. When the speed v is very small compared to the speed of light, the woman in the station and the passengers on the train measure nearly the same time between claps.

9–4 A train is traveling with speed v. A woman standing in the train station measures a time of 1 second for one train car to pass by her. (Assume that v is *not* small compared to c.)
   a. To find the length of the train car, the woman multiples the speed v by 1 second.
   b. The woman finds that the length of the moving train car is equal to the length when the train car is at rest.

c. To the passengers on the train car, the woman appears to travel past them with speed v.

d. To find the length of the train car, the passengers multiply the time for the woman to pass by the car, by the speed v.

e. The passengers find that the length of the train car is equal to the length when the train car is at rest.

f. According to the passengers, the time for the woman to pass by the length of the car is exactly 1 second.

g. Since both the woman and the passengers measure times of 1 second, the woman and the passengers find the length of the train car to be the same.

9−5 Car A, traveling at 30 mph, and car B, traveling at 20 mph, collide head-on.

a. Before the collision, the driver in car A saw car B approach at 50 mph.

b. Similarly, if the driver of car A measured the speed of light coming from the headlights of car B (with sufficient accuracy), the driver would find that the speed of light is c plus 50 mph.

c. Einstein's theory of relativity accounts for the relative speed in *all* situations.

9−6 Two people standing in a train station are separated by 20 m. Simultaneously, they fire spray cans of paint at a train that is traveling at speed v.

a. When the train stops, the paint spots are separated by a distance greater than 20 m, as the passengers and people in the station can easily see.

b. When the train is moving, passengers on the train find that the people in the station are separated by a distance greater than 20 m.

c. The passengers on the train find that the spray cans were fired simultaneously.

d. Since the spots on the train are not separated by 20 m, the passengers saw one person fire the spray can before the other person fired.

9−7 In reality, length contraction is an optical illusion.

9−8 At speeds comparable to the speed of light, Einstein's first postulate showed that the concept of relative speed must be altered, while the second postulate showed that the concept of momentum and energy must be altered.

9−9 When a twin returns from a journey into space aboard a spacecraft that traveled at one-hundredth the speed of light, the astronaut twin will have aged less than the stay-at-home twin.

9−10 Two cars approach each other traveling at speed v.

a. If the speed v is one-millionth the speed of light, then the relative speed of the two cars is very nearly equal to $2v$.

b. If the speed v is eight-tenths of the speed of light, then the relative speed of the two cars is less than $2v$.

c. The correct results in **a** and **b** require that the momentum of an object traveling with speed v is given by mv only when v is extremely small compared to the speed of light.

Questions for Thought

9−1 Just as Susan, aboard a train moving in a straight line at 2 m/sec, starts to throw a ball vertically into the air the train accelerates at 1 m/sec². Brian, standing on the station platform, observes the ball's motion. Explain why it seems reasonable to Brian that the ball moves to the rear of the train car.

9−2 Table 9−1 shows that the relative speed between Frank and Marie

a. depends only upon Frank's speed relative to the train.

b. depends only upon Frank's speed relative to the train and Marie's speed relative to the platform.

c. depends upon Frank's speed relative to the train, Marie's speed relative to the platform, and the speed and direction of the train.

d. depends only upon Marie's speed relative to the platform.

9–3 Marie, located in the train station, decides to measure the speed of light from a lamp located on a moving train. Marie, who has not previously studied Einstein's theory of relativity, expects that

a. the speed of light is *c*.

b. the speed of light will be slightly different from *c*, because of the train's motion.

c. the speed of light will be slightly different from *c*, because of the train's motion and Marie's speed relative to the platform.

In the following blanks, insert **a, b,** or **c** given above: Frank is aboard the moving train. Marie expects that Frank will measure the speed of light to be given by _____ . When Marie and Frank carry out their measurements, Marie finds that _____ and Frank finds that _____ .

9–4 Summarize the results shown in Table 9–2.

9–5 a. The R postulate is (insert: *correct* or *incorrect*) _____ and (insert: *can* or *cannot*) _____ be verified by experiment.

b. Comment on the following statement: If the R postulate were correct, then the time between two events would be different for two observers moving with relative speed *v* in a straight line.

9–6 Initially, many physicists found it very difficult to accept Einstein's second postulate. Does this difficulty seem reasonable to you?

9–7 In Fig. 9–8 Susan walks across the train car and returns. Do Lois and Margaret agree on the distance that Susan traveled? Is Susan's speed relative to Margaret the same as Susan's speed relative to Lois? Explain why Lois and Margaret agree on the time for Susan's round trip.

9–8 Suppose that there is another observer in Fig. 9–16 traveling to the right at 10 m/sec.

a. Draw a diagram similar to Fig. 9–17, showing how this observer would view the collision.

b. Does this observer find that momentum and energy are conserved?

c. Does this observer's calculation show a 5°C rise in temperature?

Questions for Calculation

9–1 Refer to Table 9–1. In case 6, Marie's motion in the station is 2 m/sec west, and Frank's motion relative to the train is 3 m/sec west. The train travels east at 1 m/sec. What is the relative speed between Frank and Marie?

9–2 Susan is at the end of a train car 24 m long, as shown in Fig. 9–21. She walks to the other end at a speed of 3 m/sec. The train has a speed of 2 m/sec.

a. How long does Susan find that it takes her to reach the other end? _____

b. It seems reasonable that Larry will find that it takes Susan _____ seconds to reach the other end.

c. During the time found in part **a**, how far has the train traveled? _____

d. Where is Larry when Susan reaches the other end? _____

e. According to Larry, Susan is initially _____ m from Larry. Larry sees Susan move toward him at _____ m/sec.

f. Use the results in part **e** to calculate the time for Susan to reach the other end, according to Larry.

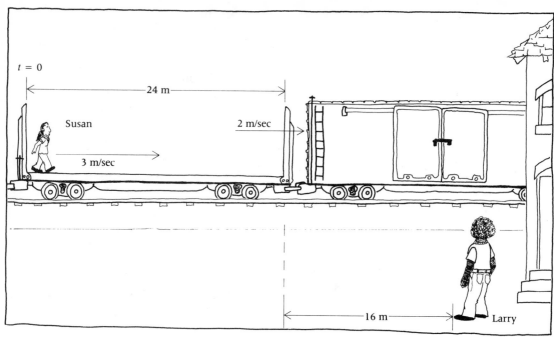

FIG. 9−21 Diagram for calculation question 9−2.

g. Is the following statement true or false: At speeds much smaller than the speed of light, the time between two events is the same for all observers. (In this example, the two events are Susan leaving one end of the car and arriving at the other end.)

9−3 Examine what would happen in calculation question 9−2 if the R postulate were correct. In that case, Susan's speed relative to the train car is 3 m/sec *and* Larry would find that Susan moves toward him at 3 m/sec.

    a. Initially, Susan is _____ m from Larry. When Susan reaches the other end of the train car, Larry is directly in front of her. Larry says that Susan travels a distance of _____ m at a speed of _____ . Therefore, Larry says that the time between two events (Susan leaving one end and arriving at the other) is _____ seconds.

    b. Susan finds that she reaches the other end after _____ seconds.

†9−4 Brian and Susan are at the center of a train car, point T, and Larry and Margaret are at point S on the station platform. The train is moving at 2 m/sec. At the instant that points T and S are adjacent, Larry walks to the left at 3 m/sec and Margaret walks to the right at 3 m/sec, relative to the station platform; Brian walks to the left at 3 m/sec, and Susan to the right at 3 m/sec, relative to the train.

    a. What are Brian's speed and Susan's speed relative to the train platform? ____
    b. Are Larry and Brian adjacent? Are Margaret and Susan adjacent? _____
    c. Suppose that the R postulate were really correct. In that case, Brian's speed relative to the train platform would be _____ m/sec, and Susan's speed relative to the train platform would be _____ m/sec. Would Larry and Brian be adjacent? Would Margaret and Susan be adjacent? Explain.
    d. Compare the results of **c** with the results for the light pulses in Fig. 9−6 and Fig. 9−7.

9-5 A light source is located at point A in Fig. 9–22(a) and emits a pulse of light that travels to point B and returns to point A. Margaret records the time t_{same} for the round-trip time, while Lois in the station records a time t_{diff}.

    a. Show that, according to Margaret, $W = ct_{same}/2$.

    b. Lois' observations are shown in Fig. 9–22(b). Show that the distance CE is $ct_{diff}/2$; the distance CD is $vt_{diff}/2$; and that $W = (t_{diff}/2)\sqrt{c^2 - v^2}$. (Hint: Find W using the Pytagorean theorem.)

    c. Equate the values of W found in parts **a** and **b** and show that

$$t_{diff} = \frac{t_{same}}{\sqrt{1 - v^2/c^2}}.$$

†9-6 Refer to Fig. 9–10. Margaret uses a meter stick and finds that the length of the train car is 24 m. The speed of the train is $0.8c$.

    a. According to Margaret, how long does it take for the car to pass Lois, an observer on the station platform?

    b. How long does Lois say that it takes the car to pass by her?

    c. What is the length of the train car according to Lois?

†9-7 A spaceship passes by an observer on earth with a speed of $0.6c$. The astronaut aboard the spaceship measures its length to be 60 m.

    a. How much time does the astronaut find that it takes the spaceship to pass the observer?

    b. Is the time in **a** t_{same} or t_{diff}? Explain.

    c. What time does the observer measure for the spaceship to pass by? Use the results of **a**.

    d. Use the results of **c** to find the length of the spaceship according to the observer.

    e. Calculate the length of the spaceship according to the observer using length contraction.

    f. The diameter of the earth is 12800 kms. What is its diameter according to the astronauts?

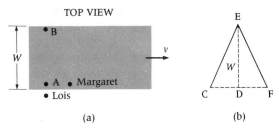

FIG. 9-22 Diagram for calculation question 9–5.

9-8 A rod traveling at $0.6c$ appears to be 0.8 m long according to a stationary observer. How long would the rod be at rest?

†9-9 This problem is an extension of Example E, illustrated in Fig. 9–5. The length L is 12 m and the train travels at $0.6c$. For simplicity, let the speed of light c be 3 m/sec.

    a. According to Frank, how much time do pulses A (or C) and B (or D) take to reach the ends of the train car?

    b. According to Larry, what is the length of the train car?

The object of the rest of the problem is to find the time for pulses C (or A) and D (or B) to reach the ends of the car, according to Larry.

    c. How far does pulse C travel from point S after 1 second? After 2 seconds? After time t? Denote the distance by x_C.

    d. How far does pulse D travel from point S after 1 second? After 2 seconds? After time t? Denote the distance by x_D.

    e. According to Larry, what is the distance d_L between the left end of the car and point S after 1 second? After 2 seconds? After time t? (Hint: At $t = 0$; d_L is not 12 m. Remember length contraction).

    f. According to Larry, what is the distance d_R between the right end of the car and point S after 1 second? After 2 seconds? After time t?

g. Find the time at which pulse C reaches the left end by letting x_C in part **c** equal d_L in part **e**.

h. Find the time at which pulse D reaches the right end by letting x_D in part **d** equal d_R in part **f**.

i. Summarizing, Frank finds that pulse A reaches the left end after _____ second(s) and pulse B reaches the right end after _____ second(s); the pulses reach the ends simultaneously. Larry, in the station, finds that pulse A reaches the left end after _____ second(s) and pulse D reaches the right end after _____ second(s); the pulses do not reach the ends simultaneously.

†9–10 A pi-meson at rest disintegrates into 2 other particles after only 2.5×10^{-8} sec. A pi-meson produced by a particle accelerator lives for 4.16×10^{-8} sec before disintegrating. Find the speed of the pi-meson produced by the accelerator. How far does it travel in the laboratory before disintegrating? How far would the pi-meson "think" that it had traveled?

†9–11 According to a stationary observer, spaceships A and B are approaching on a collision course; spaceship A has a speed of $0.6c$ and spaceship B has a speed of $0.8c$. According to an astronaut in spaceship A, how fast is B approaching?

9–12 Suppose that there is another observer in Fig. 9–18 traveling to the right at $0.8c$.

    a. Draw a diagram similar to Fig. 9–19, showing how this observer would view the collision.

    b. Use the relativistic definition of momentum and energy (Equations 14 and 15) to show that momentum and energy are conserved for this observer, too.

    c. According to this observer, what is the total energy of car A before the collision?

d. Find the rest mass of the coupled cars after the collision.

e. What are the kinetic energies of car A, car B, and the coupled cars? Show that the difference in the total kinetic energy before and after the collision results in the increase in mass of the coupled cars.

9–13 Use the values of rest mass, momentum, and energy for car A shown in Fig. 9–18 to show that the relation $E^2 = p^2c^2 + m_0^2c^4$ is valid.

9–14 Use the definition of momentum and energy in Equations 14 and 15 to show that $E^2 = p^2c^2 + m_0^2c^4$ is valid. Hint: $m_0^2c^4$ can also be written as $m_0^2c^4 (1 - v^2/c^2)/(1 - v^2/c^2)$.

†9–15 A *neutrino* is a particle that has no mass, but does have energy and momentum.

    a. Use the relationship given in calculation question 9–14 to find the relationship between momentum and energy.

    b. Use the definition of momentum in Equation 14 to show that the neutrino must travel at the speed of light if it is to have nonzero momentum. (Hint: Recall that, mathematically, zero divided by zero is not defined.)

9–16 The energy E of a particle is defined as the sum of its kinetic energy and its rest mass energy: $E = \text{KE} + m_0c^2$. When the speed of the object is much less than c, then E must equal $\frac{1}{2} m_0 v^2 + m_0c^2$.

    a. Use a hand calculator to show that
$$\frac{1}{\sqrt{1 - (.01)^2}} = 1 + \tfrac{1}{2}(.01)^2.$$

    b. Use **a** to find an approximation for the quantity $1/\sqrt{(1 - v^2/c^2)}$, when v is much less than c.

    c. Using the definition of the energy E in Equation 15 and the results of part **b**, show that $E = \frac{1}{2}m_0v^2 + m_0c^2$, when v is much less than c.

9–17 Show that, if a particle's speed were greater than the speed of light, the momentum and energy would be imaginary numbers. What do imaginary quantities indicate in physics?

Footnotes

[1] Proof of Equation 12 appears in: Margaret Stautberg Greenwood, "Relativistic Addition of Velocities Using Lorentz Contraction and Time Dilation," *American Journal of Physics*, December 1982, p. 1156.

[2] The author considers finite times to accelerate in: Margaret Stautberg Greenwood, "Use of Doppler-shifted Light Beams to Measure Time During Acceleration," *American Journal of Physics* 44(1976):259.

Max Planck

CHAPTER 10

The Dual Nature of Light

Introduction

In Chapter 8, we saw irrefutable evidence of the wave nature of light: diffraction and interference effects, which date back to Young's double-slit experiment of 1801. Thus, the wave nature of light was a cornerstone of physics when, at the turn of the twentieth century, two additional experiments concerning light were conducted. In 1900, Max Planck tried to account for the light emitted by heated objects. In another experiment called the photoelectric effect, researchers studied the ability of light to eject electrons from a metal surface.

These two experiments dealing with the emission and absorption of light were quite different from interference and diffraction experiments. Even so, it seemed reasonable and logical that they, too, would demonstrate the wave nature of light. Imagine the scientists' surprise and shock—even disbelief—when the data were, in fact, in *violent disagreement* with the wave nature of light.

In this chapter, we shall see how Planck and Einstein came to grips with this terrible dilemma. Stimulated by Planck's results, Einstein proposed a revolutionary solution to the photoelectric effect: The energy of light is concentrated in "lumps" rather than being uniformly distributed, as the wave theory holds. Because these energy lumps (today, we use the more sophisticated term *photon*) remind us of particles, light is described as having a particle nature, even though the "particles" have no mass. Furthermore, light has a *dual nature*. In some experiments, a wave nature is apparent, while in others, a particle nature is apparent.

As might be expected, this radical new theory met with a great deal of doubt. Not until 1916, some 11 years after Einstein published his solution, did Robert Millikan carry out a definitive experiment, whose data agreed completely with the predictions of Einstein's theory!

In 1923, Arthur H. Compton carried out yet another experiment that demonstrated once again the particle nature of light. Finally, the conclusions were inescapable: Since light exhibits both a wave nature and a particle nature, light must have a dual nature. This chapter covers these fascinating experiments and theories.

Today, the wave-particle duality is *the* foundation of modern physics. Chapter 15 reveals the symmetrical behavior of the wave-particle duality: Particles, such as electrons, exhibit a wave nature. Particles have a dual nature as well!

Our concluding sections are devoted to several examples of, and applications of, the photoelectric effect. For instance, everyone knows that exposure to X-rays produces harmful effects. Since X-rays and visible light are both electromagnetic radiation, why doesn't visible light produce the same effects as X-rays? We shall see that the particle nature of light provides the answer.

The many applications of the photoelectric effect are diverse: the sound track in movies, the solar battery, the television camera, and the Xerox copier, to name a few. We'll discuss these applications. Who could have possibly imagined that from a collection of cathode-ray tubes, vacuum pumps, prisms, and light sources such phenomenal technologies would eventually emerge?

Let's begin now with Planck's investigations.

Planck and the Ultraviolet Catastrophe

The German physicist Max Planck (1858–1947) was concerned with the light emitted by heated objects, but more accurately, from objects called **black bodies.** A *black body* is defined as one that absorbs *all* of the radiation that strikes it, reflecting none of it. Other objects appear colored because they reflect one particular color. For example, a leaf is green because it absorbs all other colors, but reflects green light. A black body is actually an idealized concept, but some objects approach 100 percent absorption quite closely. (An object covered with lampblack, for instance, is extremely close to being an ideal black body.)

In addition to reflecting light, objects also *emit* radiation. For example, the heating coil of an electric stove appears a dark, gray-black color when not turned on. This color is due to reflected light. However, the cold coil also emits radiation in the infrared region, to which our eyes are insensitive. But when the electric stove is turned on, the heated coil glows red. This red color is definitely due to light *emitted* by the object, and not to light reflected by it (that is, when the temperature increases, the wavelength of the predominant color shifts to a shorter wavelength—in this case, from infrared to red). Another example of this effect is a piece of metal heated in a flame. First the metal will glow red and, finally, white-hot. As it gets hotter, the predominant wavelength shifts to shorter values.

In fact, a heated solid emits a continuous *spectrum* of wavelengths, with the predominant wavelength or color being the most intense. At high temperatures, a black body is very luminous, having a predominant wavelength in the visible region of the spectrum. For example, the sun, with a surface temperature 5500°C, approximates a black body and emits an intense yellow light.

During the late nineteenth century, black-body radiation was studied experimentally by measuring the intensity of radiation for many wavelengths in the spectrum. Figure 10–1 shows this experimental data for several temperatures. Planck searched for a formula to account for these results, but he was not the first. Others had tried and failed. The English mathematician and physicist Lord Rayleigh (1842–1919) was one. His failure was quite disturbing because he used well-established principles of light and heat to derive his formula. According to Rayleigh's theory, a vibrating atom emitted light when some of its kinetic energy (heat) was transformed into light. He expected that *any* amount of kinetic energy could change into light. With this reasonable assumption, he derived the following relationship:

$$\text{Intensity of light emitted by black body} \propto \frac{1}{\lambda^4}. \qquad (1)$$

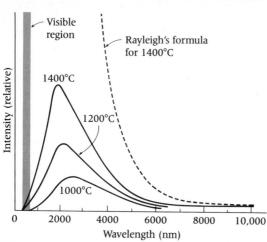

FIG. 10–1 Spectra of black-body radiation for several temperatures.

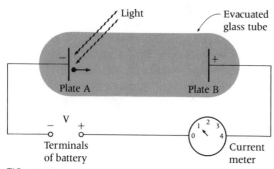

FIG. 10–2 Apparatus for observing the photoelectric effect.

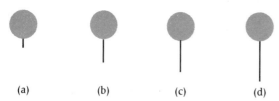

FIG. 10–3 The golf ball analogy. The golf balls are bound to the tee with varying amounts of glue. The length of the line indicates the amount of glue.

Equation 1 does fit the data at long wavelengths, as Fig. 10–1 shows, but not at short wavelengths corresponding to ultraviolet light. Scientists appropriately called this the **ultraviolet catastrophe.**

In 1900, Planck did succeed in finding a formula to predict the experimental data at *all* wavelengths, but only when he made one drastic assumption: Only *definite* amounts of energy of the vibrating atom could be changed into light, not *any* amount, as in Rayleigh's theory. When the atom vibrates with frequency *f,* the amount of energy converted into light is given by

$$E = hf, 2hf, 3hf, 4hf, \ldots \qquad (2)$$

Only one value of *h* was possible in order to fit the experimental data. Planck determined *h* to be 4.1×10^{-15} eV-sec.

While Planck's formula was successful enough in fitting the data, the underlying assumption—discrete energy transfer—made Planck and his contemporaries extremely uncomfortable. Why this restriction on the energy transfer? Nothing like it had ever been encountered before. Hopefully, some other explanation might be found. However, the *photoelectric effect*, described in the next sections, only added fuel to the fire.

The Photoelectric Effect

A piece of metal becomes quite hot when left in direct sunlight. But something else also happens, not detected so easily: Light ejects electrons from the metal. We call this the **photoelectric effect** (*photo* for light, and *electric* for electron).

We can observe the photoelectric effect using the apparatus depicted in Fig. 10–2. When light shines on plate A, the ejected electrons move toward the positively charged plate B, producing an electric current: The electrons travel through the meter, through the battery, and, finally, return to plate A. Turn off the light and the current drops to zero, as the current meter will show. Clearly, light ejects electrons from plate A. The experiment is difficult to perform because the electrical current is so extremely

TABLE 10-1 PHOTOELECTRIC EFFECT ANALOGY

EXAMPLE	ENERGY OF GOLFER'S SWING (OR ENERGY OF LIGHT)	BINDING OF GOLF BALL TO TEE (OR BINDING OF ELECTRON TO METAL)	KINETIC ENERGY OF BALL (OR KINETIC ENERGY OF ELECTRON)
1	25 units	5 units	————
2	20	————	5
3	15	15	————
4	15	20	————

small, at least one billion times smaller than the current in household appliances. Before discussing these experiments, let's see what the scientists expected to find. We'll begin by applying the principle of conservation of energy to the photoelectric effect.

Conservation of Energy

Light transmits energy. We can feel the heat energy when sunlight strikes our skin. In the photoelectric effect, the light energy overpowers the electron's attraction to the positive part of the atom, and the electron escapes. Let us apply conservation of energy to the electron's escape. First, we'll consider a simple analogy.

Suppose that a golfer, practicing long drives, hits the golf balls shown in Fig. 10-3. One of his friends, a practical joker *binds* (a term we shall continue to use) the balls to the tee with varying amounts of glue, indicated by the length of the vertical lines. Figure 10-3(a) has the smallest amount of glue, and Fig. 10-3(d), the largest.

If the golfer swings with the same amount of energy each time, the golf ball with the smallest binding (glue) will have the largest speed and hence, the largest kinetic energy. In fact, if the golf ball is too tightly bound (too much glue), the ball will remain stuck to the tee.

Suppose, now, that the golfer strikes a number of golf balls all bound by the same amount. If the golfer increases the energy of his swing each time, then the speed and kinetic energy of the golf ball increases.

The energy of the swing is divided into two parts: some of it is used to release the ball from

the tee, and the rest appears as the kinetic energy of the golf ball. Thus,

$$\begin{matrix} \text{Total energy} \\ \text{before hitting} \\ \text{golf ball} \end{matrix} = \begin{matrix} \text{total energy} \\ \text{after hitting} \\ \text{golf ball} \end{matrix} \qquad (3)$$

$$\frac{\text{Energy of}}{\text{swing}} = \frac{\text{kinetic energy}}{\text{of ball}} + \frac{\text{binding}}{\text{energy.}}$$

The binding energy is the amount of energy required to release the golf ball from the tee. Equation 3 becomes

$$\frac{\text{KE of}}{\text{golf ball}} = \frac{\text{energy of}}{\text{swing}} - \frac{\text{binding}}{\text{energy.}} \qquad (4)$$

We can write a similar equation for light striking metal. The light supplies energy B to overcome the electron's attraction to the atom and gives the electron kinetic energy:

$$\frac{\text{KE of}}{\text{electron}} = \frac{\text{energy of}}{\text{light}} - B. \qquad (5)$$

Let's use these ideas to fill in the blanks in Table 10-1. The answers to examples 1, 2, and 3 are, respectively, 20, 15, and 0 units. However, example 4 requires some discussion. In example 3, the energy is *just* sufficient to release the ball from the tee, and the golf ball does not

THE DUAL NATURE OF LIGHT

move (speed = 0). In example 4, the binding energy is *greater* than the energy of the swing! The ball will remain fastened to the tee (or the electron will not be released from the metal).

We can think of a number of interesting questions. For example, the kinetic energy of the golf ball depends upon the energy of the golfer's swing (that is, whether the shot is a long drive or a putt) and upon its binding to the tee. In the photoelectric effect, does the kinetic energy of the electron change if the wavelength of the light is changed? Does the kinetic energy change if a brighter light is used? Does it change if a different type of metal is used? Let's look first at the answers provided by the wave nature of light and compare them with experimental results.

Energy of Waves and Expected Results

A wave's amplitude, wavelength, and frequency completely define its shape (Chapter 8). But how do we define the wave's energy? Once we know this, we can find the electron's kinetic energy from Equation 5.

Picture yourself now at Waikiki Beach in Honolulu, ready to do some surfing and watching those high, magnificent waves come toward shore. Which wave has more energy—a 1-m wave or a 5-m wave? A ridiculous question! Undoubtedly, the 5-m wave has more energy. Even though these treacherous ocean waves have a different shape from the undulating water waves described in Chapter 8, we still associate the water wave's energy with its height or its amplitude. In fact, there are so many similarities between water waves and light waves (see Chapter 8), we assume that the light wave's energy also depends upon its amplitude.

Because the amplitude of a light wave refers to the strength of the electric and magnetic field, a larger amplitude means a larger electric and magnetic field. Since the force on an electron due to the electric field is given by F equals QE (Chapter 7, Equation 5), the force increases when the amplitude increases. A larger force should be more effective in ejecting an electron; thus, we associate the amplitude of a wave with its energy. Increase the amplitude of a light wave, and the energy of the wave likewise increases.

The brightness or intensity of a light beam depends upon its amplitude. For example, suppose that you switch the three-way lamp from 50 to 200 watts while reading the newspaper. The color of the light does not change, but the light is certainly brighter. What you have done is to increase the *amplitude* of the light wave. Since the light wave's energy depends on its amplitude, and its amplitude, in turn, depends upon its brightness, the energy of a light wave thus depends upon its brightness. Therefore, in the photoelectric effect, we expect the electron's kinetic energy to increase when we use a brighter light. Equation 5 shows that the kinetic energy increases when the light's energy increases.

How does the transfer of energy to the electron take place, gradually or instantaneously? Let's think about water waves for a moment.

Suppose that an oscillatory wave of amplitude A passes as you are treading water (after falling from a surfboard). As the waves go by, you *gradually* rise to a maximum height A and then fall to the minimum depth A. The potential energy you receive depends upon time: It *takes time* for you to reach the maximum height A, and time for you to fall. From this example, we arrive at a general conclusion: For a light wave (as well as a water wave), the transfer of energy occurs gradually. It takes time; it does not occur instantaneously. An electron should absorb energy gradually as the light wave passes. It can escape from the metal after it has absorbed enough energy.

Early Experimental Results

As early as 1888, scientists investigated the photoelectric effect. While these first experiments were not sophisticated (they did not

PREDICTIONS OF THE WAVE NATURE	EXPERIMENTAL RESULTS	EINSTEIN'S EXPLANATION BASED UPON PARTICLE NATURE OF LIGHT
Since the energy of light depends upon its intensity, red light and ultraviolet light of the same intensity should be *equally* effective in releasing electrons.	Ultraviolet light releases electrons, while red light does not.	An ultraviolet photon has more energy than a red photon. The energy of the ultraviolet photon is larger than the binding of the electron, and so the electron escapes. The energy of the red photon is smaller than the binding, and so the electron does not escape.
Electrons should be released if the intensity is increased sufficiently.	Electrons were not released using red light even when the intensity was greatly increased.	Increasing the intensity of red light increases the *number* of photons, but does not change the energy of the red photon.
Electrons should not be released using *very* dim light because the intensity is too small.	Electrons were released using very dim ultraviolet light, but their number was very small.	Decreasing the intensity of ultraviolet light reduces the number of photons and produces a smaller number of electrons.
There will be a time delay while the electron absorbs sufficient energy to overcome binding.	Electrons were released immediately using ultraviolet light.	Ultraviolet photons are absorbed in an instant.

measure the electron's kinetic energy), the results still proved to be quite disturbing:

1. Ultraviolet light (400 nm) of *any* brightness ejected electrons from the metal.
2. These electrons escaped *immediately* after the light was turned on. No time delay!
3. Red light (640 nm), *no matter how bright,* did not eject electrons from the metal.

These results did not agree at all with the wave nature of light, as Table 10–2 shows. *There was not a single point of agreement.* How could this be? The wave nature of light was *not* a speculation. It had been demonstrated in many experiments. Why wasn't the wave nature exhibited in the photoelectric effect?

Einstein's Solution

Einstein discerned a connection between Planck's analysis of black-body radiation and the photoelectric effect. Planck found that definite amounts of energy (nhf) must be transformed into light emitted by a heated object.

Here was a dependence upon frequency. Likewise, a dependence upon the frequency or color of light was needed to understand the curious photoelectric effect. Einstein proposed that the energy of light was concentrated into packets or "lumps" of energy rather than being distributed uniformly throughout the beam of light, as in the wave theory. Today, these packets of energy are called **photons,** a term we shall use from here on. A photon of light has an energy given by

$$E = hf.$$

When the intensity of the light increases, the *number* of photons, but not their energy, increases. The number of photons is proportional to the intensity.

When a photon strikes an electron in the metal, the entire photon is absorbed by the elec-

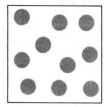

(a) Energy distribution of ultraviolet light, 100 watts

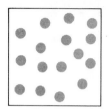

(b) Energy distribution of red light, 100 watts

(c) Energy distribution of light, 100 watts wave theory

FIG. 10−4   Light having an intensity of 100 watts strikes a metal surface. According to Einstein's theory, the light energy is subdivided into packets of energy, called *photons*. In part (a), the photons of ultraviolet light are larger than those of red light in part (b). However, since the total energy per second (wattage) is the same, there are more photons of red light than of ultraviolet light. Part (c) represents the continuous distribution of energy over the surface, as predicted by the wave theory. Note: The circles in (a) and (b) indicate the amount of energy contained in the photons, not the amount of space occupied by a photon.

tron. If the photon has sufficient energy to overcome the binding of the electron, the electron will be released *immediately* from the metal, with a kinetic energy given by

$$KE = hf - B. \qquad (6)$$

This equation is obtained by substituting hf for the energy of light in Equation 5. If hf is larger than B, the electron will escape *immediately*; if not, it won't escape at all.

We have just seen the results obtained by using red light and ultraviolet light. Now, let's see how Einstein's particle theory explains these results, where the wave nature theory does not.

Consider the light from two sources—one ultraviolet (400 nm) and the other red (640 nm)—of the same intensity, say 100 watts, striking the metal surfaces shown in Fig. 10−4. Since the frequency of light is given by f equals c/λ (Chapter 8, Equation 1), ultraviolet light has a larger frequency than red light. Because E equals hf, a photon of ultraviolet light has more energy than a photon of red light. In this case, a photon of ultraviolet light has 640/400, or 1.6 times more energy than a photon of red light. As the intensity (or wattage) of both sources is the same, the red beam has more photons of *smaller* energy than the ultraviolet beam. In Fig. 10−4, the ultraviolet photons are 1.6 times larger than the red photons. There are 16 red photons but only 10 ultraviolet photons. Thus, the *total* energy striking the surfaces in Fig. 10−4(a) and (b) is the same, but it is subdivided differently. This is in sharp contrast to the wave theory (Fig. 10−4c), where the energy of the beam is uniformly distributed over the surface. These differences in energy distribution explain why ultraviolet light releases electrons, why red light does not, and why the wave theory is inadequate. Table 10−2 gives Einstein's explanation of the experimental results.

THE CRUCIAL TEST

Even though the data showed ultraviolet photons had more energy than red photons, the kinetic energy was not measured in these early experiments. Einstein's **photoelectric equation** (as Equation 6 is now called) relates the kinetic energy of the electron to the frequency of the light. Thus, the crucial test of Einstein's theory is to find out if experimental data agree with Equation 6.

Equation 6 is plotted in Fig. 10−5 and, for the sake of comparison, the straight-line $y = 8x − 3$ is plotted in Fig. 10−6. The dashed line in Fig. 10−5 indicates that the kinetic energy can never be a negative number. If the photon energy is *less than* the binding (so that a negative number results for the kinetic energy in Equation 6), the electron simply doesn't escape. In the early experiments, this happened for red photons.

Several similarities are apparent in Figs. 10−5 and 10−6. When $x = 0$, $y = −3$. Similarly, when $f = 0$, $KE = −B$. Thus, we can find the binding energy B of the electron by extending the straight line to pass through the kinetic energy axis. We can find the coefficient of x (the number in front of x) directly from the two broken-line triangles in Fig. 10−6. In each case, the length of the vertical side divided by the length of the horizontal side equals 8. Here the "length" refers to the length along the x axis and y axes, and depends upon the scale of these axes. Similarly, we can find Planck's constant h (the coefficient of the frequency f in Equation 6) from the triangles in Fig. 10−5. The constant h equals the length of the vertical side divided by the length of the horizontal side.

The following steps constitute a crucial test of Einstein's theory:

1. Shine light of frequency f on a metal plate and measure the electrons' kinetic energy.
2. Repeat step 1 for several different frequencies.
3. Repeat step 1 for several different metal plates.
4. Plot a graph of kinetic energy versus frequency for each metal plate.
5. If this graph is a straight line, find values of Planck's constant h and binding energy B. If Einstein's theory is valid, the value of h should be the same as Planck's value.
6. The straight lines for different metal plates should be parallel.

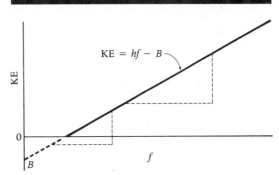

FIG. 10−5   A graph of the photoelectric equation, $KE = hf − B$.

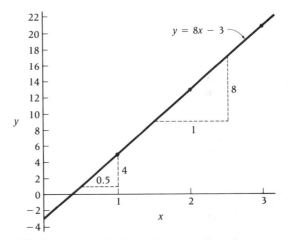

FIG. 10−6   A graph of straight-line $y = 8x − 3$.

In 1916, the American physicist Robert A. Millikan (1868–1953) performed this crucial experiment.

Millikan's Experiment

Einstein's theory certainly caused a great deal of controversy and consternation. Millikan wrote:

It was in 1905 that Einstein made the first coupling of photo effects with any form of quantum theory by bringing forward the bold, not to say the reckless, hypothesis of an electro-magnetic light corpuscle of energy hf, *which energy was transferred upon absorption to an electron. This hypothesis may well be called reckless first because an electro-magnetic disturbance which remains localized in space seems a violation of the very conception of an electro-magnetic disturbance, and second because it flies in the face of the thoroughly established facts of interference. The hypothesis was apparently made solely because it furnished a ready explanation of one of the most remarkable facts brought to light by recent investigations, viz., that the energy with which an electron is thrown out of a metal by ultraviolet light or X-rays is independent of the intensity of the light while it depends on its frequency. This fact alone seems to demand some modification of classical theory or, at any rate, it has not yet been interpreted satisfactorily in terms of classical theory.*[1]

During this time, Millikan was also carrying out experiments to measure the charge on the electron. He measured the kinetic energy of electrons released from sodium and lithium metals. Millikan's actual experiment was quite sophisticated. Figure 10–7 shows the essential features, which are the same as the apparatus depicted in Fig. 10–2 except that the voltage connections are reversed. Millikan used a mercury lamp for the light source and a prism to break up the light into a spectrum of colors. He directed only one color at a time toward the metal surface.

KINETIC ENERGY MEASUREMENTS

Figure 10–8 illustrates how Millikan determined the electrons' kinetic energy. The electrons escaping from plate A have speed v_p. As an electron travels toward plate B, it slows down due to negative charges on plate B and positive charges on plate A. This electron will stop between the plates if these forces are large enough (Fig. 10–8a); the electron will then return to plate A. In this case, the current meter reads zero.

Reduce the charges on plates A and B by reducing the voltage ($V_2 < V_1$), and the electron will travel a greater distance (Fig. 10–8b). The current meter still reads zero.

Reduce the voltage even more, and the electron will travel a still greater distance toward plate B. At some voltage, called V_{max}, the electron will *just be able* to reach plate B. "Just be able" means that the electron's speed will be zero at plate B.

Reduce the voltage still more ($V_3 < V_{max}$) and the electron will still have some kinetic energy when it bangs into plate B (Fig. 10–8d). In this case, the current meter does show a current.

The voltage V_{max} is a dividing line: For voltages less than V_{max}, Millikan observed a current on the meter; for voltages greater than V_{max}, he didn't.

There is a definite relationship between the speed v_p and the voltage V_{max} *just* required to stop the electron. For example, if the speed v_p had been larger than in Fig. 10–8, V_{max} would also have to be larger. What is the relationship between v_p and V_{max}?

In Chapter 7, we learned that the kinetic energy of the electron was related to its retarding voltage, here V_{max}. When the voltage is V_{max}, the electron *loses* V_{max} eV in traveling from plate A to plate B. Since its kinetic energy at plate B

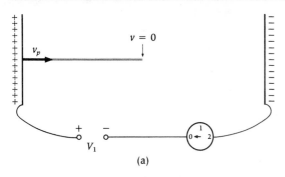

(a)

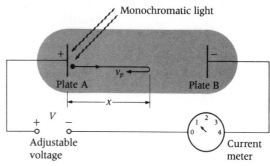

FIG. 10–7 Millikan's apparatus for measuring the kinetic energy of electrons.

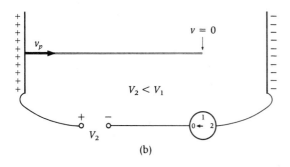

(b)

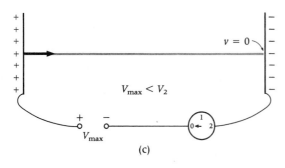

(c)

FIG. 10–8 Light strikes plate A, releasing electrons with speed v_p. In (a), the retarding force is so large that an electron stops before reaching plate B. In (b), an electron travels farther before stopping because the retarding force is smaller than in (a). In (c), an electron is just able to reach plate B, where its speed is zero, because the retarding force is less than in (b). In (d), the retarding force is less than in (c), so that the electron still has some kinetic energy left when it bangs into plate B. Its kinetic energy is $V_{max} - V_3$ eV.

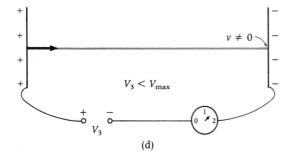

(d)

THE DUAL NATURE OF LIGHT

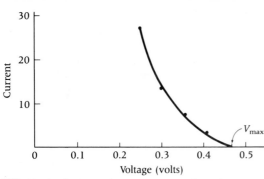

FIG. 10−9 A graph of current versus voltage for green light (546.1 nm) shining on sodium.

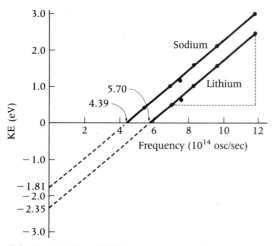

FIG. 10−10 Plot of Millikan data given in Table 10−3.

is zero, the initial kinetic energy at plate A must have been V_{max} eV, or

$$KE_A = \frac{1}{2}mv_p^2$$

$$= V_{max}eV.$$

Figure 10−9 shows how Millikan obtained the voltage V_{max} for green light (λ = 546.1 nm) shining on sodium metal. He drew a smooth curve through the four data points and extrapolated to zero current, and found that V_{max} was 0.47 volts.

DATA

Millikan repeated this experiment for several wavelengths and for sodium and lithium metals. Table 10−3 presents this data.

We use Equation 1 of Chapter 8 to find the frequency. For example, the frequency of green light is

$$f = \frac{c}{\lambda}$$

$$= \frac{3 \times 10^8\,\text{m/sec}}{546.1 \times 10^{-9}\,\text{m}}$$

$$= 5.49 \times 10^{14}/\text{sec},$$

where c equals 3×10^8 m/sec, and 1 nm equals 10^{-9} m.

The data in Table 10−3 show that light of wavelength 546.1 nm does not have enough energy to release electrons from lithium. This situation is similar to example 4 in Table 10−1.

We can now answer those crucial questions: Is the graph of kinetic energy versus frequency a straight line? If so, is the value of h the same as Planck's? Millikan's data provide a resounding yes! Figure 10−10 shows two parallel straight lines. We can calculate the value of h, as Millikan did, by using the broken-line triangle in Fig. 10−10. The constant h is given by

$$h = \frac{\text{length of vertical side}}{\text{length of horizontal side}}.$$

TABLE 10-3 MILLIKAN'S DATA

WAVELENGTH (nm)	TYPE OF LIGHT	FREQUENCY (osc/sec)	SODIUM V_{max}(volts) or KE(eV)	LITHIUM V_{max}(volts) or KE(eV)
546.1	green	5.490×10^{14}	0.47	NR*
433.9	violet	6.909	1.02	0.50
404.7	ultraviolet	7.408	1.22	0.69
365.0	ultraviolet	8.214	1.60	1.04
312.6	ultraviolet	9.590	2.13	1.62
253.5	ultraviolet	11.826	3.03	2.53

*Electrons were not released.

Using values in Table 10-3, we find

Length of vertical side $= 2.53\ \text{eV} - 0.50\ \text{eV}$

$$= 2.03\ \text{eV},$$

and

Length of horizontal side

$= 11.826 \times 10^{14}/\text{sec} - 6.909 \times 10^{14}/\text{sec}$

$= 4.917 \times 10^{14}/\text{sec}.$

Thus,

$$h = \frac{2.03\ \text{eV}}{4.917 \times 10^{14}/\text{sec}}$$

$$= 4.13 \times 10^{-15}\ \text{eV-sec}.$$

(The abbreviation "osc" refers to the *number* of oscillation, which is unitless, and so osc is omitted from the units of f). This value of h is the same as Planck's.

We could use any other triangle, and the value of h would be the same. Note that the straight lines for sodium and lithium are parallel, also indicating the same value of h. The general equation representing these two lines is

$$KE = hf - B.$$

Millikan found the binding energy B by extending the straight line through the kinetic energy axis: B equals 1.81 eV for sodium, and B equals 2.35 eV for lithium.

Millikan made another interesting comparison. Using the wave theory, Millikan *calculated* that a (standard) candle placed 3 m away from a metal plate would have to illuminate it for *4 hours* before electrons would be ejected! In his experiment, Millikan found that electrons were released immediately after the mercury light was turned on—yet another disagreement between the wave theory and experiment.

Without a doubt, Einstein's theory passed Millikan's experimental test with flying colors. Even so, Millikan wrote:

Despite then the apparently complete success of the Einstein equation, the physical theory . . . is found so untenable that Einstein himself, I believe, no longer holds to it. But how else can the equation be obtained?[2]

That is, Millikan was willing to accept the straight-line relationship between kinetic energy and frequency, but he could not accept the **particle theory:** that hf referred to an energy packet. So onerous, so repugnant, was this relatively new theory. This attitude is understandable because the wave nature of light had been accepted for an entire century! The particle theory, to repeat Millikan, "seems a violation of the very conception of an electromagnetic disturbance . . . because it flies in the face of the thoroughly established facts of interference."[3]

As time passed, the particle theory came to be accepted. In 1921, Einstein received the Nobel Prize in physics for his explanation of the photoelectric effect. In 1923, the American physicist Arthur H. Compton (1892–1962) demonstrated the particle nature of X-rays: a third experiment showing the particle nature of "electromagnetic disturbance."

Compton's Scattering of X-Rays

In his experiment, Compton directed an X-ray beam of only *one* wavelength λ_0 onto a carbon target (see Fig. 10–11). He measured the wavelength of the X-rays scattered to angle θ. The X-rays are scattered by the electrons in the carbon target and, according to the wave nature, these scattered X-rays should also have wavelength λ_0. However, Compton found that some of the X-rays had a longer wavelength. Increase angle θ and the wavelength λ increases.

Compton explained this by assuming a billiard-ball collision between a photon and an electron. The X-ray photon transfers some of its energy to the electron, giving it kinetic energy (Fig. 10–11b). Due to conservation of energy, the scattered X-ray photon must have less energy and, hence, a wavelength longer than λ_0. Since

$$E = hf$$
$$= \frac{hc}{\lambda},$$

the wavelength increases as the energy decreases.

According to Einstein's theory of relativity, a photon also has momentum. When the photon "collides" with the electron, some of its momentum is transferred to the electron.

Compton applied the conservation of energy and momentum principles to the photon-electron collision and found a formula for the wavelength at angle θ in terms of λ_0 and Planck's constant h. The wavelengths found from this formula agreed well with the experimental values. This was the third experiment that demonstrated the particle nature of light; the experimental evidence was overwhelming. There was no escape from it: Light has a particle nature.

The Nature of Light

You may well ask, What is the *real* nature of light? The interference effects demonstrate its wave nature and now three experiments show its particle nature. Is this not a contradiction? The only escape from this dilemma seems to be that light has two natures—a dual nature. Much like a schizophrenic person exhibiting multiple personalities, light has two personalities and exhibits one of them, depending upon the circumstances. Some experiments, such as the double-slit experiment, exhibit the wave nature, while others, such as the photoelectric effect, exhibit the particle nature.

When light interacts with matter (that is, light is emitted or absorbed), it exhibits its particle nature. But when light travels from place to place, it exhibits its wave nature. The two natures are, of course, entwined, because we need the *frequency* (a characteristic describing its wave nature) to obtain the energy of a photon. This dual and entwined behavior of light is called the **wave-particle duality.**

In Chapter 15, we shall examine the symmetric nature of the wave-particle duality. Just as the symmetry of a geometric design is pleasing to an artist, symmetry is pleasing to scientists as well. In 1923, Louis de Broglie speculated about the symmetry of the wave-particle duality. He wondered whether particles, such as electrons, might exhibit a wave nature.

We shall consider such fascinating speculation and experimental evidence in Chapter 15. Here, let's complete our story of light's nature by looking at two examples of, and then some applications of, the photoelectric effect.

ELECTROMAGNETIC RADIATION AND HUMAN CELLS: EXAMPLE ONE

Visible light, X-rays, and gamma rays are all forms of electromagnetic radiation (see Table 8–1, p. 181). Why do X-rays and gamma rays destroy cells of human tissue, while (fortunately for us) visible light rays do not? Since the cells absorb *photons* of radiation, we must compare the photon energies of these various rays.

The wavelength of visible light is 500 nm, and the wavelength of X-rays is about 0.1 nm. The energy of a visible light photon is

$$E = hf$$
$$= \frac{hc}{\lambda}$$
$$= \frac{(4.13 \times 10^{-15}\,\text{eV-sec})(3 \times 10^{8}\,\text{m/sec})}{500 \times 10^{-9}\,\text{m}}$$
$$= 2.48\,\text{eV}.$$

The energy of an X-ray photon is 5000 times larger, or 1.24×10^4 eV.

In contrast, the gamma rays from a cobalt-60 radioactive source (used to destroy cancer cells) have a photon energy of about 1 million electron-volts (abbreviation: MeV).

An X-ray or gamma ray photon, striking a molecule in the cell, detaches an electron from it: a process called **ionization.** An energy of 30 eV will cause the ionization of any molecule in human tissues; that is, the binding energy B of the electron is 30 eV. The remaining part of the photon's energy goes to the kinetic energy of the electron. This careening electron will then cause the ionization of other molecules, and so on. A single photon may start a chain reaction, causing destruction of many molecules.

We can now see why visible light rays do not destroy human cells: The energy of a visible light photon is simply not large enough to cause the ionization of molecules in cells. Hence, visible light poses no threat.

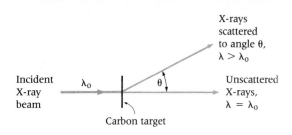

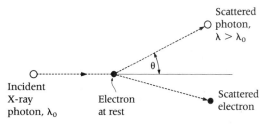

FIG. 10–11   (a) Compton directed an X-ray beam (λ_0) onto a carbon target and measured the wavelength at angle θ. (b) A billiard-ball collision between a photon and electron in the carbon target. The scattered photon must have less energy than the incoming photon and, hence, the scattered photon has a wavelength larger than λ_0.

THE DUAL NATURE OF LIGHT

Radiation causes the destruction of many molecules in the cell. How does this affect the cell's reproduction? The master molecule of the cell is the *deoxyribonucleic acid* molecule (abbreviation: DNA), which contains the genetic code. If this molecule is damaged by ionization, it will not divide properly, and a mutation will occur. If this molecule is perhaps torn into several fragments, the cell may not be able to divide at all. This leads to the "death" of the cell, since cell division cannot occur.

X-RAY PRODUCTION: EXAMPLE TWO

X-rays are produced in a cathode ray tube when electrons strike the anode, slow down, and finally stop. Here, let us suppose that the voltage connecting the cathode and anode is 35,000 volts. The kinetic energy of the electrons, just before they strike the anode, is 35,000 eV. This kinetic energy is transformed into the photon energy of the X-rays.

When only one X-ray photon is produced, the photon has an energy of 35,000 eV. The energy of an X-ray photon is

$$E = hf$$
$$= \frac{hc}{\lambda},$$

and the wavelength is

$$\lambda = \frac{hc}{E}$$
$$= \frac{(4.13 \times 10^{-15}\,\text{eV-sec})(3 \times 10^{8}\,\text{m/sec})}{35,000\,\text{eV}}$$
$$= 0.0354 \times 10^{-9}\,\text{m}$$
$$= 0.0354\,\text{nm}.$$

Suppose that 3 photons having energies of 5000 eV, 10,000 eV, and 20,000 eV are produced. The wavelengths are, respectively, 0.248 nm, 0.124 nm, and 0.0620 nm. Any number of photons can be produced when the electrons come to rest in the anode. In this example, the X-ray photons range from 35,000 eV to zero, and have wavelengths ranging from a minimum of 0.0354 nm to very long wavelengths: a *continuous* distribution of wavelengths.

In the radiologic department of many hospitals, particle accelerators generate electron beams having an energy of about 18 MeV. These electrons, aimed at a solid target, produce X-rays having a maximum energy of 18 MeV. This kind of treatment for cancerous tumors is sometimes preferred to treatment using cobalt-60 gamma rays.

Technological Applications

MOTION-PICTURE SOUND

Figure 10−12 illustrates how a movie **sound track** is made. The current from the microphone, amplified by suitable electronics, varies the *length* of the slit. (Note that the diagram shows the *width* of the slit.) As the film passes underneath the slit, its exposure depends upon the amount of current from the microphone: the length of the slit is proportional to the current.

Figure 10−13 shows the negative and positive prints of the sound track. The region corresponding to the slit is transparent in the positive print. Figure 10−14 shows a section of the movie film.

Figure 10−15 shows how the sound is reproduced. A slit of light falls onto the film. Only the light passing through the transparent part of the sound track enters the photoelectric cell, commonly called the *photocell*. As we learned, when the intensity of the light increases, the number of electrons ejected increases proportionately. Thus, the current from the photocell depends upon the amount of light passing through the sound track. This varying current is fed into the loudspeaker, producing the sound we hear at the movies.

SOLAR BATTERY

A **solar battery** is constructed of semiconducting material. The term *semiconducting* refers to material that does not conduct electrical current as readily as conductors, such as copper, nor as poorly as insulators, such as glass, which do not conduct current at all. Silicon, with very minute and controlled amounts of impurities, is a **semiconductor.** Figure 10–16 presents a schematic diagram of a solar battery. Such a battery consists of two types of silicon: *n*-type silicon, which has a small amount of arsenic added to it, and *p*-type silicon, which has a small amount of gallium added to it. When light from the sun strikes the junction between the *p*- and *n*-type silicon, electrons are freed to carry electrical current (that is, electrons are not released from the material, but are at least mobile). As a result of the sunlight, electrons are free to travel from the *n*-side through the circuit to the *p*-side. Thus, the semiconductor device acts just like a battery.

TELEVISION

Let's talk about TV in terms of its reception and its transmission.

Reception How does a TV set work? Figure 10–17 presents a diagram of a TV cathode-ray tube. The wire filament of the electron gun becomes extremely hot as a huge current passes through it: Electrons are boiled off and race toward the (positive) anode. Some of the electrons pass through a hole in the anode and are deflected by two sets of capacitor plates. The amount of charge on the *y*-plates dictates how the beam will be deflected in the vertical direction. Likewise, the charge on the *x*-plates dictates how the beam will be deflected in the horizontal direction. Thus, at one instant of time, the electron bean strikes *only one spot* on the screen. The fluorescent screen emits a speck of light when an electron strikes it. The amount of light, therefore, depends upon the number of electrons that strike the screen. The TV picture is produced as the electron beam scans the entire area.

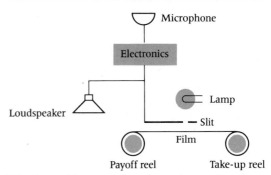

FIG. 10–12   The production of a movie sound track.

Negative   Positive

FIG. 10–13   The positive and negative prints of a sound track.

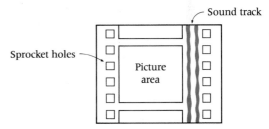

FIG. 10–14   A section of a movie film with a sound track.

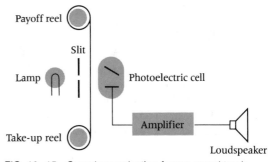

FIG. 10–15   Sound reproduction from a sound track.

THE DUAL NATURE OF LIGHT

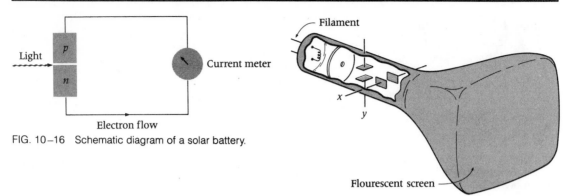

Light

p

n

Current meter

Electron flow

FIG. 10–16   Schematic diagram of a solar battery.

Filament

x

y

Flourescent screen

FIG. 10–17   A black-and-white TV cathode-ray tube.

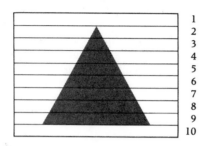

1
2
3
4
5
6
7
8
9
10

FIG. 10–18   The picture divided into 10 lines.

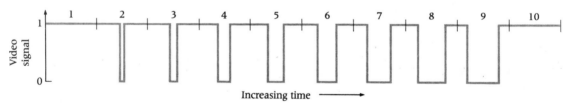

Video signal

1
0

1   2   3   4   5   6   7   8   9   10

Increasing time

FIG. 10–19   A video signal produced by a TV camera
and received by a TV set, resulting in the picture shown
in Fig. 10–18.

FIG. 10–20   The image of a triangle focused on the
plate of a TV camera.

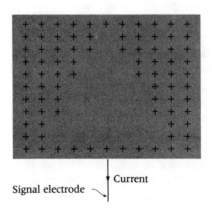

Current

Signal electrode

In commercial television, the picture area is divided into 525 lines, and the entire area is scanned at the rate of 30 times per second. Let's consider this principle by dividing the picture area into 10 lines, as shown in Fig. 10−18. The signal that will produce this picture is shown in Fig. 10−19. The signal governs the number of electrons that strike the screen at each point: the larger the signal, the larger the number of electrons.

Transmission In a TV camera, the image is focused onto the plate where the light causes electrons to be ejected (see Fig. 10−20). The *number* of ejected electrons depends upon the intensity of the light. The plate is then positively charged, and the distribution of the positive charges corresponds to the intensity distribution of the image. An electron beam aimed at the plate scans it line by line. *Each* region of the screen absorbs electrons to neutralize the positive charges, which produces a current in the signal electrode that is proportional to the intensity at that point on the screen. The video signal in Fig. 10−19 is obtained as the electron beam scans the plate.

In color TV, three video signals, corresponding to the red, blue, and green components, must be transmitted and received. With color TV, three electron guns act independently, striking red, blue, and green fluorescent material on the TV screen.

Summary

Planck tried to account for the spectrum of light emitted by a black body (an object that absorbs all radiation striking it). He could find agreement between his theory and experiment only when he assumed that the amount of heat energy converted into light equaled $hf, 2hf, 3hf, \ldots$, where h equals 4.135×10^{-15} eV-sec (the presently accepted value). Lord Rayleigh's theory failed because, at small wavelengths, he could not obtain agreement with experiment. This failure was called the ultraviolet catastrophe.

Ultraviolet light ejects electrons from a metal, while red light does not. This was shown experimentally. According to the wave nature of light, the energy of light should increase when its amplitude or brightness increased: The energy of light should not depend upon its wavelength or color. Obviously, the experimental results could not be explained by the wave nature of light.

Einstein assumed that the energy of light was subdivided into packets (now called photons) of energy. The energy of a photon is hf. When the intensity or brightness of the light increased, the *number* of photons increased, but their size *did not* change. When a photon strikes an atom, the *entire* photon is absorbed. Some energy is needed to detach the electron from the atom; this energy is called the binding energy B. The rest of the energy is converted to the electron's kinetic energy. This is expressed in the photoelectric equation ($KE = hf - B$).

If the photon energy is not large enough to overcome the electron's binding, the electron cannot escape. The frequency f is related to the wavelength λ by f equals c/λ. A photon of ultraviolet light (wavelength of 400 nm or smaller) has more energy than a photon of red light (wavelength about 600 nm). Therefore, a photon of ultraviolet light has sufficient energy to overcome the electron's binding, while one of red light does not.

Millikan showed that Einstein's photoelectric equation agreed with experimental results. Millikan measured the kinetic energy of electrons ejected by light of frequency f by determining the maximum retarding voltage V_{max} *just* needed to stop the electrons. When the voltage between the two plates of a cathode ray tube *exceeded* V_{max}, the electrons could not reach the negatively charged plate and the current meter in the circuit dropped to zero.

Millikan's data (kinetic energy on the vertical axis and f on the horizontal axis) showed a straight line in agreement with the photoelectric equation. The binding energy B is determined by extending the straight line so that it intersects the vertical axis. The constant h is determined from the steepness of the straight line. Draw a triangle with two sides parallel to axes; h is equal to the length of vertical side divided by the length of the horizontal side. This value of h agrees with Planck's value.

Compton directed X-rays of wavelength λ_0 onto a carbon target and found that scattered X-rays had a wavelength larger than λ_0. Compton accounted for the wavelength of X-rays at angle θ by considering a collision between an X-ray *photon* and an electron in the target much like a billiard-ball collision. The electron receives some energy—kinetic energy. The photon must have less energy as a result and, hence, the scat-tered photons have a wavelength greater than λ_0.

X-rays and gamma rays are harmful to human tissue because their photon energy is so much larger (greater than 5000 times larger) than the photon energy of visible light. DNA molecules, containing the genetic code, are torn apart by X-rays and gamma rays.

X-rays produced in a cathode-ray tube have a continuous distribution of energy (and, hence, wavelength). The kinetic energy of an electron striking the anode is transformed into one *or more* X-ray photons.

Some practical applications of the photoelectric effect include the motion-picture sound track, the solar battery, and the TV set.

True or False Questions

Indicate whether the following statements are true or false. Change all of the false statements so that they read correctly.

10-1 The radiation emitted by a solid object has the following properties:

    a. The light exhibits a continuous spectrum, in contrast to a gas in a cathode-ray tube, which exhibits a line spectrum.

    b. The continuous spectrum exhibits the ultraviolet catastrophe because the intensity increases as the wavelength decreases.

    c. In the continuous spectrum, the intensity is maximum at one particular value of the wavelength.

    d. When the solid is at room temperature, most of the radiation is ultraviolet.

    e. When the temperature of the solid increases, the wavelength of maximum intensity shifts to a longer wavelength.

    f. A certain class of solid objects that reflect all of the radiation striking them, and that absorb none of it, are called black bodies.

10-2 Using the principles of light, heat, and energy, Lord Rayleigh predicted the spectrum of light emitted by a black body. He predicted the following:

    a. In the continuous spectrum, the intensity of the light should be maximum for one particular value of the wavelength.

    b. In the continuous spectrum, the intensity increases when the wavelength decreases, an effect called the ultraviolet catastrophe.

10-3 a. Rayleigh and Planck agreed that light is emitted by a solid object when some kinetic energy of the vibrating atom transforms into light.

    b. Further, Rayleigh and Planck assumed that only certain amounts of kinetic energy could be transformed into light.

10-4 Around 1890, researchers directed light from a source onto a metal surface and observed whether or not electrons were ejected by the light. They observed the following:

    a. Dim red light (wavelength of 640 nm from a 25-watt source) was not able to eject electrons.

b. When the red-light source wattage was increased to 500 watts, the bright red light was able to eject electrons.

c. Dim ultraviolet light (wavelength of 320 nm from a 25-watt source) was not able to eject electrons.

d. Bright ultraviolet light from a 500-watt source was able to eject electrons.

e. When electrons were ejected, the electrons were ejected as soon as the light was turned on—immediately.

10–5 Since Young's double-slit experiment in 1801, light was known to have a wave nature. According to the wave nature of light, researchers conducting photoelectric effect experiments around 1890 *expected* the following results:

a. If no electrons are ejected, then increasing the brightness of the light (that is, increasing the energy of the light) should cause electrons to be ejected.

b. A red light source and an ultraviolet light source both having a wattage of 200 watts should be equally effective in ejecting electrons.

c. Electrons should be ejected as soon as the light is turned on—immediately.

10–6 Two light sources rated at 200 watts are placed equidistant from two identical metal squares. One source emits red light of 640 nm wavelength and the other emits ultraviolet light of 320 nm wavelength.

a. A photon of ultraviolet light contains twice as much energy as a photon of red light.

b. The square illuminated by ultraviolet light receives twice as much energy in 1 second as the one illuminated by red light.

c. In both squares, the light energy is distributed uniformly over the entire square.

10–7 Light shines on a metal surface.

a. The kinetic energy of the ejected electrons can be increased by increasing the wavelength of the light.

b. The kinetic energy of the ejected electrons does not change when the brightness of the light is increased.

c. The number of ejected electrons increases when the brightness of the light increases.

10–8 Light shines on a metal surface that has a binding energy of 5 eV.

a. Light having a photon energy of 7 eV will release electrons having a kinetic energy of 2 eV.

b. Light having a photon energy of 3 eV will release electrons having a kinetic energy of 2 eV.

10–9 An energy of 30 eV will eject electrons from molecules in cells of the human body.

a. X-rays have a photon energy greater than 30 eV.

b. Visible light has a photon energy greater than 30 eV.

10–10 Motion-picture sound utilizes the following property of the photoelectric effect: The kinetic energy of the electrons depends upon the frequency of the light.

Questions for Thought

10–1 During a hot, summer afternoon, two Cadillacs are parked side by side for two hours. One is black and one is white. At the end of two hours, the temperatures inside are measured. Which car will be hotter? Explain.

10–2 A blacksmith heats two pieces of iron in a fire. One glows white-hot, and the other glows red. Do both pieces emit light having wavelengths of 630 nm (red) and 450 nm (blue)? If

so, what can you say about the intensity of each wavelength? Which piece is hotter?

10–3 Explain what is meant by a black body and black-body radiation.

THE DUAL NATURE OF LIGHT

10–4 How did Lord Rayleigh describe black-body radiation? What relationship did he derive between the intensity and the wavelength of light? Did his theory agree with experiment?

10–5 What is the ultraviolet catastrophe?

10–6 How did Planck force agreement between theory and experiment?

10–7 A ball is thrown vertically into the air.
  a. Is there any restriction on how much of its kinetic energy can be transformed into potential energy?
  b. The ball has an "initial" kinetic energy of 6 joules. Suppose that only 2 joules of energy can be transformed from kinetic energy to potential energy, and vice versa. How would the motion of the ball appear?
  c. The ball has an initial kinetic energy of 6 joules. Suppose that only 0.0001 joules of energy can be transformed from kinetic energy to potential energy, and vice versa. How would the motion of the ball appear?

10–8 What is the photoelectric effect? Briefly describe an experiment that demonstrates the photoelectric effect.

10–9 a. Is the following statement *plausible* or *ridiculous:* The energy of a 7-m ocean wave is larger than the energy of a 1-m ocean wave. _____
  b. _____
  effects show that light has a wave nature. The amplitude of a light wave refers to the strength of_____.
  When the amplitude of light striking a metal is increased, the light exerts (insert: *a larger, a smaller,* or *the same*) _____ force on an electron in the metal.
  c. Thus, increasing the amplitude of a light wave should do which of the following?

TABLE 10–4   DATA FROM PHOTOELECTRIC EFFECT EXPERIMENT (FOR THOUGHT QUESTION 10–9e)

WAVELENGTH OF LIGHT (nm)	INTENSITY OF LIGHT SOURCE (watts)	KE OF ELECTRONS EJECTED FROM METAL (eV)
400	100	4
400	200	4
400	300	4
400	400	4

(1) Make it easier to eject an electron from the metal because the force on it is larger.
(2) Make it harder to eject an electron from the metal because the force on it is larger.
(3) Make it harder to eject an electron from the metal because the force on it is smaller.
(4) Effect no change, because the force has not changed. _____

  d. When the amplitude of a light wave increases, we observe that
  (1) its wavelength increases.
  (2) its color changes.
  (3) its brightness or its intensity increases.
  (4) the wattage of the light source increases.
  (One or more answers may be correct.)
  _____

  e. Table 10–4 shows data obtained from a photoelectric effect experiment. Are these results in agreement with answers to parts **a, b,** and **d** based upon the wave nature of light? If so, explain why. If not, explain and make a table showing the data you would have expected.

10–10 How did Einstein describe the energy of light? What is Einstein's photoelectric equation and how does it follow from conservation of energy?

10–11 Refer to Table 10–4 and answer the following questions based upon Einstein's photon theory:

a. When the intensity of the light source is changed from 100 watts to 200 watts, does the photon energy change? If so, by how much? Does the number of photons change? If so, by how much?

b. How would using a light source of longer wavelength affect the photon energy and the kinetic energy of the ejected electrons?

10–12 Make a diagram similar to Fig. 10–4(a) and (b) for yellow light (560 nm) of 100 watts intensity striking a metal surface.

10–13 Describe Millikan's experiment. How did he measure the kinetic energy of the ejected electrons? What were the results of his experiment?

10–14 Figure 10–9 shows how Millikan obtained the kinetic energy of electrons ejected from sodium by green light (λ = 546.1 nm).

a. If *all* of the electrons had a kinetic energy of 0.47 eV, make a graph of current versus voltage that you would expect to find in the experiment.

b. Figure 10–9 shows that the current is smaller when the voltage V is 0.4 volts than when V equals 0.3 volts. What happened to the other electrons?

c. Explain how Fig. 10–9 shows that the ejected electrons have a continuous range of energies, but have a *maximum* kinetic energy of 0.47 eV.

†10–15 In an experiment on the photoelectric effect, two light sources are available. One source has only frequency f_1, and the other source has only frequency f_2. Two different metal plates are used during the experiment: plate A and plate B. The following observations are recorded by the experimenters:

1. Using source with f_1 and plate A, *no* electrons are emitted.

2. Using source with f_2 and plate A, electrons are emitted.

3. Using source with f_1 and plate B, electrons are emitted.

4. Using source with f_2 and plate B, electrons are emitted.

a. Referring to observation 1, explain whether electrons will be emitted if the intensity of the light is increased, but the frequency is not changed.

b. Referring to observation 2, explain whether the kinetic energy changes as the intensity of the light is increased, but the frequency of the light is not changed.

c. In all cases where electrons are emitted, will the experimenters find more electrons emitted if the intensity of the light is increased?

d. From the observations, frequency f_1 is greater than frequency f_2. Discuss. True or false?

e. What information about the metal plates A and B is obtained by referring to observations 1 and 3?

f. Explain whether the electrons emitted in observation 3 will have a larger or smaller kinetic energy than those emitted in observation 4.

g. Explain whether the electrons emitted in observation 2 will have a larger or smaller kinetic energy than those emitted in observation 4.

10–16 Describe Compton's experiment with X-rays. What measurements did he make? How did these measurements compare with anticipated results? How did he interpret the results of his experiment?

10–17 If electromagnetic radiation possessed *only* a wave nature, explain whether X-rays and visible light would be harmful to human tissue.

10–18 The picture in a TV set is obtained by:

a. a varying number of electrons striking every part of the TV screen simultaneously.

b. an equal number of electrons striking every part of the TV screen simultaneously.

THE DUAL NATURE OF LIGHT

c. X-rays striking the screen.

d. having a beam of electrons scan the screen, line by line.

10−19 How do photocells control elevator doors? Streetlights? Explain why a bright flash of lightning occurring on a stormy night will cause streetlights (operated by photocells) to turn off.

10−20 Explain how a photocell is used in a smoke detector.

Questions for Calculation

10−1 Use conservation of energy to complete the following:

a.

Energy of golfer's swing	Binding energy of ball to tee	KE of golf ball
25 units	5 units	_____
25	_____	5 units
25	25	_____
25	30	_____
20	10	_____
30	10	_____
_____	10	5

b.

Energy of light	Binding energy of electron to metal	KE of electron
15 eV	5 eV	_____
20	5	_____
_____	5	10 eV
2	5	_____
20	10	_____
20	_____	5
20	20	_____

†10−2 a. Calculate the frequency and photon energy in eV of ultraviolet light of 400 nm wavelength.

b. Calculate the number of photons per second that strike the metal surface in Fig. 10−4(a). The intensity of the ultraviolet light is 100 watts. (Remember that 1 watt = 1 joule/sec, and 1 eV = 1.6 × 10^{-19} joules).

†10−3 Find the equations of the two parallel lines in Fig. 10−21. How is the fact that the lines are parallel reflected in the equations?

10−4 Plot the data in Table 10−5 on a graph and find the equation for the straight line.

10−5 Electrons ejected from plate A in Fig. 10−7 have a kinetic energy of 2 eV.

a. If the voltage V connecting plates A and B is 1 volt, what is the kinetic energy of electrons striking plate B? Is a current detected on the current meter?

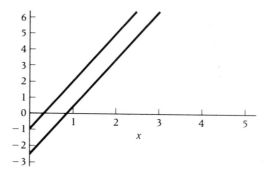

FIG. 10−21   Graph for calculation question 10−3.

TABLE 10−5   TABLE FOR CALCULATION QUESTION 10−4

FREQUENCY OF LIGHT (osc/sec)	KE OF EJECTED ELECTRONS (eV)
7.25 × 10^{14}	0
9.67	1
12.09	2
14.51	3
16.93	4

b. If V equals 1.95 volts, what is the kinetic energy of electrons striking plate B?

c. If V equals 2.05 volts, is a current detected on the current meter?

10–6 Blue light of 450 nm strikes a metal plate and ejects electrons having a kinetic energy of 2 eV.

    a. Find the photon energy of blue light in eV.

    b. What is the binding energy B?

†10–7 Ultraviolet light of wavelength 200 nm strikes a metal surface whose binding energy is 2 eV.

    a. Find the photon energy of the ultraviolet light in eV.

    b. What is the kinetic energy of the ejected electrons?

10–8 A metal has a binding energy B of 2 eV.

    a. What is minimum frequency of light that will cause electrons to be ejected?

    b. What is the wavelength of this light? What is its color?

    c. Will infrared light of wavelength 700 nm eject electrons?

10–9 The cathode and anode of an X-ray tube are connected by a voltage of 20,000 volts.

    a. What is the kinetic energy of electrons striking the anode?

    b. What is the maximum X-ray photon energy? What is the wavelength of this X-ray?

    c. How are X-ray photons below the maximum energy produced?

    d. What is the range of wavelengths produced by this X-ray tube?

Footnotes

[1] Robert A. Millikan, *Physical Review* 7 (1916):355.

[2] Ibid; p. 355.

[3] Ibid; p. 355.

Marie Curie

CHAPTER 11

Radioactivity

Introduction

Like the three Princes of Serendip in the fairy tale (who learned all kinds of things about a missing camel while merely trying to get from one city to another), Henri Becquerel was looking for something else when he discovered radioactivity. He happened to choose a uranium compound that glowed in the dark to study the phenomenon of phosphorescence. In this pursuit, he learned that this compound and *all* uranium compounds emit penetrating radiation. He had discovered radioactivity. Since Marie and Pierre Curie almost immediately made great advances in the study of radioactivity, their story logically follows Becquerel's. Their tale is such a singular one that it will be related in some detail, making this chapter somewhat different from the others. The Curies provide an interesting contrast to Becquerel in that their discoveries came after steady years of hard work in a directed search.

In the latter part of this chapter, we shall see how scientists answered the challenging question, What is the nature of these rays? Among them was none other than Ernest Rutherford, who learned that there were two components, the alpha and beta rays, in the rays from uranium. Within the next 10 years, he vigorously tracked down the nature of the alpha ray, showing that it was a helium atom with two missing electrons.

Becquerel's Discovery

Like his father and grandfather before him, Becquerel was interested in phosphorescent substances that shine in the dark. When he learned that the glass of an X-ray tube glowed green (see Chapter 1, Fig. 1–12, p. 14), he wondered whether glowing phosphorescent substances might also emit penetrating rays, such as X-rays.

To answer this question, Becquerel wrapped an unexposed photographic plate with black paper, then laid a phosphorescent substance on top of it and put the bundle in sunlight. There are many phosphorescent substances but, luckily, Becquerel chose a uranium compound, double sulfate of uranium and potassium. The sunlight would cause the phosphorescent substance to glow. If phosphorescence did involve the emission of penetrating rays, these would pass through the paper and expose the plate, even though the sunlight itself could not get through the black paper.

Sure enough, after the wrapped plate had been exposed to the sun for several hours, this happened. Becquerel reported, "When I developed the photographic plate I saw the silhouette of the phosphorescent substance in black on the negative . . . If I placed between the phosphorescent substance and the paper a coin or a metallic screen pierced with an open-work design, the image of these objects appeared on the negative."[1] The phosphorescent substance was emitting penetrating rays.

The weather intervened and Becquerel's experiments took a completely unexpected turn. As a result, on two days of his experiments, February 26 and 27, 1896, "the sun only showed itself intermittently [so] I kept my arrangements all prepared and put back the holders in the dark in the drawer of the case, and left in place the crusts of uranium salt. Since the sun did not show itself again for several days I developed the photographic plates on the 1st of March, expecting to find the images very feeble. The silhouettes appeared, on the contrary, with great intensity. I at once thought that the action might be able to go on in the dark."[2]

Continuing with his experiments, Becquerel found that the same substance kept in the dark for two months continued to emit rays and expose photographic plates. Furthermore, it was *the presence of uranium in the compound that was important and not the phosphorescence.* He wrote:

All the salts of uranium that I have studied, . . . phosphorescent or not, . . . crystallized or in solution, have given me similar results. I have thus been led to think that the effect is a consequence of the presence of the element uranium in these salts, and that the metal would give more intense effects than its compounds. An experiment made several weeks ago with powdered uranium, which has been for a long time in my laboratory, [has] confirmed this expectation: the photographic effect is notably greater than the impression produced by one of the uranium salts; and in particular by the sulphate of uranium and potassium.[3]

Becquerel found that the intensity of the rays depends upon the fraction of uranium in the compound. Since the rays are due to the uranium, then 1 g of pure uranium will emit more rays, and expose a photographic plate more strongly, than 1 g of uranium sulfate, which contains the elements sulfur and oxygen, as well as uranium.

Becquerel made another important observation about the uranium rays (see Fig. 11–1). When uranium is brought near a charged electroscope, the leaves immediately collapse; that is, the uranium rays make the air conductive and the charge on the electroscope becomes neutralized.

Becquerel, trying to find a connection between X-ray production and phosphorescence, discovered by serendipity a completely new phenomenon: the rays emitted by the element uranium. In short, he discovered radioactivity. (These rays were at first called *Becquerel rays.* When Marie Curie saw radium glow in the dark, she coined the term *radioactive* or *radioactivity,* combining the words *radiance* and *activity.*)

The Curies: Their Story[4]

In October, 1891, Marya Sklodowska's dream was finally coming true: she was on her way from Warsaw to study at the Sorbonne in Paris. She had almost despaired that it would ever happen, for eight years had slipped away since her graduation from the gymnasium. She was now 24 years old. For two years, she had given private lessons in Warsaw, and for six years she

(a) CHARGED ELECTROSCOPE

Uranium

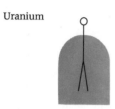

(b) ELECTROSCOPE DISCHARGED BY URANIUM RAYS

FIG. 11–1 An electroscope becomes charged by touching it with a rod charged with static electricity. For example, a rubber rod rubbed with fur becomes negatively charged, and a glass rod rubbed with silk becomes positively charged. In order to minimize the repulsion between charges, the charges move onto the thin, gold foil leaves, and the leaves separate as a result. When uranium is brought nearby, the leaves collapse, showing that the electroscope has been discharged.

had been a governess, at times in a remote village. All the while she sent money to her sister Bronya, who was studying to be a doctor in Paris. They had an agreement, which Marya had suggested: first Bronya would complete her studies with Marya's help, and then it would be Marya's turn.

One day, Bronya wrote saying that she would soon marry a doctor, and that they could provide room and board for Marya. Come. But Marya insisted on waiting yet another year, partly out of concern for her father and another sister, and partly to save more money for expenses. But, finally, she was on her way to Paris, to the Sorbonne, where political freedom and freedom of thought prevailed. Such was not the case in her beloved Poland, ruled by the Russian tsar, Alexander II. This oppressive rule had greatly influenced Marya's life.

As a child, her school was often inspected by Russian officials, and she was often called upon to recite, since she had learned Russian perfectly. If her teacher had been caught teaching Polish history and culture, the teacher could have been sent to Siberia. Her father, a teacher of mathematics and physics, had a serious disagreement with his Russian superior, and his salary was reduced. After graduating from the gymnasium, where she received a gold medal for her accomplishments, she attended the informal sessions of the Floating University, held in secret because the authorities feared independent minds.

Several times during the last century the Poles had rebelled, but had suffered terrible defeats. Now they were trying a nonviolent approach: instilling Polish nationalism in the minds of the young on the theory that an educated population was one of the best bulwarks against oppression. Toward this end, Marya had to read to workers in the garment factory, and had even gathered a small library of Polish books. While a governess in a remote village, she taught peasant children to read Polish. Marya decided to serve Poland by obtaining a master's degree and returning to Warsaw as an educator. With this resolve, she came to Paris and registered at the Sorbonne, on November 3, 1891, as Marie Sklodowska.

Marie enjoyed living with her sister and brother-in-law, and had many good times with their friends, who often came to call. But because their home was an inconvenient hour away from the Sorbonne, she finally moved to a small flat of her own. Here, she led a spartan existence devoted completely to her studies. Even though she had studied physics during the evening hours while a governess and had corresponded with her father about mathematics, she was ill prepared for the rigors of the Sorbonne, or for the lectures in rapid French. Her existence followed the bohemian stereotype: starving in an icy garret while working on a masterpiece. She was successful, for she placed first in the examinations for the master's degree in physics in 1893.

Early in 1894 Marie met Pierre Curie at a gathering of Polish friends. She asked his advice on her research on the magnetic properties of steel. Pierre, at 35, was already well known for his studies on magnetism, and was chief of laboratory at the School of Physics and Chemistry of the City of Paris.

When Marie received her second master's degree, in mathematics, Pierre urged her to return, after a summer vacation in Warsaw, to continue her studies. He also proposed marriage. Marie did return, but it was nearly a year before she accepted his proposal. Doing so meant leaving her family and giving up political activity in Poland. They were married in the summer of 1895 and spent the rest of the season roaming the French countryside on their bicycles—thereafter, their favorite way to vacation.

During the next year, Marie studied for a diploma that would allow her to teach in France. In September, 1897, their daughter Irene was born. The idea of choosing between domesticity and a scientific career did not even occur to Marie; she was successful at combining these aspects of her life.

She then decided to take her doctorate at the School of Physics. But what research should she undertake? Both Pierre and Marie were attracted to Becquerel's then-recent work. However, the only room they could find for a laboratory was a glassed-in studio that had been used as a storeroom. It was very cold and damp, and the humidity affected their sensitive instruments.

In her notebook, Marie recorded a temperature of 6°C (44°F) on one day in February. She nonetheless persisted in her research.

THE CURIES' FIRST RESEARCH IN RADIOACTIVITY

Marie decided to use an **electrical method** for measuring radioactivity, rather than the photographic method (see Fig. 11–2). Becquerel had shown that uranium rays make air conductive. Measuring the current through the air would provide a *quantitative* measure of radioactivity.

Marie's apparatus (Fig. 11–3) consisted of two metal capacitor plates about 3 cm apart, connected by a voltage of 100 volts, and a meter to measure the current. The sample to be examined was placed on the lower plate, and caused the air between the plates to conduct a current indicated on the meter. This current depends upon the strength of the radioactive sample: the number of rays emitted per second. The strength, usually called its **activity,** is defined as

$$\text{Activity} = \frac{\text{no. of rays (or disintegrations)}}{\text{time}}. \quad (1)$$

In Chapter 1, we learned that the alpha, beta, and gamma rays are emitted when the nucleus disintegrates. The unit of activity is called the **curie.** One curie, defined as the number of rays (or disintegrations) per second emitted by 1 g of radium, corresponds to 3.7×10^{10} disintegrations per second.

Today, we understand more fully why the air conducts a current. The rays from the radioactive substance ionize molecules of air. For example, the nitrogen molecule, consisting of 2 atoms of nitrogen, is ionized in one of the following ways:

$$N_2 \rightarrow N_2^+ + e^- \quad (2a)$$

$$N_2 \rightarrow N^+ + N^- \quad (2b)$$

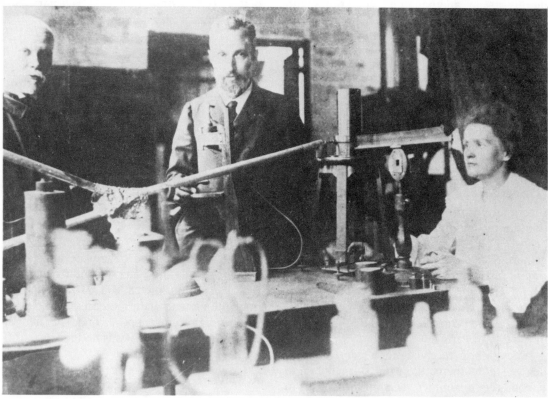

FIG. 11–2 Marie and Pierre Curie with their electrical apparatus for measuring radioactivity. (Courtesy of Argonne National Laboratory)

FIG. 11–3 The electrical method for measuring radioactivity used by Marie Curie.

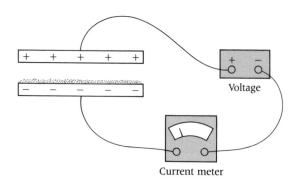

BOX 11–1

Invented by Hans Geiger in 1908, the **Geiger counter** also operates on the principle of ionization. This device consists of a central wire in a hollow metal cylinder filled with gas. A voltage of 1000 volts connects the cylinder and the central wire. Radiation enters the cylinder through the thin window and ionizes gas atoms. Due to the high voltage, the electrons *accelerate* toward the positively charged wire, and the positive ions accelerate to the negatively charged cylinder. Because the electron and positive ion gain kinetic energy, they have sufficient energy to ionize other gas atoms, which are accelerated in turn and ionize still more gas atoms, and so on. An avalanche of ionization occurs, producing a burst of current, which can be counted (see Fig. 1). Each particle passing through the thin window produces *one* burst or "one count."

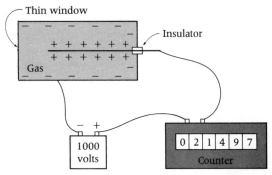

FIG. 1   The Geiger counter.

In some models, each burst activates a speaker producing a "beep." The Geiger counter is still the most widely used instrument for detecting radiation.

Each ray may ionize 100,000 air molecules before its energy is depleted and it comes to a halt. The free electrons and negative ions move toward the positive plate, while the positive ions travel toward the negative plate. When the charges reach their respective plate, the amount of charge on the plate is reduced. In Chapter 7, we learned that the amount of charge on the capacitor plates depends upon the voltage connecting the plates. Here, ionization reduces this charge; the battery acts to maintain the status quo by transferring electrons from one plate to the other, producing a current shown on the meter.

The first thing Marie undertook was to examine the various chemical compounds of uranium with her electrical radiation detector. Table 11–1 shows the results of this study.

Marie next examined all known chemical elements. In the two years since Becquerel's discovery, no one else had thought to make such a systematic survey. She soon found that the element thorium, which had, next to uranium, the highest atomic mass then known, was also radioactive. Almost simultaneously, a German

TABLE 11–1 RADIOACTIVITY OF URANIUM COMPOUNDS

SUBSTANCE	CURRENT (amperes)
Metallic uranium (containing a little carbon)	2.3×10^{-11}
Green oxide of uranium, U_3O_4	1.8
Hydrated uranic acid	0.6
Uranate of sodium	1.2
Uranate of potassium	1.2
Uranate of ammonium	1.3
Uranium sulphate	0.7
Sulphate of uranium and potassium	0.7
Nitrate of uranium	0.7
Phosphate of copper and uranium	0.9
Oxysulphide of uranium	1.2

Source: Marie Curie's thesis, in *Chemical News and Journal of Physical Science* (London) 88 (1908):88.

RADIOACTIVITY

TABLE 11-2 RADIOACTIVITY OF MINERALS

SUBSTANCE	CURRENT (amperes)
Uranium	2.3×10^{-11}
Pitchblende from Johanngeorgenstadt	8.3
Pitchblende from Joachimsthal	7.0
Pitchblende from Pzibran	6.5
Pitchblende from Cornwallis	1.6
Cleveite	1.4
Chalcolite	5.2
Autunite	2.7
Orangite (native oxide of thorium)	2.0

Source: Marie Curie's thesis, in *Chemical News and Journal of Physical Science* (London) 88 (1903):99.

scientist, G. C. Schmidt, also discovered that thorium was radioactive.

When two elements were found to be radioactive, rather than just one, Marie suggested a new explanation for radioactivity. All elements have distinct chemical and physical properties, such as boiling point, melting point, color, and formation of chemical compounds. Radioactivity had been considered to be a property of the *element* uranium. Finding yet another radioactive element removed uranium from its unique position and led Marie to conclude that radioactivity was not merely a chemical property of a given element, but something far more fundamental: *Radioactivity was a property of the atom.*

Next, Marie examined samples of minerals from the School of Physics' collection (see Table 11-2). She expected that the minerals would be radioactive only if they contained uranium or thorium, and that their activities would depend on how much of these two elements was present. Her expectation proved true except in the case of pitchblende (which contains uranium, barium, bismuth, and other elements) and chalcolite (which contains copper and uranium). In these two cases, the activity was *much greater* than expected. In fact, the activity of pitchblende was about 4 times that of *pure* uranium. This discovery was so astounding that she repeated the experiment many times to be sure there was no error.

Since all known chemical elements had been tested for radioactivity, Marie concluded that the minerals must contain *a new element* more active than either uranium or thorium. To have escaped detection by the chemists, this new element must be an exceedingly small fraction of the pitchblende ore, which she estimated at about 1 percent. (It turned out to be 1 part in a million!) Pierre was so fascinated with Marie's conclusions that he gave up his own very different research and joined her efforts to find this new element.

They proceeded with a chemical separation to obtain the element contained in the pitchblende. Using the electrical method, they tested each separation to see where the radioactivity resided, and found that it resided with the separated bismuth and also with the separated barium. In a report to the French Academy of Science, they wrote:

By carrying on these different operations [chemical separations] we obtained products which were more and more active. Finally, we obtained a substance whose activity is about 400 times greater than that of uranium.

We have attempted to discover among bodies which are already known if there are any which are radioactive. We have examined compounds of almost all the simple bodies; thanks to the kindness of several chemists we have received specimens of very rare substances. Uranium and thorium were the only ones which were evidently active. Tantalum perhaps is very feebly so.

We believe, therefore, that the substance which we removed from pitchblende contains a metal which has not yet been known, similar to bismuth in its chemical properties. If the existence of this new metal is confirmed, we propose to call it polonium, after the name of the native country of one of us.[5]

The Curies reported their results with the barium sample as follows:

We believe . . . that this substance, although for the most part consisting of barium, contains in addition a new element which gives it its radioactivity and which furthermore is very near barium in its chemical properties. These are the reasons which speak in favor of this view.

1. Barium and its compounds are not ordinarily radioactive. Now, one of us has shown that radioactivity seems to be an atomic property, persisting in all the chemical and physical states of matter. . . . the radioactivity of our substance, which does not arise from barium, ought to be attributed to another element.

2. The first substances which we obtained, in the state of hydrated chlorides, had a radioactivity 60 times greater than that of metallic uranium. . . . We have . . . obtained chlorides which have an activity 900 times greater than that of uranium. We have been stopped by the lack of material, but . . . we may assume that the activity would have . . . increased if we had been able to continue. These facts can be explained by the presence of a radioactive element of which the chloride is less soluble in alcoholic solution than is barium chloride.

3. . . . M. Demarcay has found in the spectrum a ray which seems not to belong to any known element. . . . The intensity of this ray increases at the same time as the radioactivity and this, we think, is a strong reason for attributing it to the radioactive part of our substance.

The various reasons which we have presented lead us to believe that the new radioactive substance contains a new element, to which we propose to give the name radium.[6]

The Curies had discovered two radioactive elements in the pitchblende: *polonium* and *radium*. Because the existence of these new elements drastically changed the status quo, and because radioactivity was not understood, many scientists adopted a wait-and-see attitude.

Next, the Curies set out to obtain a pure radium sample and determine that all-important property: its atomic mass. Because the amount of polonium in the bismuth sample was so small,

they could not separate it *completely* from the bismuth, so they could not determine its atomic mass. By determining the atomic weight of radium, they hoped to convince everyone that they had discovered two new elements. The Curies could not even get a suitable laboratory at the Sorbonne. They continued their research at Marie's original storeroom-laboratory.

Since they now needed *tons* of pitchblende, and the ore was costly because it contained valuable uranium, the Curies arrived at an ingenious solution: they would work with the useless *residue* of the ore, from which the uranium had already been extracted. The polonium and radium should be in the residue. They arranged to get their residue from a mine in Joachimsthal, Austria, for just the shipping cost, which the Curies themselves paid.

Outside Marie's storeroom-laboratory was a dirt-floored shed that could be used for the extraction work. The couple decided to divide their efforts. Marie would continue with the chemical separation process to obtain a *pure* sample of radium and to determine its atomic mass, while Pierre would investigate the properties of radium and the nature of radioactivity.

For four years, Marie persisted in her work. First, she separated barium (plus radium) from the pitchblende, and then separated the radium from the barium. Finally, after treating nearly a ton of pitchblende residue, she obtained one-tenth of a gram of pure radium that was about a million times more active than uranium. She knew that it was *pure* radium because the spectrum of light from the sample showed no contamination from barium.

ATOMIC MASS OF RADIUM

Marie determined the atomic mass of radium by following certain steps.

Step One. She converted radium to radium chloride ($RaCl_2$), a molecule with 1 atom of

TABLE 11–3   DATA FOR DETERMINING THE ATOMIC MASS OF RADIUM

MASS OF RADIUM CHLORIDE (grams)	MASS OF SILVER CHLORIDE (grams)	ATOMIC MASS OF RADIUM (amu)
0.09192	0.08890	225.3
0.08936	0.08627	225.8
0.08839	0.08589	224.0

Source: Marie Curie's thesis, in *Chemical News and Journal of Physical Science* (London) 88 (1903):159.

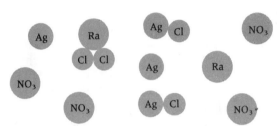

FIG. 11–4   The formation of silver chloride (AgCl) molecules.

radium and 2 atoms of chlorine, and weighed this sample. (See Table 11–3). The mass of the radium chloride sample depends upon the number of radium chloride molecules in the sample (denoted by the symbol N) and the mass of each molecule:

$$\text{Mass of RaCl}_2 \text{ sample in grams} = (N)\left(\begin{array}{c}\text{mass of a}\\ \text{RaCl}_2 \text{ molecule}\\ \text{in grams}\end{array}\right). \quad (3)$$

The *relative* masses of atoms are given by their atomic masses (Chapter 1). We can find the relative mass of a molecule, called the **molecular mass,** by adding the atomic mass of each atom. If we use the symbol R for the atomic mass of radium and 35.4 amu for the atomic mass of chlorine, then the molecular mass of radium chloride is $R + 2(35.4 \text{ amu})$. Since the mass of a molecule in grams is proportional to its molecular mass, Equation 3 becomes

$$\text{Mass of RaCl}_2 \text{ sample in grams} \propto (N)\left(\begin{array}{c}\text{molecular}\\ \text{mass of}\\ \text{RaCl}_2 \text{ in amu}\end{array}\right)$$

$$\propto (N)\,(R + 70.8 \text{ amu}). \quad (4)$$

Step Two.   She dropped the radium chloride sample into a solution of silver nitrate ($AgNO_3$), a molecule with 1 atom of silver, 1 atom of nitrogen, and 3 atoms of oxygen. When chemical compounds are in solution, the molecules split apart into positively and negatively charged ions: Ra^{2+} ions and two Cl^- ions for radium chloride, and Ag^+ and $(NO_3)^-$ ions for silver nitrate. Then one silver ion joins with one chlorine ion to form a molecule of silver chloride (AgCl), as shown in Fig. 11–4. Marie started out with N molecules of $RaCl_2$ and ended up with $2N$ molecules of AgCl. (Note that the total number of chlorine atoms in both cases is $2N$.)

Silver chloride is a crystalline power and precipitates out of solution. Marie separated the silver chloride precipitate from the solution and weighed it:

Mass of
AgCl sample $\propto (2N)$ $\begin{pmatrix} \text{molecular} \\ \text{mass of} \\ \text{AgCl in amu} \end{pmatrix}$ (5)
in grams

$$\propto (2N)(107.8\,\text{amu} + 35.4\,\text{amu}),$$

where $2N$ is the number of silver chloride molecules in the sample, and the atomic mass of silver is 107.8 amu.

Dividing Equation 4 by 5, she found

$$\frac{\text{Mass of } RaCl_2}{\text{Mass of } AgCl} = \frac{R + 70.8\,\text{amu}}{2(143.2\,\text{amu})}. \quad (6)$$

We now see the rationale for Marie's procedure. The mass of each chemical compound depends upon its molecular mass. When the chlorine atoms switch partners (from radium to silver), the change in the mass of the chlorine compound reflects the difference in the atomic mass between radium and silver. Thus, by weighing the compounds before and after the chlorine atoms had switched partners (and knowing the atomic masses of chlorine and silver), Marie could determine the atomic mass of radium R. Solving Equation 6 for R, she found

$$R = \frac{(286.4\,\text{amu})(\text{mass of } RaCl_2)}{(\text{mass of } AgCl)} - 70.8\,\text{amu}.$$

Table 11–3 shows the data from Marie Curie's thesis. The average value of three measurements is 225 amu. Finally, after four years of work, she had determined the atomic mass of radium.

THE CURIES' SUCCESS AND THEIR RECOGNITION

After the birth of their daughter, the Curies needed more money. Pierre tried several times to get a chair at the Sorbonne, so that he could teach less and devote more time to research. Certainly, he was qualified, but he was not enough of a political creature to get into favor with the right committees. He did get a better-paid position at an annex of the Sorbonne, but it still required a lot of teaching. Likewise, Marie obtained a professorship at the Higher Normal School for Girls at Sevres. Beyond their teaching duties, Marie and Pierre continued to work long hours in the shed. Driving themselves in this way, they were always exhausted, and their health was, of course, affected.

After finally taking time to write a thesis, Marie received her doctorate in physics from the Sorbonne in June, 1903.

At one point in their work, Pierre had exposed his arm to radium rays and received a burn that took more than two months to heal. He collaborated with medical doctors, and they found that the radium rays could be used to destroy diseased tissue and thus treat certain forms of cancer. This was the first effective nonsurgical weapon used against cancer.

By further hard work, Marie and Pierre then separated 1 gram of radium from 8 tons of pitchblende ore. Needless to say, they received many requests for their chemical method of purifying radium. A decision had to be made: Should they patent the process? If they did, they could become very wealthy and could build the laboratory of their dreams. They decided, however, that profiting from their discovery was contrary to the scientific spirit and that they should not take advantage of the fact that radium had a medical application. They answered all who wrote, and published their method openly. The radium industry flourished. At that time, radium sold for $150,000 per gram.

In 1903, Marie and Pierre Curie, along with Henri Becquerel, received the Nobel Prize in physics for their discoveries in radioactivity.

Finally, Pierre was offered a chair at the Sorbonne, although the couple still had no adequate laboratory. Pursued by journalists and photographers, they remained untouched, and even bewildered by such admiration. They wished only to work.

In 1904, their second child, Eve, was born.

On April 19, 1906, Pierre, perhaps lost in thought, was killed in a street accident. Here, the words of Eve, their daughter and biographer, are fitting:

At the moment when the fame of the two scientists and benefactors was spreading through the world, grief overtook Marie; her husband, her wonderful companion, was taken from her by death in an instant. But in spite of distress and physical illness, she continued alone the work that had been begun with him and brilliantly developed the science they had created together.[7]

Marie was appointed to succeed Pierre as professor at the Sorbonne, the first woman to hold such a position in France. In 1911, she received the Nobel Prize in chemistry for discovering radium and polonium—the first person to receive it twice. Still, the most prestigious accolade of France, election to the Academy of Science, was not presented to her.

Finally, in July, 1914, the much-delayed and long-awaited Radium Institute, built according to Marie's plans, was finished. At last, she had a proper laboratory.

When war broke out, she immediately sought a way to help her adopted country. Although X-rays had been discovered in 1895, only large hospitals had X-ray equipment. Marie developed the first mobile X-ray unit, an ordinary car with X-ray equipment and a dynamo driven by the car motor. She outfitted 20 cars and drove one herself, going to outlying areas to take X-rays of wounded soldiers. In addition, Marie equipped 200 X-ray rooms. At the Radium Institute, she taught 250 technicians how to take X-rays. The 1917 armistice brought a special victory to Marie, for Poland was now a free country.

Now, she resumed the work interrupted by the war. From 1919 to 1935, some 483 publications resulted from the work of students and scientists at the Radium Institute. At Marie's side was her daughter Irene, also a researcher in physics. Irene married another student, Frederic Joliot. Later, they, too, would also receive the Nobel Prize.

In 1920, Marie was interviewed by an American magazine editor, Mrs. William Brown Meloney. After visiting the magnificent laboratories in the United States, Mrs. Meloney found the Radium Institute quite modest. Mme Curie was asked what she would like most to have, and she replied that she needed a gram of radium. Struck by the irony of this wish, Mrs. Meloney returned to New York, organized a committee, and launched a national campaign that raised $100,000 in less than a year for the Marie Curie Radium Fund. The committee asked that Marie and her daughter come to America to receive this gift. They did so, and went on to tour the country, greeted everywhere by huge crowds.

The words of her daughter Eve best describe the latter part of Marie Curie's life:

The rest of her life resolves itself into a kind of perpetual giving. To the war wounded she gave her devotion and her health. Later on she gave her advice, her wisdom, and all the hours of her time to her pupils, to future scientists who came to her from all parts of the world.

When her mission was accomplished she died exhausted (July 6, 1934), having refused wealth and endured her honors with indifference.[8]

We shall have more to say about the work of the remarkable Curies in the section on induced radioactivity.

Properties of the Rays

The attention and curiosity of many scientists were focused on these new discoveries. Now, we shall see *how* they determined the properties of these rays.

Among the interested scientists was Ernest Rutherford, of whom we shall hear a great deal more. In 1895, Rutherford came from New Zealand to the Cavendish Laboratory, where he studied with J. J. Thomson. In 1898, he obtained a professorship at McGill University in Montreal, where he carried out the pioneering research for which he received the Nobel Prize in 1908. His first research at McGill was studying uranium rays, using the same electrical method as the Curies (see Fig. 11–3).

The plates were 20 cm square and 4 cm apart; the voltage connecting them was 50 volts. Powdered uranium oxide was spread uniformly over the center of the bottom plate, and successive layers of aluminum foil were placed over it. If the aluminum foil absorbs some of the rays, Rutherford thought, then the ionization of the air, and likewise the current, should be reduced. In this way, he hoped to obtain information about how the rays were absorbed. The graph in Fig. 11–6 shows Rutherford's data.[9]

With the first few layers of aluminum foil, the current is greatly reduced. In fact, four layers reduce the current by a factor of 20, from 182 to 9.4 units. But after 10 layers, further layers cause only very small reductions. Rutherford concluded that the uranium radiation is complex, consisting of at least two distinct types of radiation—a readily absorbed one, which he called for convenience the **alpha radiation,** and a more penetrating one, which he called the **beta radiation.**

Immediately scientists asked, Are these alpha rays and beta rays charged particles? If so, what is their charge? What is their mass? Or, are they similar to X-rays? Let's briefly review Thomson's work with cathode rays to see how such questions can be pursued (see Chapter 7 for more details).

Thomson had found the mass-to-charge ratio *m/e* of cathode rays by using a magnetic field and also sending a beam of rays between charged capacitor plates. The first step was to send the beam through a magnetic field of known strength

FIG. 11–5 Marie Curie with daughter Irene carrying out an experiment at the Radium Institute. (AIP Niels Bohr Library, William G. Myers Collection)

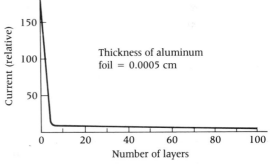

FIG. 11–6 A graph of Rutherford's data that shows the current versus the number of layers of aluminum foil placed over the bottom plate of the capacitor in Fig. 11–3.

and measure the radius of its circular path.* The direction in which the beam bent would show whether it was positively or negatively charged. If we use the force on a charged particle in a magnetic field ($F = qvB$) and the acceleration ($a = v^2/r$), Newton's second law becomes

$$evB = \frac{mv^2}{r},$$

where e is the charge on the electron. The mass-to-charge ratio is

$$\frac{m}{e} = \frac{rB}{v}. \qquad (7)$$

The only unknown on the right-hand side is the speed v. Thomson determined this speed in a second step: sending the beam between charged capacitor plates and using a magnetic field as well. Thus, he could obtain a value for m/e because all terms on the right side of Equation 7 were known.

BETA RAYS

Several scientists, including Becquerel,[10] Pierre Curie,[11] and a German scientist named Giesel,[12] investigated the beta rays using a magnetic field. They found that the beta rays bent in the direction for negatively charged particles. Becquerel's apparatus appears in Fig. 11–7. Beta rays emerging from a radioactive source in a small lead container are bent in a semicircle by the magnetic field, exposing the photographic plate when they strike. The record on the photographic plate shows the direction in which the beta rays are bent and the radius of their path.

Next, Becquerel used charged capacitor plates and a magnetic field to find the speed v and the

*For the sake of simplicity, we consider the deflection of the beam produced by a magnetic field rather than by charged capacitor plates, as Thomson did in his experiment. See p. 154 for additional comments.

mass-to-charge ratio m/e. He found that their speed was larger than one-half the speed of light, and the mass-to-charge ratio close to that for the cathode ray (electron). Therefore, beta rays were identified as electrons.[13]

ALPHA RAYS

The first experiments using a magnetic field showed that, while the beta rays are easily deflected, the alpha particles continue to move in a straight line. You might jump to the conclusion that the alpha particles are not charged, that they are neutral. However, there is another possibility. If the mass of the alpha particle is *much* larger than the mass of the beta ray, then, in the same magnetic field, the alpha particle will bend much less than the beta ray. It might, in fact, bend too little to be observed by available equipment. The alpha and beta deflections can be compared by using Equation 7, which can be rewritten as

$$r = \frac{mv}{qB}, \qquad (8)$$

where q has been substituted for e.

Let's substitute values for m, v, and q into Equation 8 and compare the radii of the paths of the alpha particle and the beta ray. The mass of the alpha particle is about 7000 times the mass of the electron ($M = 7000\ m_e$); the speed of the alpha particle is 5 percent of the speed of light ($0.05c$), compared to one-half the speed of light ($0.5c$) for the beta ray; the charge of the beta ray is $-e$, and the charge of the alpha particle is $+2e$. The substitution of values into Equation 8 gives r equals $0.5\ m_e\ c/Be$ for the beta ray, and r equals $175\ m_e\ c/Be$ for the alpha particle. The radius of the alpha particle's path is 300 times the radius of the beta particle's path. As Fig. 11–8 shows, a large radius corresponds to a small deflection from a straight line.

The tiny deflection of the alpha particle can be increased by using a much stronger magnetic field. This is what Rutherford set out to do, but not until 1903 could he get a strong enough magnetic field. As soon as he did, he showed that a beam of alpha particles bent like a beam of positively charged particles.

Next, Rutherford used charged capacitor plates as well as a magnetic field to determine the mass-to-charge ratio m/q. He found it to be about 3500 times *larger* than for the beta ray. There were several reasonable possibilities. For example, the alpha particle might be a hydrogen molecule with one of its two electrons missing (that is, a hydrogen gas consists of molecules containing 2 atoms of hydrogen, each atom 1800 times as massive as an electron). Therefore, a hydrogen molecule with one missing electron would have a single positive charge and a mass about 3600 times larger than an electron or beta ray. Or the alpha particle might be a helium atom (mass about 7000 times larger than an electron) with 2 electrons missing. This particle would have two positive charges. In both cases, the mass-to-charge ratio m/q would be about 3500 times larger than for a beta ray, as required.

In 1909, Rutherford, with T. D. Royds,[14] demonstrated conclusively the nature of the alpha particle, using the apparatus in Fig. 11–9. Rutherford and Royds put a gaseous radioactive element, radon, in a glass tube A, surrounded by glass tube T. The walls of tube A were so thin (only one-hundredth of a millimeter) that the alpha particles could pass right through because of their large kinetic energy. After waiting six days for the alpha particles to accumulate in the larger glass tube T, Rutherford and Royds let mercury in to compress the gas into the upper part of the tube, where electrodes had been inserted into the glass. These electrodes were connected by a high voltage, which caused the gas to glow. The light given off, when passed through a prism, showed a spectrum identical to that of a helium gas.

This experiment showed that an alpha particle is a helium atom with 2 electrons removed. (An alpha particle has two positive charges and, when it passes through the walls of tube A, it picks up two electrons to become a helium atom. This explains why its spectrum is identical to that of a helium gas.) In modern terms, an alpha particle is the nucleus of a helium atom.

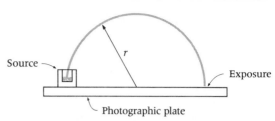

FIG. 11–7 Photographic technique for determining the radius of the path of beta rays in a magnetic field. [*Source:* Richtmyer, Kennard, and Lauritsen, *Introduction to Modern Physics*, 5th ed. (New York: McGraw-Hill, 1955), p. 434.]

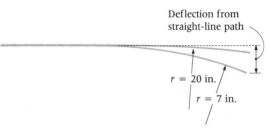

FIG. 11–8 A comparison of the radius of a circular path with its deflection from a straight-line path.

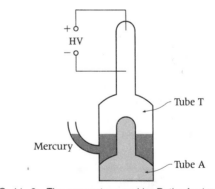

FIG. 11–9 The apparatus used by Rutherford and Royds to show that an alpha particle is a helium atom with 2 electrons removed: the nucleus of a helium atom.

RADIOACTIVITY

In 1900, the French scientist P. Villard[15] discovered yet another type of ray, using the experimental set-up shown in Fig. 11–10. Rays emerging from a radium source are subjected to a strong magnetic field. Two photographic plates wrapped in black paper are placed one over the other in the path of the rays. When the plates are developed, the upper plate shows two regions of exposure, one due to the negatively charged beta rays that have been bent by the magnetic field, and the other due to rays traveling in a straight line from the source. The alpha particles travel in a straight line because the magnetic field is not strong enough to bend them. But the lower plate shows only one region of exposure—from the "straight-line" rays. The alpha and beta rays were absorbed by the glass in the upper plate. This new ray is very penetrating since it easily travels through glass. When a lead sheet 0.3 mm thick was placed on the upper plate, these new rays were still able to penetrate it and expose the bottom photographic plate, though less intensely. This new ray, called a **gamma ray,** resembles X-rays: uncharged and very penetrating.

Visible light, X-rays, and gamma rays are all forms of electromagnetic radiation. Visible light has a wavelength between 400 nm and 700 nm; X-rays, between 5×10^{-4} nm and 1 nm; and gamma rays, less than 0.1 nm.

Figure 11–11 recaps the response of alpha, beta, and gamma rays to a magnetic field, and Table 11–4 lists some basic properties of the three types of radiation.

Induced Radioactivity and Half-Life

The Curies found that _any_ substance placed near a tube of radium becomes radioactive, even though ordinarily it is not radioactive. Their laboratory was soon contaminated by this induced radioactivity. Marie wrote:

When one studies strongly radioactive substances, special precautions must be taken if one wishes to be able to continue taking delicate measurements. The various objects used in a chemical laboratory, and those which serve for experiments in physics, all become radioactive in a short time and act upon photographic plates through black paper. Dust, the air of the room, and one's clothes, all become radioactive. The air in the room is a conductor. In the laboratory where we work, the evil has reached an acute stage, and we can no longer have any apparatus completely isolated.[16]

Pierre Curie and Rutherford both examined this induced radioactivity, using the electrical method. A sample was exposed to the rays from a radioactive source, removed from the source, and then the induced activity was measured by placing the sample on the bottom plate shown in Fig. 11–3. As before, the current is proportional to the activity (rays emitted per time). They both found that the current or activity decreased with time in a very precise way. For example, Pierre Curie found that the radioactivity induced in glass by radium decreased by a factor of 2 after each 4-day time interval (that is, after 4 days, the activity was one-half its original intensity; after 8 days, one-fourth of its original intensity; after 12 days, one-eighth, and so on).

Rutherford found a similar behavior for the radioactivity induced in zinc by thorium, but it decreased by a factor of 2 after 11 hours.[17] _The time period required for the activity of a radioactive sample to decrease by a factor of 2 is called the_ **half-life** _and it is different for every radioactive substance._ This relationship between activity and half-life is shown in Fig. 11–12.

In their research, the Curies found no drop in activity of their radium sample. There is good reason for this: The half-life of radium is 1620 years!

Heat Generation in Radium

As the Curies accumulated more purified radium, they found, to their surprise, that it stayed 1.5°C _above_ room temperature. The Curies asked,

TABLE 11–4   PROPERTIES OF ALPHA, BETA, AND GAMMA RAYS

	MASS (amu)	CHARGE	NATURE	PENETRABILITY
Alpha particle	4	+2	Nucleus of helium atom	Easily absorbed. The range of a 5 MeV alpha particle in air is only 3.5 cm.
Beta ray	$\dfrac{1}{1822}$	−1	Electron	More penetrating than an alpha particle. The range of a 1 MeV beta ray in air is 328 cm, and in aluminum the range is 0.16 cm. An aluminum absorber 0.015 cm thick will cut the intensity of 1 MeV beta rays in half.
Gamma ray	0	0	Electromagnetic radiation, wavelength less than 0.1 nm	More penetrating than X-rays. An aluminum absorber with a thickness of 4.3 cm or a lead absorber with a thickness of 0.91 cm will cut the intensity of 1 MeV gamma rays in half.

Note: Since an alpha particle is the nucleus of a helium atom, its mass is given as 4 amu—the atomic mass of helium. The mass of the beta ray is given in atomic mass units. The mass of a helium atom is 6.67×10^{-27} kg. The charge is given in terms of the charge on the electron. In the last column, the term *range* refers to the maximum distance traveled by the ray. The unit of energy is MeV, where 1 MeV equals 10^6 eV. (1 MeV equals 1.6×10^{-13} joules.)

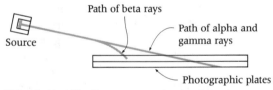

FIG. 11–10   Villard's apparatus, showing the existence of penetrating rays called *gamma rays*.

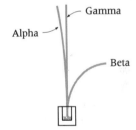

FIG. 11–11   The behavior of alpha, beta, and gamma rays in a magnetic field.

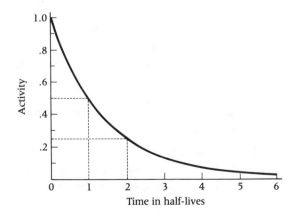

FIG. 11–12   A graph showing how the activity of a radioactive substance decreases with time: The activity is cut in half after a time called the *half-life*.

What is the source of the heat? Where is the energy coming from? Pierre and a student, A. Laborde,[18] showed that 1 g of radium gives off 100 calories of heat each hour. One **calorie** (abbreviation: cal) of heat is defined as the amount of heat required to raise the temperature of 1 g of water 1°C.

Rutherford also became interested in the problem of heat generation (see Chapter 6). Knowing that alpha particles are easily absorbed, he suggested that not all of them escape from radium, that some get absorbed by the radium itself. When a fast-moving alpha particle is stopped in the radium, its kinetic energy is changed into heat. Rutherford explained:

This emission of heat is, in reality, directly connected with the radioactivity of radium, and is not an independent phenomenon. There is little doubt that the heat emission is a direct consequence of the continuous expulsion of alpha particles from the radium. The alpha particles are readily absorbed in their passage through matter; and in a pellet of radium, only a small percentage of the total number emitted are able to escape into the surrounding gas (air). The rest are absorbed in the mass of the radium itself. The radium is thus subjected to an intense and continuous bombardment by the alpha particles projected from its own mass. Just as a target is heated by the impact of bullets upon it, so the radium becomes hot in consequence of its self-bombardment.[19]

To say that the heat results from the absorption of the rays still does not explain the source of energy for the rays. How can a substance throw out alpha and beta rays at speeds near the speed of light? In 1903, Rutherford discussed the matter:

As far as experiments have gone, the rate of heat emission of radium is independent of external conditions. Curie and Laborde found that the radium gave out the same amount of heat when immersed in ice as under ordinary conditions. J. J. Thomson

has pointed out that it is impossible to suppose that under such conditions the radium borrowed the heat from the surrounding air.

It has been pointed out that the radio elements have no special chemical characteristics except their high atomic weight which distinguish them from the other chemical elements. It is thus probable that the energy resident in the atoms of the elements is enormous compared with that released or absorbed in chemical reactions. This energy has not been observed on account of the difficulty of breaking up the atoms by the physical and chemical processes at our disposal.[20]

Scientists concluded that the energy from radioactivity must be due to the energy *somehow* stored in the atom—atomic energy. Today, we know that the energy is stored in the nucleus, and we call it nuclear energy.

Rutherford also speculated that the energy of the sun is due to the energy of the atom:

The maintenance of solar energy, for example, no longer presents any fundamental difficulty if the internal energy of the component elements is considered to be available, i.e., if processes of sub-atomic change are going on.[21]

Later (Chapter 19), we shall see that the energy of the sun is due to the atomic energy released when 4 hydrogen atoms combine to form 1 atom of helium.

At the turn of the century, so many questions remained unanswered: How can radioactivity be induced in substances not normally radioactive? What is atomic energy? How is it stored in the atom? How can it be released? How does our sun use atomic energy? Undoubtedly, the most fundamental question is: What does the atom look like? In Chapter 12, we shall see how Rutherford, in yet another experiment with alpha particles, answered this question.

Summary

In 1896, Becquerel accidentally discovered radioactivity when he found that *all* uranium compounds—not only the phosphorescent uranium compound he initially used—emitted penetrating radiation. He found that the rays from uranium could discharge a charged electroscope.

Madame Curie used an electrical method in pursuing the study of radioactivity. The device she employed (a capacitor connected to a voltage source and current meter) was the forerunner of the Geiger counter. Madame Curie systematically measured the activity of many uranium compounds, testing all known elements for radioactivity. Outside of uranium, she found that only thorium was radioactive. She then checked minerals for radioactivity and discovered that pitchblende was far more active than uranium or thorium, indicating undiscovered radioactive elements in it. With chemical separation procedures, she and her husband Pierre uncovered two such elements in pitchblende. These they named radium and polonium.

In the separation, radium appeared in the barium sample, showing that barium and radium have similar chemical properties. When the light from this sample was analyzed, the spectrum showed many lines from barium, but the wavelength of one line had not been found before in *any* element. This line was due to radium. As they purified their radium sample (increasing the percentage of radium compared to barium), the Curies found that the spectral line in question became stronger and also the activity became stronger, providing evidence for a new element. To determine the atomic mass of radium, Marie extracted one-tenth of a gram of pure radium from one ton of pitchblende. Then she produced a radium chloride compound and weighed it carefully. She chemically replaced the radium with silver, forming silver chloride, which she weighed again. The change in mass of the chlorine compound (from radium to silver) reflects the difference in atomic mass of radium and silver. With this maneuver, she was able to determine the atomic mass of radium to be 225 amu (compared to 226 amu accepted today).

Pierre Curie discovered that rays from radium could be used to treat cancer. Soon radiation therapy was adopted all over the world.

The Curies became famous, but not wealthy. They did not seek a patent on their procedure to extract radium from pitchblende.

The rays of uranium consist of three different types: alpha, beta, and gamma. The properties of these rays were determined by many scientists, but a principal investigator was Rutherford. He placed uranium on the bottom plate of the electrical apparatus and observed the current. He then placed thin sheets of aluminum foil on top of the uranium and watched the current drop because some of the rays had been absorbed by the foil. Rutherford concluded that the sharp drop in the current was due to absorption of one kind of ray, which he called the alpha rays. The more penetrating rays that could pass through 100 foils he called the beta rays. The mass-to-charge ratio of the beta rays was the same as that of the electrons, and the beta rays were identified as electrons. While beta rays were easily deflected by a magnetic field, alpha rays were not, indicating that they were very massive compared to beta rays.

When Rutherford obtained a sufficiently strong magnetic field, he found that alpha particles deflected as positively charged particles would. The mass-to-charge ratio of the alpha particle was consistent with it being the nucleus of a helium atom (a helium atom with 2 electrons missing). This was confirmed by Rutherford and Royds, who collected an alpha particle gas in what was essentially a discharge tube. The light from the alpha particle gas had the same spectrum as helium. Villard discovered the third ray—gamma rays—which were extremely penetrating compared to beta rays. The rays from a radioactive source were subjected to a magnetic field and the beta rays were deflected. Passing the beam through glass caused the alpha rays to be absorbed. Only the gamma rays were left in the beam, which caused exposure on a photographic plate, showing evidence of their existence.

The Curies found that any substance placed near radium becomes radioactive, but the activity dropped off in a precise way, decreasing by a factor of 2 during a time called the half-life. Much to the Curies' surprise, the purified radium remained 1.5°C above room temperature. Rutherford explained that the heat production is due to energetic alpha particles (produced within the radium sample) being absorbed before they can reach the surface of the sample. The kinetic energy of the alpha particles, transferred to the radium atoms with which the alpha particles collide, appears as heat. Scientists concluded that the energy of the rays (such as the kinetic energy of the alpha particles from radium) must be due to the energy somehow stored in the atom, which they called atomic energy. Today, we know that this energy is stored in the nucleus as nuclear energy.

True or False Questions

Indicate whether the following statements are true or false. Change all of the false statements so that they read correctly.

11-1 In his experiments with uranium compounds, Becquerel found
    a. only uranium compounds that were phosphorescent emitted penetrating rays.
    b. sunlight caused phosphorescent uranium compounds to emit rays.
    c. sunlight caused all uranium compounds to emit rays.
    d. all uranium compounds spontaneously emitted penetrating rays.
    e. 1 gram of pure uranium emitted more rays per second than 1 gram of uranium sulfate.
    f. the penetrating rays could pass through paper.
    g. the penetrating rays could expose a photographic plate.
    h. a connection between X-ray production and phosphorescence.
    i. the penetrating rays from uranium caused the leaves of an electroscope to separate because the rays produced charged particles in the air.

11-2 Madame Curie developed an electrical method for studying radioactivity. In this method,
    a. the two plates of a capacitor were linked to a battery and a current meter.
    b. the substance to be examined was placed on the bottom plate of the capacitor.
    c. the rays from a radioactive substance traveled through the current meter, producing a reading on this meter.
    d. the rays from the radioactive substance ionized the air between the capacitor plates.
    e. the ionized air neutralized charges on the capacitor plates.
    f. the battery acted to restore charges that were neutralized on the capacitor plates, and a current was produced.
    g. a stable substance produced a small current, while a radioactive substance produced a large current.
    h. the current was proportional to the activity of the substance on the bottom plate.

11-3 Madame Curie's electrical method for studying radioactivity and the Geiger counter operate on different principles.

11-4 Madame Curie tested all elements that were known at that time and found that only uranium was radioactive.

11-5 Madame Curie concluded that the mineral pitchblende must contain a new element (or new elements) because the activity of pitchblende was less than the activity of uranium.

11-6 In the chemical separation of pitchblende,
    a. the Curies discovered only one new radioactive element: radium.

b. the radium and barium were difficult to separate because their chemical properties were quite different.

c. the radium and barium were difficult to separate because radium and barium are in the same row of the periodic table of the elements.

11–7 The light emitted by a sample containing barium and radium showed evidence of radium because the spectrum contained lines from barium and helium. (Radium emits alpha particles, which are helium nuclei.)

11–8 Using the electrical method developed by the Curies, Rutherford placed uranium on the bottom plate of the capacitor and then placed thin aluminum foils on top of the uranium.

> **a.** When foils were added, the current decreased because the foils deflect the rays backwards.
>
> **b.** When 4 foils had been added, the current dropped slightly; when 100 foils had been added, the current dropped sharply.
>
> **c.** The alpha rays were absorbed by the first 4 foils, but some beta rays could even penetrate 100 foils.
>
> **d.** Gamma rays were not detected in this experiment.

11–9 The following are properties of the alpha, beta, and gamma rays:

> **a.** The alpha ray is the same as an electron.

b. The gamma ray is similar in nature to X-rays.

c. The beta ray is the nucleus of a helium atom.

d. The alpha ray is about 7000 times more massive than a beta ray.

e. The gamma ray has no mass and no charge.

f. The gamma ray penetrates matter most easily, and the beta ray is most easily absorbed by matter.

g. In a magnetic field, the gamma ray travels in a straight line, showing that it is neutral.

h. In a magnetic field, the alpha ray is deflected in the same direction as a beta ray, showing that the alpha and beta rays are oppositely charged.

i. In a magnetic field, the beta ray is more easily deflected (smaller radius path) than the alpha ray, showing that the alpha ray is more massive than the beta ray.

j. The alpha particle has a larger mass-to-charge ratio than the beta ray.

11–10 Radium caused glass to become radioactive with a four-day half-life.

> **a.** This means that the activity drops by a factor of 2 after four days and the activity drops to zero after eight days.

Questions for Thought

11–1 Describe how Becquerel discovered radioactivity.

11–2 Which would be more radioactive: a 1-g sample of uranium tetrachloride (UCl_4), a molecule with 1 atom of uranium and 4 atoms of chlorine, or a 1-g sample of uranium tetrafluoride (UF_4), a molecule with 1 atom of uranium and 4 atoms of fluorine? Explain.

11–3 In Fig. 11–1 the electroscope discharges when uranium is brought nearby. Explain this effect in terms of ionization.

11–4 Describe Marie Curie's electrical method for measuring radioactivity. Explain why the current is proportional to the activity of the radioactive sample.

11–5 Why did Marie Curie conclude that pitchblende contained new element(s)?

RADIOACTIVITY

11–6 Does the mineral chalcolite, listed in Table 11–2, contain new element(s) also? Explain.

    a. How many new elements did the Curies find in the pitchblende?

    b. How did the chemical separation of the elements in pitchblende show that they had found new elements?

    c. How did they describe the chemical properties of these new elements?

    d. Locate these new elements in the periodic table of the elements (p. 6) and comment about the Curies' conclusion about their chemical properties.

    e. If you know that the chemical formula for calcium chloride is $CaCl_2$, can you draw any conclusions about the chemical formula for barium chloride and radium chloride? Explain.

11–7 Barium was separated from the pitchblende and the spectrum of light from it was examined. How did this spectrum prove the existence of a new element?

11–8 In order to determine the atomic mass of radium, Marie needed a *pure* radium sample. How did Marie determine that the separated sample contained *only* radium?

11–9 Briefly describe how Marie Curie determined the atomic mass of radium.

11–10 Describe how alpha, beta, and gamma rays are affected by a magnetic field. How does this provide information about their mass and charge?

11–11 The Curies' purified radium sample remained 1.5°C above room temperature. How did Rutherford explain this? What theories were proposed to account for the production of heat so that the principle of conservation of energy would not be violated?

11–12 Marie Curie's measurements of radioactivity using the electrical method are shown in Tables 11–1 and 11–2. These tables do not list the thickness of the layer or the mass of the material in grams. In fact, Marie did examine the effect of the layer thickness as shown by the following data from her thesis:

	Thickness of layer (mm)	Current (amperes)
Uranium oxide	0.5	2.7×10^{-11}
Uranium oxide	3.0	3.0×10^{-11}

She says, "It may be concluded from this that the absorption of uranium rays by the substance which generates them is very great, since the rays proceeding from deep layers produce no significant effect."[22]

    a. In light of Rutherford's absorption experiments, is the current shown in Tables 11–1 and 11–2 due primarily to alpha or beta rays?

    b. Explain Marie's conclusion about the effect of the layer thickness in your own words.

11–13 Recall the photographic method that Becquerel used to discover radioactivity. Were any of the rays absorbed by the paper in which the photographic plate was wrapped? What rays caused the exposure on the plate?

11–14 Suppose that early in the evening you had read this chapter and then watched the late horror movie on TV. Later, as you sleep, you dream of taking a trip to Hawaii. The bellhop carrying your luggage to your room is called Drake U. Lar. As you walk down the hotel corridor, a horrendous bolt of lightning flashes and a guest shrieks, supposedly from surprise at the lightning. You turn around and scream. The bellhop has turned into Dracula. Dracula points to three rooms labeled alpha, beta, and gamma. He says each one contains a particular type of radioactive source. He begins to shove you toward the rooms. Which one would you try to claim as your room and why?

11–15 In Chapter 10, we learned that X-rays cause damage to cells of human tissue by ionizing molecules and, in particular, ionizing the

DNA molecule, which governs cell division. Alpha, beta, and gamma rays are harmful because they cause ionization.

a. Which ray is the most highly ionizing; that is, which produces the largest number of ion per centimeter of travel?

b. Radiation can be used to destroy cancerous cells. *If* the radiation could reach the cells, which one would be most effective in destroying cancerous cells in a localized tumor—alpha, beta, or gamma rays?

c. Assuming that the radioactive source is outside the patient's body, which ray(s) could be used in treating skin cancer?

In treating a tumor just below the surface of the skin? A tumor deep within the body? Explain.

d. Explain (in a general way) how the ionization of water molecules in a cell leads to the formation of hydrogen peroxide (H_2O_2), a deadly poison.

11–16 After graduation from college, you are employed by a steel company. During a tour of the operations, you notice that the thickness of steel is monitored by using a radioactive source. Explain the set-up of the radioactive source and detector. If the steel is supposed to be 0.02 cm thick, what kind of source could be used—one emitting alpha, beta, or gamma rays?

Questions for Calculation

11–1 When Marie Curie placed metallic uranium on the bottom plate of the capacitor in Fig. 11–3 the current meter read 2.3×10^{-11} amperes (Table 11–1). Since 1 unit of charge equals 1.6×10^{-19} coulombs, how many units of charge reach each capacitor plate in 1 second? (1 ampere is defined as 1 coulomb/sec.)

11–2 Brian and Susan, playing with Tinker Toys, assemble 5 of the structures shown in Fig. 11–13(a), which have a total mass of 400 g. Shortly thereafter, they assemble 10 of the structures shown in Fig. 11–13(b), which have a total mass of 350 g. Block S has a mass of 20 g. Find the mass of block R and block C.

†11–3 The alpha particles and beta rays from uranium have several different kinetic energies, but the alpha particles' kinetic energies are close to 5 MeV and the beta rays', to 1 MeV. Here, consider how these particles are stopped as they travel through air—78 percent nitrogen (N_2) and 21 percent oxygen (O_2). When an alpha particle (or a beta ray) collides with a nitrogen molecule, for example, the nitrogen molecule becomes ionized, as shown in Equation 2. This collision produces a pair of ions (hereafter called *ion-pairs*, which refers to the fragments on the right-hand side of Equation 2a and 2b); since

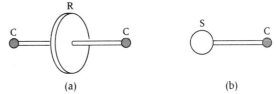

FIG. 11–13 Diagram for calculation question 11–2.

35 eV is needed to ionize the air molecule, the kinetic energy of the alpha particle (or beta ray) is reduced by 35 eV.

a. How many molecules of air are ionized as a 5 MeV alpha particle comes to a stop?

b. If the alpha particle comes to a stop in 3.5 cm (see Table 11–4), how many ion pairs on the average are produced per cm?

c. How many molecules of air are ionized as a 1 MeV beta ray comes to a stop?

d. If the beta ray comes to a stop in 328 cm (see Table 11–4), how many ion pairs on the average are produced per cm?

TABLE 11-5   DATA FOR CALCULATION QUESTION
11-6

COUNTING TIME INTERVAL IN MINUTES	COUNTS RECORDED DURING ONE-HALF MINUTE INTERVALS
0-0.5	2878
1-1.5	2370
2-2.5	1831
3-3.5	1390
4-4.5	1047
5-5.5	810
6-6.5	633
7-7.5	547
8-8.5	402
9-9.5	320

†11-4 Rutherford placed powdered uranium oxide in the center of the bottom plate of a capacitor. The plates were 20 cm square and 4 cm apart. He measured the current produced by rays and how it changed when the rays were absorbed by aluminum foil placed over the uranium oxide. Figure 11-6 shows Rutherford's data. The object of this problem is to compare the number of alpha particles emitted per second (N_a) with the number of beta rays emitted per second (N_b) from Rutherford's uranium sample. The results of calculation question 11-3 are needed.

   a. The data in Fig. 11-6 shows the current produced by the alpha particles and beta rays streaming in all directions from the uranium. The current measures the number of ion pairs per second reaching the capacitor plates. Denote N_{ipa} and N_{ipb} as the number of ion pairs per second caused by the alpha particles and beta rays, respectively. From the data in Fig. 11-6, estimate the ratio N_{ipb}/N_{ipa}.
   b. If N_a is the number of alpha particles per second from uranium, use the results from calculation question 11-3 to find the total number of ion pairs per second N_{ipa}, due to the alpha particles (in terms of N_a).
   c. The beta rays are not stopped as they travel between the capacitor plates. Since the beta rays are emitted in all directions, they travel a greater distance than the 4 cm distance between the capacitor plates. Assume that 10 cm is the average distance traveled by the beta rays before the beta ray either strikes one of the plates or escapes through the open sides. Use the results of calculation question 11-3 to find the number of ion pairs produced by a 1 MeV beta ray traveling a distance of 10 cm.
   d. If N_b is the number of beta rays per second from uranium, what is the total number of ion pairs per second due to the beta rays, N_{ipb}?
   e. What is the ratio of the number of beta rays per second to the number of alpha particles per second, N_b/N_a?

11-5 Use the data in Fig. 11-6 to find an approximate value for the range of 5 MeV alpha particles in aluminum. Compare it with the accepted value of 0.0022 cm.

11-6 A Geiger counter is used to measure the decay of a radioactive source. A counting interval of one-half minute is followed by a one-half minute interval in which no counts are recorded, as shown in Table 11-5.

Using semilog graph paper, plot counts along the vertical axis and time along the horizontal axis. Use the time at the start of the counting period. Label the vertical axis as shown in Fig. 11-14(a). Draw a straight line that seems to be the best fit through the data points. (The data points will not fall exactly on the line because of uncertainties in any measurement.) Find the half-life of the radioactive source by finding the time required to reduce the activity (counts per minute) by a factor of 2.

11-7 To study the absorption of gamma rays from a cesium-137 source, an absorber is placed between the source and a Geiger counter. The gamma rays have an energy of 0.66 MeV. When

TABLE 11–6 DATA FOR CALCULATION QUESTION 11–7			
THICKNESS OF ALUMINUM ABSORBER (cm)	COUNTS PER MINUTE	THICKNESS OF LEAD ABSORBER (cm)	COUNTS PER MINUTE
0.0	5009	0.0	4999
0.4	4579	0.2	3936
0.8	4207	0.4	3055
1.0	4115	0.5	2677
1.5	3699	0.6	2323
2.0	3355	0.8	1877
2.5	3000	1.0	1404
3.0	2738	1.5	788
3.5	2466	2.0	405
4.0	2200		
4.5	2000		
8.0	989		

a gamma ray strikes the Geiger counter, it gives a pulse of current. These pulses or counts are counted for a period of time. The number of counts per minute for each absorber thickness is shown in Table 11–6.

a. Plot counts per minute versus absorber thickness on semilog graph paper. Label the vertical axis, as shown in Fig. 11–14(a).

b. Use the graph to find the thickness of lead or aluminum required to reduce the intensity of gamma rays from 5000 counts per minute to 2500 counts per minute.

c. Pick two other points on each straight line and find the additional thickness required to reduce the intensity by 2. Compare these answers to those in part **b.**

d. What thicknesses of aluminum and lead are required to reduce the gamma ray intensity by 4, by 8, and by 16?

e. Explain how the photoelectric effect and Compton effect explain the scattering and absorption of 0.66 MeV gamma rays.

11–8 Table 11–7 shows the data obtained by placing an aluminum absorber between a thallium-204 radioactive source, emitting 0.76 MeV beta rays, and a Geiger counter.

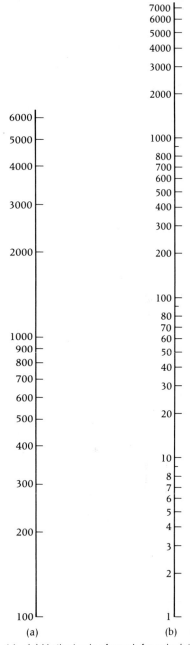

FIG. 11–14 (a) Vertical axis of graph for calculation questions 11–6 and 11–7. (b) Vertical axis of graph for calculation question 11–8.

TABLE 11–7 DATA FOR CALCULATION QUESTION 11–8	
THICKNESS OF ALUMINUM ABSORBER (cm)	COUNTS PER MINUTE
0.0000	6701
0.0033	5436
0.0044	5168
0.0058	4493
0.0083	4239
0.0109	3717
0.0160	2645
0.0225	1788
0.0290	1109
0.0337	756
0.0447	287
0.0534	114
0.0667	28
0.0852	5

a. Plot the counts per minute versus absorber thickness on semilog graph paper. Label the vertical axis as shown in Fig. 11–14(b). Plot a straight line through the data for the first seven points in Table 11–7 and a curved line through the rest of the data.

b. Using the straight-line portion of the graph, find the thickness of aluminum required to reduce the intensity of beta rays by 2.

c. Compare the answer in **b** with the thickness of aluminum required to reduce the intensity of 0.66 MeV gamma rays by a factor of 2 in calculation question 11–7**b**.

d. Using the graph, *estimate* the range of 0.76 MeV beta rays in aluminum. Compare with the accepted value of 0.11 cm.

†11–9 The density of lead (atomic mass number $A = 207$; atomic number $Z = 82$) is 11.4 g/cm^3. The density of aluminum ($A = 27$; $Z = 13$) is 2.70 g/cm^3.

a. Use the fact that the atomic mass in grams contains 6×10^{23} atoms to find the number of atoms in 1 cm^3 of lead and of aluminum.

b. How many electrons does an atom of aluminum and one of lead contain?

c. How many electrons are contained in 1 cm^3 of lead and aluminum?

d. Assuming that beta rays and gamma rays interact with electrons, use part **c** to explain why the penetration of beta and gamma rays through aluminum is greater than for lead.

†11–10 In 1957, a committee of the National Academy of Sciences first proposed the burial of radioactive wastes from nuclear reactors in geologically stable rock formations or salt formations 600 m below the earth's surface. Experiments show that the intensity of gamma rays is reduced by a factor of 10 in passing through one-third of a meter of rock or soil.

a. What is the chance that alpha and beta rays will reach the earth's surface?

b. By what factor (expressed in powers of 10) will the intensity of gamma rays be reduced in passing through 600 m?

c. How long will it take a gamma ray to travel 600 m? (A gamma ray travels with the speed of light. Assume a speed of 2×10^8 m/sec through rock.)

†11–11 A Geiger counter is located 0.2 m from a radioactive source emitting gamma rays *uniformly in all directions*. The counting rate is 1000 counts per minute. If the Geiger counter is moved to 0.4 m, what do you expect the counting rate to be? What would the counting rate be if the counter is moved to 0.8 m? (Imagine the radioactive source at the center of a sphere. Compare the number of gamma rays per cm^2 striking the sphere, where the radius of the sphere is 0.2 m and 0.4 m. The surface area of a sphere is $4\pi r^2$.)

†11–12 Radioisotope power generators (nuclear batteries) power the *Voyager II* spacecraft probing the planet Saturn (August, 1981), powered experiments during the *Apollo* lunar landings, now power arctic weather stations and weather satellites, and are currently available for heart pacemakers that can last up to 20 years. The basic idea behind such generators involves converting the heat generated in radioactive isotopes to electricity. Polonium-210 (Po-210), the radioactive isotope discovered by Marie Curie, has been used in some generators. Po-210 emits alpha particles having a kinetic energy of 5.30 MeV and has a half-life of 138 days. One gram of Po-210 emits 1.65×10^{14} alpha particles per second.

a. What is the activity of 1 g of Po-210 in curies?

b. Assuming that the kinetic energy of all the alpha particles is converted into heat, find the heat energy generated in watts by 1 g of Po-210. (1 MeV = 1.6×10^{-13} joule; 1 watt = 1 joule/sec.)

c. In 1959, the demonstration model used Po-210 and generated 2.5 watts of electrical power. If only 10 percent of the heat is converted into electricity, how many grams of Po-210 are needed to generate this electrical power? After 138 days, what electrical power will be generated by this battery? (Note: When the two ends of two dissimilar metal wires are joined to form a loop, and the two junctions are kept at different temperatures, a current flows through the loop. This principle to generate electricity is used in some nuclear batteries: one junction is near the radioactive isotope, and the other is at the cooler surface of the battery.)

d. Strontium-90 (Sr-90) emits beta rays having a kinetic energy of 0.54 MeV and has a half-life of 28 years. In what sense might Sr-90 make a better battery?

11–13 One gram of radium gives off 100 calories of heat each hour. (Remember that 1 cal is the amount of heat required to raise the temperature of 1 g of water by 1°C.) How long would it take for the heat given off by 100 g of radium to heat a cup of water (240 g) from 20°C (68°F) to the boiling point of 100°C? (Assume no heat losses due to conduction or radiation.)

Footnotes

[1] W. F. Magie, *Source Book in Physics* (New York: McGraw-Hill, 1935), p. 610. (Reprinted by permission.)

[2] Ibid; p. 610.

[3] Ibid; p. 610.

[4] Biographical information regarding the Curies has been obtained from: Eve Curie, *Madame Curie* (New York: Doubleday & Company, Inc., 1937).

[5] Magie, p. 614. (Reprinted by permission.)

[6] Ibid; p. 615. (Reprinted by permission.)

[7] Eve Curie, *Madame Curie* (New York: Doubleday & Company, Inc., 1937), p. ix. (Excerpt from *Madame Curie* by Eve Curie, p. ix. Copyright 1937 by Doubleday & Company, Inc. Reprinted by permission.)

[8] Ibid; p. x. (Excerpt from *Madame Curie* by Eve Curie, p. x. Copyright 1937 by Doubleday & Company, Inc. Reprinted by permission.)

[9] *The Collected Papers of Lord Rutherford of Nelson*, Vol. 1, (London: George Allen & Unwin, 1962), p. 175. [This article originally appeared in *Philosophical Magazine* 47, Ser. 5 (1899):109.]

[10] *Science Abstracts*, Section A, No. 1260 and No. 1261, (1900).

[11] *Science Abstracts*, Section A, No. 830, (1900).

[12] *Science Abstracts*, Section A, No. 61, (1900).

[13] F. K. Richtmyer, E. H. Kennard, and T. Lauritsen, *Introduction to Modern Physics*, 5th ed. (New York: McGraw-Hill, 1955), p. 454.

(*Note:* Marie Curie's thesis, *Radioactive Substances*, has been reprinted by Greenwood Press, Westport, Connecticut, 1971.)

[14] *Collected Papers, Lord Rutherford,* Vol. 2, p. 163. [E. Rutherford and T. D. Royds, *Philosophical Magazine* 17, Ser. 6 (1909):281.]

[15] *Science Abstracts,* Section A, No. 1648 (1900).

[16] E. Curie, p. 196. (Excerpt from *Madame Curie* by Eve Curie, p. 196. Copyright 1937 by Doubleday & Company, Inc. Reprinted by permission.)

[17] *Collected Papers, Lord Rutherford,* Vol. 1, p. 245. [E. Rutherford, *Philosophical Magazine,* 49, Ser. 5 (1900):161.]

[18] P. Curie and A. Laborde, *Comptes Rendus* 136(1903):673.

[19] *Collected Papers, Lord Rutherford,* Vol. 1, p. 646. (E. Rutherford, *Technics,* July 1904, p. 11.)

[20] Ibid; p. 620. (E. Rutherford, *Transactions of the Australasian Association for the Advancement of Science,* Dunedin, January 1904, p. 87.)

[21] Ibid; p. 608. [E. Rutherford, *Philosophical Magazine,* Ser. 6 (1903):576.]

[22] Marie Curie's thesis, in *Chemical News and Journal of Physical Science* (London) 88, No. 2283 (1903):98.

Introduction

After J. J. Thomson's discovery of the electron, scientists wondered, What does the atom look like? Since an atom is neutral, it must contain equal amounts of positive and negative charge. How are the electrons and positive charges arranged in the atom?

In 1904, J. J. Thomson himself and H. Nagoaka proposed models of the atom. Thomson's model was generally accepted, while Nagoaka's was not. We shall, in this chapter, learn the reason for this. We'll open our discussion with some brief words about oscillating charged particles, which will provide the background we need to understand the models of the atom.

Next, we'll see how Rutherford got his idea for the alpha-scattering experiments and why Thomson's model could not account for some experimental results. Not only did Rutherford propose a new model of the atom, he also described an experiment to test it. In conclusion, we'll consider Geiger and Marsden's experimental test by comparing their results with the predictions of Rutherford's model.

Ernest Rutherford

CHAPTER 12

Rutherford Discovers the Nucleus

Oscillating Charges and Electromagnetic Waves

Radio station WGN in Chicago (AM 720) transmits electromagnetic waves at a frequency of 720,000 oscillations per second. The electrons in the broadcast antenna (a large metal tower whose height is comparable to the wavelength of the waves being emitted) oscillate at a frequency of 720,000 oscillations per second; electromagnetic waves of the same frequency result. This example illustrates a basic principle of electricity: When a charged particle oscillates, electromagnetic waves, having the same frequency as the oscillation, are produced.

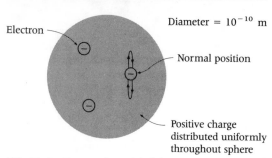

Electron ─

Diameter $= 10^{-10}$ m

─ Normal position

─ Positive charge distributed uniformly throughout sphere

FIG. 12–1   Thomson's model of the atom.

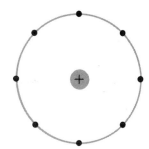

FIG. 12–2   Nagoaka's model of the atom.

Thomson's Model of the Atom

Each chemical element emits a characteristic spectrum of light (Chapter 1). How does an atom emit light? Why is the spectrum of light different for each element? Thomson and Nagoaka both tried to answer these questions.

Thomson suggested that the atom consists of a sphere of positive charge with electrons embedded inside it, as shown in Fig. 12–1. The diameter of the sphere was about 10^{-10} m, corresponding to the known size of the atom (Chapter 1, Equation 2). Usually, the electrons are at rest, but when they are disturbed or excited, the electrons oscillate about their equilibrium position and emit electromagnetic radiation. In the hydrogen atom, the single electron is located at the center and, when disturbed, oscillates back and forth with a *single* frequency, producing electromagnetic waves of that frequency. This frequency corresponds to a wavelength of 120 nm. (Remember that the wavelength equals the speed of light divided by the frequency.) But the hydrogen spectrum doesn't have a single wavelength. It has 4 wavelengths in the visible region: 656 nm (red), 486 nm (blue), 434 nm (violet), and 410 nm (violet). Obviously, Thomson's inability to account for these wavelengths was a serious deficiency. Even so, Thomson's model gained acceptance, while Nagoaka's model did not.

Nagoaka's Model of the Atom

The Japanese physicist Hantaro Nagoaka (1865–1950) proposed a model of the atom that consisted of a positively charged center with electrons arranged in a circle at equal angular intervals.[1] (See Fig. 12–2.) The electrons revolve in this circular orbit. The atom emits light when an electron makes small oscillations perpendicular to the plane of this page.

Nagoaka's model was not accepted because it didn't seem to depict an atom that was stable. As an electron *revolves* in a circle, electromagnetic radiation of the same frequency as the number of revolutions per second should be emitted. Since electromagnetic waves carry off energy, the energy of the electron should decrease, and the electron should move inward. Its path would then be a spiral, and it would crash into the positively charged center (see Fig. 12–3). Thus, the atom would not be stable, and the size of the atom would shrink from 10^{-10} m to the size of the positive center.

You can see that options for creating models were extremely limited. According to the principles of electricity, an unexcited electron (that is, one not emitting electromagnetic waves) must be at rest. For this reason, Thomson's model seemed the only choice, and it was accepted, even though it could not explain the spectra of the elements.

Scattering Experiments

AN IDEA GERMINATES

While he was studying the absorption of alpha particles, Rutherford noticed something very curious indeed. Some of the alpha particles, after passing through mica (a mineral found in the structure of rock) only 0.003 cm thick, were bent by *as much* as 2° from a straight-line path.[2] (See Fig. 12–4a.) From his earlier research, Rutherford knew that a huge magnetic field was needed to bend alpha particles by the same amount (Fig. 12–4b). The force on alpha particles in the mica must be 1000 times (which equals 3 cm/0.003 cm) larger than the force due to the magnetic field! Rutherford concluded:

Such a result brings out clearly the fact that atoms of matter must be the seat of very intense electrical forces—a deduction in harmony with the electronic theory of matter.[3]

An idea began to germinate: Alpha particles could probe these electrical forces. Recall that this idea was presented by analogy in Chapter 1 (see Fig. 1–18, p. 22). In observing the scattering of balls by an obstacle, one can determine

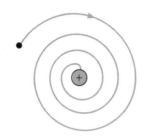

FIG. 12–3 The spiral path of an electron revolving around a positively charged center.

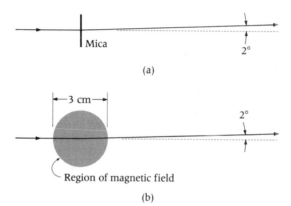

FIG. 12–4 (a) A beam of alpha particles scattered by a piece of mica. (b) Alpha particles deflected by a magnetic field. (*Source:* H. Geiger and E. Marsden, *Proceedings of the Royal Society*, Vol. 82, A, 1909, p. 495.)

RUTHERFORD DISCOVERS THE NUCLEUS

whether the obstacle is square, circular, or triangular. In like manner, alpha particles could probe the shape of the atom.

In 1907, Rutherford accepted the Chair of Physics at the University of Manchester in England. To a list of possible research topics, he added the scattering of alpha particles.[4]

Hans Geiger, an inventive German scientist, was one of Rutherford's associates, and together they found a way to count alpha particles. An alpha particle, upon striking a zinc sulfide screen, produces a tiny flash of light that can be seen with a magnifying glass.

Chapter 1 reviewed this episode in the history of physics: Geiger directed a beam of alpha particles onto a gold foil and counted the number of them scattered to various angles between 0° and nearly 20°. Then, Ernest Marsden, a young graduate student, started to work with Geiger. As his first research project, Rutherford suggested that Marsden look for the backward-scattering of alpha particles. With Thomson's model as a base, however, Rutherford didn't really expect Marsden to see anything.

Figure 12−5 shows the apparatus used by Geiger and Marsden. The zinc sulfide screen S was placed behind the lead plate, so that no alpha particles from a radioactive source could strike it directly. The alpha particles struck the platinum foil R at an *average* angle of 90°. Those reflected or scattered backwards were counted at different points of screen S. "To find the whole number of reflected particles, it was assumed that they were distributed uniformly round a half sphere with the middle of the reflector (R) as centre. Three different determinations showed that of the incident alpha particles about 1 in 8000 was reflected. . . ."[5] Similar results were obtained with other foils such as gold.

Rutherford referred to the backward-scattering as the most incredible event in his life. Let's see why.

Predictions of Thomson's Model

Let's assume for an instant that Thomson's model of the atom is correct and see what to expect for alpha particles scattered by the atoms of a gold foil.

Only the positive part of the atom causes scattering. For example, when a massive alpha particle does interact with a puny electron, it simply "pushes" the electron out of the way and continues unaffected. (We used conservation of momentum to examine this in Chapter 4.)

Several experiments, such as the scattering of X-rays, suggested that the number of electrons in an atom was roughly equal to one-half the atomic mass number, or $A/2$. Since the atom is neutral, the amount of positive charge was also taken to be $A/2$. The atomic mass number A of gold is 197; the amount of positive charge in the atom and the number of electrons was thus 100. This theory was quite accurate for helium: Helium has an atomic mass number of 4 and the alpha particle (a helium atom with 2 electrons removed) has 2 positive charges.

We begin with the scattering due to 1 atom. The alpha particle at point B in Fig. 12−6 is repelled by both the positive charge below it (force F_1) and by the charge above it (force F_2). The net force ($F_1 - F_2$) causes the alpha particle to bend upward, as shown. Similarly, the other alpha particle is bent downward. Calculations show that the maximum scattering angle will be 1/30° (assuming the sphere of positive charge has a diameter of about 10^{-10} m and contains 100 positive charges). For the sake of illustration, the scattering angle in Fig. 12−6 is 5°, or 150 times larger than the maximum.

According to Thomson's model, the alpha particle encounters many atoms as it travels through the foil, scattering from each one. Because of this multiple scattering, an alpha particle emerges from the foil bent at an angle to its initial direction.

A mechanical analog on display at the Museum of Science and Industry in Chicago (see Fig. 12–7) illustrates multiple scattering.* As the balls fall, they strike pegs, arranged in a regular fashion. Many of the balls are found at the center (the point from which the balls were dropped), and fewer are found as the distance from the center increases. A similar kind of behavior (even to the shape of the curve) is expected for the scattering of alpha particles.

Suppose that a beam of alpha particles is directed toward a gold foil only 400 atoms thick. Since the scattering from 1 atom is only 1/30° (according to Thomson's model), how can an alpha particle emerge from the foil with a scattering angle θ larger than this? To do so, the alpha particle must be scattered in one direction (upward) more often than in the opposite direction (downward), as shown in Fig. 12–8.

A scattering of 1° would require an *excess* of 30 scatterings in one direction. This could happen if the alpha particle were scattered by 400 atoms, with 215 scatterings in one direction (upward) and 185 in the other direction (downward). A 1° scattering would also occur if the alpha particle were scattered by 500 atoms, with 265 in one direction and 235 in the other direction. There are many combinations that would give a scattering of 1°.

A scattering of 10° would require an excess of 300 scatterings in one direction. One possibility would be the scattering from 500 atoms, with 400 in one direction and 100 in the opposite direction.

The most likely situation is an equal number of scatterings in either direction, where the excess in one direction is zero.

The difficulty in obtaining a scattering angle much greater than a few degrees may be understood when we consider the tossing of a coin. If a coin were tossed 500 times, the *most likely* result would be the same number of heads and tails. It is not unreasonable that 500 tosses would yield 265 heads and 235 tails. Obtaining 400

*The analogy between the multiple scattering of alpha particles and falling balls striking pegs was suggested by E. N. da C. Andrade, *Rutherford and the Nature of the Atom* (New York: Doubleday, 1964), p. 112.

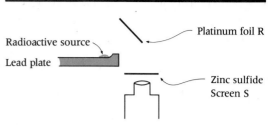

FIG. 12–5 Apparatus used by Geiger and Marsden to look for backward scattering of alpha particles. (*Source:* H. Geiger and E. Marsden, *Proceedings of the Royal Society*, Vol. 82, A, 1909, p. 495.)

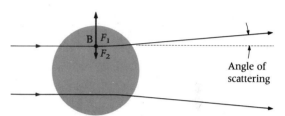

FIG. 12–6 The scattering of alpha particles predicted by Thomson's model. For the sake of illustration, the angle of scattering of 1/30° is drawn as 5°.

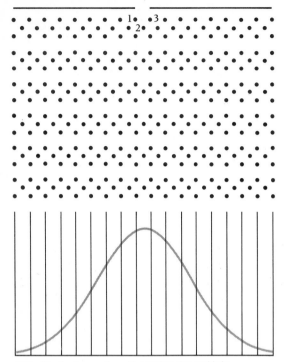

FIG. 12–7   An analogy illustrating the multiple scattering of alpha particles. Balls, which have the same diameter as the pegs, fall through the opening at the top of the diagram with various directions, so that they initially strike pegs 1, 2, or 3. The balls strike many pegs as they fall, and then drop into the slots at the bottom. The height of the column of balls in each slot is indicated by the curve. This is a mathematical curve called a *Gaussian*. Such a curve is more commonly called a *bell-shaped curve*. (*Source:* Museum of Science and Industry, Chicago.)

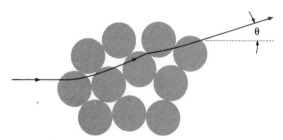

FIG. 12–8   The multiple scattering of alpha particles by gold atoms, based upon Thomson's model.

tails and 100 heads, however, would be extremely improbable.

Taking the probabilities into account, we can see that the chance of an alpha particle being scattered to angles larger than 90° (or reflected from the foil like a ball bouncing against a wall) is very nearly zero—only 1 in 10^{3500} alpha particles would be scattered backward! Impossible to observe. *If* the atom really "looks like" Thomson's model, then an experiment should show the following:

1. Most of the alpha particles will scatter less than 1°. (The most probable angle is 0°.)
2. Scattering to an angle greater than 20° is *extremely* improbable.
3. Scattering to angles greater than 90° is impossible.

Rutherford's Model of the Atom

Rutherford realized that Thomson's model could not be correct. The backward-scattering was certainly extraordinarily small (only 1 in 8000), but it was *not zero*. Rutherford pondered, What must the atom look like to produce this effect? He wrestled with this dilemma for nearly two years before he arrived at a solution.

Throughout the text, we've learned how Rutherford thought of a new model of the atom. Let's summarize this knowledge briefly.

Most of the alpha particles striking a gold foil are scattered by only a few degrees and their acceleration is very small. However, those scattered backwards must come to a stop in the foil and, as a result, their deceleration is extremely large. This deceleration was calculated to be about 10^{21} m/sec² (Chapter 2). Rutherford realized that the force needed to produce this large deceleration must be due to the alpha particle's repulsion to the positive part of the atom. He could account for the backward-scattering if the nucleus—the modern term for the sphere of positive charge—had a size of about 10^{-14} m (Chapter 3). In that case, the alpha particle would be so close to the center of the nucleus that the force of repulsion would be very large. Rutherford used conservation of energy to find the radius of the gold nucleus to be about 3×10^{-14} m (Chapter 6).

Based upon all of these ideas, Rutherford proposed the model of the atom shown in Fig. 12-9. Because the diameter of the atom was about 10^{-10} m, Rutherford concluded that the nucleus was surrounded by a sphere of that diameter containing the electrons. How the electrons moved (or did not move) inside that sphere Rutherford did not specify.

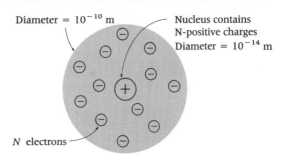

FIG. 12-9   Rutherford's model of the atom.

PREDICTIONS

Figure 12-10 shows a narrow beam of alpha particles striking a gold foil. Note that the nuclei are separated by a distance of about 10^{-10} m. Because the size of the nucleus is so very small compared to the size of the atom, only an extremely small number of alpha particles will come close to the nucleus itself. What happens to these few alpha particles? Let's find out by investigating the scattering from one *nucleus* of gold, as shown in Fig. 12-11.

The trajectories of several alpha particles are shown. The letter H represents the perpendicular distance between the original direction of the alpha particle and the center of the nucleus. Rutherford found that the angle of scattering (designated by θ) is directly related to the value of H. Table 12-1 shows the results of his calculations.[6]

Some understanding of why alpha particles are scattered to different angles can be gained by comparing the trajectories of the alpha particles scattered to 53° and 152°. Because the alpha particle scattered to 152° has the smaller value of H, it approaches the nucleus more closely than the one scattered to 53°. As a result, the alpha particle scattered to 152° has the larger force exerted on it and is, hence, scattered to the larger angle.

We can visualize the vast emptiness of the atom by considering that, using the scale in Fig. 12-11, the nearest nucleus would be 63 m away in any direction. The atom is mostly empty space!

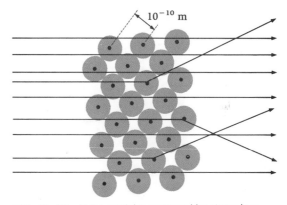

FIG. 12-10   Alpha particles scattered by atoms in a gold foil, as predicted by Rutherford.

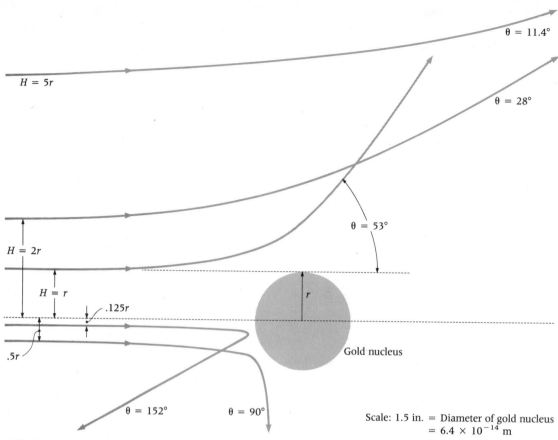

FIG. 12–11 The scattering of alpha particles approaching a gold nucleus. On the scale of this drawing, another gold nucleus would be 63 m away in any direction. (Adapted from E. N. da C. Andrade, *Rutherford and the Nature of the Atom,* p. 119. Copyright © 1964 by Doubleday & Company, Inc. By permission of the publisher.)

Scale: 1.5 in. = Diameter of gold nucleus
= 6.4×10^{-14} m

TABLE 12–1   CLOSENESS OF APPROACH TO GOLD NUCLEUS VERSUS SCATTERING ANGLE

H	θ
0.0	180°
0.125r	152°
0.25r	127°
0.5r	90°
r	53°
2r	28°
5r	11.4°
10r	5.7°

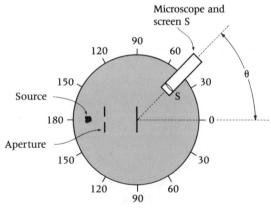

FIG. 12–12   The apparatus used by Geiger and Marsden to test Rutherford's predictions.

How, exactly, did Rutherford predict the number of alpha particles scattered to various angles θ? Suppose that the thickness of the foil in Fig. 12−10 corresponds to 1000 atoms placed end to end. Table 12−1 shows that any alpha particle, which is aimed directly at the nucleus (that is, H is either equal to, or less than, r), would be scattered to angles larger than 53°. Because the area of the nucleus is 100 million times smaller than the area of the atom, the probability that an alpha particle will hit any nucleus is very small. Therefore, only a miniscule number of alpha particles will be scattered to angles larger than 53°. In fact, due to the vast emptiness of the atom, most of the alpha particles will not be scattered at all. For these reasons Rutherford concluded that, if an alpha particle *is* scattered by one nucleus, the chance that it will now strike another nucleus is nearly zero. Therefore, the scattering of an alpha particle is due to only one nucleus in the foil.

Based upon such logic, Rutherford calculated the number of alpha particles that should be scattered at angle θ. He quickly told Geiger about his calculations. Geiger related:

One day Rutherford, obviously in the best of spirits, came into my room and told me that he now knew what the atom looked like and how to explain the large deflections of alpha particles. On the very same day I began an experiment to test the relations expected by Rutherford between the number of scattered particles and the angle of scattering.[7]

The Crucial Test: Comparison Between Theory and Experiment

Figure 12−12 shows the experimental apparatus used by Geiger and Marsden to test Rutherford's predictions. A narrow beam of alpha particles, produced by alpha particles passing through a circular aperture, strikes the gold foil. Those scattered to angle θ strike the zinc sulfide screen, producing flashes, which Geiger or Marsden counted. They rotated the microscope and screen to detect alpha particles at various

TABLE 12−2   DATA FROM SCATTERING EXPERIMENT

θ	NUMBER OF FLASHES IN TIME T
15°	132,000
22.5°	27,300
30°	7800
37.5°	3300
45°	1435
60°	477
75°	211
105°	69.5
120°	51.9
135°	43.0
150°	33.1

Source: M. Geiger and E. Marsden, *Proceedings of the Royal Society,* 25, A, (1913), p. 604.

angles from 0° to 150°. They found that the number of alpha particles decreases very rapidly with increasing angle, as shown in Table 12−2.

Figure 12−14 shows the beautiful correspondence between the predictions of the Rutherford model and the experiment. Note that the vertical axis is divided in an unusual way: the numbers 10° (10° = 1), 10^1, 10^2, 10^3, and so on are separated by equal distances on the axis. This permits the plotting of such widely separated numbers as 132,000 and 33 on the same graph.

SOME EXPERIMENTAL DETAILS

Imagine performing this experiment. You might wonder how anyone would have the patience and persistence to count such a huge number of flashes. Look at the entry for 15°—132,000 flashes! How long did it take to count that incredible number? Geiger and Marsden reported that they did count a *total* of about 100,000 flashes in the course of their experiment, using gold and silver foils. Actually, far

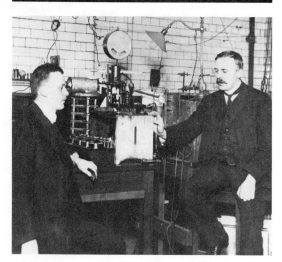

FIG. 12−13 Rutherford (on right) and Geiger with their apparatus for counting alpha particles. (By kind permission of the Physics Department of the University of Manchester)

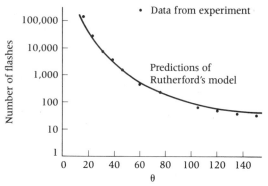

FIG. 12−14 Comparison of Rutherford's predictions (shown by the curved line) with Geiger's and Marsden's experimental data points.

less than 132,000 flashes were observed at 15° because the data have been corrected for several effects.

Geiger and Marsden found that no more than 90 flashes per minute could be counted with accuracy, and this certainly seems reasonable. How is it possible to ensure that no more than 90 alpha particles per minute *will* strike the screen?

First of all, the intensity of alpha particles (or number of alpha particles per second) striking the foil could be adjusted by changing the diameter of the aperture. The number of alpha particles reaching the screen is directly proportional to the number striking the foil. For this reason, an intense beam was used when observing at large angles (where the number of flashes per second was very low), and a very weak beam was used when observing at small angles (where the number of flashes per second was very high). The data were corrected for this effect.

Secondly, because the radioactive source had a half-life of 3.75 days, the intensity of the alpha particle beam striking the foil decreased by a factor of 2 after 3.75 days. Therefore, Geiger and Marsden obtained the data at large scattering angles first, when the source was strong, and then, as the activity of the source grew weaker, they gathered the data at small scattering angles. The radioactive decay law (Chapter 11) was used to correct the data.

Also, the distance of the zinc sulfide screen to the foil could be changed. When the counting rate was very small at large scattering angles, the screen was moved closer to the foil. When the counting rate was very high at small angles, the screen was moved farther away.

The experiment was carried out with these adjustments, which permitted the flashes to be counted with accuracy. The data in Table 12−2 then represent the results that would have been obtained if

1. the flashes at each angle θ had been observed for the same length of time *T.*
2. the zinc sulfide screen had been at the same distance from the foil for all measurements.

3. the same diameter aperture had been used for all measurements.

4. the activity of the radioactive source had been the same for all measurements.

Rutherford's model had passed the experimental test with flying colors, as we can see by the excellent agreement between theory and experiment presented in Fig. 12–14. You might expect that this crucial test of Rutherford's model would have made headlines (at least in the physics community), but this did not happen, perhaps because Rutherford had still not solved the problem of the electrons. In Rutherford's model, the electrons could not be initially at rest or they would be attracted to the positive nucleus. If they moved, they should emit electromagnetic radiation and then spiral into the nucleus, as we mentioned.

Rutherford's model could not represent a stable atom, and this aspect of it caused a big problem regarding its acceptance. However, a young Danish physicist, Niels Bohr, was working with Rutherford at the University of Manchester and was completely captivated with Rutherford's new model. Bohr used another clue—the light emitted by a hydrogen atom—to describe the motion of the electron in the hydrogen atom, solving the stability problem of the Rutherford model in the bargain. Bohr's model of the hydrogen atom will be our major focus in the next chapter.

Summary

In 1904, Thomson and Nagoaka proposed models of the atom. According to Thomson's model, the atom consists of a sphere of positive charge about 10^{-10} m in diameter (consistent with the known size of the atom), with electrons embedded inside this sphere. Light is emitted when electrons oscillate about their equilibrium positions. No light is emitted when the electron is at rest. This agreed with a basic principle of electricity: When a charged particle oscillates (or accelerates in any manner), electromagnetic waves are produced. Nagoaka's

model, in which equally spaced electrons rotated around a central positive charge, did not gain acceptance. The electrons are accelerating and should emit electromagnetic radiation and spiral into the central positive charge.

When Rutherford observed that alpha particles were deflected by as much as 2° in passing through mica, he realized that there were very strong electrical forces in the atom. He could probe these forces with alpha particles. At first, Geiger directed alpha particles toward a gold foil and counted the number scattered to angles between 0° and 20°. Rutherford suggested that Marsden, a young graduate student, look for alpha particles scattered between 90° and 180°. The most astounding result was that 1 in 8000 alpha particles was scattered backwards. The Thomson model could not account for this result. (According to the Thomson model, an alpha particle would be scattered by a single atom only by one-thirtieth of a degree. After encountering many atoms, an alpha particle would be scattered by a few degrees at most, but backward-scattering was impossible.) Rutherford wondered what the atom must look like to produce backward-scattering and, after two years, arrived at an answer: The atom must consist of a central positive sphere (nucleus) having a diameter of only 10^{-14} m. The electrons were contained within a sphere, having a diameter of 10^{-10} m. An alpha particle aimed at the nucleus directly would experience such a large force of repulsion that it would stop and then be scattered backwards. Since the nucleus was 10,000 times smaller than the size of the atom, the chance that an alpha particle would hit the nucleus head-on was extremely small. Most of the alpha particles would pass through empty space in the gold foil and either not be scattered at all, or be scattered (by one nucleus) to a small angle. On the basis of his model, Rutherford calculated the number of alpha particles that should be scattered to various angles by the gold foil. This prediction could be compared with an experiment, which was performed by Geiger and Marsden. Their experimental results agreed beautifully with Rutherford's predictions. The atom does, in fact, look like Rutherford's model.

True or False Questions

Indicate whether the following statements are true or false. Change all of the false statements so that they read correctly.

12–1 A basic principle of electricity is as follows: When a charged particle oscillates, electromagnetic waves, having the same frequency as the oscillation, are produced.

12–2 Nagoaka proposed a model of the atom with electrons revolving around a central positive charge.

   a. This model was rejected because it did not explain the spectrum of the hydrogen atom.

12–3 The hydrogen spectrum contains four lines in the visible region of the spectrum.

   a. Calculation based upon Thomson's model of the hydrogen atom predicted only one line in the ultraviolet region of the spectrum.

12–4 When alpha particles passed through a thin piece of mica, the alpha particles were scattered by 2°.

   a. This 2° angle seemed very small to Rutherford because he knew how easily alpha particles are deflected.

   b. The atoms in the mica must exert a very weak force on the alpha particle.

12–5 Alpha particles strike a gold foil. The following are predictions based upon the Thomson model:

   a. The sphere of positive charge of the gold atom has a diameter of 10^{-10} m.

   b. The electrons of the gold atom are embedded in the sphere of positive charge.

   c. In passing through the gold foil, the alpha particle scatters from many gold atoms.

   d. An alpha particle scattered to 0° is the result of multiple scattering from many gold atoms.

   e. An alpha particle scattered to 5° is the result of multiple scattering from many gold atoms.

   f. The force of repulsion between an alpha particle and one gold atom is very weak.

   g. Scattering to angles larger than 90° is impossible.

12–6 Alpha particles strike a gold foil. The following are predictions based upon the Rutherford model:

   a. The sphere of positive charge of the gold atom has a diameter of 10^{-10} m.

   b. The electrons of the gold atom are embedded in the sphere of positive charge.

   c. In passing through the gold foil, the alpha particle scatters from many gold atoms.

   d. An alpha particle scattered to 0° is the result of multiple scattering from many gold atoms.

   e. An alpha particle scattered to 5° is the result of multiple scattering from many gold atoms.

   f. The force of repulsion between an alpha particle and one gold atom is very weak.

   g. Scattering to angles larger than 90° is impossible.

12-7 A beam of alpha particles strikes a gold foil.

    a. An alpha particle scattered to an angle of 53° comes closer to a gold nucleus than one scattered to 28°.

    b. An alpha particle aimed directly at the center of a gold nucleus is scattered to an angle between 90° and 180°.

12-8 Rutherford showed that the atom consists mostly of empty space.

12-9 To test Rutherford's predictions based upon his model of the atom, Geiger and Marsden aimed alpha particles at a gold foil and measured the number of alpha particles scattered to various angles.

    a. The data obtained by Geiger and Marsden showed that 17 times as many alpha particles were scattered to an angle of 15° as were scattered to 30°.

    b. The data also showed that 277 times as many alpha particles were scattered to an angle of 15° as were scattered to 60°.

    c. The data from the experiment agreed with predictions based upon Rutherford's model of the atom.

Questions for Thought

12-1 Describe Thomson's model of the atom. How did it explain the emission of light?

12-2 Describe Nagoaka's model of the atom. How did it explain the emission of light?

12-3 Explain why Thomson's model of the atom was accepted and Nagoaka's was not.

12-4 What observations caused Rutherford to think about alpha-scattering experiments?

12-5 Suppose that a beam of alpha particles strikes a gold foil, and its thickness corresponds to 1000 gold atoms placed end to end. How would Thomson's model account for scattering of 0°? For 6°? For 13°?

12-6 Flip a coin 10 times and count the number of heads and tails. Repeat this experiment many times, or ask others to help you out. How many times did the coin come up 5 heads and 5 tails, 6 heads and 4 tails (or vice versa), and 7 heads and 3 tails (or vice versa)? Discuss the probability of scattering to 0°, 6°, and 13° in thought question 12-5.

12-7 Describe Rutherford's model of the atom. How did it explain the emission of light?

12-8 How did Rutherford propose that his model of the atom be tested?

12-9 Briefly describe Geiger's and Marsden's experiment to test Rutherford's model.

12-10 Explain why Rutherford could not account for the behavior of electrons in an atom from the results of alpha-scattering experiments.

Questions for Calculation

12-1 What is the wavelength of electromagnetic waves that have a frequency of 720,000 oscillations per second? (The speed of light equals 3×10^8 m/sec.)

12-2 a. Suppose that the alpha particles striking the mica and entering the magnetic field in Fig. 12-4 have a speed of 1.5×10^{-7} m/sec. How long does it take the alpha particles to pass through the mica (0.003 cm thick) and leave the magnetic field (width = 3 cm)?

b. Suppose that the force on the alpha particle in the mica (F_M) is directed vertically upward and the force on the alpha particle in the magnetic field (F_B) is also directed vertically upward. Let m_a be the mass of the alpha particle. Use Newton's second law to find the acceleration of these alpha particles, a_M and a_B, in the vertical direction.

c. Initially, the alpha particles have no velocity in the vertical direction; only a velocity in the *horizontal* direction. But the forces F_M and F_B produce an *acceleration* in the *vertical* direction, a_M and a_B, and, hence, velocities in the vertical direction. The alpha particles in Fig. 12–4 have the same deflection because their velocities in the vertical direction, after leaving the mica and magnetic field, are identical. Using the following equation, show that F_M equals 1000 F_B.

$$\begin{pmatrix} \text{Velocity in} \\ \text{vertical} \\ \text{direction} \end{pmatrix} = \begin{pmatrix} \text{acceleration} \\ \text{in vertical} \\ \text{direction} \end{pmatrix} (\text{time}).$$

12–3 Figure 12–15 shows a design on a linoleum floor. If 30 balls are dropped at random, how many will hit black squares and how many will hit white squares? If a ball is dropped at random onto this pattern, what is the probability that the ball will hit a black square? A white square? Show that the probability of striking a black square is given by

$$\frac{\text{Total area of black squares}}{\text{Total area of all squares}}.$$

†12–4 Alpha particles uniformly strike a section of a gold foil 5 mm wide and 5 mm long. This section contains 3×10^{18} gold atoms. In order to find the number of alpha particles scattered to various angles, calculate the area of the

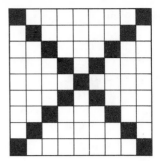

FIG. 12–15 Diagram for calculation question 12–3.

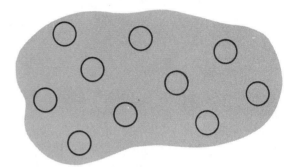

FIG. 12–16 Small part of a gold foil, showing the nuclei of gold atoms. It doesn't make any difference whether the nuclei are at the front or rear of the foil. The nuclei are so small that they will not overlap.

nuclei intercepting the alpha particles and compare it to the total area of the foil. This approach is similar to that in calculation question 12–3.

a. What is the area of the gold foil struck by alpha particles?

b. Figure 12–16 shows a very small part of the gold foil. Each circle indicates the *nucleus* of a gold atom. Imagine that a beam of alpha particles is aimed at this foil in a direction perpendicular to the plane of this paper. Using Table 12–1 draw a shaded area on each nucleus, such that alpha particles aimed at this area will be scattered to angles between 90° and 180°. Calculate the area of one such shaded area, assuming the radius of the gold nucleus to be 3.2×10^{-14} m. Calculate the total area due to 3×10^{18} gold nuclei.

c. Calculate the fraction of the incident alpha particles scattered to angles between 90° and 180° by determining

$$\frac{\text{Total area of all nuclei in (b)}}{\text{Area of gold foil in (a)}}.$$

d. Compare the fraction in part **c** with the fraction 1/8000.

12−5 Repeat calculation question 12−4, but calculate the fraction of the incident alpha particles scattered to angles between 5.7° and 180°, and the fraction scattered between 0° and 5.7°.

12−6 Show that the section of the gold foil 5 mm wide and 5 mm long (calculation question 12−4) and 20.4×10^{-6} m thick contains 3×10^{18} atoms. The density of foil is 1932 kg/m^3; the atomic mass of gold is 197 amu. (Remember that 197 g of gold contains 6×10^{23} atoms.)

Footnotes

[1] *Science Abstracts*, Section A, Vol. 7, No. 961, (1904).

[2] *The Collected Papers of Lord Rutherford of Nelson*, Vol. 1, (London: George Allen & Unwin, 1962), p. 867. [E. Rutherford, *Philosophical Magazine*, 12, Ser. 6 (1906):134.]

[3] Ibid; p. 867.

[4] N. Feather, *Lord Rutherford* (London: Priory Press Limited, 1973), p. 117.

[5] H. Geiger and E. Marsden, *Proceeding of the Royal Society*, 82, A, (1909), p. 495.

[6] *Collected Papers, Lord Rutherford*, Vol. 2 [E. Rutherford, *Philosophical Magazine*, 21, Ser. 6 (1911):238.]

[7] Excerpt from *Rutherford and the Nature of the Atom* by E. N. da C. Andrade. Copyright © 1964 by Doubleday & Company, Inc. Reprinted by permission of the publisher.

Niels Bohr

CHAPTER 13

Atoms, Light, and Lasers

Introduction

How do atoms emit light? Scientists had been pondering that question for decades when, in 1913, Niels Bohr proposed his model of the hydrogen atom. We have learned a great deal about the Bohr model in Part I and Part II. Here, we'll go further. We'll see how Bohr calculated the wavelengths of light emitted by hydrogen and how his calculations agreed with wavelength measurements obtained experimentally. Bohr had shown that the electron, traveling only in allowed orbits, can have only certain well-defined energies. We now express this by saying that the hydrogen atom has quantized energy levels. Hydrogen and *all* chemical elements have characteristic line spectra. Bohr's success in accounting for the hydrogen line spectrum led directly to the idea that, like hydrogen, all atoms have quantized energy levels. In an ingenious experiment carried out in 1914, J. Franck and G. Hertz showed that this was true for mercury.

Atoms cannot only emit light, they can also absorb light. For example, when a beam of yellow light passes through a sodium vapor, the sodium atoms absorb some of the yellow light. However, sodium atoms will not absorb blue light. In this chapter, we shall see why atoms absorb light of some wavelengths, but not others. Phenomena such as these provide astronomers with the "tools" to determine the composition of our sun and other stars, and even their surface temperatures.

In the hydrogen atom, light is emitted when the electron jumps from one allowed orbit to a smaller one—from one quantized energy level to another. In a laser, also, light is emitted when the electron jumps from one quantized energy level to another. Lasers share this feature with other light sources. However, in contrast to conventional light sources, laser light is monochromatic and is emitted along one direction. We'll conclude our discussion by learning about the operation of a laser, about the characteristics that differentiate it from ordinary light, and about the laser's fantastic technological applications.

Bohr's Postulates

After receiving his doctorate from the University of Copenhagen in 1911, Niels Bohr came to the Cavendish Laboratory of Cambridge University to study with its illustrious director, J. J. Thomson. Bohr had become fascinated with the recent, albeit revolutionary, theories of Planck and Einstein: The energy of light is subdivided into photons of energy, $E = hf$. Because Thomson could not account for the wavelengths of light emitted by chemical elements, Bohr thought these new ideas of Planck and Einstein should—somehow—be incorporated into Thomson's model. Maybe agreement with experiment might result.

Thomson could not accept these revolutionary ideas. After several disagreements, Bohr decided to leave the Cavendish Laboratory, and he went to work with Rutherford at the University of Manchester. He arrived at a most opportune time; Rutherford had recently proposed the nuclear model of the atom, and Geiger and Marsden were carrying out the alpha particle scattering experiments to test this theory. These were exciting times! However, Bohr's time abroad was coming to an end, and he returned to Copenhagen in 1913. But, his interest in the nuclear model continued. Bohr felt that there was one additional clue—the light emitted by elements—that could be used to glean information about the behavior of the electrons in an atom. Recall from Chapter 12 that Rutherford did not discuss in detail the behavior of the electrons. Bohr chose to investigate the hydrogen atom because it contained only one electron.

Bohr pointed to the discovery of X-rays, Rutherford's model of the atom, and the photoelectric effect. He concluded that classical theory of electricity was not adequate "in describing systems of atomic size. Whatever the alteration in the laws of motion of the electrons may be, it seems necessary to introduce . . . Planck's constant. . . ."[1] Bohr proposed a planetary model of the hydrogen atom in which the single electron revolved around the nucleus in a circular orbit. According to the established

FIG. 13–1   Niels Bohr. (AIP Niels Bohr Library)

principles of electricity, this electron should emit electromagnetic radiation, lose energy, and spiral into the nucleus (see Chapter 12). Bohr simply *ignored* this objection. Instead, he adopted quantization from Planck by requiring that the electron could travel only in circular orbits, so-called allowed orbits, where the angular momentum was an integral multiple of $h/2\pi$. Bohr's postulates are as follows:

1. No electromagnetic radiation is emitted when the electron travels in one of the allowed orbits.
2. When an electron jumps from one allowed orbit to a smaller allowed orbit of lower energy, electromagnetic radiation is emitted. The energy of a photon of light is given by

$$hf = E_i - E_f, \qquad (1)$$

where E_i is the energy of the initial larger orbit, and E_f is the energy of the final smaller orbit.
3. The angular momentum of an allowed orbit is given by

$$mvr = \frac{nh}{2\pi}, \qquad (2)$$

where $n = 1, 2, 3$, and so forth.

Using these postulates, Bohr calculated the wavelengths of light emitted by hydrogen.

Bohr Accounts for the Hydrogen Spectrum

Let's briefly review what we have already learned about Bohr's theory. In Chapter 3, we saw that the attraction between the negatively charged electron and the positively charged nucleus caused the electron to travel in a circular orbit around the nucleus. For a given value of the radius, we could find the speed of the electron in that orbit, but we couldn't select the so-called allowed orbits. However, an orbiting electron has angular momentum, and Bohr's third postulate requires that the electron can have only certain values of the angular momentum: $h/2\pi$, $2(h/2\pi)$, $3(h/2\pi)$, and so on. Equation 7 in Chapter 4 (p. 94) is actually Bohr's third postulate.

An electron traveling in a circular orbit has two kinds of energy: kinetic energy due to its motion and potential energy due to its distance from the nucleus. Its energy E is the sum of the kinetic and potential energies. Equation 12 of Chapter 6 (p. 132) shows that the energy E is given by:

$$E = \frac{-13.6\,\text{eV}}{n^2}. \qquad (3)$$

Figure 13-2 shows an energy-level diagram in which the energies E are obtained by using Equation 3.

According to Bohr's second postulate, the hydrogen atom emits a photon of light when the electron jumps to a smaller orbit. Now, we want to find the wavelength of this light and compare it with experimental values. We can do this in three short steps:

Step one Find the photon energy:

Photon energy = difference in energy of two orbits.

For example, when the electron jumps from the third orbit to the second, the difference in energy is equal to $(-1.51\ \text{eV}) - (-3.4\ \text{eV})$, or 1.89 eV. The photon energy of the light is 1.89 eV.

Step two Find the frequency of the light:

Photon energy = hf

$$f = \frac{\text{photon energy}}{h}.$$

In this example, the frequency is 1.89 eV/h.

Step three   Find the wavelength λ:

$$\lambda = \frac{c}{f},$$

where c is the speed of light. (See Chapter 8, Equation 1, p. 170.)

Substituting for the frequency f from above, the wavelength becomes

$$\lambda = \frac{hc}{\text{photon energy}}$$

$$= \frac{hc}{\text{energy difference}}.$$

Since h equals 4.1354×10^{-15} eV-sec and c equals 2.99793×10^{8} m/sec, the wavelength is found to be

$$\lambda = \frac{12.40 \times 10^{-7}\,\text{m-eV}}{\text{energy difference}}.$$

Changing the units, this becomes

$$\lambda = \frac{1240\,\text{nm-eV}}{\text{energy difference}}, \qquad (4)$$

where 1 nm equals 10^{-9} m. In this example, the wavelength of light emitted when the electron jumps from the third to second orbit is 1240 nm-eV/1.89 eV, which equals 656 nm. This is the red line in the spectrum.

Remember that Balmer found a formula for the wavelengths of the four lines in the visible spectrum (Chapter 1). When the initial orbit is the sixth, fifth, fourth, or third orbit and the final one is the second orbit, Equation 4 is equivalent to Balmer's formula. (Calculation question 13-5 outlines the simple steps to show this equivalence.)

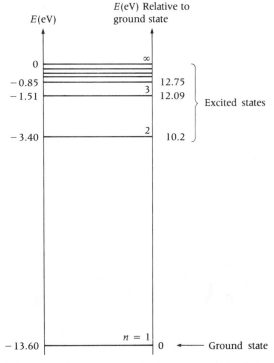

FIG. 13–2   An energy-level diagram for hydrogen. The allowed or quantized energy levels, given by Equation 3, are plotted on the vertical axis of this graph. When the electron is in the lowest energy level ($n = 1$), we say that the hydrogen atom is in the *ground state*. When the electron is in a higher energy level, the hydrogen atom is in an *excited state*. The energies on the right indicate the energy of the level relative to the ground state. For example, the excited state of n equals 2 is 10.2 eV larger than the ground state. This is found by subtracting -13.6 eV from -3.4 eV, or $(-3.4\,\text{eV}) - (-13.6\,\text{eV}) = +10.2$ eV.

Hydrogen Emission Spectrum

Figure 13–3 shows how the spectrum of hydrogen is obtained. Light from a cathode ray tube containing hydrogen gas passes through a

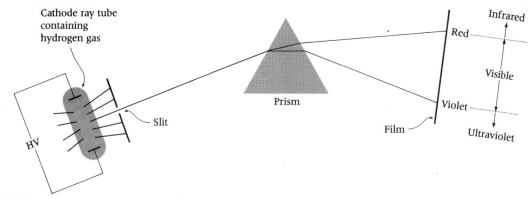

FIG. 13–3   Experimental apparatus used to obtain the hydrogen emission spectrum.

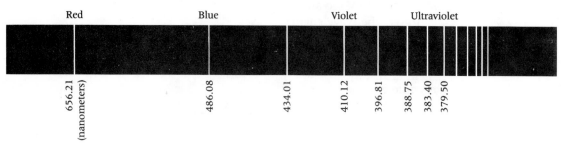

FIG. 13–4   The hydrogen emission spectrum.

slit to form a beam of light. This beam strikes a prism (or a diffraction grating) where the light is broken up into its spectrum and captured on film.

HOW LIGHT IS EMITTED

The cathode ray tube contains hydrogen gas at a pressure of about 1/100 of atmospheric pressure. Before the high voltage is switched on, the electron of a hydrogen atom is in the smallest $n = 1$ orbit; this is also the smallest energy orbit, called the **ground state.** When the high voltage is switched on, hydrogen atoms near the cathode and anode are ionized. Let's con-

sider those hydrogen atoms near the cathode. The electron of a hydrogen atom is strongly repelled by the cathode, but the nucleus is greatly attracted to it. These forces are so strong that the hydrogen atom is simply ripped apart; the freed electron rushes toward the anode, and the hydrogen nucleus rushes toward the cathode. (A similar process occurs near the anode.)

Because there are about 3×10^{17} hydrogen atoms per cubic centimeter at this pressure, the freed electron soon collides with another hydrogen atom. During this collision, the freed electron loses kinetic energy, but the hydrogen atom gains this amount of energy, and its electron jumps from the $n = 1$ orbit to a larger and higher energy orbit. Almost immediately, this electron of the hydrogen atom jumps to a smaller and lower-energy orbit, emitting a photon of light. This process of "ionization plus collision" explains why light is emitted by a cathode ray tube, whose anode and cathode are connected to a high voltage.

BOHR'S THEORY COMPARED WITH EXPERIMENT

What appears on the film in Fig. 13–3, then, are many images of the slit: lines of many different wavelengths called **spectral lines**. Figure 13–4 shows a diagram of the hydrogen emission spectrum, where the white lines indicate exposure of the film. Table 13–1 compares Bohr's theory (Equation 4) with experimental values of the wavelength. The agreement is excellent!

Bohr's theory was a tremendous breakthrough, for which he received the Nobel Prize in 1922. Shortly after his research had been published, the Royal Danish Academy of Science funded Bohr to build his own institute for atomic studies and to provide fellowships for students of theoretical physics. (In turn, the Academy received much of its funding from the Carlsberg Brewery, whose founder had willed income from the brewery to be used for the development of science.) The Bohr Institute of Theoretical Physics soon became the center for the study of the revolutionary quantum physics. Many scientists from all over the world came to study with Bohr and to be inspired by the intellectual atmosphere of this Institute. Among the scientists were Paul Dirac, Wolfgang Pauli, Werner Heisenberg, George Gamow, and J. Robert Oppenheimer, to name only a few. Today, the Bohr Institute is still a renowned center for the study of theoretical physics. In 1975, Bohr's son Aage, along with B. R. Mottelson and J. Rainwater, received the Nobel Prize for their theories concerning the nucleus of the atom. Much of this research was carried out at the Institute.

Hydrogen Absorption Spectrum

When a hydrogen atom absorbs a photon of light, the electron jumps to a *larger* orbit. For example, when a hydrogen atom absorbs a 10.2-

TABLE 13–1   COMPARISON OF BOHR'S THEORY WITH EXPERIMENT

n_i	n_f	BOHR'S PREDICTIONS USING EQUATION 4 (WAVELENGTH IN nm)	EXPERIMENTAL VALUES (WAVELENGTH IN nm)
3	2	656.30	656.21
4	2	486.18	486.08
5	2	434.09	434.01
6	2	410.22	410.12
7	2	397.05	396.81
8	2	388.95	388.75
9	2	383.58	383.40
10	2	379.83	379.50
11	2	377.02	376.75
∞	2	365.00	
2	1	121.54	121.57
3	1	102.55	102.57
4	1	97.24	97.25
5	1	94.96	94.97
∞	1	91.20	

eV photon, the electron jumps from the first to the second orbit (see Fig. 13–5), since 10.2 eV is equal to the difference in energies of these two orbits. Similarly, when a 12.09-eV photon is absorbed, the electron will jump from the first to the third orbit. However, a 12-eV photon will pass right by the hydrogen atom and not be absorbed because its energy is not sufficient enough to raise it to the third, but too high to raise it merely to the second. The photon energy must *exactly* match the energy difference between the two orbits if the photon is to be absorbed. Hence, *hydrogen atoms absorb the same wavelengths of light that they emit.*

When a hydrogen atom absorbs a 13.6-eV photon, the electron will jump from the first orbit to a very large one, where its energy E is zero. What does this mean?

When the electron's energy E is negative, the electron travels in a circular orbit around the nucleus: It is bound to the nucleus by an attractive force. When the electron's energy E is zero,

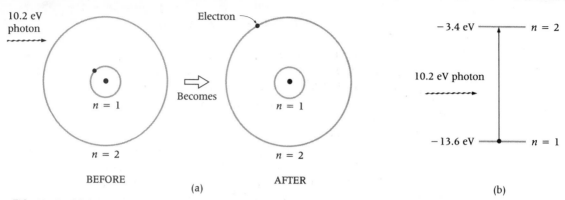

FIG. 13–5 (a) A 10.2-eV photon approaches a hydrogen atom with the electron in the first orbit. This photon is absorbed (disappears), and the electron jumps to the second orbit. (b) An energy-level diagram that shows the same process happening in (a).

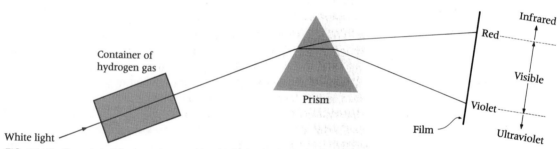

FIG. 13–6 Experimental apparatus used to obtain the hydrogen absorption spectrum.

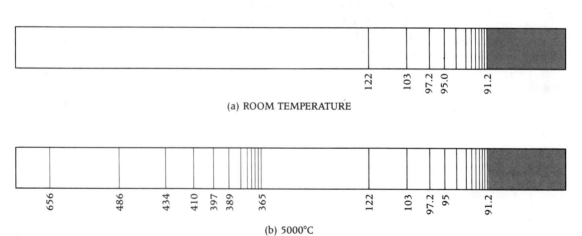

FIG. 13–7 (a) The hydrogen absorption spectrum at room temperature. (b) At 5000°C.

Equation 3 shows that $-13.6 \text{ eV}/n^2$ must also be zero, and n must approach infinity. This, in turn, means that the radius of the "orbit" must be infinitely large and, therefore, the electron is no longer attracted to the nucleus. The electron, no longer bound to the nucleus, becomes a free particle. When the electron is freed from the atom, we say that the atom is *ionized*. An energy of 13.6 eV is required to ionize a hydrogen atom.

A photon having an energy greater than 13.6 eV will ionize the hydrogen atom, and the remaining energy will appear as kinetic energy of the electron. For example, when a 15-eV photon is absorbed by a hydrogen atom, the freed electron has a kinetic energy of 1.4 eV (15 eV minus 13.6 eV).

Equation 4 reveals that an energy of 13.6 eV corresponds to a wavelength of 91.2 nm, and energies larger than 13.6 eV correspond to wavelengths smaller than 91.2 nm. Therefore, *all* wavelengths of light 91.2 nm or less will be absorbed and will ionize hydrogen atoms. However, wavelengths larger than 91.2 nm will be absorbed only if they will cause the electron to jump to a larger orbit. For this reason, only certain wavelengths larger than 91.2 nm will be absorbed.

Figure 13–6 illustrates how we can obtain the hydrogen absorption spectrum. A beam of white light (having a wide range of wavelengths from ultraviolet, visible, through infrared) enters a container filled with hydrogen gas. Some photons are absorbed by the gas, but the remaining ones strike the prism; the spectrum of light is captured on the photographic film. Figure 13–7 shows such a spectrum, where the gas is at room temperature in part (a), and at a temperature of 5000°C in part (b).

From Fig. 13–7(a), we see that, as expected, all wavelengths less than 91.2 nm are absorbed, while only certain ones with wavelengths greater than 91.2 nm are absorbed. Table 13–1 shows that light of wavelength 122 nm is absorbed because the electron can jump from the first to the second orbit (and so on, for the other absorption lines in the spectrum).

Of course, the electrons in these higher-energy orbits jump back very soon to the ground state, emitting photons in *all directions,* as shown in Fig. 13–8. Very few photons are directed to the right (the same direction as the incident beam). Light that is not absorbed by the hydrogen gas passes through the container as though the container were empty and exposes the film. This sharp contrast in exposure produces a line on the film.

When the temperature of the hydrogen gas in Fig. 13–6 is raised to 5000°C, a few of the hydrogen atoms (actually, much less than 1 percent) have electrons in the second orbit. When the temperature increases, so does the kinetic energy of the atoms. When these more-energetic hydrogen atoms collide, there is enough energy so that one (or both) of the electrons can jump into the second orbit. Such an atom can then absorb a photon of light, and the electron will jump from the second orbit to a larger one. Table 13–1 shows that a hydrogen gas at a temperature of 5000°C can absorb light having wavelengths of 656 nm, 486 nm, 434 nm, . . . to a limit of 365 nm. Such a spectrum is illustrated in Fig. 13–7(b).

Notice the faint lines appearing in Fig. 13–7(b). Consider, for example, the 656-nm line due to the electron in the second orbit absorbing a 1.89-eV photon and jumping to the third orbit. Because there are so few electrons in the second orbit, not all of the 1.89-eV photons entering the container are absorbed. Some of them reach the film, producing a partial exposure, a gray line: part way between white (no absorption) and dense black (strong absorption).

At a temperature greater than 5000°C, more hydrogen atoms would be in the second orbit, and those absorption lines would be blacker.

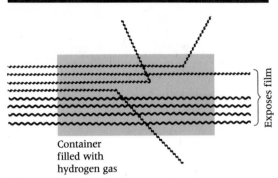

Container
filled with
hydrogen gas

FIG. 13–8 Photons of two different energies and, hence, different wavelengths enter a container of hydrogen gas. One photon energy is exactly that required to raise the electron to a higher-energy level, and these photons are absorbed. Because these photons are re-emitted in many different directions, they do not expose the film. Since the other photons are not absorbed and their direction is unchanged, they do expose the film.

TABLE 13–2   FOUR WAYS TO REACH THE $n = 1$ LEVEL

TRANSITIONS	PHOTON ENERGIES (eV)	TOTAL ENERGY OF ALL PHOTONS (eV)
$4 \rightarrow 1$	$hf_1 = 12.75$	12.75
$4 \rightarrow 2$ $2 \rightarrow 1$	$hf_2 = 2.55$ $hf_3 = 10.20$	12.75
$4 \rightarrow 3$ $3 \rightarrow 1$	$hf_4 = 0.66$ $hf_5 = 12.09$	12.75
$4 \rightarrow 3$ $3 \rightarrow 2$ $2 \rightarrow 1$	$hf_4 = 0.66$ $hf_6 = 1.89$ $hf_3 = 10.20$	12.75

The reverse also holds. By comparing the relative intensity of the absorption lines, you can determine the temperature of the hydrogen gas.

Quantized Energy Levels

During the late nineteenth century, scientists studied the spectra of chemical elements, searching for regularities in these spectra. While each chemical element had a characteristic spectrum, it was, however, a *line* spectrum. Examining the spectrum of a chemical element, researchers found that the frequency of one spectral line was equal to the sum of the frequencies of several other spectral lines. Such relationships could be found in the spectra of all chemical elements. Bohr's theory of the hydrogen atom provided marvelous insight into these regularities.

Let's now focus our attention on those hydrogen atoms with electrons in the $n = 4$ level. Table 13-2 shows that there are four ways to reach the $n = 1$ level. Each time an electron jumps to a lower-energy level, the hydrogen atom emits a photon of light. We can use the values in Fig. 13–2 to calculate the photon energies. For example, when the electron jumps from the $n = 4$ orbit to the $n = 3$ orbit, the photon has an energy given by Equation 1, or

$$hf_4 = 12.75 \, \text{eV} - 12.09 \, \text{eV}$$
$$= 0.66 \, \text{eV}.$$

Table 13–2 shows clearly that the photon energies sum to 12.75 eV:

$$12.75 \, \text{eV} = hf_1$$
$$hf_1 = hf_2 + hf_3.$$

Dividing this equation by h, we obtain

$$f_1 = f_2 + f_3,$$

and, similarly,

$$f_1 = f_4 + f_5$$

and

$$f_1 = f_3 + f_4 + f_6.$$

This was exactly the feature found in the spectra of all chemical elements! It followed that each element had a distinctive set of quantized energy levels. When one of its electrons jumps to a lower-energy level (Fig. 13−9a), the atom emits a photon of light having an energy hf equal to E_i minus E_f. An atom absorbs a photon of light when the photon energy is exactly that needed to raise the electron to a higher-energy level (Fig. 13−9b), or to ionize the atom. As for hydrogen, atoms of a chemical element absorb light that they are able to emit.

The Franck-Hertz Experiment

MERCURY HAS QUANTIZED ENERGY LEVELS

Only one year after Bohr had published his paper on the hydrogen atom, J. Franck (1882−1964) and G. Hertz (1887−1975) showed, in 1914, that mercury has an energy level 4.9 eV above its ground state (Fig. 13−10). Basically, they found that electrons colliding with mercury atoms would not lose any of their kinetic energy unless that kinetic energy was larger than 4.9 eV. Their experiment is described in Fig. 1, Box 13−1.

CALCULATING THE WAVELENGTHS OF LIGHT: AN EXAMPLE

Using the energy-level diagram for mercury in Fig. 13−10, we can calculate the wavelengths of light when the electron jumps from the levels of

(a) 9.53 eV to 7.73 eV

(b) 7.73 eV to 4.89 eV

(c) 7.73 eV to 5.46 eV

(d) 8.84 eV to 6.70 eV

(e) 6.70 eV to the ground state

(f) 4.89 eV to the ground state

The photon energy is equal to the difference in energy of the two orbits. In (a), the energy difference is 9.53 eV minus 7.73 eV, or 1.80 eV.

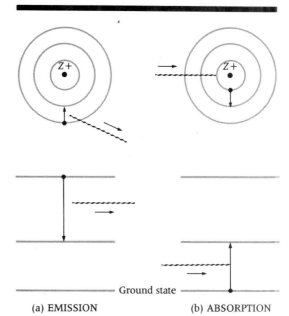

(a) EMISSION   (b) ABSORPTION

FIG. 13−9 (a) Light is emitted by an atom when an electron jumps to a lower-energy level. (b) Light is absorbed by an atom when the photon energy is exactly equal to the difference in energy between the two energy levels. The electron jumps to a higher-energy level.

ATOMS, LIGHT, AND LASERS

BOX 13–1   OBTAINING DATA FROM THE FRANCK-HERTZ EXPERIMENT

The cathode ray tube contains mercury vapor. The cathode is heated, so that electrons are simply "boiled off" the cathode. (Cathode rays are produced spontaneously only if a high voltage—such as 10,000 volts—connects the cathode and anode.) These electrons are accelerated by the positive charge on the grid G. As indicated, there are openings in the grid G, so that electrons can pass through it. Electrons that *are able* to reach the anode A produce a current detected by the current meter.

Franck and Hertz changed V_G (keeping V_A always 1.5 volts smaller) and looked for a decrease in current, which happened when V_G was 4.9 volts. The essence of their results is indicated in part (a) and part (b). When the voltage connecting cathode C and grid G was greater than 4.9 volts, the current dropped dramatically. This showed that mercury has its first energy level 4.9 eV above its ground state. Let's see why.

Look at part (b). Suppose that an electron makes a collision with a mercury atom just in front of grid G. This electron had a kinetic energy of 5.0 eV. The mercury atom absorbs 4.9 eV, and one of its electrons is raised to an energy level 4.9 eV above its ground state. The kinetic energy of the electron is reduced to 5.0 eV minus 4.9 eV, which equals 0.1 eV. This electron passes through the grid G and enters the region between the grid G and the anode A. Because the voltage on anode A is 3.5 volts, while that on grid G is 5.0 volts, grid G has more positive charge on it than anode A. The electron is more strongly attracted to grid G. And in traveling from grid G toward anode A, the electron decelerates. Since the electron has a kinetic energy of only 0.1 eV (and the voltage difference between grid G and

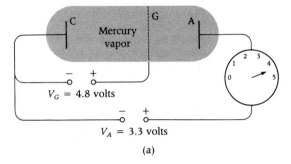

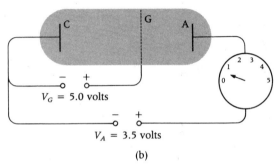

FIG. 1   A diagram of the Franck-Hertz equipment.

anode A is 1.5 volts), the electron quickly comes to a stop. Due to the stronger attraction, the electron travels to grid G. The electron does not reach the anode A and, hence, doesn't cause a current in the meter.

In part (a), an electron colliding with a mercury atom doesn't lose any of its kinetic energy because it simply doesn't have enough energy. This electron passes through the grid with a kinetic energy of 4.8 eV. It, too, decelerates as it travels toward anode A, but it has enough kinetic energy to reach anode A, where its kinetic energy is 4.8 eV minus 1.5 eV, which equals 3.3 eV.

From Equation 4, the wavelength is given by:

$$\lambda = \frac{1240 \text{ nm-eV}}{1.80 \text{ eV}}$$

$$= 689 \text{ nm}.$$

This wavelength corresponds to red light. Similarly, the wavelength in (b) is 437 nm (blue); in (c), 546 nm (green); in (d), 579 nm (yellow); in (e), 185 nm (ultraviolet); and, in (f), 254 nm (ultraviolet). (The spectrum of mercury appears in Fig. 1–11, p. 13.)

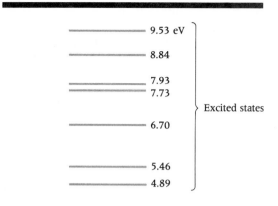

9.53 eV

8.84

7.93
7.73

} Excited states

6.70

5.46

4.89

Ground state ———— 0

FIG. 13–10   An energy-level diagram of mercury.

Problems with Bohr's Model

It wasn't long before difficulties with Bohr's theory began to emerge. While the theory showed that all atoms have quantized energy levels, Bohr could not calculate their energies, as he had for hydrogen. He could not even calculate the energies for helium, which has only two electrons. Then, too, the spectral lines of hydrogen did not all have the same intensity, and Bohr could not explain this phenomenon.

Indeed, his theory was a brilliant beginning, but he had discovered only the "tip of the iceberg"; much remained to be uncovered. In Chapter 15, we shall see that Bohr's theory is in fact a special case of a yet more general theory based upon the idea that particles, like light, have a dual nature. However, this general theory retained, as it must, the concept of quantized energy levels. This more general theory shows *how* these quantized energy levels can be calculated.

Now, let's consider several illustrations of quantized energy levels.

The Solar Spectrum

You know what the **solar spectrum** looks like, right? You see it every time you look at a rainbow. It's a *continuous* distribution of colors from red, orange, and so on through blue. That's what

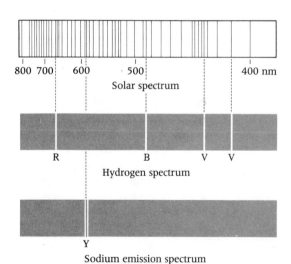

800 700   600   500   400 nm
Solar spectrum

R   B   V   V
Hydrogen spectrum

Y
Sodium emission spectrum

FIG. 13–11   The solar spectrum; the emission spectrum of hydrogen; the emission spectrum of sodium. Compare the three.

Joseph von Fraunhofer (1787–1826), a Bavarian optician, thought in 1814. But when he examined the sun's spectrum *in detail,* he found numerous black lines superimposed on the expected continuous spectrum (see Fig. 13–11). Fraunhofer was so baffled by these observations (could there be something wrong with his apparatus or methods?) that he did not announce his results until 1817.

No explanation was offered until 1859 when two German scientists—Gustav Kirchhoff (1824–1887) and Robert Bunsen (1811–1899), inventor of the Bunsen burner—compared the emission and absorption spectra of elements. These scientists established two laws already familiar to us: (1) Each element has a characteristic line spectrum, and (2) each element absorbs the light that it emits. These laws enabled them to explain the black lines in the sun's spectrum, an interpretation still valid today.

Light from the sun's hot surface (5800°C) passes through the cooler vapor extending for about 2500 miles from the sun's surface. Some of this light is absorbed by this vapor, producing about 20,000 absorption lines. Since the spectrum is characteristic of each element, the elements in the sun (or in any star) can be determined. A comparison of the lines in Fig. 13–11(b) and 13–11(c) with those in Fig. 13–11(a) shows that hydrogen and sodium are present in the sun. In this way, 65 elements have been identified in the sun's spectrum. By comparing the intensity or blackness of the various absorption lines, we can establish the relative proportion of the different elements. In the sun, 90 percent of the atoms are hydrogen, and about 10 percent are helium, with only an extremely small percentage of heavier elements. For example, for every 10,000 atoms of hydrogen, there are 1000 atoms of helium, 6 of oxygen, 3 of nitrogen, 2 of carbon, and 1 of neon.

We can even determine the surface temperature of the sun and other stars from their spectra. Because the earth's atmosphere absorbs ultraviolet light, only the visible region, from 400 nm to 700 nm, can be obtained. The presence or absence of certain absorption lines indicates the surface temperature of a star. For example, the sun has a surface temperature of 5800°C, and a small percentage of the hydrogen atoms in the gaseous layer will have electrons in the $n = 2$ orbit. In this case, the hydrogen absorption lines are quite faint. In comparison, these hydrogen lines for a star with a surface temperature of 10,000°C are quite intense.

The Laser

The word **laser** is an acronym for "*l*ight *a*mplification by *s*timulated *e*mission of *r*adiation." In 1916, Einstein published a paper describing the process of stimulated emission. Figure 13–12 compares spontaneous emission with **stimulated emission.** In Fig. 13–12(a), the electron *spontaneously* jumps from E_H to the ground state ($E = 0$) and emits a photon of energy E_H. There is absolutely no restriction on its direction. In Fig. 13–12(b) a photon of energy E_H passes by the atom and *stimulates* the electron in E_H to jump to the ground state, emitting a photon identical to the original: same direction and same phase. Where there was initially one photon, there are now two identical photons. These two can pass by 2 more atoms (with electrons in E_H), resulting in 4 photons; these 4 photons result in 8 photons; and so on. The light can be amplified.

Einstein did not mention any possible applications. We can easily see why applications were not immediately evident. First of all, electrons usually occupy the *ground* state. How do you get *large numbers* of atoms in E_H? Secondly, spontaneous emission occurs very soon after the electron jumps to E_H. For example, in hydrogen, the electron in the $n = 2$ orbit will jump to the $n = 1$ orbit before 10^{-8} seconds has passed! How can you "force" stimulated emission to occur before spontaneous emission has a chance to happen? In 1960, T. H. Maiman, at Hughes Aircraft Co., solved these problems and developed the ruby laser.

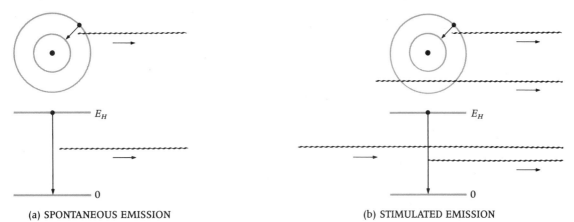

(a) SPONTANEOUS EMISSION   (b) STIMULATED EMISSION

FIG. 13–12   A comparison of stimulated and spontaneous emissions. (a) Spontaneous emission. (b) Stimulated emission. In (b), the two light waves have the same direction and the same phase (meaning that a peak of one wave overlaps a peak of the other wave).

THE RUBY LASER

Ruby is an aluminum oxide crystal in which 0.05 percent of the aluminum atoms have been replaced by chromium atoms, which are responsible for the ruby's red color and for the laser action. Figure 13–13(a) shows a photograph of the ruby laser, and Fig. 13–13(b) illustrates it schematically.

The ends of the ruby rod, parallel to a very high degree of accuracy, are silvered. One end is only partially silvered, so that some light can pass through it, while some is reflected.

Pumping   The chromium atoms are normally in their ground state. An electron will jump to higher-energy levels when a chromium atom absorbs a photon of exactly the right energy. For example, in Fig. 13–14, the electron jumps to the 2.27-eV energy level when the chromium atom absorbs a 2.27-eV photon (yellow light with a wavelength of 545 nm). The purpose of the flash lamp is to provide photons which the chromium atoms absorb and cause electrons to jump from the ground state to energy levels of 2.27 eV and higher. We say that "the chromium atoms are pumped to excited states." Some of these electrons jump to the 1.79-eV energy level, which has an extremely important property for laser light.

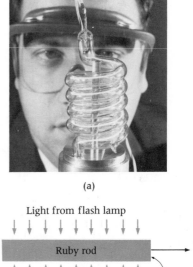

(a)

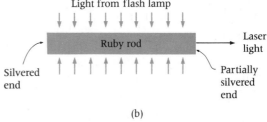

(b)

FIG. 13–13   (a) Dr. Theodore H. Maiman of Hughes Research Laboratory, where the scientific breakthrough was achieved. Dr. Maiman studies the laser's main parts, a light source surrounding a rod of synthetic ruby crystals. (Courtesy of Hughes Aircraft Company) (b) A diagram of a ruby laser.

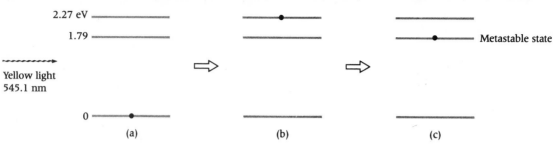

2.27 eV

1.79

Yellow light
545.1 nm

Metastable state

0

(a)   (b)   (c)

FIG. 13–14   (a) The chromium atom in the ground state
absorbs a 2.27-eV photon. (b) The electron jumps to
the 2.27-eV energy level. (c) The electron jumps to the
1.79-eV energy level.

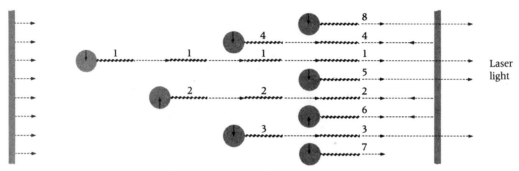

Laser
light

FIG. 13–15   Laser light—a chain reaction. Atom 1 spontaneously emits a photon along the axis of the ruby rod.
This photon stimulates atom 2 to emit a photon. Two
photons cause four, four cause eight, and so on.

Metastable State (almost stable)   The 1.79-eV state
is called a **metastable state** because an electron in this energy level will remain there for
as long as 10^{-3} seconds before it spontaneously
jumps to the ground state. This doesn't seem
very long, but it's 100,000 times longer than in
most atoms.

Recall that, for the hydrogen atom, the electron remains only 10^{-8} seconds in the $n = 2$
orbit before it jumps to the ground state. Because
the chromium atoms can be pumped to the

metastable state, there will be *more atoms in the
metastable state than in the ground state.* This is
most significant for laser action.

Laser Action   Figure 13–15 shows a ruby rod
in which the chromium atoms have been
pumped to the metastable state. **Laser action**
begins when an electron spontaneously jumps
to the ground state and—importantly—emits a
photon along the axis of the rod. This red light
has a photon energy of 1.79 eV and a wavelength of 694.3 nm. For the sake of illustration,
suppose that the initial spontaneous emission
(photon 1 in Fig. 13–15) occurs only 10^{-5} seconds after the electron has been pumped to the

1.79-eV energy level. This is 100 times shorter than its life expectancy of 10^{-3} seconds. The majority of the other chromium atoms, therefore, will still be in the 1.79-eV state for some time, permitting the original photon to pass by them and cause stimulated emission. These 2 photons then cause stimulated emission in 2 more atoms, resulting in 4 photons, and so on. A "chain reaction" occurs. Laser action is possible because stimulated emission occurs *before* the electron spontaneously jumps to the ground state. Light is reflected at each end of the ruby rod and laser action continues until all of the atoms have reached the ground state. Some of the light passes through the partially silvered end. For more laser light, the flash lamp must be flashed again. The light from a ruby laser is pulsed light.

There are now many types of lasers, such as the helium-neon laser, producing red light, and the carbon dioxide laser, producing infrared radiation. All have these features in common: some method to pump atoms to an excited metastable state, and the process of stimulated emission.

Technological Applications of the Laser

Due to the special properties of its light, the laser is one of the most significant technological advances of the 1960s. Its monochromatic light is emitted in one direction, and the light waves are emitted in phase. A laser beam is narrow and stays narrow. Its light can be concentrated into an exceptionally small area, producing beams that have attained 500 million watts per square centimeter. (In contrast, the power of the sun at the earth's surface is 6500 watts per square centimeter.)

Lasers may be used in a large variety of ways. Some examples: measuring the earth-to-moon distance (by bouncing a laser beam from the moon's surface); welding a detached retina in place (through special laser surgical instruments); aligning the two-mile long accelerator tube at Stanford University (by sighting along a laser beam); reproducing color photographs (by using blue, green, and red laser light to produce three negatives); identifying diamonds (by employing the distinctive pattern of reflected laser light). The applications for diverse fields are virtually endless.

Right now, much research is aimed at using the laser in the area of communications. Like the radio transmitter, laser light produces electromagnetic waves of a single frequency and waves that are in phase. But there is one exciting difference: the frequency of visible laser light (about 10^{15} osc/sec) is 1 billion times larger than the frequency of radio waves, which is 10^6 osc/sec.

Why exciting? The transmission of the human voice requires a *range* of frequencies, called the *bandwidth*, of about 4000 osc/sec. Because visible light has such large frequency compared to radio waves, many more channels of communication—more telephone calls, more radio stations, more TV channels—could be transmitted by laser light than are presently available at radio frequencies. In Chicago, during 1977, telephone calls were transmitted by sending laser light through thin, specially designed glass fibers placed beneath the city's streets. A cable containing 144 glass fibers is less than one-half inch in diameter and has the capacity to carry 50,000 telephone conversations. Figure 13−16 shows a laser beam passing through a short section of this glass. Note that you can see the beam entering and leaving, but do not see the beam in the glass. If this had been ordinary glass instead, so much scattering of light would have occurred that the beam would be clearly visible in the glass, too. Since the laser light must travel over a long distance, the loss of light through scattering must be made as small as possible. The glass fibers were designed to have this important no-scattering property. In 1979, Southern New England Telephone began the first service

FIG. 13–16   The passage of a laser beam through a
short section of lightguide glass. (Photo courtesy of *The
Western Electric Engineer.* Copyright © 1980 by Western
Electric Company, Inc.)

FIG. 13–17   A lightguide cable containing 12 glass
fibers carries color TV, voice, and data signals in the
form of light pulses for the 1980 Winter Olympics at Lake
Placid, New York. The lightwave system, manufactured
by Western Electric for New York Telephone, runs along
a 2½-mile route linking the Lake Placid telephone
switching office, the Olympic ice arena, and the
broadcast center serving 26 mass-media agencies.
(Photo by Nancie Battaglia, Lake Placid, New York.
Photo courtesy of *Western Electric*)

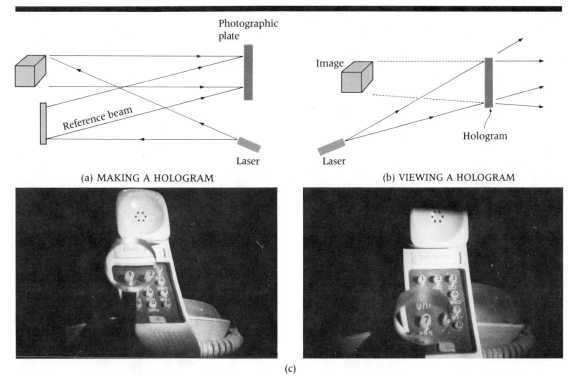

Photographic plate

Reference beam

Laser

(a) MAKING A HOLOGRAM

Image

Laser

Hologram

(b) VIEWING A HOLOGRAM

(c)

FIG. 13–18 (a) Constructing a hologram. (b) Viewing a hologram. (c) Two photographs, taken at different viewing positions, of the light emerging from the hologram in (b). In this hologram, a magnifier is mounted in a *fixed* position in front of a telephone. The three-dimensional nature of the light from the hologram is evident here because the magnifier enlarges different numbers. It is just as though you were looking at the actual object from two different angles. (Hologram by T. H. Jeong, Lake Forest, Ill.)

over a permanent route using the lightwave technology in Connecticut. In the future, we shall certainly see many telephone companies shifting to this new technology. The lightwave technology was also used in the 1980 Winter Olympics at Lake Placid, New York (Fig. 13–17).

THREE-DIMENSIONAL IMAGES

Laser-produced three-dimensional images called **holograms,** have captured the imagination of the American public. Perhaps one day holographic motion pictures and TV may be a reality, although this would require an incredible amount of information to be transmitted. (In the film *Star Wars*, a foot-high holographic image

of Princess Leia was beamed from the head of the droid R2D2.)

The holographic process was discovered in 1947 by the Hungarian-born physicist Denis Gabor (1900–1979) of the Imperial College of Science and Technology in London. Gabor received the Nobel Prize for his work in 1971.

Figure 13–18(a) shows, in essence, how a hologram is produced. The laser beam is split into two parts. One part strikes the object, where it is reflected, and then travels to the film. The other part, called the *reference beam*, does not

ATOMS, LIGHT, AND LASERS

strike the object at all, but is reflected by a mirror, and then it too travels to the film. The interference between the two beams of light is recorded on the photographic film. Where constructive interference occurs, the film is exposed—blackened; where destructive interference occurs, the film is not exposed and is clear when developed. Thus, on the photographic plate, a series of fringes, oftentimes circular, appears in contrast to the image of the object that is on the negative of a film in conventional photography.

Figure 13–18(b) shows how a three-dimensional image is produced from a hologram. Laser light strikes the hologram in the same direction as the reference beam. The fringes on the plate diffract the light, so that a three-dimensional image appears. Figure 13–18(c) shows two photographs, taken at different angles, of the light emerging from the hologram in (b). As the viewer's head or the camera is moved, the perspective of the image changes, just as it does for a three-dimensional object. The word *hologram* derives from two Greek words meaning "whole information," or "whole picture."

Summary

Niels Bohr calculated the radius and speed of the electron's orbit using Coulomb's law and $F = ma$. He found the radii of allowed orbits by requiring that the electron's angular momentum be restricted to integral amounts of $h/2\pi$. By summing the electron's kinetic energy and potential energy, Bohr found the energy sum E to be -13.6 eV/n^2. When the electron jumps to a lower-energy level, the atom emits a photon having an energy that is given by $hf = E_i - E_f$. Bohr was able to derive Balmer's formula for the wavelengths in the visible region of hydrogen's spectrum.

Bohr's theory could be generalized: All atoms have quantized energy levels; light is emitted when an electron jumps to a lower-energy level. That each chemical element has a distinct set of energy levels explains the characteristic spectra of chemical elements. The Franck-Hertz experiment showed that mercury has quantized energy levels.

The absorption lines in the solar spectrum occur because photons of certain energies are absorbed by the gas extending for miles above the sun's surface. The chemical elements in the sun can be identified by comparing the solar absorption spectrum with the emission spectra of elements.

In a ruby laser, chromium atoms absorb photons of light from a flash lamp and are pumped to a metastable state, a state with a comparatively long lifetime (10^{-3} seconds). Laser action begins when one chromium atom spontaneously emits a photon along the axis of the ruby rod, and stimulates another atom to emit an *identical* photon, resulting in two photons of same phase and same direction. This process, called stimulated emission, is responsible for the laser light. The laser light passes through the partially silvered end of the ruby rod. Laser action continues until all chromium atoms have reached the ground state.

The technological applications of the laser are numerous and growing. One, the holographic process, produces three-dimensional images that may revolutionize the film industry.

True or False Questions

Indicate whether the following statements are true or false. Change all of the false statements so that they read correctly.

13–1 The following statements concern Bohr's model of the hydrogen atom:

    a. According to Bohr's first postulate, no radiation is emitted when the electron travels in an allowed circular orbit around the nucleus.

b. Bohr first postulate conformed to established principles of electricity.

c. The speed of an electron in an allowed orbit was found using an established principle of electricity: Coulomb's law.

d. The radii of the allowed orbits are found from Bohr's second postulate.

e. The wavelengths of light emitted by the hydrogen atom are found from Bohr's third postulate.

f. Conservation of energy appears in Bohr's second postulate.

13–2 An electron in the hydrogen atom jumps between two allowed orbits that have energies E equal to -0.54 eV and -0.38 eV.

a. The orbit having energy E equal to -0.38 eV will have the larger radius.

b. The larger value of energy E is -0.54 eV.

c. Since the energies E are negative, the photon energy will be a negative number.

d. The photon energy will be 0.16 eV.

e. The energy of the fifth allowed orbit is -0.54 eV.

f. The light emitted will be in the ultraviolet region of the spectrum.

13–3 In hydrogen atom A, the electron jumps from the second to first orbit ($2 \rightarrow 1$). In hydrogen atom B, the electron jumps from the third to second orbit ($3 \rightarrow 2$).

a. Atom A emits a photon of larger energy.

b. Atom A emits light of longer wavelength.

c. Atom A emits ultraviolet light.

d. Atom B emits visible light.

13–4 Photons of ultraviolet light having energy of 12.5 eV and 14 eV strike a container of hydrogen gas, where electrons are in the first orbit.

a. The 12.5-eV photons will surely be absorbed by the hydrogen gas because an energy of only 10.2 eV is needed to raise the electron to the second orbit and 12.09 eV is needed to raise the electron to the third orbit.

b. The 14-eV photons will surely be absorbed by the hydrogen gas because an energy of 13.6 eV is needed to ionize the hydrogen atom.

13–5 Hydrogen gas at room temperature will not absorb light in the visible region of the spectrum.

13–6 In the spectrum of light emitted by a hydrogen atom, the frequency of one line plus the frequency of a second line equals the frequency of a third line.

a. This frequency-addition characteristic is found in the spectra of some elements, but not in others.

13–7 Each element emits light having a characteristic line spectrum because each element has a characteristic set of excited energy levels.

13–8 The first evidence of the element helium (Greek *helios:* the sun) was a new line in the solar spectrum.

a. This line in the absorption spectrum of the sun had a wavelength not found in any other element.

b. Without a doubt, the emission spectrum of helium (in a discharge tube) would contain the same line as found in the solar spectrum.

13–9 In a laser,

a. the atoms must have an excited state that is metastable.

b. laser action can begin when more atoms are in the ground state than in the metastable state.

c. one atom emits light by the process of stimulated emission, causing other atoms to emit light by the process of spontaneous emission.

13–10 A hologram, produced by laser light, utilizes a basic property of light, called diffraction.

ATOMS, LIGHT, AND LASERS

Questions for Thought

Questions 13–1 through 13–4 are multiple-choice questions.

13–1 The radii of the allowed orbits in the hydrogen atom are found by considering

   a. only the effect of Coulomb's law.
   b. only the restriction on the angular momentum.
   c. conservation of energy.
   d. the effect of Coulomb's law along with a restriction on the angular momentum.

13–2 When the light from the sun is put through a prism,

   a. a continuous spectrum of colors is found.
   b. some wavelengths of light are missing, due to elements in the sun emitting only certain wavelengths.
   c. some wavelengths of light are missing, due to absorption in passing through the earth's atmosphere.
   d. some wavelengths of light are missing, due to absorption of the light in passing through the surface gas on the sun.

13–3 The purpose of the flash lamp in the ruby laser is to

   a. cause the electron to reach a higher-energy state.
   b. cause the electron to fall to a lower-energy state.
   c. heat up the ruby so that it will glow.
   d. none of the above.

13–4 In the process of stimulated emission in a laser, a photon of light of frequency f_0 strikes an atom and the result is

   a. the photon is absorbed causing the electron to go to a higher orbit.
   b. two photons of different energy are emitted.
   c. two photons having a frequency f_0

are obtained and travel in different directions, since the process is random.
   d. two photons having a frequency f_0 are obtained and travel in the same direction with the same phase.

13–5 Consider the electron in the hydrogen atom traveling in a circular orbit around the nucleus. In the following list, state whether each item represents a so-called classical concept, or a departure from classical physics—a quantum effect—by inserting a C or Q in the blank:

_____ Use of $F = ma$ and Coulomb's law to calculate the speed of the electron at a given radius.

_____ The electron has angular momentum.

_____ In an allowed orbit, the electron can have only angular momentum given by $nh/2\pi$, where $n = 1, 2, 3, \ldots$.

_____ The electron can travel only in allowed orbits.

_____ The electron traveling in allowed orbits does not emit electromagnetic radiation.

_____ The electron has kinetic energy, potential energy, and an energy equal to their sum.

_____ The energy sum E of the electron can have only certain values.

_____ The hydrogen atom emits light when the electron jumps to a smaller allowed orbit.

13–6 The following questions refer to Bohr's model of the hydrogen atom: Insert: *larger, smaller,* or *equal to* in the blanks.

   a. The radius of the $n = 4$ orbit is _____ than the radius of the $n = 1$ orbit.
   b. When the electron is in the $n = 4$ orbit, the force is _____ than when in the $n = 1$ orbit.
   c. The speed of the electron is _____ in the $n = 4$ orbit than in the $n = 1$ orbit.
   d. The total energy of the electron is _____ for $n = 4$ than for $n = 1$.

e. The angular momentum of the electron is _____ for $n = 4$ than for $n = 1$.

13–7 In a cathode ray tube containing hydrogen, explain how ionization occurs near the anode. How does this ionization ultimately result in light being emitted by the cathode ray tube?

13–8 Insert *increases* or *decreases* in the blanks.
  a. As an object falls to the ground due to gravity, its potential energy _____ and its kinetic energy _____ .
  b. As a positively charged particle moves toward a negatively charged particle, the potential energy _____ and their total kinetic energy _____ .
  c. From these two examples, can you conclude that, in general, a system tends to move so that its potential energy has the smallest possible value?
  d. Extend part **c** to the hydrogen atom and explain why the electron is normally found in the $n = 1$ orbit. Why is this called the ground state?

13–9 Given sufficient engineering technology, it is possible to place a satellite in a circular orbit around the sun with *any* desired radius. However, the electron in Bohr's hydrogen atom can travel only in allowed orbits. The satellite is governed by Newton's law of universal gravitation and the electron, by Coulomb's law, but both forces are inversely proportional to r^2. What is the reason for "*any* desired radius" for the satellite versus "allowed orbits" for the electron?

13–10 a. In the Franck-Hertz experiment, explain how mercury atoms are raised from the ground state to the 4.9-eV excited state.
  b. Franck and Hertz also examined the light emitted by the mercury vapor in their experiment. What wavelength of light is emitted when the electron jumps from the 4.9-eV energy level to the ground state? Did they see this wavelength when V_G in Fig. 1, Box 13–1, was 4.8 volts? When V_G was 5.0 volts? Explain.

13–11 A mercury vapor street lamp is in essence a cathode ray tube containing mercury vapor. Explain how connecting a high voltage between the cathode and anode causes light to be emitted.

13–12 Examining the spectrum of a star, an astronomer finds that it does not contain hydrogen absorption lines in the visible region of the spectrum. Lines in the ultraviolet cannot be observed because the earth's atmosphere absorbs ultraviolet radiation.
  a. Which of the following conclusions is possible, as well as reasonable: (1) The star does not contain hydrogen. (2) The star does contain hydrogen, but the surface temperature is so small that the hydrogen atoms are all in the ground state. (3) The star does contain hydrogen, but the surface temperature is so high that all of the hydrogen is ionized, so that an absorption spectrum for hydrogen is not observed.
  b. Explain how the astronomer can determine the surface temperature of the star by assuming that it is a black body and examining the star's spectrum. (Black body radiation was discussed in Chapter 10.) How does this affect the choices in part **a**?

13–13 Consider a hypothetical atom that has an excited metastable state at 2 eV above ground state and another excited state at 4 eV above ground state.
  a. Sketch the energy-level diagram.
  b. Suppose that the hypothetical atom is in the metastable state and a photon of 2 eV interacts with this atom. Describe what happens: (1) if the process of stimulated emission occurs; (2) if absorption of the photon occurs.

13–14 Describe the basic principles of the operation of a laser. What is pumping? A metastable state? Why must the laser material have a metastable state? How does laser light differ from light produced by ordinary lamps?

Questions for Calculation

13-1 The energy sum E of the electron in the $n = 1$ orbit is _____ eV. If the potential energy of the electron in the $n = 1$ orbit is -27.2 eV, then the kinetic energy of the electron is _____ eV. The energy sum E of the electron in the $n = 2$ orbit is _____ eV. When the electron jumps between the $n = 2$ and $n = 1$ orbits, a photon is emitted having an energy of _____ eV.

13-2 a. Show that Equation 1 ($hf = E_i - E_f$) is an application of the conservation of energy principle, TE_{before} equals TE_{after}, where the total energy TE means the sum of all types of energies.
b. When the electron of a hydrogen atom is in the $n = 2$ energy level, its energy sum E is _____ . When this electron jumps to the ground state, its energy sum E is _____ . In order to conserve energy, the photon must have an energy of _____ .
c. When the electron of a hydrogen atom is in the $n = 2$ energy level, its angular momentum is _____ . When this electron jumps to the ground state, its angular momentum is _____ . In order to conserve angular momentum, the photon must have an angular momentum of _____ .
d. Use part **c** to show that the concept of angular momentum must be extended beyond the standard definition of angular momentum for a massive particle mvr. (Does a photon have mass?)

13-3 How much energy is needed to ionize a hydrogen atom in which the electron is in the $n = 2$ energy level? What wavelength of light will supply this energy?

13-4 A beryllium (Be) atom contains 4 electrons and a nucleus containing 4 positive charges. A Be^{3+} ion, missing three of its electrons, consists of a single electron circling a nucleus with 4 positive charges.
a. Consider this electron circling the nucleus with a radius r and write an equation for the force of attraction between the electron and the beryllium nucleus. Write a similar equation for the hydrogen atom. For the same radius, which force is larger?
b. Consider the electron in a hydrogen atom in the $n = 1$ orbit. Suppose that it were possible to add 3 more positive charges to this hydrogen nucleus, so that the nucleus then contained 4 positive charges. What would happen to the electron of this hypothetical atom? Would the $n = 1$ allowed orbit be larger or smaller than before?
c. Do you expect that the speed of the electron in the $n = 1$ orbit of a Be^{3+} ion will be larger or smaller than that of the electron in the $n = 1$ orbit of a hydrogen atom? Why?
d. Construct a table similar to Table 6-2 (p. 131) for the Be^{3+} ion and show that:
(1) The radius of an allowed orbit is given by

$$r = \frac{(0.528 \times 10^{-10} \, m)n^2}{Z} \, ;$$

(2) The speed in an allowed orbit is given by

$$v = \frac{(2.19 \times 10^6 \, m/sec)Z}{n} \, ;$$

and (3) the energy E is given by

$$E = \frac{(-13.6eV)Z^2}{n^2} \, ,$$

where $Z = 4$ for beryllium. The angular momentum will still be $h/2\pi$ in the first allowed orbit. How much energy is needed to ionize a Be^{3+} ion?

e. Construct an energy-level diagram for Be^{3+} and calculate the wavelength of light emitted when the electron jumps from the $n = 2$ level to the $n = 1$ level.

13–5 Show that Equation 4 is equivalent to Balmer's formula, given by

$$\lambda = 364.56\,nm \times \frac{n^2}{n^2 - 2^2},$$

where $n = 3, 4, 5,$ and 6. Follow these steps:

a. Denote the initial larger-energy orbit by n_i and the final smaller-energy orbit by n_f. Since E equals -13.6 eV/n^2, write down the energy difference E_i minus E_f in terms of n_i and n_f. Show that the energy difference is given by

$$E_i - E = \frac{(13.6eV)(n_i^2 - n_f^2)}{n_f^2 n_i^2}.$$

(Think about how you add two numerical fractions by finding the same common denominator and follow the same procedure here.)

b. Substitute the formula for the energy difference found in **a** into Equation 4. Set n_f equal to 2 and show that Balmer's formula results.

13–6 Consider the hydrogen atom with its electron in the first allowed orbit.

a. Compare the force of attraction between the electron and the nucleus with the attraction between the masses, using Newton's law of universal gravitation. The radius of the first allowed orbit is 0.53×10^{-10} m. The mass of the electron is 9.11×10^{-31} kg, and the mass of the hydrogen nucleus is 1.67×10^{-27} kg. The charge on the electron is 1.6×10^{-19} coulombs. K is 9×10^9 N-m^2/C^2 and G is 6.67×10^{-11} N-m^2/kg^2.

b. Show that the ratio of the electric and gravitational forces does not depend upon the distance between the electron and the nucleus.

c. Explain why gravity has not been mentioned in this chapter as a possible force holding the electron in orbit around the nucleus.

13–7 In the Franck-Hertz experiment, an electron from the cathode C collides with a mercury atom just in front of the grid G.

a. Answer the following questions when V_G in Fig. 1, Box 13–1, is 5.5 volts and V_A is 4 volts: What is the electron's kinetic energy before and after collision with a mercury atom? Will this electron be able to reach the anode A? If so, what is its kinetic energy just before reaching anode A?

b. Answer the questions in part **a** when V_G is 4.5 volts and V_A is 3 volts.

13–8 An electron in the 7.93-eV energy level of mercury (see Figure 13–10) can reach the ground state in two ways: (1) by jumping to the 4.89-eV energy level and then to the ground state, or (2) by jumping to the 6.70-eV energy level and then to the ground state. Find the photon energies of these 4 photons and find a simple relationship among them. What is the relationship among the frequencies of photons of light?

†13–9 The intense yellow line in the sodium spectrum has a wavelength of 589 nm. This yellow light is emitted when an electron jumps from an excited state to the ground state. Find the energy of this excited state.

13–10 The glass tube of a fluorescent lamp is filled with mercury vapor and coated with a mixture of fluorescent materials that control the color of the light. When the lamp is turned on, the mercury atoms are raised from the ground state to the 4.89-eV excited state.

a. What wavelength of light is emitted when the mercury atoms return to the ground state? Is this light visible or ultraviolet light?

b. The light emitted by the mercury atoms is absorbed by atoms of the fluorescent coating, raising them to a 4.89-eV excited state. Show that, if these atoms of the fluorescent material return to the ground state by emitting two photons, one of the photons can be in the visible region of the spectrum. Visible light ranges from 400 nm (blue) to 700 nm (red.)

Footnote

[1]Niels Bohr, *Philosophical Magazine* 26 (1913): 1.

Introduction

Shortly after his discovery in 1895, Roentgen showed that X-rays are not charged particles: They did not bend when passing through a magnetic field. Are they neutral particles? Or, are they like light, electromagnetic waves? Although early research suggested that X-rays are waves, controversy raged about the nature of X-rays until 1912 when Max von Laue settled the question once and for all. This chapter begins with the early research and continues with Laue's dramatic demonstration of the interference of X-rays.

The next step was taken by Sir William Henry Bragg, professor at the University of Leeds, and his son, William Lawrence Bragg. The Braggs developed a method to measure the wavelength of X-rays and to study the arrangement of atoms in a crystal. Lawrence describes their interest:

At first the main interest in von Laue's discovery was focused on its bearing on the controversy about the nature of X-rays; it proved that they were waves not particles. It soon became clear to some of us, however, that this effect opened up a new way of studying matter, that in fact man had been presented with a new form of microscope, several thousand times more powerful than any light microscope, that could in principle resolve the structure of matter right down to the atomic scale.[1]

The Braggs found that some wavelengths from the X-ray tube were very intense. In 1913, H. G. J. Moseley systematically studied the X-ray lines, as they are called, by using all known elements for the anode of the X-ray tube. (Remember that X-rays are produced in an X-ray tube when electrons strike the anode and slow down). These studies showed that the positive charge on the nucleus is equal to the atomic number Z. Also, an amazing connection exists between the hydrogen spectrum and X-ray lines.

Finally, we shall see how researchers used X-rays to probe the structure of biochemical molecules: vitamin B-12, proteins, and DNA (the molecule that carries the genetic code in cells).

H. G. J. Moseley

CHAPTER 14

X-rays, Crystals, and the Atomic Number

Early Research

If X-rays are electromagnetic waves, then researchers should be able to observe diffraction and interference effects, as they had for visible light. We can observe the diffraction of light through a single slit, *but* the width of the slit and the wavelength must be about the same size. Similarly, in order to observe the diffraction of X-rays by a single slit, the slit size must be on the order of the wavelength.

Not knowing the wavelength of X-rays, or even whether X-rays had a wave nature, the early workers had virtually no ideas regarding the width of the slit that might show diffraction effects. In 1899, H. Haga and C. H. Wind of the University of Groningen in Holland thought of a clever way around this difficulty. They directed an X-ray beam through a V-shaped slit 4 cm long and only 0.025 mm wide at its wider end.[2] A photographic plate behind the slit showed a broadening of the X-ray beam after passing through the narrowest part of the V. Attributing this fuzzy image to diffraction, they estimated[3] the wavelength to be about 10^{-10} m. But these results were not convincing—not dramatic enough—because the effects were so small.

Laue and Interference Effects

In a flash of insight, the German physicist Max von Laue (1879–1960) realized that interference effects should be apparent when X-rays struck a crystal. His earlier work had prepared him for this great discovery, for which he received the Nobel Prize in 1914. Laue had investigated the diffraction of light by a cross-grating (see Fig. 14–1). Just as an ordinary diffraction grating produces regions of constructive interference along a single *line*, a cross-grating produces regions of constructive interference over an *area*. Further, Laue was familiar with the concept of crystal structure, as he explains in his Nobel lecture:

There was yet another important circumstance. Mineralogists since Hauy (1774) and Bravais (1848) had explained (crystal structure) simply and clearly as due to the arrangement of the atoms in a space lattice. [See Fig. 14–2]. But these ideas led to no further physical deduction and so continued to exist as a doubtful hypothesis comparatively unknown to physicists. In Munich, however, where models of the . . . space-lattices were to be seen . . . I too became acquainted with it. . . . That the lattice constant (distance between centers of atoms) in crystals is of the order of magnitude of 10^{-10} m . . . was easily confirmed from the density, the molecular weight, and the mass of the hydrogen atom*. . . . The order of magnitude of the wavelength of X-rays was 10^{-11} meters. Hence, if X-rays were passed through a crystal, the ratio of the wavelength and the lattice constant was extraordinarily favorable. . . . At once . . . I expected interference effects with X-rays.[4]

To understand the interference effects shown in Fig. 14–3, let's have another look at the diffraction effects for water waves in Fig. 8–7 (p. 172). There, we saw that circular diffracted waves are produced when plane water waves strike a peg. The diameter of the peg is slightly smaller than the wavelength of the water waves. Since the wavelength of X-rays is comparable to the diameter of an atom, we let the waves in Fig. 8–7 represent X-rays and the peg represent an atom. When the X-rays strike the crystal, circular diffracted waves emerge from each atom struck by the X-rays. Since the atoms in the crystal are arranged in a regular pattern, as are the slits in a diffraction grating for light, regions of constructive interference occur. These regions are seen on the photographic film. The unexposed regions on the film are due to destructive interference of the many waves.

---

*In Chapter 1, we calculated the size of the atom (and, hence, the distance between centers of atoms) to be about 10^{-10} m.

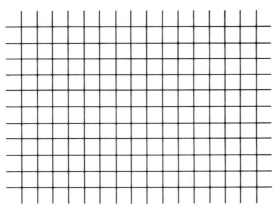

FIG. 14−1   A diagram of a cross-grating, which has two sets of slits, one at right angles to the other.

| Cubic | Orthorhombic | Trigonal |

FIG. 14−2   Three types of Bravais lattices. In 1774, Rene Hauy proposed that the known angles between the faces of a crystal would be accounted for if the crystal were made up of a small structure, now called a *unit cell*, repeated throughout the crystal. In 1848, Auguste Bravais proposed that molecules in a crystal are arranged in regular three-dimensional lattices. This would explain the existence of the 32 types of crystalline substances found in nature.

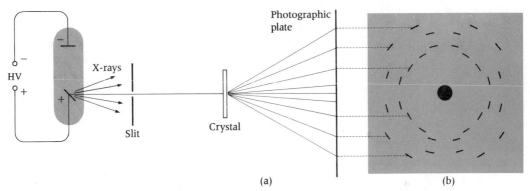

(a)   (b)

FIG. 14−3   (a) The experimental apparatus used by Friedrich and Knipping to observe the interference of X-rays diffracted by atoms in a crystal. (b) An artist's rendering of a photograph obtained by using a zinc sulfide crystal. The regions of constructive interference are indicated by the exposed (or black) regions on the film. The regions of destructive interference are indicated by the unexposed (or gray) regions on the film.

The Braggs and Crystal Structure

In 1915, the father-and-son English physicists Henry Bragg (1862−1942) and Lawrence Bragg (1890−1971) were awarded the Nobel Prize. In a passage from his Nobel lecture, Lawrence tells how he began his work:

Professor Laue had made some of his earliest experiments with a crystal of zinc sulphide. . . . Whereas there were a large number of directions in which one would expect to find a diffractional beam, only a certain number of these appeared on the photographic plate. . . . In studying his work, it occurred to me that perhaps we ought to look for the origin of this selection of certain directions of diffraction in the peculiarities of the crystal structure. . . . I tried to attack the problem from a slightly different point of view.[5]

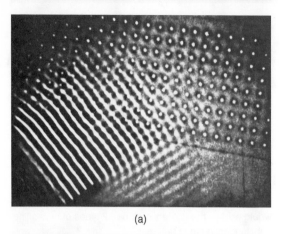

(a)

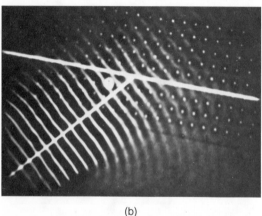

(b)

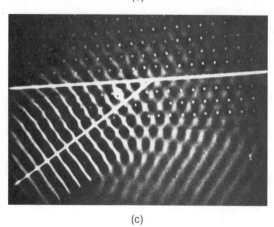

(c)

FIG. 14–4   Ripple-tank photographs showing plane water waves striking a lattice of pegs. Circular, diffracted waves are produced by each peg struck by the waves. In part (a) and part (c), the waves interfere constructively, and a reflected plane wave appears. In part (b), destructive interference occurs, and no reflected wave is observed. The angle that the incident beam makes with the lattice is the same in part (a) and part (b), but smaller in part (c). The wavelengths are the same in part (b) and part (c), but smaller in part (a). (Courtesy of Education Development Center, Newton, Ma)

We can understand the Braggs' approach most easily by examining ripple-tank photographs. Figure 14–4 shows what happens when plane water waves strike a regular lattice of pegs. A circular diffracted wave emerges from each peg struck by the waves.

In Fig. 14–4(a), the circular, diffracted waves from each peg interfere constructively, which results in a plane wave making the same angle θ with the lattice as the incident beam. It seems as if the X-ray beam is reflected from the surface of the crystal, similar to the manner in which light is reflected by a mirror.

In Fig. 14–4(b), angle θ for the incident beam is the same as in part (a), but the wavelength is larger. There is now no evidence of a "reflected" beam. This is due to the destructive interference of the diffracted waves from each peg.

In Fig. 14–4(c), the wavelength is the same as in part (b), but the reflected beam reappears when the angle θ is changed.

If we had repeated the experiments in Fig. 14–4 using a lattice with a different distance between the pegs, the results (whether or not a reflected beam appeared) would have changed.

Imagine now that we superimpose part (b) on part (a), matching the lattices. We would have an incident beam with *two* wavelengths, but a reflected beam with only *one* wavelength.

The similarity of Fig. 14–4 to X-rays striking a crystal is obvious. Summarizing the results of Fig. 14–4 in terms of X-rays, rather than water waves, we conclude:

1. The wavelength of the X-ray beam reflected by the crystal depends upon the angle θ and the distance d between atom planes.

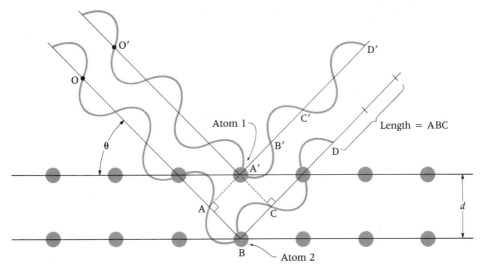

FIG. 14–5 Waves diffracted by atoms 1 and 2. At the instant of time shown, the front of the waves are at points D and D'. Initially, the waves were at point O and O' and subsequently moved to points A and A', B and B', C and C', and finally to D and D'. The waves interfere constructively because the distance ABC is equal to one wavelength. (In the crystal, the atoms are actually side by side. Here, the distance between centers of atoms is greatly enlarged for clarity.)

2. By comparing part (b) and part (c), we see that a reflected beam will appear for some angle θ.

3. If the X-ray beam striking the crystal contains many wavelengths, we can use the crystal to obtain a spectrum of X-rays. The superimposition of part (a) and part (b) shows that *only one* wavelength is reflected by the crystal. To select other wavelengths from the beam, we must rotate the crystal.

Let's examine each of these conclusions more closely. We shall see how the Braggs used their wavelength measurements as a "microscope, several thousand times more powerful than any light microscope" to examine crystal structure.

In conclusion 1, we see that the wavelength depends upon the angle θ and the distance d between atom planes. Let's see how this wavelength can be determined.

Figure 14–5 shows waves diffracted by atoms 1 and 2. Initially the fronts of the waves are at points O and O'. At later times, the fronts are at points A and A', then at B and B', at C and

C', and finally at D and D'. Constructive interference occurs because the distance ABC is equal to one wavelength. Find ABC, and you find the wavelength:

$$\text{Wavelength} = ABC. \qquad (1)$$

Figure 14–6 shows that the distance ABC changes with the angle, so there is only one angle where the distance ABC is equal to the wavelength. If you draw a diagram to scale (knowing the distance d and angle θ for constructive interference), and measure the distance ABC, you can find the wavelength.*

---

*Constructive interference will also occur if the distance ABC equals 2λ or 3λ, and so on. We'll neglect this complication here.

X-RAYS, CRYSTALS, AND THE ATOMIC NUMBER

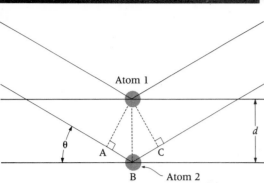

FIG. 14—6 Waves diffracted by atoms 1 and 2. Here, the angle θ is smaller than in Fig. 14—5 and, likewise, the distance ABC is smaller, too. There is only one value of angle θ where the distance ABC is equal to the wavelength of the incident beam.

If all of this has a familiar ring, it is because we used the same symbols d and θ to discuss a diffraction grating. Figure 14—7 compares these two ways to measure wavelength.

CONTINUOUS AND LINE SPECTRA

Figure 14—7 describes how to obtain a spectrum of X-rays. The Braggs designed an instrument, called an **X-ray spectrometer,** to do this conveniently (see Fig. 14—8). The intensity of the reflected beam is proportional to the current in the ionization chamber. It is quite similar to a Geiger counter, but it gives a *value* for the current, not just a large burst of current, as the Geiger counter does.

First, Sir Henry used an X-ray tube with a platinum anode and a sodium chloride (salt) crystal.[6] Rotating the crystal and the ionization chamber so that the reflected beam would enter it, he obtained the spectrum in Fig. 14—9. He found six strong peaks—*X-ray lines,* as they are now called—superimposed on a continuous spectrum indicated by the dashed line.[7] He had expected the continuous spectrum. But what, however, caused the lines?

X-rays are produced when electrons come to a stop in the anode. The kinetic energy of an electron is transformed into one or more X-ray photons. If only one photon is produced, that photon energy equals the kinetic energy of the electron. If several X-ray photons are produced, they share the electron's kinetic energy. Thus, in an X-ray tube, there is a continuous distribution of photon energies. Since the photon energy and wavelength are related (photon energy = hc/λ), there is a continuous distribution of X-ray wavelengths (see Chapter 10, p. 244).

When the Braggs used different elements for the anode, they found different sets of X-ray lines superimposed upon a continuous spectrum. Therefore, the X-ray spectrum consists of two parts: (1) the continuous spectrum and (2) a characteristic line spectrum that depends upon the anode material.

In his Nobel lecture, Lawrence describes research with the spectrometer:

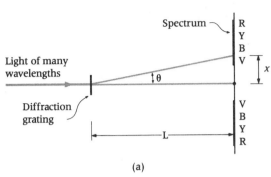

Spectrum

R
Y
B
V

x

Light of many
wavelengths

Diffraction
grating

θ

L

V
B
Y
R

(a)

FIG. 14–7 (a) When visible light strikes a diffraction grating, a spectrum appears on the screen. Each position corresponds to a different wavelength. To determine the wavelength of a given color, you must measure the angle θ (or equivalently, the distances L and x) and know the distance d between two consecutive slits in the grating.

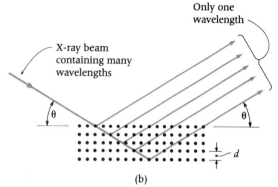

Only one
wavelength

X-ray beam
containing many
wavelengths

θ

θ

d

(b)

FIG. 14–7 (b) When an X-ray beam strikes a crystal, a diffracted beam emerges at angle θ. While the incident beam contained many wavelengths, the *reflected beam contains only one.* This wavelength can be found because you know angle θ and the distance d between atom planes. If you change the angle at which the incident beam strikes the crystal, the reflected beam will contain *another* single-wavelength component of the incident beam. To obtain an X-ray spectrum, continuously change the angle and observe the intensity of the reflected beam.

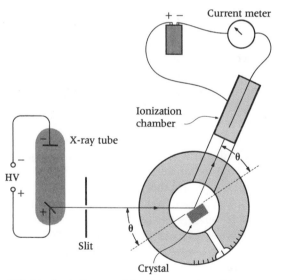

+ − Current meter

Ionization
chamber

X-ray tube

HV

Slit

Crystal

θ

θ

FIG. 14–8 A diagram of the X-ray spectrometer.

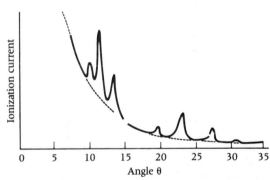

Ionization current

0   5   10   15   20   25   30   35
Angle θ

FIG. 14–9 The X-ray spectrum obtained by Sir Henry Bragg using a sodium chloride crystal and an X-ray tube with a platinum anode.

X-RAYS, CRYSTALS, AND THE ATOMIC NUMBER

In examining the effect for varying angles of incidence my father discovered that a very strong reflection appeared when a given crystal face was set at certain definite angles. . . . This constitutes the first evidence of the existence of characteristic 'lines' in the radiation given off by the anode. By using tubes with anodes of platinum, osmium, tungsten, nickel, and other metal, it became clear that each gave off X-rays containing characteristic lines . . . These . . . lines . . . proved a most powerful method of finding the arrangement of atoms in the crystal. The structure of sodium chloride and potassium chloride, which had been suggested by the Laue photographs, were established on a firm basis by means of the spectrometer . . .[8]

THE CRYSTAL STRUCTURE OF SODIUM CHLORIDE

We return now to Lawrence Bragg's statement in the Introduction, "(X-rays) opened up a new way of studying matter . . . right down to the atomic scale." So far, we have seen how to measure X-ray wavelengths by determining θ and knowing d. However, turning the tables, you can determine the distance d between atom planes if you know the wavelength of an X-ray *line* and measure the angle θ. In this way, you can glean information about how the atoms are oriented in the crystal and about their spacing. The Braggs quickly set out to study the sodium chloride crystal.

Laue's X-ray photographs had suggested a cubic structure for the sodium chloride crystal. The cubic sodium chloride crystal in Fig. 1−6 (p. 9) surely suggests this too. Laue suspected that this crystal contained a sodium chloride molecule at each corner of a cube. However, the Braggs' measurements showed that "sodium and chlorine ions occur alternately in three directions at right angles, like a chessboard in three dimensions." We shall see how the Braggs arrived at this important conclusion.

Figure 14−10 shows this chessboard arrangement. An electron from a sodium atom moves over to a chlorine atom, forming Na^+ and Cl^- ions. The crystal is held together by the attraction between negative and positive ions. Figure 14−10, a realistic representation of the sodium chloride crystal (in that the ions are stacked one right next to the other), shows two sets of atom planes. Figure 14−11 shows a third set in a three-dimensional diagram, where the atoms are separated (unrealistically) from each other.

Both sodium ions and chlorine ions diffract X-rays, producing circular waves, but the heights or amplitudes of these waves are different. Therefore, if you want to consider interference effects, you must consider waves of the same height. For example, *complete* destructive interference occurs when a peak of one X-ray wave overlaps a trough of another wave, provided their amplitudes are the same. This means that the wave must come from identical atoms. Therefore, the distance between atom planes refers to the distance between *adjacent but identical* planes. Since atom planes of sets A and B contain both sodium ions and chlorine ions, adjacent planes are also identical planes. Calling D the distance between adjacent sodium and chlorine ions, the distance between atom planes for set A is D; for set B, the distance is $D/\sqrt{2}$; and, for set C, the distance is $2D/\sqrt{3}$.

Now, let's see what the distances would be if Laue's educated guess—molecules at the corner of a cube—had been correct. In Fig. 14−11, imagine a sphere representing a sodium chloride molecule where ever you see a sodium or chlorine ion. The distances between atom planes for sets A and B would be the same, as we found above. However, for set C, the distance between identical atom planes would now be $D/\sqrt{3}$.

The Braggs oriented the crystal in their spectrometer so that the X-ray beam was directed toward set C atom planes. Experimentally, they found the angle θ for constructive interference. From this angle, and from the known wavelength of the X-ray line, they determined that the distance between set C atom planes was $2D/\sqrt{3}$. From this, they concluded that the sodium chloride crystal must look like the diagrams shown in Figs. 14–10 and 14–11.

Moseley's Experiment with X-rays and the Significance of the Atomic Number

Let's return to the research at the University of Manchester in progress during 1913. Rutherford, Geiger, and Marsden had just completed their alpha-scattering experiments, which proved the existence of the nucleus. Bohr's theory of the hydrogen atom had triumphantly predicted the hydrogen spectrum; the Braggs had shown how to use a crystal to measure the wavelength of X-rays and determine crystal structure. It was in this climate (one major achievement following another) that the English physicst H. G. J. Moseley (1882–1915) began his work with X-rays.

Moseley had come to the University of Manchester in 1910, attracted by the work of Rutherford and his collaborators, and his first studies dealt with radioactivity. He had also worked with Charles Galton Darwin (grandson of Charles Darwin, who had written *Origin of the Species*), using a crystal to study the *continuous* spectrum of X-rays. Immediately after hearing about the Braggs' research with X-ray lines, Moseley decided to study the characteristic X-ray lines of the elements in a systematic way. His skill as an experimenter was well known:

It is typical of the scarcity of apparatus, even in Manchester, in these times that the Gaede (vacuum) pump which Moseley used to evacuate his (cathode ray) tube was borrowed from Balliol College, Oxford. The apparatus was a troublesome set-up to handle, but Moseley was a man of great experimental skill and prodigious industry, who produced his results in surprisingly short time. He worked late at night: in fact, it was said of him that his specialized attainments included a knowledge of where in Manchester to get a meal at three o'clock in the morning. C. G. Darwin states that "he was without exception, the hardest worker I have ever known."[9]

Moseley used many different elements for the anode in an X-ray tube and measured the wave-

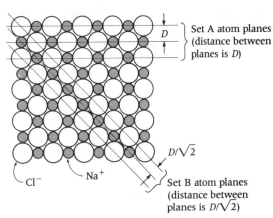

FIG. 14–10 A top view of a sodium chloride crystal. (See, for example, the top view of the three-dimensional crystal shown in Fig. 14–11.) The distance between adjacent sodium and chlorine ions is D. The distance between set A atom planes is D. The distance between set B atom planes is $D\sqrt{2}$.

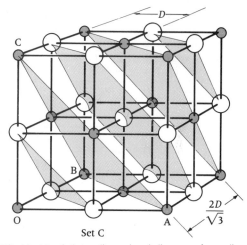

FIG. 14–11 A three-dimensional diagram of a sodium chloride crystal, where the distance between ions has been greatly exaggerated for clarity. Set C atom planes are shown. Each plane contains only one kind of ion. For this reason, the distance between adjacent but identical atom planes is $2D/\sqrt{3}$, where D is the distance between adjacent sodium and chlorine ions.

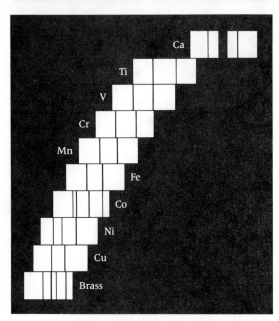

FIG. 14–12 Artist's reproduction of Moseley's original X-ray photographs.

TABLE 14–1 WAVELENGTH OF ONE X-RAY LINE IN THE X-RAY SPECTRA OF THE ELEMENTS

ELEMENT	Z	ATOMIC MASS (amu)	WAVELENGTH (nm)
Calcium	20	40.09	0.3357
Scandium	21	44.1	
Titanium	22	48.1	0.2766
Vanadium	23	51.06	0.2521
Chromium	24	52.0	0.2295
Manganese	25	54.93	0.2117
Iron	26	55.85	0.1945
Cobalt	27	58.97	0.1794
Nickel	28	58.68	0.1664
Copper	29	63.57	0.1548
Zinc	30	65.37	0.1446

Source: H. G. J. Moseley, "The High-Frequency Spectra of the Elements," *Philosophical Magazine* 27(1913):1024.

length of the characteristic X-ray lines. He used essentially the same method as the Braggs (Fig. 14–8), but replaced the ionization chamber with a photographic plate "which makes the analysis of the X-rays as simple as any other branch of spectroscopy."[10] Figure 14–12 shows a diagram of some of his X-ray photographs. From one element to the next, the X-ray lines seem to shift in a regular manner. Table 14–1 shows the wavelength of one of these X-ray lines for each element.

Figure 14–13(a) plots the wavelengths in Table 14–1 versus the atomic number Z of the anode material and Fig. 14–13(b) plots the wavelengths versus the atomic mass. Immediately, one notes the smooth curve in part (a), but sees the impossibility of drawing a smooth curve through the data points in part (b). Obviously, the atomic number Z has a greater significance than merely indicating the location of the element in the periodic table. What is it?

Moseley answered this question when he established an amazing connection between the X-ray line wavelengths and Bohr's model of the hydrogen atom.

Bohr's theory can be applied to the so-called one-electron atom (really an ion) where one electron circles a nucleus with N-positive charges. (See Calculation question 13–4.) Figure 14–14 compares a one-electron atom with a hydrogen atom.

The radius of the first allowed orbit is smaller for the one-electron atom, but the speed is larger, which seems reasonable because of the greater force of attraction of the electron to the nucleus. In the hydrogen atom, the angular momentum mv_1r_1 in the first allowed orbit must equal $h/2\pi$. Figure 14–14 shows that the electron in the one-electron atom also has an angular momentum equal to $h/2\pi$, as it must. We can see that the kinetic energy ($\frac{1}{2}mv^2$) of the electron in the one-electron atom is N^2 times the kinetic energy in hydrogen. The potential energy is proportional to Q_Ne/r, where Q_N is the charge on the nucleus, e is the charge on the electron, and r is the radius. Comparing the charge on the nucleus and the radius in both cases, we

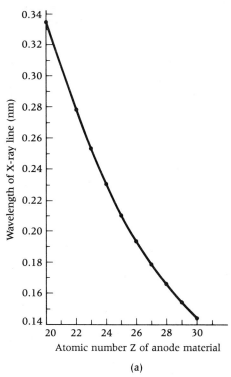

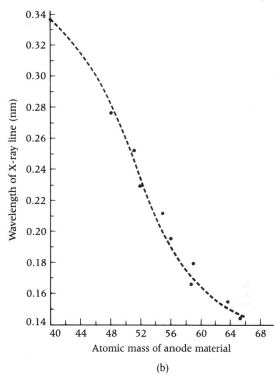

FIG. 14–13 (a) A graph showing the wavelength of X-ray line versus atomic number Z of the anode material of an X-ray tube. (b) A graph showing the wavelength of X-ray line versus the atomic mass of the anode material of an X-ray tube.

find that the potential energy for the one-electron atom is N^2 times that in hydrogen. The energy E (the sum of the kinetic and potential energies) equals -13.6 eV$/n^2$ for hydrogen, and $-13.6\ N^2$ eV$/n^2$ for the one-electron atom.

Consider the electron jumping from the second orbit to the first orbit in both atoms. For hydrogen, the photon energy of the light is equal to the difference in energy E of the two orbits:

$$\begin{aligned}\text{Photon energy} \atop \text{for hydrogen} &= E_i - E_f\\ &= (-3.4\,\text{eV}) - (13.6\,\text{eV})\\ &= 10.2\,\text{eV}.\end{aligned}$$

In the one-electron atom, the photon energy is given by

Photon energy
for one-electron $= 10.2\ N^2$ eV.
atom

Equation 4 of Chapter 13 shows that the wavelength of the emitted light is given by

$$\lambda = \frac{1240\,\text{nm-eV}}{(\text{energy difference})}.$$

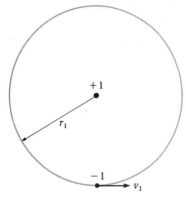

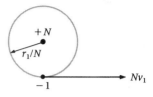

Hydrogen atom

One electron atom
N-positive charges
on nucleus

Speed v_1
Radius r_1
KE_1
PE_1
$E = \dfrac{-13.6\,\text{eV}}{n^2}$

Speed Nv_1
Radius r_1/N
$N^2 \times KE_1$
$N^2 \times PE_1$
$E = \dfrac{-13.6\,N^2\,\text{eV}}{n^2}$

Angular momentum $= mv_1r_1$

Angular momentum $= m(Nv_1)(r_1/N)$
$\qquad\qquad\qquad\quad = mv_1r_1$

FIG. 14–14   A comparison of the properties of a one-electron atom with those of a hydrogen atom. The one-electron atom has a single electron circling a nucleus with N positive charges. Both atoms show the first allowed orbit, where the angular momentum mv_1r_1 is equal to $h/2\pi$.

For hydrogen, the wavelength is given by

$$\lambda = \frac{1240\,\text{nm-eV}}{10.2\,\text{eV}}$$

$$= 121.5\,\text{nm}.$$

For the one-electron atom, the wavelength is given by:

$$\lambda = \frac{1240\,\text{nm-eV}}{10.2\,N^2\,\text{eV}}$$

$$\lambda = \frac{121.5\,\text{nm}}{N^2}$$

The important conclusion is: *When an electron in a one-electron atom jumps from the second orbit to the first, the wavelength of the light is equal to 121.5 nm/N², where N is the number of positive charges in the nucleus.*

How did Moseley establish the relationship between his X-ray line wavelengths and those of the one-electron atom? For each element in Table 14–1, calculate the wavelength given by

$$\lambda = \frac{121.5\,\text{nm}}{(Z - 1)^2}, \tag{2}$$

and you will find that it is the same as the X-ray wavelength. For example, for calcium ($Z = 20$), we find 121.5 nm/19², which equals 0.3366 nm; and for zinc we find 121.5 nm/29², which equals 0.1445 nm. These two calculations show excellent agreement with the X-ray wavelengths in Table 14–1. Thus, the wavelength of the X-ray lines is given by Equation 2. Comparing Equation 2 with the results of the one-electron theory, we see that the quantity Z minus 1 is equivalent to N, the number of charges on the nucleus. *Moseley concluded that the atomic number Z represents the number of positive charges in the nucleus and the number of electrons circling the nucleus.*

But why is the number of positive charges in the nucleus Z, rather than Z minus 1. One simple answer is that the hydrogen nucleus must contain one positive charge, showing that the number of charges here is equal to Z, not to Z minus 1. Secondly, the anode material does not really consist of one-electron atoms, but rather atoms with many electrons circling the nucleus. This fact plays an important role in understanding X-ray lines.

Figure 14–15 illustrates how the 20 electrons of calcium revolve in orbit around the nucleus. (See Chapter 15 for additional discussion.) When the electron beam in an X-ray tube strikes the calcium anode, an orbital electron may be knocked out of the atom. Suppose, for example, that an electron in the first orbit is knocked out and an electron in the second orbit jumps down to take its place. This electron "sees" the Z-positive charges of the nucleus *and* the electron still in the first orbit, yielding a total of Z minus 1 positive charges. This accounts for the Z minus 1 term, rather than Z, in Equation 2.

Another X-ray line occurs when an electron is knocked out of the second orbit and one from the third orbit jumps down to fill the vacancy, and so on.

Moseley's life was short. As Segre writes:

Moseley's story is poignant and tragic. . . . He was obviously a major rising star among British scientists. . . . In an experimental career of very few years he obtained results of permanent value and everlasting fame. At the beginning of the First World War he enlisted as a volunteer and, in spite of attempts by Rutherford and others to protect him from mortal danger in order to save him for English science, he insisted on doing combat duty. He was killed in action at age twenty-seven in the Dardanelles expedition conceived by Churchill.[11]

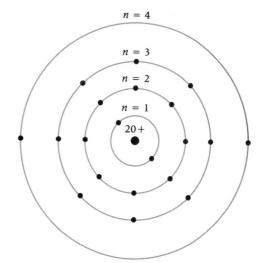

FIG.14–15 The structure of the calcium atom.

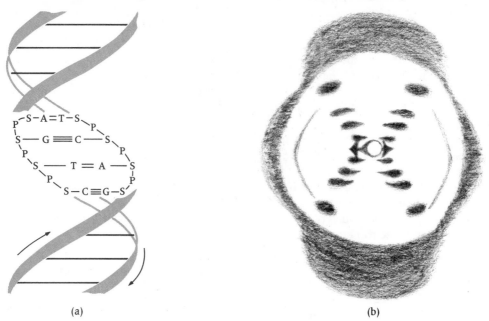

FIG. 14–16 (a) The helical structure of the DNA molecule. Each helix consists of two components: sugar units S alternating with phosphate units P. A unit called a *nitrogen base* (consisting of atoms of hydrogen, carbon, nitrogen, and oxygen) is attached to each sugar unit. There are four kinds of nitrogen bases: adenine A, thymine T, guanine G, and cytosine C. (b) Artist's reproduction of the key X-ray diffraction photograph of DNA.

X-ray Diffraction and the Structure of Biochemical Molecules

Following the Braggs' pioneering work, scientists developed many ingenious methods for using X-rays to study crystal structure. Chemical compounds of every type were investigated, and the nature of metal alloys was explained. Attention was also directed to biochemical molecules. Of this development, Sir Lawrence writes:

We now come to a most dramatic turning point in the history of X-ray analysis. When vitamin B-12 was analyzed (Dorothy Crowfoot Hodgkin was the leader of this investigation at Oxford), with 181 atoms in the molecule, it seemed hard to imagine that much more complex structures could ever be tackled; it had taken eight years to complete and the difficulties increase as a high power of the number of atoms. And then, as the result of an investigation that had lasted for some 20 years, a way was finally found to solve the structure of the immensely more complicated molecules of living matter, the proteins. The first of these to be analyzed, by John C. Kendrew in 1955, was myoglobin, which has 2,500 atoms in its molecule.[12]

Also, in the 1950s, F. H. C. Crick (b. 1916), an English biophysicist, and J. D. Watson (b. 1928), an American geneticist, proposed the model of **deoxyribonucleic acid,** called **DNA,** the molecule that carries the genetic code in a cell (see Fig. 14–16a). The key X-ray diffraction photograph is shown in Fig. 14–16(b). J. D. Watson describes this photograph:

This photograph, taken at Kings College, London, in the winter of 1952–53, by Rosalind Franklin, experimentally confirmed the then current guesses that DNA was helical. The helical form is indicated by the crossways pattern of X-ray reflections (photographically measured by darkening of the X-ray film) in the center of the photograph.[13]

J. D. Watson recalls showing the completed paper to Sir Lawrence:

The solution to the structure was bringing genuine happiness to Bragg. That the result came out of the Cavendish and not Pasadena (where Linus Pauling was also investigating DNA) was obviously a factor. More important was the unexpectedly marvelous nature of the answer, and the fact that the X-ray method he had developed forty years before was at the heart of a profound insight into the nature of life itself.[14]

Summary

Early research showed that X-rays diffracted slightly after passing through a very narrow slit, indicating a wavelength of about 10^{-10} m. Because atoms had a diameter on the order of 10^{-10} m, Laue realized that atoms would diffract X-rays, producing circular waves. If atoms in a crystal had an orderly arrangement, analogous to slits in a diffraction grating, then these diffracted waves should interfere constructively in some directions. When X-rays were aimed at a crystal, the photographic plate behind the crystal revealed exposed regions due to constructive interference.

The Braggs designed a spectrometer to measure the intensity and wavelength of X-rays reflected by the crystal. Although the incident beam contained many wavelengths, the reflected beam (making the same angle θ with the atom planes as the incident beam) contained only one wavelength. This wavelength, determined by Equation 1, depends upon the distance d between atom planes and the angle θ. The X-ray spectrum (X-ray intensity, or the number of X-ray photons per second entering an ionization chamber, versus angle θ) showed a line spectrum superimposed on a continuous spectrum. The features of the line spectrum depended upon the anode material.

The Braggs used their spectrometer measurements to show that sodium chloride had a cubic crystal structure. Further, it was an ionic compound with a Na^+ ion at one corner of a cube with a Cl^- ion at the adjacent corner. They investigated the crystal structure of many compounds.

In 1913, Moseley showed the fundamental significance of the atomic number Z. He measured the wavelengths of the characteristic X-ray lines of many elements. Comparing them to the light emitted by a hydrogen atom when the electron jumps from the second to first orbit (121.5 nm), he found that the X-ray wavelengths were given by $121.5 \text{ nm}/(Z-1)^2$, which is the wavelength emitted by a one-electron atom. Moseley concluded that Z represented the number of positive charges in the nucleus and, since the atom is neutral, the number of electrons in the atom.

The research of the Braggs opened up a new field of research: using X-rays to study crystal structure. For example, in 1953, X-ray diffraction photographs showed that DNA had a helical structure.

True or False Questions

Indicate whether the following statements are true or false. Change all of the false statements so that they read correctly.

14–1 In 1899, researchers directed an X-ray beam toward a V-shaped slit. A sharp image of the V-shaped slit appeared on the photographic plate behind the slit, showing the wave nature of X-rays.

14–2 Just as Becquerel had discovered radioactivity quite accidentally, Laue was looking for another effect when he discovered the wave nature of X-rays.

X-RAYS, CRYSTALS, AND THE ATOMIC NUMBER

14−3 Laue and his colleagues directed an X-ray beam toward a crystal and placed a photographic plate behind the crystal.

   a. The photographic plate showed not only a central exposed spot, but many exposed regions, or spots.

   b. These exposed spots show that atoms in the crystal diffract X-rays from their straight-line path.

   c. These exposed spots are due to constructive interference of X-rays diffracted by many atoms.

   d. The unexposed regions on the photographic plate show that a single atom diffracts X-rays in only certain directions.

14−4 To measure the wavelength of X-rays, the Braggs directed the X-ray beam toward a crystal and measured the intensity of the X-ray beam after it had passed through the crystal.

14−5 An X-ray beam having a wavelength of 5 nm strikes a crystal so that angle θ is 30°. A reflected beam of X-rays appears when angle θ is 30°. [Angle θ is the angle that the incident (or reflected) beam makes with the surface of the crystal.]

   a. When an X-ray beam having a wavelength of 3 nm strikes the same crystal at the same angle of 30°, then the reflected beam will still appear at an angle of 30°.

   b. When the X-ray beam having a wavelength of 5 nm strikes another type of crystal at the same angle of 30°, then the reflected beam will still appear at an angle of 30°.

   c. The wavelength of the X-ray beam can be measured by finding the angle θ at which a reflected beam appears and knowing the distance between atoms in the crystal.

14−6 The Braggs showed that the spectrum of X-rays from an X-ray tube consists of extremely intense X-ray lines superimposed upon a continuous spectrum.

14−7 In an X-ray tube, the X-ray lines are produced by electrons coming to a stop in the anode of the tube; the kinetic energy of the electrons is transformed into X-ray photons.

14−8 The Braggs studied the crystal structure of sodium chloride.

   a. They used the continuous X-ray spectrum from an X-ray tube.

   b. They measured the intensity of different X-ray wavelengths.

   c. They measured the distance between atom planes of the crystal.

   d. They showed that the sodium chloride crystal did *not* consist of molecules at the corners of a cube because the distance between atom planes in this model did not equal that found in their experiment.

   e. They showed that the sodium chloride crystal consisted of alternating sodium and chlorine ions because the distance between atom planes in this model did equal that found in their experiment.

14−9 Moseley used many different elements for the anode in an X-ray tube and measured the wavelength of the characteristic X-rays lines.

   a. When he plotted a graph of wavelength versus the atomic mass of the anode material, he found a smooth curve.

   b. When he plotted a graph of wavelength versus the atomic number of the anode material, he found that a smooth curve did not pass through the points on the graph.

   c. Moseley found that the wavelength of the X-ray lines was related to the light emitted by the hydrogen atom when the electron jumps from the second to the first orbit. (In this case, the hydrogen atom emits light of 121.5 nm.)

   d. The wavelength of an X-ray line is found by dividing 121.5 nm by $(Z - 1)^2$, where Z is the atomic number of the anode material.

14-10 X-ray lines are produced when an electron strikes the anode and an electron in an atom of the anode is ejected from the first orbit. When an electron in the second (or larger orbit) jumps to fill the vacancy in the first orbit, radiation having a wavelength in the X-ray region of the spectrum is emitted.

    a. As the atomic number of the anode material increases, the photon energy of the X-ray increases, because the energy difference between the second and first orbits increases.

    b. As the atomic number of the anode material increases, the wavelength of the X-ray line increases.

Questions for Thought

14-1 How did Hauy and Bravais explain crystal structure?

14-2 Why did Laue expect interference effects to occur when X-rays were aimed at a crystal?

14-3 In what way was Laue prepared for this great discovery?

14-4 Describe the experimental setup used to obtain Laue-type photographs.

14-5 Why doesn't a diffraction grating show interference effects with X-rays?

14-6 Figure 14-5 shows waves emerging at an angle of 45°, producing constructive interference. Figure 14-17 shows an incident beam at 45°, but the direction of the emerging wavelets at 58°. Draw emerging wavelets from atoms 1 and 3 and show that destructive interference occurs for waves emerging from atoms in the *same plane,* unless the waves have the same angle as the incident beam.

14-7 Figure 14-18 shows X-rays reflected by one set of atom planes. Draw two more sets of atom planes and show how the incident beam is reflected by these planes. Use this diagram to explain Laue-type photographs.

14-8 a. Describe the Braggs' X-ray spectrometer.

    b. Explain how they obtained an X-ray spectrum.

    c. Explain how they found the X-ray wavelength.

    d. What unusual features did the X-ray spectra exhibit? How did these features depend upon the X-ray tube?

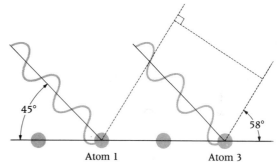

FIG. 14-17 Diagram for thought question 14-6.

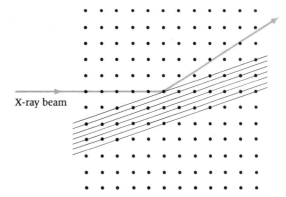

FIG. 14-18 Diagram for thought question 14-7.

14−9 What did Moseley find when he plotted a graph of X-ray wavelength versus the atomic number Z for the element used as the anode in the X-ray tube? When he plotted another graph of wavelength versus atomic mass? What did he conclude?

14−10 Describe the connection between X-ray lines and Bohr's theory of the hydrogen atom.

14−11 Describe the origin of X-ray lines.

Questions for Calculation

14−1 **a.** Consider a 30,000-eV X-ray photon entering an ionization chamber. If 30 eV is required to ionize a molecule of gas in the ionization chamber, how many molecules does this X-ray photon ionize?
b. If the X-ray intensity is defined as the number of photons per second entering the ionization chamber, show that the current from the chamber is proportional to the X-ray intensity.

†14−2 **a.** Explain how X-rays are produced in an X-ray tube.
b. Explain why the wavelengths of X-rays have a continuous range of values.
c. If the voltage between cathode and anode of an X-ray is 40,000 volts, what is the maximum frequency and minimum frequency of an X-ray photon? What is the minimum wavelength?

14−3 Refer to Fig. 14−5 and write an equation, similar to Equation 1, that governs the destructive interference between wavelets emerging from the atoms.

14−4 The distance D in Fig. 14−10 is the distance between adjacent sodium and chlorine ions.

a. Use the Pythagorean theorem to show that the shortest distance between two sodium ions and between two chlorine ions is $\sqrt{2}D$.
b. Show that the distance between atom planes of set B is $D/\sqrt{2}$.

Footnotes

[1] From "X-ray Crystallography," Sir Lawrence Bragg. *Scientific American* (July 1968):58. Copyright © 1968 by Scientific American, Inc. All rights reserved.

[2] Niels H. de V. Heathcote, *Nobel Prize Winners in Physics, 1901–1950* (New York: Henry Schuman, Inc., 1953), p. 120.

[3] J. D. Stranathan, *The Particles of Modern Physics* (Philadelphia: The Blakiston Company, 1942), p. 263.

[4] Heathcote, p. 121. (© The Nobel Foundation 1920.)

[5] Heathcote, p. 134. (© The Nobel Foundation 1923.)

[6] W. H. Bragg, *Nature* (January 23, 1913).

[7] F. K. Richtmyer, E. H. Kennard, and T. Lauritsen, *Introduction to Modern Physics* (New York: McGraw-Hill, 1955), p. 363.

[8] Heathcote, p. 134. (© The Nobel Foundation 1923.)

[9] Excerpt from *Rutherford and the Nature of the Atom* by E. N. da C. Andrade. (New York: Doubleday, 1964), p. 130. (Copyright © 1964 by Doubleday & Company, Inc. Reprinted by permission of the publisher.)

[10] H. G. J. Moseley, "The High-Frequency Spectra of the Elements," *Philosophical Magazine* 27(1913):1024.

[11] Emilio Segre, *From X-rays to Quarks* (San Francisco: W. H. Freeman and Co., 1980), p. 133.

[12] From "X-ray Crystallography," Sir Lawrence Bragg. *Scientific American* (July 1968):70. (Copyright © 1968 by Scientific American, Inc. All rights reserved.)

[13] J. D. Watson, *Molecular Biology of the Gene*, 3rd ed. (New York: W. A. Benjamin, Inc., 1976), p. 209.

[14] J. D. Watson, *The Double Helix* (New York: Antheneum, 1968), p. 194.

Introduction

In 1924, about 28 years after the discovery of radioactivity, Paris was the scene of yet another epoch-making discovery: Louis de Broglie, a student at the University of Paris, wrote a doctoral thesis proposing that electrons behave like waves. His theory completed the symmetry: Light has a particle nature as well as a wave nature and, now, de Broglie claimed, particles also have a wave nature. Two years later, Erwin Schroedinger developed a wave equation capable of describing the wave nature of particles in much greater detail. In 1927, two experiments demonstrated the wave nature of particles. Just as interference effects demonstrate the wave nature of light and X-rays, so do they demonstrate the wave nature of particles.

In this chapter, we'll examine de Broglie's theory, as well as the comprehensive theories of Schroedinger, Born, and Heisenberg, which followed. We'll also examine some of those ingenious experiments that proved these theories correct. These new theories provided marvelous insight into the structure of the atom, the periodic table of the elements, the nature of alpha decay, and they made possible technological applications, such as the electron microscope.

Erwin Schroedinger

CHAPTER 15

The Wave-Particle Duality and the Structure of the Atom

De Broglie: Electrons Have a Wave Nature

The investigations of the French physicist Louis de Broglie (b. 1892) were prompted in part by conversations with his older brother Maurice, an experimentalist. The Paris laboratory of Maurice concentrated on X-ray studies, and it was here that Louis became initially interested in experimental research and focused, with his brother's guidance, on the dual nature of particles and waves. In his autobiographical notes, Louis de Broglie wrote:

I had long discussions with my brother on the interpretation of his beautiful experiments on the pho-

FIG. 15-1 Louis de Broglie. (AIP Niels Bohr Library)

toelectric effect and corpuscular (particle) spectra. . . . These long conversations with my brother about the properties of X-rays . . . led me to profound meditations on the need of always associating the aspect of waves with that of particles.[1]

De Broglie received the Nobel Prize in 1929. In his Nobel lecture, he describes the dual nature of light and X-rays, and how he arrived at the conclusion that electrons, too, must have a wave nature:

When I began to think about these difficulties, two things struck me particularly. On the one hand, the theory of light quanta could not be considered satisfactory, for it defined the energy of a light-corpuscle by the relation $E = hf$,* in which the frequency f occurs. Now a purely corpuscular theory does not contain any element permitting a frequency to be defined. If only for this reason, it is necessary in the case of light to introduce simultaneously the idea of a corpuscle and the idea of periodicity. On the other hand, the determination of the stable motions of the electron in the (hydrogen) atom brings in whole numbers, and so far the only phenomena in physics where whole numbers come in are the phenomena of interference[†]. . . . This gave me the idea that electrons, too, could not be represented as simple corpuscles, but that to them also must be attributed a periodicity.

Thus I arrived at the following general idea, which has guided my researches: For matter, just as . . . for . . . light, we must introduce at one and the same time the corpuscle concept and the wave concept. In other words, in both cases we must assume the existence of corpuscles accompanied by waves. But corpuscles and waves cannot be independent since, according to Bohr, they are complementary to each other; consequently it must be possible to establish a certain parallelism between the motion of a corpuscle and the propagation of the wave which is associated with it. The first thing to be done, then, was to establish this correspondence.[2]

*Symbols have been changed to conform to those used in this text.

†See Equation 2 of Chapter 8.

In his doctoral thesis, de Broglie established "this correspondence." The wavelength λ, associated with a particle of mass m moving with speed v, is given by

$$\lambda = \frac{h}{mv} \cdot \qquad (1)$$

How can we obtain this relationship?

First, let's consider light. A photon of light, even though it has no mass, has both energy and momentum. (Remember from our study of the Compton effect that the collision of a photon with an electron is reminiscent of a billiard ball collision. A photon, like a billiard ball, has momentum.) The theory of relativity specifies that the energy E of a photon is related to its momentum p by

$$E = pc. \qquad (2)$$

(This relationship is derived in calculation question 9–13. Let the rest mass equal zero.) Since the photon's frequency f and its wavelength λ are related by f equals c/λ, the energy of a photon is given by

$$E = hf \qquad (3)$$
$$= \frac{hc}{\lambda} \cdot$$

Setting Equation 2 equal to Equation 3, we find

$$pc = \frac{hc}{\lambda},$$

or

$$\lambda = \frac{h}{p} \cdot \qquad (4)$$

Let us find the wavelength of a particle of mass m moving with speed v. Since the momentum p of a particle is given by p equals mv, an equation analogous to Equation 4 is

$$\lambda = \frac{h}{mv} \cdot$$

This is the same as Equation 1.

APPLICATION TO BOHR MODEL

In Bohr's model of the hydrogen atom, the requirement for an allowed orbit is

$$mvr = \frac{nh}{2\pi},$$

where mvr is the angular momentum of the electron and n can only be an integer. De Broglie showed that this same relationship could be found starting with Equation 1 and using the notions of constructive and destructive interference that apply to all types of waves. The allowed orbits correspond to constructive interference of the wave associated with the electron; all other orbits correspond to destructive interference.

In Fig. 15–2, consider the electron making two complete revolutions around the circular orbit. On the second trip, the wave must have exactly the same phase as the wave from the first trip in order for constructive interference to occur, as shown in Fig. 15–2(a). Constructive interference occurs when the circumference is a whole number of wavelengths. Figure 15–2(b) shows that destructive interference occurs when the circumference is a half number of wavelengths. On the first trip around, a peak occurs near the end of the circumference. On the start of the second trip, a trough must occur. This produces destructive interference when the circumference is a half number of wavelengths. When the circumference is equal to other fractions of a wavelength, the many waves from the many trips around the circular orbit will also produce destructive interference.

The requirement for constructive interference is

$$2\pi r = n\lambda, \qquad (5)$$

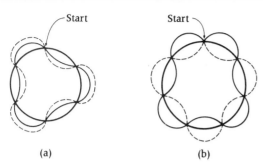

(a)   (b)

FIG. 15–2 (a) The third allowed orbit where the circumference is equal to three wavelengths. The solid line indicates the waves from the first revolution around the circular orbit. The dashed line indicates the waves from the second revolution. The peaks of the "solid" and "dashed" waves overlap, so that constructive interference occurs. (b) This orbit, slightly larger than the third orbit, is not allowed because the circumference is equal to 3.5 wavelengths. Destructive interference occurs because the peaks of the "solid" waves from the first trip around the circumference overlap the troughs of the "dashed" waves from the second trip.

FIG. 15–3 Lester Germer, holding glass tube containing the nickel target, and Clinton J. Davisson. (AIP Niels Bohr Library)

where $n = 1, 2, 3, \ldots$ With the circumference $2\pi r$ identified as the path difference, Equation 5 is the same as Equation 2 of Chapter 8 (see p. 173), which governs the constructive interference of waves from two sources. The integer n in Bohr's theory led de Broglie to recall this equation and to associate waves with this electron. Since λ equals h/mv, Equation 5 becomes

$$2\pi r = \frac{nh}{mv},$$

or   (6)

$$mvr = \frac{nh}{2\pi}$$

This is exactly the angular momentum restriction stated in Bohr's third postulate!

WAVELENGTH OF OBJECTS

Wavelength of an Electron An electron having a kinetic energy of 54 eV has a speed of 4.36×10^6 m/sec. The mass of an electron is 9.11×10^{-31} kg; h equals 6.63×10^{-34} joules-sec. (Remember that a joule is a unit of energy, and that 1 joule is shorthand for 1 kg-m^2/sec^2; h equals 6.63×10^{-34} kg-m^2/sec.) Then,

$$\lambda = \frac{6.63 \times 10^{-34}\,\text{kg-m}^2/\text{sec}}{(9.11 \times 10^{-31}\,\text{kg})(4.36 \times 10^6\,\text{m/sec})}$$
$$= 1.67 \times 10^{-10}\,\text{m}$$
$$= 0.167\,\text{nm}.$$

Wavelength of a Golf Ball A 45-g golf ball moving with a speed of 30 m/sec has a wavelength given by

$$\lambda = \frac{6.63 \times 10^{-34}\,\text{kg-m}^2/\text{sec}}{(0.45\,\text{kg})(30\,\text{m/sec})}$$
$$= 4.9 \times 10^{-34}\,\text{m}$$
$$= 4.9 \times 10^{-25}\,\text{nm}.$$

The wavelength associated with the golf ball is incredibly small. The wavelength of the electron is comparable to that of X-rays and to the spacing of atoms in a crystal. Chapter 14 taught us that X-rays were diffracted by crystals precisely because the wavelength of X-rays and the distance between atoms of a crystal were about the same size.

On November 29, 1924, de Broglie presented his doctoral thesis "Recherches sur le theorie des Quanta" to the Faculty of Science at the University of Paris. . . . Although the examining committee highly appraised the originality of the work, it did not believe in the physical reality of the newly proposed waves. When asked by Perrin whether these waves could be experimentally verified, de Broglie replied that this should be possible by diffraction experiments of electrons by crystals.[3]

In 1925, W. Elsasser, a German scientist, suggested the same experiment, but these suggestions were not immediately heeded.

Davisson-Germer Experiment

In 1927, the experiments of American physicist Clinton J. Davisson (1881–1958) and his associate, Lester Germer (1896–1971), at Bell Telephone Laboratories, clearly demonstrated the wave nature of electrons (see Fig. 15–3). Davisson recalls:

It would be pleasant to tell you that no sooner had Elsasser's suggestion appeared than the experiments were begun in New York which resulted in a demonstration of electron diffraction—pleasanter still to say that the work was begun the day after copies of de Broglie's thesis reached America. The true story contains less of perspicacity and more of chance![4]

In 1925, Davisson and Germer were investigating the reflection of a beam of electrons from a nickel target when an accident occurred. Their apparatus was contained in an evacuated glass bulb, which was accidentally broken. The air caused the nickel target to oxidize and blacken. Germer relates:

Here again good fortune intervened. The oxidized and blackened (nickel) reflector might very well have been thrown away and a shiny new one substituted for it. The course of the experiment was otherwise. It was decided to clean the oxide from the surface by heating it for a long time in hydrogen.[5]

A nickel crystal consists of the atoms arranged in a cubic structure. Ordinary nickel, such as that used for their reflector, is a **polycrystalline** material, which means that there are many very small crystals oriented *randomly*. However, the prolonged heating of the nickel reflector caused the crystals to rearrange, so that it consisted of only a few *large* crystals. Figure 15–4(a) shows the experimental apparatus. The number of electrons per unit time scattered to angle ϕ was measured. The detector was moved to many scattering angles. When the experiment was continued, Davisson and Germer found that the distribution in angle of electrons scattered from the nickel was *quite different* from that found before the accident. In fact, at some angles, the number of electrons was *very much* larger than before the accident. There were now "peaks" in the data. What was the cause of these dramatic effects? Davisson recalls:

Thus the New York experiment was not, at the inception, a test of the wave theory. Only in the summer of 1926, after I had discussed the investigation in England with Richardson, Born, Franck and others, did it take on this character.[6]

We have learned that the wavelength of light can be determined by using a diffraction grating (Chapter 8) and that the wavelength of an X-ray can be determined by using a crystal (Chapter 14). In each case, the angle of scattering determines the wavelength. The distance between atoms in a crystal plays the same role as the distance between slits in a diffraction grating.

If electrons really have a wave nature and obey de Broglie's relationship in Equation 1, then

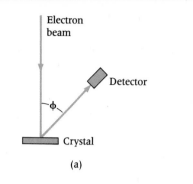

Electron beam

Detector

φ

Crystal

(a)

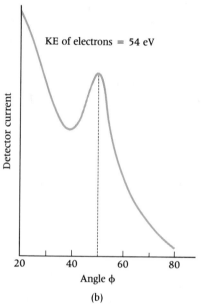

KE of electrons = 54 eV

Detector current

20   40   60   80

Angle φ

(b)

FIG. 15–4   A schematic diagram of the Davisson-Germer experimental apparatus. The apparatus was enclosed in a glass tube and evacuated to an extremely low pressure—100 billion times smaller than atmospheric pressure. The detector current measures the relative number of electrons scattered to angle φ. (b) A graph showing detector current versus scattering angle φ, defined in (a).

the scattering of electrons by a crystal should be no different than the scattering of X-rays.

Figure 15–4(b) shows how electrons, having a kinetic energy of 54 eV, are scattered by a nickel crystal. A peak, attributed to the constructive interference of electron waves, appears at 50°. Using the known distance between atoms in the nickel crystal and the angle of 50°, we can find the de Broglie wavelength to be 0.165 nm. The following "example" section displays the details of this calculation.

In the last section, we calculated the wavelength of 54-eV electrons using de Broglie's relationship (Equation 1) to be 0.167 nm. The agreement between theory and experiment is surely excellent!

When the kinetic energy of the electron beam was changed, the angle of peak changed as well. In all, Davisson and Germer found 19 cases where the wavelength extracted from the scattering experiments agreed very well with de Broglie's theoretical value.

DETERMINING THE DE BROGLIE WAVELENGTH: AN EXAMPLE

Experiments with X-rays showed that a nickel crystal had a cubic structure and that the distance between atoms is 2.15×10^{-10} m.

In the Davisson-Germer experiment, the electron beam struck the nickel surface at a right angle. The reinforcement occurred at 50°, as shown in Fig. 15–5. The atom planes responsible for this scattering are likewise shown in this figure. The distance between these atom planes is 0.909×10^{-10} m. The Bragg angle θ is 65°.

According to Equation 1 of Chapter 14, constructive interference occurs when

$$\lambda = ABC, \qquad (7)$$

where ABC is defined in Fig. 14–5 of Chapter 14.

According to de Broglie's theory, electrons behave like waves and should act just like X-rays striking a crystal. We should therefore be able to determine the de Broglie wavelength from the angle for constructive interference and the distance between atom planes, as we did for X-rays in Chapter 14. Figure 15−6 shows how to determine the distance ABC by drawing a triangle to scale:

$$ABC = 1.64 \times 10^{-10} \text{ m.}$$

Using Equation 7, we find that the de Broglie wavelength is 1.64×10^{-10} m, or 0.164 nm. (Some inaccuracy is introduced by drawing the triangle to scale.)

Davisson and Germer succeeded where others had failed . . . G. P. Thomson, who did find that evidence by a very different method, testified to the magnitude of the technical achievement as follows:[7] [David and Germer's work] *was indeed a triumph of experimental skill. The relatively slow electrons* [they] *used are most difficult to handle. If the results are to be of any value the vacuum has to be quite outstandingly good. Even now (1961) . . . it would be a very difficult experiment. In those days it was a veritable triumph. It is a tribute to Davisson's experimental skill that only two or three other workers have used slow electrons successfully for this purpose.*'[8]

Diffraction of Electron Waves by Foils

In 1927, English physicist George P. Thomson (1892−1975), J. J. Thomson's son, demonstrated the wave nature of electrons in a completely different way than Davisson and Germer. He directed a beam of electrons toward a metallic foil so thin that electrons could pass through it. Behind the foil, he placed a photographic plate to see how the electrons were scattered. This experiment is similar to Laue's experiment with X-rays (see Chapter 14, Fig. 14−3). Figure 15−7 shows the striking similarity between electrons and X-rays. Without a doubt, X-rays and electrons both have a wave nature. Because the metal foils were polycrystalline (that is, many small crystals randomly oriented), you see concentric rings showing constructive interference rather than dots found using a single crystal.

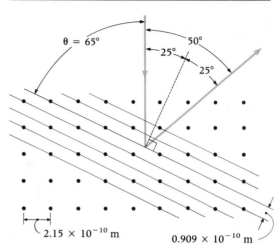

FIG. 15−5 The scattering of electrons by nickel crystal. An intense beam occurs at 50° from the incident electron beam. The atom planes responsible for their reflection are indicated.

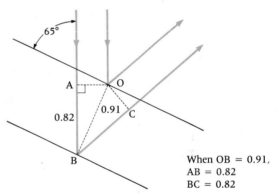

When OB = 0.91,
AB = 0.82
BC = 0.82

FIG. 15−6 A right triangle drawn to scale to find the distance ABC.

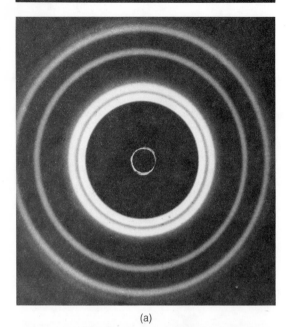

(a)

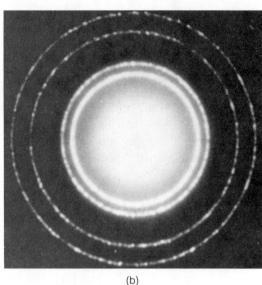

(b)

FIG. 15-7 (a) Diffraction pattern produced by electrons having a de Broglie wavelength of 0.05 nm passing through an aluminum foil. (b) Diffraction pattern produced by X-rays of wavelength 0.071 nm passing through an aluminum foil. (Courtesy of Education Development Center, Newton, Ma.)

C. J. Davisson and G. P. Thomson shared the Nobel Prize in 1937 for demonstrating the wave nature of electrons.

The Electron Microscope

The **electron microscope** was used to obtain the photograph of uranium atoms in Fig. 15-8. The electron microscope is also an invaluable tool in the field of biology. It is, for example, capable of resolving individual virus molecules about 20 nm in diameter. An ordinary light microscope is not capable of defining such fine details. The reason for the electron microscope's capability has to do with diffraction. Turn to Fig. 8-7 (Chapter 8, p. 172), which shows the circular diffracted waves produced by water waves in a ripple tank striking a small peg. Imagine that the peg is an object being viewed by light waves in a microscope and you can see the problem. The object appears very large because of the diffracted waves. Diffraction effects, which occur when the object and wavelength are on the same order, limit the effectiveness of the light microscope. Diffraction, however, *does not* occur when the object being viewed is much larger than the wavelength of the waves striking it. This is the reason the electron microscope was invented.

In a typical electron microscope, the electron beam is accelerated by a voltage of 75,000 volts. The de Broglie wavelength of these electrons is only 0.004 nm, about 100,000 times smaller than the wavelength of visible light. Because the de Broglie wavelength is so small compared to the size of the virus molecules (20 nm), diffraction effects do not come into play.

The basic principles of an electron microscope resemble those of a light microscope. Because electrons are influenced by electric and magnetic fields, they are focused by electrostatic and magnetic lenses, just as light waves in an ordinary microscope are focused by glass lenses. In both cases, the lenses serve to magnify the object being examined. In an electron microscope, the lenses focus the electron beam onto a fluorescent screen, so that the images can be viewed and photographed.

Schroedinger's Wave Equation

Erwin Schroedinger (1887–1961) carried out his monumental work at the University of Zurich in 1926. He had given a colloquium on the then-recent work of Louis de Broglie and this preparation set in motion his research, which culminated in his wave equation.[9] Only a few months after his colloquium, his paper on the wave equation appeared in the same journal that had published the revolutionary discoveries of Planck and Einstein.

There are many wave equations in physics, and Schroedinger turned to them for inspiration. For example, a wave equation describes the vibration of a taut wire, such as a plucked violin string. Starting with the initial shape of the wire, the wave equation specifies the amplitude of the wire at any point on it at a given time.

In Chapter 8, we considered light—electromagnetic waves. The wave equation for a light wave specifies the electric field E and the magnetic field B at any point in space at a specified time. Figures 8–30 and 8–31 (Chapter 8, p. 190) show how an electromagnetic wave travels through free space. The wave equation describes this behavior. But the wave equation does more than this—it describes more complicated situations; for example, the reflection and refraction that occurs when the light beam strikes a surface.

Similarly, Schroedinger developed a wave equation for a particle, in which the symbol Ψ (the capital Greek letter psi) specified the height of the wave at any point in space. The symbol Ψ is called the **wave function.**

For a free particle traveling with constant speed v, Schroedinger's wave equation yielded the de Broglie wavelength (Equation 1), as it certainly must do to be a valid equation. More importantly, however, this wave equation also described the wave for a particle with a force

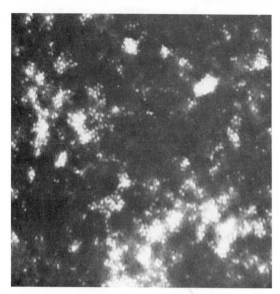

FIG. 15–8 The electron microscope reveals individual uranium atoms. (Courtesy of Albert Crewe, University of Chicago)

FIG. 15–9 Erwin Schroedinger. (AIP Niels Bohr Library)

347

acting on it—something de Broglie didn't do. The force acting on a particle causes its kinetic energy and potential energy to change, but their sum must remain constant:

$$E = KE + PE. \qquad (8)$$

This conservation of energy principle is imbedded in (and is, in fact, the very fabric of) Schroedinger's wave equation.

THE HYDROGEN ATOM

Immediately, Schroedinger solved the wave equation for the hydrogen atom, using the potential energy of the electron, given by

$$PE = \frac{-Ke^2}{r}.$$

He found the allowed energy levels to be the same as Bohr's:

$$E = \frac{-13.6\,eV}{n^2}. \qquad (9)$$

However, also emerging from the wave equation were two other **quantum numbers** l and m, which, like n, are integers. The values of l and m are governed by n, and are given by

$$l = 0, 1, 2, 3, \ldots, n-1 \qquad (10)$$
$$m = -l, \ldots -2, -1, 0, 1, 2, \ldots +l. \qquad (11)$$

For example, when $n = 3$, l can equal 0, 1, or 2. If $l = 2$, then m can be -2, -1, 0, 1, or 2.

While n specifies the energy E, it *does not* specify the angular momentum as it did in Bohr's theory, where the angular momentum was given by $nh/2\pi$. According to Schroedinger's theory, the angular momentum of the electron is given by

$$\text{Angular momentum} = \sqrt{l(l+1)}\,\frac{h}{2\pi}.$$

Thus, $\sqrt{l(l+1)}$ replaces the integer n in the description of angular momentum. When $l = 2$, the angular momentum is 2.45 $h/2\pi$.

Quite puzzling was the height of the wave Ψ. What did it mean? Scientists discussed this at length. In 1926, the German physicist Max Born (1882–1970), a professor at the University of Gottingen, provided the interpretation accepted today:

$$\begin{array}{l}\text{Probability of finding a}\\ \text{particle at a given point} = \Psi^2. \qquad (12)\\ \text{in space}\end{array}$$

Ψ is the height of the wave at that point.

According to the Bohr model of the atom, the electron is found in only certain well-defined orbits and nowhere in between. The Schroedinger-Born theory completely demolished the notion of orbits and replaced it with the notion of **probability.** The electron can be found at any distance from the nucleus, and Ψ^2 gives the probability of finding it there. The following "example" section illustrates this idea.

THE PROBABILITY OF FINDING ELECTRONS: AN EXAMPLE

Consider 1000 hydrogen atoms having $n = 1$, $l = 0$, $m = 0$, and $E = -13.6$ eV. Figure 15–10 compares the Schroedinger-Born theory with Bohr's theory.

In Fig. 15–10(a), we see that the electron of each atom has the radius r_1, the radius of the first allowed orbit in Bohr's theory of the hydrogen atom. Figure 15–10(b) shows a series of 12 equally spaced concentric circles covering the distance $r = 0$ to $r = 6r_1$. We must extend this diagram to a three-dimensional point of view and imagine 12 nested *spheres* centered on the nucleus. The number between two consecutive spheres indicates the number of hydrogen atoms with electrons between these two imaginary spheres. For example, of the 1000 hydrogen atoms, 253 electrons are between r_1 and 1.5 r_1; 113 electrons are between $2r_1$ and $2.5r_1$, and so on.

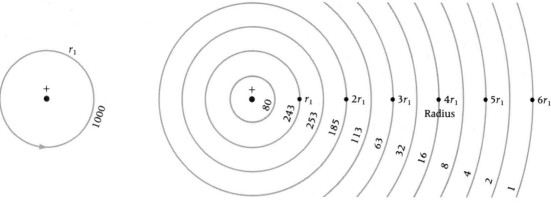

(a) BOHR MODEL   (b) WAVE NATURE DESCRIPTION

FIG. 15–10   A comparison of Bohr's theory with the Schroedinger-Born theory. (a) Bohr's orbit model. (b) Wave-nature model for 1000 hydrogen atoms in the ground state, $n = 1$; r_1 is the radius of the first allowed orbit in Bohr's theory.

Next, we consider *one* hydrogen atom and ask, What is the *probability* that the electron will be found between two imaginary spheres in Fig. 15–10(b)? The probability that the electron will be between $r = 0$ and $r = 0.5r_1$, is 80 in 1000, or 0.08; between $r = 0.5r_1$ and r_1, the probability is 0.243; and so on. Note that if you sum all of these probabilities, the total probability is 1. This means that the electron is always found somewhere between $r = 0$ and $r = 6r_1$. The probability that the electron has a radius larger than $6r_1$ is less than 1 in 1000 and is taken as zero in this example. Figure 15–11(a) shows a graph of probability versus radius, and a smooth curve drawn through these probability values.

Now, picture two imaginary spheres centered on the nucleus: one having a radius r and the other having a radius just slightly larger, $r + \Delta r$. For example, let $r = r_1$ and $r + \Delta r = 1.001r_1$. Figure 15–11(b) shows the probability of finding the electron in a hydrogen atom between the radius r and $r + \Delta r$. Curves are shown for three different sets of n and l quantum numbers.

The curve for $n = 1$ and $l = 0$ has the same shape as the dashed curve in Fig. 15–11(a). We see that the probability is largest between $r = r_1$ and say, $r = 1.001r_1$, an interesting similarity to the Bohr model. While the electron can be found anywhere between $r = 0$ and $r = 6r_1$, it is most likely to be found at r_1. The determinism of the Bohr model has been replaced by a statistical model.

For hydrogen's second energy level, $n = 2$ and $E = -3.4$ eV. As the restriction on the quantum numbers in Equations 10 and 11 shows, l can be either 1 or 0. Figure 15–11(b) shows that a unique probability distribution is associated with each set of quantum numbers. For $n = 2$ and $l = 1$, the probability is largest for r equal to $4r_1$—the same radius as the second allowed Bohr orbit. Note also that the electron with n equal to 2 will often have a radius larger than $6r_1$.

The curves shown in Fig. 15–11(b) are obtained by solving Schroedinger's wave equation for the hydrogen atom. Born provided the statistical interpretation, which is illustrated, for example, in Fig. 15–10.

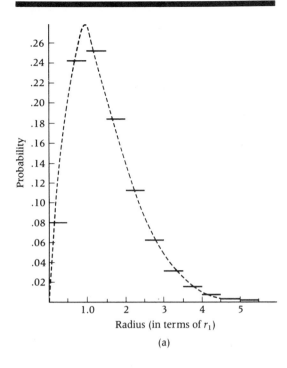

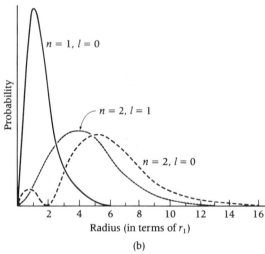

FIG. 15–11   (a) The probability of finding an electron of the hydrogen atom between r and $r + 0.5$ versus r. (b) The probability of finding an electron between r and $r + \Delta r$ versus r.

Quantum Numbers and the Periodic Table

Schroedinger, in fact, was not the first to introduce the quantum numbers n, l, and m. Scientists had introduced them earlier to "patch up" Bohr's theory and to understand the spectra of chemical elements. By 1925, these quantum numbers and a fourth one, called the *spin*, were used to explain the structure of the periodic table of the elements (see Fig. 1–3, p. 6). Let's look at these developments.

As diffraction gratings improved (see Chapter 8), so did the resolution of the spectra. In the hydrogen spectrum, what had originally appeared as one line, split into several closely spaced lines. How could Bohr's theory be extended to account for this so-called fine structure?

In 1915, Arnold Sommerfeld (1868–1951), professor of theoretical physics at Munich, tackled this problem. In Bohr's theory, light is emitted when the electron jumps from one energy level to another of lower energy. Since the spectral lines have several components, the energy levels must, likewise, have several closely spaced components. To explain this so-called fine structure, Sommerfeld introduced the quantum number l. In the Bohr-Sommerfeld theory, the electron moves in an elliptical orbit, and the quantum numbers n and l specify its shape. The quantum number l was restricted to values $l = 0, 1, 2, 3 \ldots n - 1$, which is the same as Equation 10. In addition, each set of (n, l) quantum numbers gave a different energy level. For example, the $n = 3$ energy level in Bohr's theory split into three levels, one for each permitted value of l: 2, 1, and 0.

Another complication arose when the light source for the spectrum (hydrogen or any other element) was placed in a magnetic field: The spectral lines split into several lines, a phenomenon called the **Zeeman effect,** after Pieter Zeeman (1865–1943), a Dutch physicist who first observed this effect for sodium in 1896.

In 1916, Sommerfeld and the Dutch-born physicist Peter Debye (1884–1966) introduced a third quantum number m to account for the

Zeeman effect. The quantum number m was restricted to values given by: $m = -l, \ldots, -2, -1, 0, 1, 2, \ldots, + l$, which is the same as Equation 11. In a magnetic field, each set of quantum numbers (n, l, m) gave a different energy level. For example, the $n = 3$ energy level splits into nine levels.

Figure 15–12 shows the meaning of these quantum numbers. The angular momentum of the electron is represented by an arrow perpendicular to the plane of the orbit, where the length of the arrow indicates the amount of angular momentum. The projection of this arrow on the vertical direction (defined by the magnetic field in that direction) must equal $mh/2\pi$. When m is positive, the angular momentum arrow is pointed upward; when negative, downward.

The Schroedinger theory replaced the notion of orbits with the idea of probability, as we have learned. In the hydrogen atom, the electron can be found at any distance from the nucleus. The quantum number n dictates the electron's energy, and the quantum numbers n, l, and m describe the probability of finding the electron at any given point in space. However, we also saw that remnants of Bohr's theory remained in Schroedinger's theory. Recall that when $n = 1$, the electron was *most likely* to be found at a radius equal to that of the first allowed orbit. When $n = 2$ and $l = 1$, the electron was most likely to be found at the radius of the second allowed orbit. Similarly, there is a grain of truth in the diagrams of Fig. 15–12. When $l = 2$ and $m = 0$, the probability of finding the electron at a specified radius is *largest* in the orbit plane shown in part (c). When $l = 2$ and $m = \pm 2$, there is a large probability of finding the electron at a specified radius in the orbit planes shown in part (a) and part (d). Thus, while we must now consider probabilities rather than well-defined orbits to describe the electron in the hydrogen atom, the diagrams in Fig. 15–12 give us a mental image of where the probabilities are large.

In 1925, two Dutch physicists, George Uhlenbech (b. 1900) and Samuel Goudsmit (1902–1978) proposed still another quantum number: **electron spin.** As the electron revolves around

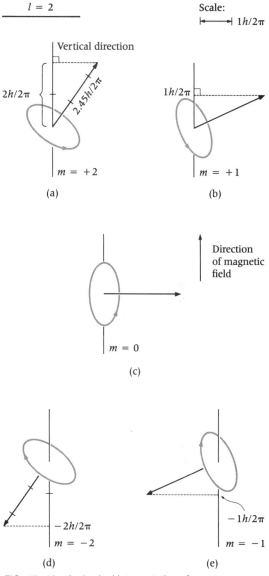

FIG. 15–12 A physical interpretation of quantum numbers l and m. When $l = 2$, the angular momentum, given by $\sqrt{l(l + 1)}\, h/2\pi$, is 2.45 $h/2\pi$. The angular momentum is represented by an arrow perpendicular to the plane of the orbit. The scale shows a length representing an angular momentum of $h/2\pi$. The quantum number m specifies the projection of the arrow on the vertical axis. When $m = 2$, the projection has a length equal to $2h/2\pi$. Thus, the values of m indicate the orientation of the orbit plane in space.

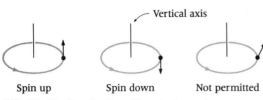

FIG. 15–13 Quantization of the electron spin.

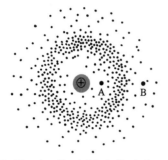

FIG. 15–14 The density of dots indicate the probability of finding $n = 1$ electrons of lithium at that location.

the nucleus, the electron spins on its axis, just as the earth spins on its axis once a day as it revolves around the sun. The direction of the "spin" is pointed either up or down in the vertical direction (Fig. 15–13). The electron spin is quantized.

In the same year, Austrian-born Wolfgang Pauli (1900–1958) realized that there was a connection between the periodic table and the four quantum numbers (n, l, m, and spin). He proposed the following principle, now known as the **Pauli exclusion principle:** *No two electrons in an atom can have the same set of quantum numbers* (n, l, m, spin).

Consider the helium atom, which has 2 electrons, and the lithium atom, which has 3 electrons. (Recall that the number of electrons is equal to the atomic number Z.) When $n = 1$, $l = 0$, and $m = 0$, according to Equations 10 and 11. There are only two sets of quantum numbers: $(1,0,0,+)$ and $(1,0,0,-)$, where $+$ indicates spin up and $-$, spin down. The 2 electrons of helium have these two sets of quantum numbers. In lithium, the third electron must have another set, so it must have $n = 2$. When $n = 2$, l can be either 1 or 0. The following "example" section shows why the third electron has $l = 0$.

DETERMINING WHY THE $n = 2$ ELECTRON IN HELIUM HAS $l = 0$: AN EXAMPLE

Here, we want to see why the third electron in helium has $n = 2$, $l = 0$, rather than $n = 2$, $l = 1$. To solve this problem, we must consider the most probable location of the first two electrons, as shown by Fig. 15–14. The density of dots indicates the probability of finding the electrons at that location. If the third electron is located at point A, for example, then it "sees" the three positive charges of the lithium nucleus. But, if the third electron is at point B, then it "sees" the 3 positive charges of the nucleus *and* the 2 negative electrons, yielding a net charge of only 1 positive charge. Thus, the electron at point A experiences a larger force than at point B due to the larger net charge and also due to its closer distance to the nucleus.

We know that the n and l quantum numbers govern the probability of finding the electron at a given location. We now ask, Does $l = 0$ or $l = 1$ have the larger probability of being found at point A? The probability distributions shown in Fig. 15–11(b) are for hydrogen, but lithium would have similar diagrams. Comparing the curves for $n = 2$, when $l = 1$ and $l = 0$, we see that the $l = 0$ curve has a (small) peak at a smaller radius than the $l = 1$ curve. Thus, $l = 0$ more closely corresponds to point A, while $l = 1$ corresponds to point B. So, the third electron will have $l = 0$ because of the larger force associated with that value of l.

In lithium, 2 electrons have quantum number sets given by: $(1,0,0,+)$ and $(1,0,0,-)$. The third electron has $n = 2$, $l = 0$, and also $m = 0$, but the electron spin can be either up or down. Table 15–1 shows one of many possibilities for the quantum number of electrons in atoms. For lithium, the spin for the third electron is chosen as spin up. Thus, the third (or last) electron has the quantum numbers shown, and the other electrons have the values preceding it in the table.

The 10 electrons of neon complete the second shell. The third shell, beginning with $n = 3$, ends with the element argon. When $n = 3$, l can be either 2, 1, or 0. We see that $n = 3$, $l = 2$ is in the fourth shell, not the third. The reason for this has to do with angular momentum considerations discussed in our last "example" section. An $n = 4$, $l = 0$ electron has more force exerted on it on the average than an $n = 3$, $l = 2$ electron because of the lower angular momentum. So, the nineteenth electron of potassium will have $n = 4$, $l = 0$.

In the fourth shell, you see that the element chromium ($Z = 24$) and copper ($Z = 29$) are not listed. They are somewhat of an anomaly in that they do not follow the sequence in Table 15–1. Rather, chromium has 1 electron with $n = 4$, $l = 0$, and 5 electrons, with $n = 3$, $l = 2$; copper has 1 electron with $n = 4$, $l = 0$ and 10 electrons with $n = 3$, $l = 2$.

Mendeleev, remember, placed the alkali elements lithium, sodium, potassium, rubidium,

TABLE 15–1 POSSIBLE QUANTUM NUMBERS FOR ELECTRONS OF ATOMS

n	l	m	SPIN	ELEMENT	NUMBER OF ELECTRONS	SHELL
1	0	0	+	H	1	1
1	0	0	−	He	2	
2	0	0	+	Li	3	
2	0	0	−	Be	4	
2	1	1	+	B	5	
2	1	1	−	C	6	
2	1	0	+	N	7	2
2	1	0	−	O	8	
2	1	−1	+	F	9	
2	1	−1	−	Ne	10	
3	0	0	+	Na	11	
3	0	0	−	Mg	12	
3	1	1	+	Al	13	
3	1	1	−	Si	14	
3	1	0	+	P	15	3
3	1	0	−	S	16	
3	1	−1	+	Cl	17	
3	1	−1	−	Ar	18	
4	0	0	+	K	19	
4	0	0	−	Ca	20	
3	2	2	+	Sc	21	
3	2	2	−	Ti	22	
3	2	1	+	V	23	
3	2	1	−			
3	2	0	+	Mn	25	
3	2	0	−	Fe	26	
3	2	−1	+	Co	27	
3	2	−1	−	Ni	28	4
3	2	−2	+			
3	2	−2	−	Zn	30	
4	1	1	+	Ga	31	
4	1	1	−	Ge	32	
4	1	0	+	As	33	
4	1	0	−	Se	34	
4	1	−1	+	Br	35	
4	1	−1	−	Kr	36	

and cesium in the same column because they had similar chemical properties. We now see (from Table 15–1) that these atoms are all similar because, in each case, there is 1 electron beyond closed (or filled) shells. Similarly, the inert gases (helium, neon, argon, krypton, xenon

WAVE-PARTICLE DUALITY AND THE ATOM

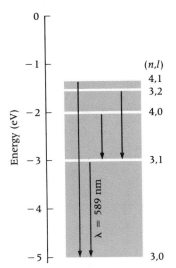

FIG. 15–15   An energy-level diagram of sodium.

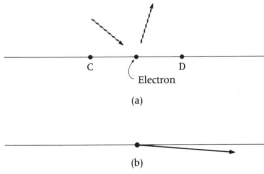

FIG. 15–16   (a) An electron has entered the region between points C and D, where it is struck by a photon, which then enters the observer's eye. (b) The electron's momentum is changed as a result.

and radon) in the last column of the periodic table, all correspond to closed shells. The halogens (fluorine, chlorine, bromine, and iodine) all correspond to 1 electron *missing* from a closed shell.

Molecules are formed when electrons are shared by the atoms. For example, in the sodium chloride molecule, the electron from sodium becomes attached to the chlorine atom, and each atom then has closed shells; similarly, so it is with the compounds KCl, KBr, NaBr, NaI, etc. Oxygen ($Z = 8$) has 2 electrons missing from a closed shell. In water (H_2O), 1 electron from each hydrogen atom is shared by oxygen so that oxygen has a closed shell. The inert gases are inactive or inert because their shells are already closed. Thus, the chemical properties are determined by the number of electrons beyond closed shells. Elements belonging to a column of the periodic table have the same number of electrons beyond closed shells.

EXCITED STATES

Light is emitted by an atom when the electron jumps from an excited state to another of lower energy. The difference in energy of the two orbits is converted into a photon of light. Figure 15–15 shows an energy level diagram for sodium. Sodium has one electron above two closed shells. The lowest energy, or ground state energy, occurs when the electron has $n = 3$, $l = 0$. The next highest energy occurs when the electron has $n = 3$, $l = 1$ and so on, as shown in Fig. 15–15. The yellow light of sodium vapor lamps, frequently used for highway illumination, occurs when the electron jumps between these two states (see Fig. 15–15). Several other transitions are also shown in Fig. 15–15.

Heisenberg Uncertainty Principle

In 1927, Werner Karl Heisenberg (1901–1976), a German physicist who had studied with Max Born and Niels Bohr, showed that there is a limit to how accurately one can measure *simul-*

taneously the position and momentum of a particle. While the **uncertainty principle** applies equally well to an electron and a 1-kg ball, more startling consequences result for the electron.

In Fig. 15–16, the electron (mass m) travels with speed v along the straight line. When it enters the region between points C and D, we want to measure its momentum. In order to observe the electron, at least one photon of light must bounce off it and enter the observer's eye. From Equation 4, the momentum of a photon is given by p equals h/λ. In bouncing off the electron, the photon transfers some of its momentum to the electron. The maximum amount transferred can certainly be no more than the photon's momentum. Therefore, the *change* in the electron's momentum, which we shall call Δp, is *at most* h/λ. Figure 15–16(b) shows that the measurement has disturbed the electron, which consequently changes direction. As a result of the measurement, the electron's momentum in part (b) is given by

$$\text{Momentum after observation} = p \pm \Delta p \qquad (13)$$
$$= mv \pm \frac{h}{\lambda}.$$

This shows that we can measure the electron's momentum to any desired accuracy if we use a very long wavelength. Our goal was not only to measure the electron's momentum, but to locate it between points C and D as well.

The electron diffracts light. This spreading out of the light produces an uncertainty in the electron's position. To understand this, consider 2 electrons separated by the distance λ and illuminated with light of wavelength λ. The diffraction produced by each electron would be so pronounced that we would not be able to tell whether there were one electron or two. That is, the uncertainty in the electron's position Δx is approximately equal to the wavelength of light shining on it:

$$\Delta x = \lambda. \qquad (14)$$

If we want only to measure the electron's position, this shows that *decreasing* the wavelength decreases the uncertainty in position.

When we measure *both* the momentum and the position, the uncertainty in position Δx is decreased by using a short wavelength, but the uncertainty in Δp is increased. If a long wavelength of light is used, then Δx is large, but Δp is small.

Multiplying the uncertainty in position by the uncertainty in momentum, we find the approximate relationship

$$\Delta p \, \Delta x \cong \frac{h}{\lambda} \times \lambda = h.$$

(The symbol $\cong$ is read "is approximately equal to.")

More accurately, the Heisenberg uncertainty principle states that the uncertainty in momentum multiplied by the uncertainty in the momentum caused by the measuring process itself must be greater than or equal to $h/2\pi$, or

$$\Delta p \, \Delta x \geqslant \frac{h}{2\pi}. \qquad (15)$$

As an example, suppose that the electron is traveling with a speed of 1.5×10^6 m/sec when it is somewhere to the left of point C. Our goal is to measure its speed when it is between points C and D, a distance of 10^{-10} m. Since Δx is 10^{-10} m, Equation 15 becomes

$$\Delta p = \frac{h}{2\pi \, \Delta x} = \frac{6.63 \times 10^{-34} \, \text{m}^2\text{-kg/sec}}{2\pi \times 10^{-10} \, \text{m}}$$
$$= 1.05 \times 10^{-24} \, \text{m-kg/sec}.$$

Since Δp equals $m \, \Delta v$,

$$\Delta v = \frac{\Delta p}{m} = \frac{1.05 \times 10^{-24} \, \text{m-kg/sec}}{9.11 \times 10^{-31} \, \text{kg}}$$
$$= 1.15 \times 10^6 \, \text{m/sec}.$$

Carrying out the measurement, we will obtain a speed given by $(1.50 \pm 1.15) \times 10^6$ m/sec. This corresponds to an uncertainty in the measurement of 77 percent. Even with more sophisticated equipment, it is simply not possible to obtain a better measurement than this.

Next, let us consider a 1-kg ball falling from rest. After the ball has traveled a distance of 1 m, its speed is 4.427 m/sec. We want to check this out by measuring its position and speed simultaneously. Our experiment shows that the distance the ball has traveled is between 1.0007 m and 0.9997 m: Δx equals 0.0010 m. The uncertainty in momentum is given by

$$\Delta p = \frac{1.05 \times 10^{-34} \, \text{m}^2\text{-kg/sec}}{2\pi \times 0.0010 \, \text{m}}$$
$$= 1.05 \times 10^{-31} \, \text{m-kg/sec},$$
$$\Delta v = \frac{1.05 \times 10^{-31} \, \text{m-kg/sec}}{1 \, \text{kg}}$$
$$= 1.05 \times 10^{-31} \, \text{m/sec}.$$

The speed is $4.427 \pm (1.05 \times 10^{-31})$ m/sec. Thus, in order to see the consequences of Heisenberg's uncertainty principle, we would have to measure our ball's speed accurately to 31 decimal places. Here, the result of applying the uncertainty principle agrees with our instinct: we will not be able to detect any noticeable effect.

Photons bouncing off the ball disturb it hardly at all, but photons disturb the electron greatly. The uncertainty principle embodies both of these effects. Both the smallness of Planck's constant h and the mass of the electron combine to emphasize the unusual effects that we have just seen for the electron. If Planck's constant h were a much larger value, we would observe the uncertainty principle operating in our daily lives.

In a delightful little book called *Mr. Thompkins in Wonderland,* the author George Gamow[10] explores what would happen if Planck's constant were very large. Box 15–1 displays an extract and a drawing from Gamow's book.

Alpha Decay

In 1928, three physicists, George Gamow and, independently, R. W. Gurney, and E. U. Condon succeeded in explaining alpha decay using Schroedinger's wave equation. It was, to be sure, a resounding triumph for the new quantum theory.

These investigators viewed alpha decay as it is depicted in Fig. 15–17. The alpha particle exists as a separate entity inside the nucleus, striking the nuclear surface many times per second. During one such encounter, the alpha particle escapes. They realized that the alpha particle must have an attractive force keeping it inside the nucleus. In fact, this attractive force must be larger than the repulsion of the alpha particle to the protons' positive charge. The exact nature of this force was not really understood until the Japanese physicist Hideki Yukawa (1907–1981) proposed the nuclear force in 1934: a very strong attractive force between protons and neutrons that acts only when the distance between them is about 10^{-14} m or less.

Due to a net attractive force, the alpha particle had remained inside that nucleus since the formation of the nucleus. How could it be, that, all of the sudden, this force was no longer able to keep the alpha particle inside? Once outside, of course, the alpha particle would move away from the nucleus due to the repulsion of the positive charge on the nucleus.

To try to understand this dilemma, we shall look first at a less abstract situation and examine the role played by Planck's constant h.

In Fig. 15–18, the toy truck is at rest at point B, where its potential energy mgh is 9.4 joules ($m = 1$ kg). We shall assume that no energy is lost to friction, or to any other form. The truck will not be able to cross over either hill, but will

BOX 15–1 QUANTUM JUNGLES

George Gamow explains in the introduction:

The hero of the present stories is transferred, in his dreams, into several worlds . . . where the phenomena, usually inaccessible to our ordinary senses, are so strongly exaggerated that they could easily be observed as the events of ordinary life. He was helped in his fantastic but scientifically correct dream by an old professor of physics . . . who explained to him in simple language the unusual events which he observed . . .

And now to the story of Chapter 8, "Quantum Jungles":

Next morning Mr. Tompkins was dozing in bed, when he became aware of somebody's presence in the room. Looking round, he discovered that his old friend the professor was sitting in the armchair, absorbed in the study of a map spread on his knee.

"Are you coming along?" asked the professor, lifting his head.

"Coming where?" said Mr. Tompkins, still wondering how the professor had got into his room.

"To see the elephants, of course, and the rest of the animals of the quantum jungle. . . . You see this region which I've marked with red pencil on the map? It seems that everything within it is subject to quantum laws with a very large quantum constant. . . ."

Mr. Tompkins agreed to go. On the way they pick up Sir Richard, a famous tiger hunter. The story continues:

The sea journey was nothing remarkable, and Mr. Tompkins scarcely noticed the time until they came ashore in a fascinating oriental city, nearest populated place to the mysterious quantum regions.

"Now," said the professor, "we have to buy an elephant for our journey inland. . . ."

Mr. Tompkins inspected the elephant from all sides; it was a very beautiful, large animal, but there was no marked difference in its behavior from the elephants he had seen in the Zoo. He turned to the professor—"You said that this was a quantum ele-

FIG. 1 Drawing from Chapter 8, "Quantum Jungles," of George Gamow's book, *Mr. Tompkins in Wonderland.*

phant, but it looks just like an ordinary elephant to me . . ."

"You show a peculiar slowness of comprehension," said the professor. "It is because of its very large mass. I told you some time ago that the uncertainty in position and velocity depends on the mass; the larger the mass, the smaller the uncertainty . . . Now, in the quantum jungle, the quantum constant is rather large, but still not large enough to produce striking effects in the behavior of such a heavy animal as an elephant. . . . But I expect that all smaller animals will show very remarkable quantum effects."

"Isn't it nice," thought Mr. Tompkins, "that we are not doing this expedition on horseback? If that were the case, I should probably never know whether my horse was between my knees or in the next valley."

. . .

BOX 15–1 continued

Mr. Tompkins, trying to keep his balance between the elephant's ears, decided to make use of the time by learning more about quantum phenomena from the professor.

"Can you tell me, please," he asked, turning to the professor, "why do bodies with small mass behave so peculiarly, and what is the commonsense meaning of this quantum constant that you are always talking about?"

"Oh, it is not so difficult to understand," said the professor. "The funny behavior of all objects you observe in the quantum world is just due to the fact that you are looking at them." . . . In making any observation of the motion you will necessarily disturb this motion. In fact, if you learn something about the motion of a body, this means that the moving body delivered some action on your senses or the apparatus you are using. Owing to the equality of action and reaction (Newton's Third Law) we must conclude that your measuring apparatus also acted on the body and, so to speak, 'spoiled' its motion, introducing an uncertainty in its position and velocity." . . . .

At this moment a terrible roar filled the air and their elephant jerked so violently that Mr. Tompkins almost fell off. A large pack of tigers was attacking their elephant, jumping simultaneously from all sides. Sir Richard grabbed his rifle and pulled the trigger, aiming right between the eyes of the tiger nearest to him. The next moment Mr. Tompkins heard him mutter a strong expression common among hunters; he shot right through the tiger's head without causing any damage to the animal.

"Shoot more!" shouted the professor. "Scatter your fire all round and don't mind about precise aiming! There is only one tiger, but it is spread around our elephant and our only hope is to raise the Hamiltonian."

The professor grabbed another rifle and the cannonade of shooting became mixed up with the roar of the quantum tiger. An eternity passed, so it seemed to Mr. Tompkins, before all was over. One of the bullets 'hit the spot' . . .

"Who is this Hamiltonian?" asked Mr. Tompkins after things had quieted down.

"Oh!" said the professor, "I am so sorry. In the excitement of the battle I started to use scientific language. . . . (The term) is named after an Irish mathematician, Hamilton, who first used this mathematical form. I just wanted to say that by shooting more quantum bullets we increase the probability of the interaction between the bullets and the body of the tiger. In the quantum world, you see, one cannot aim precisely and be sure of a hit. Owing to the spreading out of the bullet, and of the aim itself, there is always only a finite chance of hitting, never a certainty. In our case we fired at least thirty bullets before we actually hit the tiger." . . .

Passing farther through the quantum land our travellers met quite a lot of other interesting phenomena, such as quantum mosquitoes, which could scarcely be located at all, owing to their small mass, and some very amusing quantum monkeys. Now they were approaching . . . a native village. . . . The ringing of the bells became unbearable to Mr. Tompkins' ears. He stretched his hand out, grabbed something, and then threw it away. The alarm clock hit the glass of water standing on his night table and the cold stream of water brought him to his senses. He jumped up, and started to dress rapidly. In half an hour he must be at the bank.

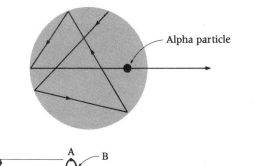

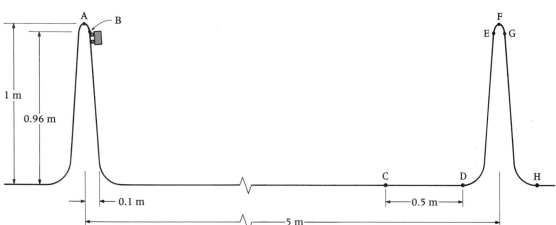

FIG. 15–17 An alpha particle will strike the nuclear surface many times before it escapes.

Alpha particle

FIG. 15–18 Example used to study alpha decay.

travel indefinitely between points B and E. The truck is trapped between B and E.

This example is analogous to alpha decay. As the truck climbs the hill, the force of gravity slows it down. Similarly, as the alpha particle attempts to cross the nuclear surface, the net attractive force slows it down and it cannot escape. Discussing it in terms of energy, we say that, as the truck climbs the hill, its kinetic energy decreases and its potential energy increases. The truck is trapped if its kinetic energy at the bottom of the hill is less than the potential energy at the top of the hill (points A and F). Similarly, an alpha particle is trapped in the nucleus because its energy inside the nucleus is less than its potential energy at the surface. That is, as the alpha particle approaches the nuclear surface, it encounters a barrier, a *potential-energy* barrier, just as the truck approaches a hill, which is also a potential-energy barrier.

Suppose, now, that, when you release the truck at point B, you give it a slight push, so that its kinetic energy at point B is 1 joule. Its total energy there is then 10.4 joules. The potential energy at points A and F is 9.8 joules. Therefore, when the truck climbs the second hill, it will not stop at point E, nor at point F, but will cross over the hill. In this case, the truck will be able to escape.

In this example, we see no way for the truck to escape "sometimes." Either it is trapped, or it escapes. One or the other. It is not trapped for a while (as the alpha particle in the nucleus is) and then after perhaps a thousand years is able to escape (as the alpha particle does).

WAVE-PARTICLE DUALITY AND THE ATOM

To see why the truck behaves so differently from the alpha particle, we are going to use Gamow's device and let *h* be a very large number. From our discussion of the Heisenberg uncertainty principle, recall that large effects occurred for the electron because the ratio h/m was so much larger for the electron than for the 1-kg ball. To make this ratio larger here, we are going to let *h* equal 1 joule-sec. First, we shall develop a plausibility argument* based upon the uncertainty principle and then see the results of a quantative treatment, using Schroedinger's equation.

PLAUSIBILITY ARGUMENT

Suppose that we release the truck from rest at point B. At the bottom of the hill, its kinetic energy is 9.4 joules and the truck has a speed of 4.36 m/sec. (KE = ½(1 kg)(4.36 m/sec)^2 = 9.4 joules.) When the truck is between points C and D, we measure its momentum. The uncertainty in the momentum is

$$\Delta p = \frac{h}{2\pi \Delta x}$$
$$= \frac{1 \text{ joule-sec}}{2\pi(0.5 \text{ m})}$$
$$= 0.32 \text{ kg-m/sec.}$$

The uncertainty in the speed Δv is $\Delta p/m$ equals 0.32 m/sec, since the mass *m* is 1 kg. Thus, the speed is between 4.68 m/sec and 4.04 m/sec. This yields an upper limit on the kinetic energy of ½(1 kg)(4.68 m/sec)^2, or 10.95 joules, and a lower limit of 8.16 joules. The upper limit on the kinetic energy is larger than the potential energy at point F, but the lower limit is less. Therefore, if we repeat this experiment many times, sometimes the truck will cross over the hill and sometimes it won't. This mimics the behavior of the alpha particle in the nucleus.

*A similar argument is found in R. D. Evans, *The Atomic Nucleus* (New York: McGraw-Hill, 1955), p. 61.

PREDICTIONS OF SCHROEDINGER'S EQUATION

Now, we take up the quantitative treatment, using Schroedinger's equation. Since the analogy between alpha decay and our example has been established, we shall hereafter use the term *particle,* for which you can substitute equally well *truck* or *alpha particle.* We shall use the term *barrier* for potential energy barrier in both cases.

Suppose that a particle approaches the right barrier in Fig. 15–18. The solution of Schroedinger's equation, shown in Fig. 15–19, describes the behavior of the particle. The **wavefunction** Ψ_{inc} describes the particle approaching the barrier; Ψ_b, inside the barrier (between points D and H); Ψ_c to the right of the barrier; and Ψ_{refl}, reflected by the barrier. (When light strikes a window pane, some of the light *waves* are reflected and some are transmitted.)

The wavefunctions Ψ_{inc} and Ψ_c move to the right, while Ψ_{refl} moves toward the left. To the right and left of the barrier, the particle is subjected to no forces, and its wavelength, therefore, is simply given by the de Broglie wavelength

$$\lambda = \frac{h}{mv}$$
$$= \frac{1 \text{ joule-sec}}{(1 \text{ kg})(4.36 \text{ m/sec})}$$
$$= 0.23 \text{ m.}$$

Since the wavefunction Ψ_b, obtained by solving Schroedinger's wave equation, is not zero, there is a definite probability of finding that the particle has crossed the barrier. Remember that the probability of finding a particle at any point is given by its wavefunction squared (Equation 12).

Note that the amplitude of the wavefunction Ψ_c is less than Ψ_{inc}. Schroedinger's theory shows that the probability that the particle has crossed the barrier is given by

$$\begin{array}{l}\text{Probability of}\\\text{crossing}\\\text{barrier}\end{array} = \frac{(\text{amplitude of } \Psi_c)^2}{(\text{amplitude of } \Psi_{inc})^2}$$

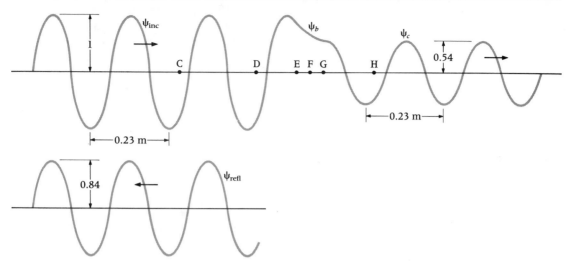

FIG. 15–19 The wavefunction describing a particle's behavior as it encounters a potential-energy barrier. At points E and G, the potential energy is equal to the particle's kinetic energy before it encountered the barrier.

The probability that the particle does not cross the barrier, or that it is trapped, is

$$\text{Probability of being trapped} = \frac{(\text{amplitude of } \Psi_{\text{refl}})^2}{(\text{amplitude of } \Psi_{\text{inc}})^2}$$

In this example, the amplitude of Ψ_{inc} is 1; that of Ψ_c is 0.54; and that of Ψ_{refl} is 0.84. Therefore, the probability that the particle crosses the barrier is $(0.54)^2$, or 0.29; the probability that the particle is trapped behind the barrier is $(0.84)^2$, or 0.71. As the particle approaches the barrier, we cannot say whether it will cross or not, but we can give odds on the results: 29 percent of the time it will cross, and 71 percent of the time it will not. If the particle approaches the barrier 100 times, 29 times it will cross and 71 times it will not. *Probability has replaced determinism.*

Suppose that we carry out the following experiment 100 times. We release the particle from rest at point B in Fig. 15–18 and record the time it takes to cross the barrier. That is, the particle will travel back and forth between points B and E until, one time, when it approaches the barrier, it will cross and escape. Our observations will agree with the predictions of Schroe-dinger's equation. Of the 100 particles approaching the barrier for the first time, 29 of them will cross, leaving 71 trapped. These 71 particles will now approach the other hill and 29 percent of them—21 particles—will cross. The 50 remaining particles will approach the barrier for a third time, and so on. Since the particle has a speed of 4.36 m/sec, it travels the distance of 5 m between the hills in 1.15 sec. Table 15–2 shows how many particles cross on each approach, as well as how long they were trapped behind the barrier.

There are some interesting effects that can be seen from analyzing Table 15–2. In the second column, we see that the number approaching the barrier (or the number trapped) drops by a factor of 2 after 2.30 seconds: 100 to 50, 71 to 36, 18 to 9, etc. Similarly, the number crossing the barrier drops by a factor of 2 after 2.30 seconds: 29 to 14, 14 to 7, 11 to 5, etc. We call this time the **half-life.** *The important conclusion here*

TABLE 15-2 HALF-LIFE BEHAVIOR OF PARTICLES APPROACHING A BARRIER

APPROACH TO BARRIER	NUMBER APPROACHING BARRIER	NUMBER CROSSING BARRIER	TIME BEHIND BARRIER (seconds)
1st time	100	29	1.15
2nd	71	21	2.30
3rd	50	14	3.45
4th	36	11	4.60
5th	25	7	5.75
6th	18	5	6.90
7th	13	4	8.05
8th	9	3	9.20
9th	6	1	10.35
10th	5	2	11.50
11th	3	1	12.65

is that the half-life depends upon the percentage of the particles crossing the barrier. In this example, that percentage was 29 percent, or the probability was 0.29. If the percentage is greater than 29 percent, then the half-life is shorter than 2.30 seconds; if smaller than 29 percent, the half-life is longer. In general, to find the half-life, plot the number approaching the barrier (or the number crossing the barrier) versus time. Find the time required to drop this number by a factor of 2.

We turn now to alpha decay and the great success of Gamow, Gurney, and Condon in applying Schroedinger's equation to this problem. An alpha particle nearing the nuclear surface approaches a potential-energy barrier. Using Schroedinger's equation, these physicists calculated the probability that an alpha particle would be able to cross this barrier. With this number, they could also *calculate* the half-life, much as we have done with the information in Table 15-2. They could then compare this theoretical value of the half-life with the experimental value. The activity is the number of alpha particles per second emitted by the radioactive source. (This is analogous to the third column in Table 15-2.) This half-life is the time required for the activity to drop by a factor of 2. Comparing their theoretical values for the half-life for many radioactive isotopes with experimental values, these investigators found *excellent* agreement. As mentioned, this was a resounding triumph for the new quantum theory: Schroedinger's wave equation was essential in understanding the structure of the atom *and* the nucleus.

Summary

In 1924, de Broglie proposed that particles have a wave nature, where the wavelength is given by λ equals h/mv.

Two experiments demonstrated the wave nature of electrons: (1) Directing an electron beam toward a nickel crystal, Davisson and Germer measured the intensity of electrons scattered to various angles and found one angle where the intensity was very large. This was due to constructive interference of the electrons' waves diffracted by atoms in the crystal. From the distance between crystal planes and the angle, they determined the wavelength, which agreed with the theoretical value obtained from de Broglie's theory. (2) Thomson directed a beam of electrons toward a thin, metal foil. On a film placed behind the foil, he found regions of constructive and destructive interference, just as Laue had observed for X-rays.

The electron microscope utilizes the wave nature of electrons. Since the de Broglie wavelength of these electrons is small compared to the objects being viewed, undesirable diffraction effects are eliminated.

Schroedinger developed a wave equation for a particle, which described the particle's behavior when a force acted on it. Solving this wave equation for the hydrogen atom, he found the same energy levels as Bohr, and introduced the quantum numbers n, l, and m. In contrast to the height of a wave on a plucked violin string, the height of the electron wave Ψ has less tangible physical significance: Ψ^2 gives the probability of finding the electron at a given point in space.

An electron in an atom has four quantum numbers: n, l, m, and spin. As the electron revolves around the nucleus, the electron spins on its axis, pointed either up or down in the vertical direction. According to the Pauli exclusion principle, no two electrons can have the same set of quantum numbers. This limits the number of electrons that can occupy the first shell ($n = 1$) to two and the number in the second shell ($n = 2$) to eight, and so on. The structure of the periodic table of the elements then became evident. In atoms of inert gases, the shells are completely filled, while in the alkali metals (lithium, sodium, potassium, etc.) there is one electron above closed shells.

Heisenberg showed that there was a limit to how closely you could simultaneously measure the position and momentum of a particle. In order for an electron to be observed, a photon of light must bounce off the electron and enter the observer's eye. This causes the electron to recoil, producing an uncertainty in the electron's momentum Δp. The electron diffracts the light, which produces an uncertainty in its position, designated by Δx. According to the Heisenberg uncertainty principle, the measuring process induces uncertainties in momentum and position that are related by $\Delta p \, \Delta x \geq h/2\pi$.

Before 1928, alpha decay had presented a serious dilemma to physicists. According to classical concepts, the alpha particle should be either bound to the nucleus indefinitely, or it should not be found in the nucleus at all. Why should an alpha particle suddenly be emitted? In 1928, scientists, applying Schroedinger's wave equation to an alpha particle existing in the nucleus, explained the basic nature of the decay, *and* showed why this decay should follow the well-known half-life behavior. As it roams around inside the nucleus and approaches the surface, an alpha particle encounters a potential-energy barrier. Classically, the alpha particle should not be able to cross this barrier, just as a particle having a kinetic energy of 94 joules should not be able to cross a hill, where the particle potential energy at the top would be 98 joules. However, according to Schroedinger's theory, the behavior of the alpha particle is governed by its wave nature. This theory showed that there was a small probability that the alpha particle could cross the potential energy barrier, if not on the first approach, then perhaps on the second, or the third, and so on. This probability of "crossing" leads directly to the half-life phenomenon.

True or False Questions

Indicate whether the following statements are true or false. Change all of the false statements so that they read correctly.

15–1 When the speed of an electron increases, the de Broglie wavelength decreases.

15–2 In Bohr's model of the hydrogen atom, the radius of an allowed orbit must be an integral multiple of the de Broglie wavelength associated with the electron.

15–3 Due to its much larger mass, a baseball has a much larger de Broglie wavelength than an electron.

15–4 The experiment carried out by Davisson and Germer for electrons was very similar to the experiment for X-rays carried out by Laue.

15–5 To view virus molecules about 20 nm in diameter with an electron microscope, the de Broglie wavelength of the electrons must be much less than 20 nm.

15–6 Schroedinger's description of the behavior of the electron in the hydrogen atom corresponds to nature more closely than does Bohr's description.

15–7 An electron in a hydrogen atom has $n = 1$, and $l = 0$. According to Schroedinger's description of the hydrogen atom,

---

WAVE-PARTICLE DUALITY AND THE ATOM

a. the electron has energy E equal to -13.6 eV only some of the time.

b. the electron will have a radius between zero and r_1 (r_1 is the radius of the first allowed Bohr orbit) 50 percent of the time.

c. an electron with $n = 1$ will be found more often with a radius between r_1 and $1.5r_1$ than an electron with $n = 2$.

15–8 The following statements concern Schroedinger's description of atoms:

a. The quantum number n specifies the energy E and the angular momentum of the electron.

b. When the quantum number n equals 2, the quantum number l can equal 2, 1, or 0.

c. When the quantum number l equals 1, the quantum number m can equal 1, 0, or -1.

d. The quantum number m indicates the orientation of the orbit plane in space where the electron is most likely to be found.

e. In an atom, 8 electrons can have n equal to 3.

f. According to the Pauli exclusion principle, only two electrons can have the same set of n, l, m, and spin quantum numbers.

15–9 According to Heisenberg's uncertainty principle,

a. the position of an electron can be measured to any desired degree of accuracy.

b. the momentum of an electron can be measured to any desired degree of accuracy.

c. the uncertainty in the electron's position multiplied by the uncertainty in the electron's momentum can be no smaller than $h/2\pi$.

15–10 An alpha particle in the nucleus approaches the nuclear surface.

a. Schroedinger's wave equation describing the behavior of the alpha particle indicates whether or not the alpha particle will escape from the nucleus on that approach.

Questions for Thought

15–1 Explain why the photoelectric effect exhibits the *dual* nature of light.

15–2 Explain the connection between the quantum number n in Bohr's model of the hydrogen atom and another integer associated with the interference effects between two waves.

15–3 Describe the essential features of the Davisson-Germer experiment.

15–4 Suppose that you have two nickel crystals. One is a single crystal and the other is polycrystalline. Make a sketch showing the difference between these two crystals.

15–5 What did Davisson and Germer observe in their experiment when they used a nickel crystal containing only a few large crystals (compared to using a polycrystalline nickel target)?

15–6 What limits the effectiveness of an optical microscope? How does an electron microscope get around these difficulties? How would you decrease the wavelength of the de Broglie waves of electrons in an electron microscope?

15–7 a. What information does the wave equation for a plucked violin string yield?

b. What information does the Schroedinger equation for a hydrogen atom yield?

15–8 a. Does the height of the wave Ψ in Schroedinger's wave equation have *physical* significance, just as the amplitude *A* of a plucked violin string does?

b. What is the significance of Ψ?

15–9 State the Pauli exclusion principle.

15–10 Write the electron configurations for oxygen ($Z = 8$) and calcium ($Z = 20$).

15–11 In terms of the structure of the atoms, explain why there are families of elements with similar chemical properties.

15–12 In order to observe an electron, at least one photon must strike the electron and then enter the observer's eye. Heisenberg showed that uncertainties in position and momentum result. How would these uncertainties be affected if light had only a wave nature, where the energy of the wave could be as small as desired?

Questions for Calculation

15–1 Show that the right-hand side of Equation 1, h/mv, has units of length.

15–2 A 10-g bullet has a speed of 500 m/sec. What is its de Broglie wavelength?

†15–3 **a.** If an electron has a de Broglie wavelength of 1×10^{-10} m, what is its speed? ($m_e = 9.1 \times 10^{-31}$ kg, $m_{proton} = 1.67 \times 10^{-27}$ kg).

b. If a proton has a de Broglie wavelength of 1×10^{-10} m, what is its speed?

15–4 For added confirmation of de Broglie's theory, Davisson and Germer aimed an electron beam having a kinetic energy of 65 eV at a nickel crystal, perpendicular to the surface, as shown in Fig. 15–20. The beam was "reflected" from the crystal planes as shown, and a maximum intensity was observed at an angle ϕ of 44°. The distance between these atom planes is 0.805×10^{-10} m.

a. From the diagram in Fig. 15–20(b) show that the distance AB is about 0.75×10^{-10} m.

b. Using the results of **a,** find the wavelength of the de Broglie waves.

c. From the definition of kinetic energy, show that the speed of 65-eV electrons is 4.78×10^6 m/sec. (1 eV = 1.6×10^{-19} joules).

d. Use Equation 1 to find the de Broglie wavelength, and compare it to that found in **b.** What does this show?

†15–5 Consider 1000 hydrogen atoms in the ground state, $n = 1$. If it were possible to mea-

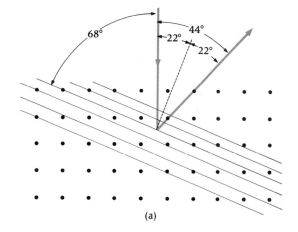

(a)

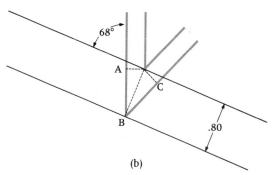

(b)

FIG. 15–20 Diagram for calculation question 15–4.

sure the position of the electrons, how many of them

    a. would have the radius r less than r_1?

    b. would have the radius r greater than r_1?

    c. would have a radius between r_1 and $2r_1$?

15–6 Consider two hydrogen atoms: one with $n = 2, l = 1$, and the other with $n = 2, l = 0$.

    a. Is it possible that the electron in both of these atoms can be found to have $r = 8r_1$?

    b. Is the probability that the electron will be found at $r = 8r_1$ the same for both atoms?

15–7 The electron of the hydrogen atom has quantum number $n = 4$.

    a. What are the allowed values of the quantum number l?

    b. For each allowed value of the quantum number l, what are the allowed values of the quantum number m?

15–8 Show that the maximum number of electrons having $l = 4$ is 18.

†15–9 Beta rays (or electrons) are emitted from the nucleus with a kinetic energy of several MeV (1 MeV $= 10^6$ eV). For a while, physicists thought that electrons were contained in the nucleus.

    a. Supposed that an electron is located within the nucleus that has a diameter of 10^{-15} m. Use Heisenberg's uncer-

tainty principle and find the uncertainty in the momentum when the electron is located within the nucleus. ($h/2\pi = 6.58 \times 10^{-16}$ eV-sec.)

    b. Set the momentum equal to the uncertainty in momentum found in **a.** Use the relativistic relationship, $E^2 = p^2c^2 + m_0^2c^4$, to find the energy of the electron located within the nucleus. The rest mass energy m_0c^2 of an electron is 0.511 MeV. Compare this energy with the kinetic energy of beta rays.

†15–10 Suppose that the distance between the two hills in Fig. 15–18 were 10 m instead of 5 m. Find the half-life of particles trapped between the two hills. (See Table 15–2.)

†15–11 If the height of the hills and/or the thickness of the hills in Fig. 15–18 is increased, but the kinetic energy at the bottom of the hill is still 9.4 joules, then the probability that the particle will cross the hill is reduced. Let this probability be 0.16. Make a table similar to Table 15–2 and find the half-life. ($h = 1$ joule-sec.)

15–12 As Table 15–2 shows, 29 particles were trapped for 1.15 seconds, 21 for 2.30 seconds, etc.

    a. Calculate the average lifetime behind the barrier by calculating [29(1.15) + 21(2.30) + $\cdots$ + 1(12.65)]/98.

    b. Is the average lifetime larger or smaller than the half-life? Why does this seem reasonable?

Footnotes

[1] Max Jammer, *The Conceptual Development of Quantum Mechanics* (New York: McGraw-Hill, 1966) p. 239 (and references cited therein).

[2] Niels H. de V. Heathcote, *Nobel Prize Winners in Physics, 1901–1950* (New York: Henry Schuman, Inc., 1953) p. 290. (© The Nobel Foundation 1930.)

[3] Jammer, p. 246.

[4] *Nobel Prize Lectures: Physics* (Amsterdam, New York: Elsevier, 1964). (© The Nobel Foundation 1964.)

[5] Karl K. Darrow, "Davisson and Germer," *Scientific American* 178 (May 1948):51.

[6] *Nobel Prize Lectures.*

[7] G. P. Thomson, *The Inspiration of Science* (London: Oxford University Press, 1961), p. 163. (Reprinted by Doubleday, Garden City, New York, 1968).

[8] Richard K. Gehrenbeck, "Electron diffraction: fifty years ago," *Physics Today* 31 (January 1978):34.

[9] Jammer, p. 257.

[10] George Gamow, *Mr. Tompkins in Wonderland* (London: Cambridge University Press, 1940), Chapter 8.

Introduction

We have, by now, learned a great deal about the general structure of the atom and about the properties of radioactivity. In this chapter, we'll focus on the nucleus and answer some fundamental questions we have raised (Chapter 11): What is nuclear energy? How is it stored in the atom? How can it be released?

We begin this chapter by discussing the work of the prolific physicist Ernest Rutherford and his collaborator Frederick Soddy who, in 1902, determined the nature of radioactivity. Rutherford and Soddy arrived at the earth-shaking conclusion that atoms disintegrate into fragments. For example, radium is radioactive because a radium nucleus breaks up into an alpha particle (helium nucleus) and a radon nucleus. In the course of their work, they uncovered many radioactive substances, too many to fit within the confines of the periodic table. Soddy suggested the existence of isotopes: Atoms of the same element that have different atomic masses.

Next, we'll turn to the nature of radioactivity, raising (and examining) additional questions: Why does one nucleus disintegrate before another? What kind of energy is transformed into the kinetic energy of the alpha and beta rays? Why are some nuclei radioactive and others stable?

In the years between Rutherford's discovery of the proton (1919) and Chadwick's discovery of the neutron (1932), physicists grappled with the question, Does the nucleus contain electrons? Since beta rays are electrons, it seemed quite reasonable to think so. However, evidence kept mounting that, in fact, the nucleus could *not* contain electrons. We shall follow these fascinating arguments and see how Chadwick's discovery of the neutron and the alleged existence of the neutrino (a particle accompanying beta decay) settled this question.

Finally, we'll consider what happens when the nucleus emits a gamma ray. Just as an atom emits light when an electron jumps from one orbit to another, the nucleus gets rid of excess

James Chadwick

CHAPTER 16

The Nucleus

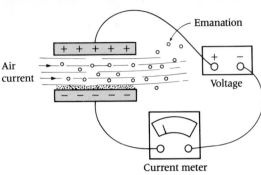

FIG. 16–1 The electrical apparatus used by Rutherford to measure the activity of thorium, which emits an emanation in addition to alpha, beta, and gamma rays. The emanation also emits an alpha particle. The current is erratic because a draft of air carries off the emanation.

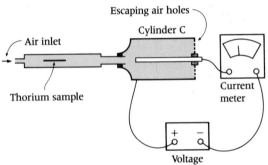

FIG. 16–2 Apparatus for measuring the half-life of the emanation.

energy by emitting a gamma ray. Just as the atom has energy levels, the nucleus has energy levels, too.

Disintegration of the Atom

Let's now visit the famous Rutherford as he carries out research at McGill University in Montreal, Canada, in 1899. Like Madame Curie, Rutherford used the apparatus shown in Fig. 16–1 to study radioactivity. Because the rays from the radioactive material on the bottom plate ionize the air, a current flows through the circuit. This current indicates the activity (the number of rays per second) of the radioactive material. At first, Rutherford used uranium, but after thorium was found to be radioactive, he tried it, too. In sharp contrast to uranium compounds, where the current was always steady, the thorium compounds produced an erratic current "being sometimes five times as great as at other"[1] times. Why this difference? In his very next research project, Rutherford compared the properties of uranium and thorium.

He found that the rays from thorium were much more penetrating than those from uranium: "Another anomaly that thorium compounds exhibit is the ease with which the radiation apparently passes through paper."[2] He tracked down the reason for the erratic electrical current: "The movement of the air caused by opening or closing of a door at the end of the room opposite to where the apparatus is placed, is often sufficient to diminish (the electrical current)."[3]

Suppose that thorium emits a fourth type of radiation (which uranium does not) and that this **emanation** passes through paper, while ordinary rays do not. Further, the emanation itself must be radioactive, because, when a draft of air blows it from the apparatus, the electrical current falls off. A fourth type of radiation would explain all of the differences between thorium and uranium. Rutherford tested these ideas, using the apparatus shown in Fig. 16–2.

Rutherford wrapped the thorium sample in paper, which stopped the ordinary rays, but not the emanation. A current of air carried the emanation into the large cylinder C, where it accumulated for a short while. Then Rutherford shut off the flow of air and measured the activity of the emanation in cylinder C at regular intervals. This activity was due to alpha particles emitted by the emanation. Figure 16−3 shows that the emanation has a half-life of 1 minute. That is, the activity drops by a factor of 2 after each minute.

COLLABORATION WITH FREDERICK SODDY

In 1901, Rutherford enlisted the help of the English chemist Frederick Soddy (1877–1956), who had just recently taken the position as laboratory instructor in McGill's chemistry department.

Soddy soon learned that the emanation was an inert gas, because it did not form chemical compounds or change in any way, even when subjected to high temperatures or to the most active chemical reagents, such as sulfuric acid, red hot magnesium powder, and red hot palladium.

Thus, the thorium sample gives off an inert gas, which is radioactive. That gas—the emanation—can easily pass through paper. In fact, all inert gases have this ability to diffuse through matter. A commonplace example of this is a helium-filled balloon, which loses its buoyancy as the helium diffuses through the balloon's skin.

Rutherford and Soddy discussed what to do next. They decided that Soddy should look for a chemical impurity in the thorium sample. Perhaps, they reasoned, it was an impurity that gave off the emanation.

After several false starts, Soddy found a chemical separation procedure that worked.* He

*To thorium nitrate dissolved in water, Soddy added ammonia. A precipitate or powder of thorium hydroxide appeared in the solution. Soddy separated the precipitate from the solution. The impurity remained in the solution, which was then evaporated to dryness.

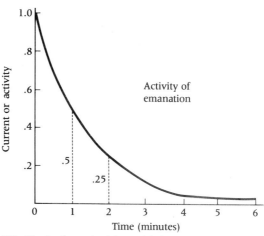

FIG. 16−3 A graph of the activity of the emanation versus time. The emanation has a half-life of 1 minute because the activity drops by a factor of 2 after each minute.

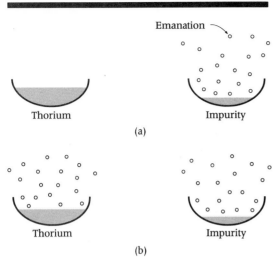

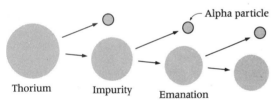

FIG. 16−4 (a) Immediately after the chemical separation procedure, *only* the impurity element emits the emanation. (b) Several weeks after the chemical separation procedure, the thorium also emits the emanation.

FIG. 16−5 The steps in the radioactive disintegration of thorium.

then isolated an impurity from the thorium sample. This impurity (a chemical element distinct from thorium) emitted the emanation, while the pure thorium did not, though the pure sample still retained a slight activity. This is where matters stood at Christmas in 1901.

When Rutherford and Soddy returned to the laboratory in January, they were astonished to find that the activity of the thorium sample had greatly increased. They also found that it, too, now gave off the emanation (see Fig. 16−4). Could it be that the thorium, from which all of the impurity element had been removed, contained the impurity once again? To answer this question, they subjected the thorium sample to the chemical separation procedure and, indeed, did find the impurity in it again.[4] This meant that the impurity *was* generated in the thorium.

In another experiment, Rutherford and Soddy concentrated their attention on the impurity, discovering that it emitted alpha particles and that its activity decreased according to a four-day half-life. Interestingly, the rate at which the impurity throws off the emanation also drops by a factor of 2 after 4 days. They felt that these two facts had to be related.

Rutherford and Soddy concluded that the results of all of their experiments could be explained by the disintegration of the atom. A thorium atom disintegrates into an impurity atom and an alpha particle. In turn, the impurity atom breaks up into an alpha particle and an atom of the emanation. The emanation atom likewise disintegrates into an alpha particle and an atom of another element (see Fig. 16−5). First of all, this theory would explain how the impurity element is generated in the thorium sample and, secondly, why the production of the emanation and the alpha particle activity of the impurity element both have a four-day half-life.

Here, we have discussed three successive disintegrations, but there are actually four more steps in the series. The disintegrations continue, ejecting alpha particles and beta rays, until an atom of lead is reached. At this point, the disintegrations stop and we say that lead is stable: It does not emit rays.

Rutherford and Soddy formulated two so-called **displacement laws** that describe the radioactive transformations:

1. Whenever a substance disintegrates by emitting an alpha particle, the new element has an atomic mass less by 4 amu, and an atomic number less by 2 units, than its parent.

2. Whenever a substance disintegrates by emitting a beta ray, the new element has the same atomic mass as its parent, but an atomic number greater by 1.

Isotopes

Scientists found that uranium, too, disintegrates in a series of steps similar to those depicted in Fig. 16–5. With each radioactive disintegration, an atom of a *different* element results. Radium, the element that Madame Curie spent four years separating from pitchblende ore, was formed after five successive disintegrations, beginning with uranium. This explains why radium is always found in uranium samples. The situation became even more complicated when two more series of transformations were discovered. When these four series were completely worked out, it developed that there were over 40 new substances, all of which had an atomic mass less than uranium.

There seemed to be two conflicting pieces of evidence. First of all, there were not that many empty spaces in the periodic table (which is to say, not that many undiscovered elements). In 1910, only six elements with an atomic number Z less than 92 ($Z = 43, 44, 72, 75, 87, 91$) had not yet been discovered. Secondly, these 40 substances could certainly be distinguished because each one had a distinct half-life that differed from the others. This would indicate that there were over 40 *different* substances produced and that there were no duplications in the four series of transformations. How could this conflict be resolved?

ISOTOPES: A SOLUTION

A third piece of evidence caused Soddy to propose a radical solution in 1910. The chemical properties of the many radioactive substances were investigated and, in several cases, two substances having *different* values for the half-life were found to have *identical* chemical properties. If these two substances were mixed together, it would have been impossible to separate them by any chemical methods. The phrase "identical chemical properties" means that both substances were the *same* element (that is, had the same atomic number Z). Until 1910, all atoms of an element were thought to be *identical in every respect*. If all atoms of an element were identical, how could two samples of one element have different half-lives?

Soddy's solution was indeed radical and toppled the cherished identical-atom concept. He proposed that the atoms of an element were *not* identical, but could differ in a basic property: the mass of the atom. Thus, an element could have several **isotopes,** or several groups of atoms of different atomic masses. For example, suppose that there are two radioactive samples, A and B, which have identical chemical properties, but different half-lives. The atoms of sample A are all identical, and the atoms of sample B are all identical. However, an atom of sample A has a mass that differs from an atom of sample B. Also, atoms of sample A have a half-life that differs from atoms of sample B. Samples A and B would then be two isotopes of the same element.

In 1921, Soddy received the Nobel Prize in chemistry for his contributions to the chemistry of radioactive substances and for predicting the existence of isotopes.

TABLE 16–1 MASSES OF ISOTOPES

ISOTOPE	MASS (amu)	ISOTOPE	MASS (amu)
$^{1}_{0}n$	1.008665	$^{23}_{11}Na$	22.98977
$^{1}_{1}H$	1.007825	$^{26}_{12}Mg$	25.98259
$^{4}_{2}He$	4.00260	$^{27}_{13}Al$	26.98153
$^{7}_{3}Li$	7.01600	$^{30}_{14}Si$	29.97376
$^{8}_{4}Be$	8.0053	$^{60}_{27}Co$	59.93381
$^{9}_{4}Be$	9.01218	$^{60}_{28}Ni$	59.9302
$^{11}_{5}B$	11.00931	$^{90}_{36}Kr$	89.91959
$^{12}_{6}C$	12.00000	$^{143}_{56}Ba$	142.92054
$^{13}_{6}C$	13.00335	$^{210}_{83}Bi$	210.04975
$^{14}_{6}C$	14.00324	$^{210}_{84}Po$	209.9829
$^{14}_{7}N$	14.00307	$^{222}_{86}Rn$	222.0175
$^{16}_{8}O$	15.99491	$^{226}_{88}Ra$	226.0254
$^{17}_{8}O$	16.99913	$^{234}_{90}Th$	234.0436
$^{19}_{9}F$	18.99840	$^{235}_{92}U$	235.04395
$^{20}_{10}Ne$	19.99244	$^{238}_{92}U$	238.0508
$^{22}_{10}Ne$	21.99138		

TABLE 16–2 HALF-LIFE BEHAVIOR

TIME (sec)	NUMBER OF ELEMENT X NUCLEI	NUMBER OF ELEMENT Y NUCLEI	NUMBER OF ALPHA PARTICLES EMITTED
0	10,000	0	0
10	5000	5000	5000
20	2500	7500	7500
30	1250	8750	8750
40	625	9375	9375

SEPARATION OF ISOTOPES:
THE MASS SPECTROGRAPH

In 1913, J. J. Thomson observed isotopes of neon by using a modified version of a cathode ray tube that contained a small amount of neon. He found that most of the atoms had an atomic mass of 20 amu, but a small fraction had 22 amu.

In 1922, the English physicist F. W. Aston (1877–1945), a former student of Thomson's, made additional improvements in the **mass spectrograph,** which is the device that obtains a "spectrum" of the masses. Aston showed that there are two isotopes of chlorine having atomic masses 35 amu and 37 amu.

In 1933, K. T. Bainbridge (b. 1904), an American scientist, developed a mass spectrograph of great accuracy. In this model, singly charged ions, all having the same speed, enter a perpendicular magnetic field, which causes the ions to travel in circular paths. Because the magnetic field exerts the same force on all ions, the more massive ion bends less and, hence, travels in a circle of larger radius than that of a lighter ion. The ions of different isotopes, therefore, travel in circles of different radii and then strike a photographic plate, producing a spectrum of masses. (See Chapter 7, thought question 7–25).

The atomic mass shown in the periodic table of the elements is an average of the atomic masses of the isotopes of an element. For example, 91 percent of the atoms of neon have an atomic mass of 20 amu; 9 percent have an atomic mass of 22 amu; and only a trace has an atomic mass of 21 amu. The atomic mass is given by

Atomic mass of neon
$$= (0.91)(20 \text{ amu}) + (0.09)(22 \text{ amu})$$
$$= 20.2 \text{ amu}.$$

Table 16–1 shows the masses of some isotopes, where an atom of carbon-12 has a mass of exactly 12.00000 amu.

Let's now raise some of the basic questions concerning radioactive disintegrations: Can the half-life of an isotope be understood in a fundamental way? Why are some atoms stable and others radioactive? How is energy stored in an atom?

The Nature of Radioactivity

ALPHA DECAY

After Rutherford discovered the nucleus in 1913, it became apparent that alpha decay was due to the disintegration of the *nucleus*. Since an alpha particle is the nucleus of a helium atom, it carries away 2 positive charges and 4 units of mass. The remaining nucleus then has 2 fewer units of positive charge and a mass smaller by about 4 amu. For example, when radium-226 emits an alpha particle, the remaining nucleus is radon-222:

$$^{226}_{88}\text{Ra} \rightarrow {}^{4}_{2}\text{He} + {}^{222}_{86}\text{Rn}. \quad (1)$$

Remember that the superscript to the left of the chemical symbol gives the atomic mass number, which is the atomic mass rounded off to an integer (see Chapter 1). For example, the atomic mass number of radium is 226.

Note that Equation 1 agrees with Rutherford and Soddy's first displacement law.

THE STATISTICAL NATURE OF RADIOACTIVE DECAY

In a sample of radium, for instance, why does one nucleus decay after 1 second and another not until, say, 3000 years later? If all radium nuclei are identical, why don't they all behave in the same way?

In 1928, Gamow, Gurney, and Condon used Schroedinger's wave equation to explain this puzzling effect (see Chapter 15). They imagined the alpha particle as a separate entity moving around inside the nucleus. When it strikes the nuclear surface, there is a probability that the alpha particle will escape. These researchers calculated this probability. There is also a probability that the alpha particle will not escape! If it does not escape on one bounce, perhaps it

will the next time, or the next time . . . and so on. Schroedinger's equation, you will recall, replaced determinism with statistical probability.

It stands to reason that the more radioactive nuclei in a sample, the more alpha particles per second will be emitted. That is, the activity of a radioactive sample is directly proportional to the number of nuclei in that sample:

$$\text{Activity of sample} \propto \text{number of radioactive nuclei in sample}. \quad (2)$$

Experimentally, we know that the activity drops by a factor of 2 after a time called the half-life. Equation 2 shows that the number of radioactive nuclei in the sample likewise drops by a factor of 2 after one half-life. Let's use an example to illustrate these effects.

Suppose that a radioactive sample contains 10,000 atoms of element X and has a half-life of 10 seconds. The nucleus of element X emits an alpha particle, and the nucleus of element Y remains.

$$^{A}_{Z}\text{X} \rightarrow {}^{4}_{2}\text{He} + {}^{A-4}_{Z-2}\text{Y}$$

After 10 seconds, there are only 5000 nuclei of X remaining, but 5000 nuclei of Y. There are always 10,000 nuclei of *some* kind. Table 16−2 shows how the nuclei of elements X and Y vary with time.

From 0 to 10 seconds, 5000 alpha particles were emitted; from 10 to 20 seconds, 2500; from 20 to 30 seconds, 1250. Thus, we see that the activity (the number of alpha particles per second) also follows the half-life behavior. Equation 2 shows that, if the number of nuclei drops by a factor of 2 during one half-life, then the activity should also drop by a factor of 2 during the same time. Table 16−2 confirms that this is true.

THE NUCLEUS

In 1905, Einstein stated that mass and energy are interchangeable. $E = mc^2$ is the famous equation governing such transformations. **Nuclear energy**—the energy stored in the atom—is due to the transformation of mass into energy as the nucleus disintegrates. In 1932, an experiment showed that mass and energy are truly equivalent.

To review a moment: The masses of atoms are often expressed in atomic mass units, which are abbreviated as amu. For example, the mass of a hydrogen atom is 1.007825 amu and the mass of a carbon atom is 12.00000 amu (see Table 16–1). This means that a carbon atom is very nearly 12 times more massive than a hydrogen atom. We know (Chapter 1) that 12 g of carbon contain 6.02×10^{23} atoms. Therefore, 1 carbon atom has a mass of

$$\frac{12\,g}{6.02 \times 10^{23}} = 1.992 \times 10^{-23}\,g$$
$$= 1.992 \times 10^{-26}\,kg.$$

That is, 12 amu corresponds to 1.992×10^{-26} kg and, hence, 1 amu corresponds to 1.66×10^{-27} kg.

To find out how much energy is produced when 1 amu is converted completely into energy, we use $E = mc^2$, where c is the speed of light, 3.00×10^8 m/sec:

$$E = (1.66 \times 10^{-27}\,kg)(3.00 \times 10^8\,m/sec)^2$$
$$= 14.9 \times 10^{-11}\,kg\text{-}m^2/sec^2$$
$$= 1.49 \times 10^{-10}\,joules.$$

To convert this energy to electron volts (because that unit is easier to work with), we use the following:

$$1\,eV = 1.60 \times 10^{-19}\,joules,$$

or, in terms of 1 million electron volts,

$$1\,MeV = 1.60 \times 10^{-13}\,joules.$$

The energy contained in 1 amu is, therefore,

$$1\,amu = \frac{1.49 \times 10^{-10}\,joules}{1.60 \times 10^{-13}\,joules/MeV} \quad (3)$$
$$= 931\,MeV.$$

Let us now see how $E = mc^2$ applies to the alpha decay of radium.

ALPHA DECAY OF RADIUM

In Fig. 16–6 the radium nucleus is drawn larger than the alpha particle and the radon nucleus. This emphasizes the idea that the mass before the decay is larger than the total afterwards. According to conservation of energy, the total energy is equal to the kinetic energy plus the mass energy:

$$TE = ME + KE.$$

If the mass energy ME decreases, the kinetic energy KE must increase, so that the total energy TE remains constant.

Table 16–3 clearly shows the conservation of energy principle. (The masses of the isotopes are given in Table 16–1.) Because the total energy must be the same before and after the decay, the kinetic energy of the alpha particle plus the kinetic energy of the radon nucleus must be 226.0254 amu minus 226.0201 amu, or 0.0053 amu. We now use the fact that 1 amu is equivalent to 931 MeV (Equation 3) to convert this kinetic energy to million electron-volts:

KE of alpha particle + KE of radon nucleus $= 0.0053\,amu \times \dfrac{931\,MeV}{amu}$

$= 4.9\,MeV.$

The following considerations show that the alpha particle has a kinetic energy of 4.8 MeV, and the radon nucleus has a kinetic energy of 0.1 MeV.

The momentum of the radium nucleus is zero. After the disintegration, the momentum of the alpha particle in one direction must be equal to the momentum of the radon nucleus in the opposite direction, so that the total momentum remains zero. Since the momentum is defined as mass multiplied by velocity, the alpha particle's speed must be 222/4 times larger than that of the radon nucleus. (See Chapter 4, p. 90.) For this reason, the alpha particle has a kinetic energy equal to 98 percent of the disintegration energy, or 4.8 MeV, while the radon nucleus has kinetic energy of only 0.1 MeV. (These calculations are outlined in calculation question 6−17.) In some of the examples used later on, we may neglect the small recoil kinetic energy of the nucleus.

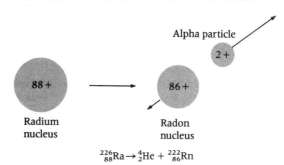

$$^{226}_{88}Ra \rightarrow \, ^{4}_{2}He + \, ^{222}_{86}Rn$$

FIG. 16−6 The radioactive decay of radium. The radium nucleus is drawn larger than that of the radon nucleus and the alpha particle to show that mass disappears during the disintegration.

NATURAL RADIOACTIVITY VERSUS STABILITY

Why are atoms of one element stable, while the atoms of another element are naturally radioactive? What criterion determines this? The phenomenon of **natural radioactivity** has been limited to elements of large atomic mass, such as uranium, radium, thorium, and so forth. Why is this so? Let's consider a lightweight element, such as carbon.

Atoms of carbon-12 are stable; a nucleus of carbon-12 *does not* break up into an alpha particle and a nucleus of beryllium-8. Table 16−1 gives us the reason. Compare the mass of carbon-12 shown on that table with that of beryllium-8 plus helium-4. The mass energy *after* the hypothetical delay is *larger* than that before. This violates conservation of energy and, of course, cannot occur. For this reason, carbon-12 atoms are stable.

In conclusion, atoms of an isotope are naturally radioactive when

$$\frac{\text{Total mass}}{\text{before decay}} > \frac{\text{total mass}}{\text{after decay}}. \qquad (4)$$

The difference in mass is converted into the kinetic energy of the particles after the decay. Atoms of an isotope are stable when

$$\frac{\text{Total mass before}}{\text{hypothetical decay}} < \frac{\text{total mass after}}{\text{hypothetical decay}}. \qquad (5)$$

TABLE 16−3   THE ALPHA DECAY OF RADIUM-226

	KE (amu)	ME (amu)	TE (amu)
Before decay	0	226.0254	226.0254
After decay	KE of alpha + KE of radon	222.0175 +4.0026 = 226.0201	226.0254

TABLE 16-4 ARTIFICIAL DISINTEGRATION OF THE NITROGEN NUCLEUS

	KE (amu)	ME (amu)	TE (amu)
Before collision	0.00827	14.00307 + 4.00260 = 18.00567	18.01394
After collision	KE of proton + KE of oxygen nucleus	16.99913 + 1.00783 = 18.00696	18.01394

Artificial Disintegration of the Nucleus and the Discovery of the Proton

Rutherford discovered the proton[5] by placing an alpha particle source inside a container of nitrogen gas (Chapter 1). When the alpha particle collided with a nitrogen nucleus, a proton from either the alpha particle or from the nitrogen nucleus was knocked loose during the collision. The nuclear reaction equation for this process is

$$^{14}_{7}\text{N} + ^{4}_{2}\text{He} \rightarrow ^{1}_{1}\text{H} + ^{17}_{8}\text{O}.$$

The proton is a nucleus of a hydrogen atom and, hence, has the symbol $^{1}_{1}\text{H}$. (Again: the superscript denotes the atomic mass number A which is the atomic mass rounded off to an integer; the subscript denotes the atomic number Z.) There are a total of 9 positive charges before and after the collision. The atomic mass numbers sum to 18 on both sides of the equation. Due to the rounding, this *does not* say that the total mass before the collision is equal to that afterwards. However, we can say that, to the *nearest integer,* there are 18 amu before and after the collision. As Table 16-4 shows, there is a total mass of 18.00567 amu before the collision, but 18.00696 amu after the collision.

The alpha particles in this experiment had a kinetic energy of 7.7 MeV. In Table 16-4, this kinetic energy is expressed in amu's:

$$\frac{\text{KE of alpha}}{\text{particle}} = \frac{7.7 \text{ MeV}}{931 \text{ MeV/amu}}$$
$$= 0.00827 \text{ amu}.$$

The mass of each particle is taken from Table 16-1, and then the total mass before and after the collision is obtained. The total energy before the collision is found by summing the kinetic energy of the alpha particle and the total mass before the collision, which yields 18.01394 amu. Since energy is conserved, the total energy after the collision must also be 18.01394 amu. Note immediately that the total mass after the collision is larger than before, but, in order to conserve energy, the kinetic energy afterwards must be smaller:

$$\frac{\text{KE of}}{\text{proton}} + \frac{\text{KE of}}{\text{oxygen nucleus}}$$
$$= 18.01394 \text{ amu} - 18.00696 \text{ amu}$$
$$= (0.00698 \text{ amu})(931 \text{ MeV/amu})$$
$$= 6.50 \text{ MeV}.$$

In 1919, Rutherford was elected to the Cavendish Chair of Physics at Cambridge, the most prestigious professorship in England. James Chadwick joined him in these investigations. By bombarding with alpha particles, they changed boron into carbon, fluorine into neon, and so on. Is this not reminiscent of the ancient alchemist's dream of turning lead into gold? Because the alpha particle must come close to that incredibly small nucleus, the transformation is not efficient. Only about one proton is produced for every one million alpha particles.

Rutherford and Chadwick also noted the release of energy when alpha particles struck aluminum.

$$^{27}_{13}\text{Al} + ^{4}_{2}\text{He} \rightarrow ^{1}_{1}\text{H} + ^{30}_{14}\text{Si}$$

The protons had *more* kinetic energy than the alpha particles: "This additional energy must come from the atom in consequence of its disintegration."[6] From the data in Table 16–1, we can see that the total mass on the right side of this reaction equation is less than that on the left side. Hence, mass must be converted to kinetic energy. This explains the observations of Rutherford and Chadwick.

Cloud Chamber

The study of nuclear physics was greatly aided by use of the **cloud chamber,** which was invented by the Scottish physicist C. T. R. Wilson (1869–1959) in 1912 at the Cavendish Laboratory (see Fig. 16–7). The cylinder of the chamber has a transparent glass window at one end and a movable piston at the other end. The cylinder contains air (or a gas), water vapor, and excess water, so that the water vapor is saturated. In order to understand how the cloud chamber operates, let's consider a familiar situation first.

The term *relative humidity* used in weather forecasting refers to the amount of water vapor that the air contains at a given time. When the relative humidity is 90 percent, the air contains 90 percent of the moisture that it can normally hold at that particular temperature. At a lower temperature, the air can hold a smaller amount of water. If a cold front moves in suddenly, then rain droplets condense on dust particles in the air.

Great care is taken to remove all dust particles from a cloud chamber. When the piston is rapidly lowered, the gas expands, dropping greatly in temperaure as a result. If there are no dust particles present, then the air or gas becomes supersaturated (that is, it contains more water vapor than it can normally hold at that temperature). If a charged particle passes through the chamber when the gas is supersaturated, then water droplets condense on that particle, which forms a track that can be photographed through the glass window.

The cloud chamber was used right away to study the scattering of alpha particles. Figure 16–8 illustrates a cloud chamber filled with oxygen. You can see that the alpha particle is

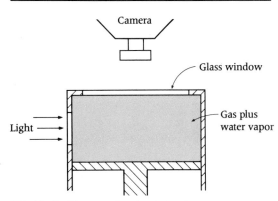

FIG. 16–7   Diagram of a cloud chamber.

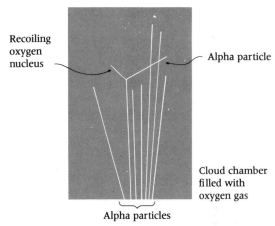

FIG. 16–8   A cloud chamber photograph showing an alpha particle scattered by an oxygen nucleus. (Artist's drawing based on a photograph in J. D. Stranathan, *The Particles of Modern Physics*, Philadelphia: The Blakiston Co., p. 408.)

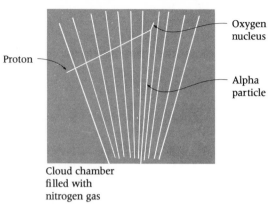

Cloud chamber
filled with
nitrogen gas

FIG. 16–9 A cloud chamber photograph showing the proton and oxygen nuclei that result when an alpha particle strikes a nitrogen nucleus. (Artist's drawing based on a photograph in P. M. S. Blackett and D. S. Lee, *Proceedings of the Royal Society*, 136 A, 1932, p. 325.)

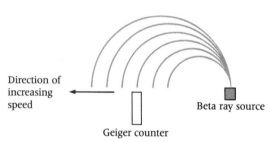

FIG. 16–10 A perpendicular magnetic field causes beta rays to travel in circular paths. The radius of the circle is dictated by the speed of the beta ray. Fast-moving beta rays are harder to bend and, hence, have a larger radius than slow-moving beta rays.

scattered through a large angle in a *single* collision with an oxygen atom, in perfect agreement with Rutherford's theory of alpha scattering. The path of the recoiling oxygen atom is also visible.

Figure 16–9 illustrates a cloud chamber filled with nitrogen gas. The alpha particle strikes a nitrogen nucleus, and a proton and oxygen nucleus are formed as a result. The long, thin track is the proton, and the much shorter track is the oxygen nucleus. Blackett obtained this photograph using an apparatus that took photographs automatically every 15 seconds. In order to obtain these results, Blackett took 23,000 photographs, which each contained about 20 alpha particle tracks. In all of these photographs, he found *only eight* tracks similar to the one shown in Fig. 16–9.

Does the Nucleus Contain Electrons?

After the discovery of the proton, many physicists thought it reasonable that the nucleus was composed of protons and electrons. Since the electron's mass is about 2000 times smaller than the proton's, it was felt that the mass of the nucleus must be largely due to the protons. The nucleus must contain A protons. Since the nucleus contains Z positive charges, there would then be A minus Z electrons in the nucleus because

Number of
positive charges = $A(+1) + (A - Z)(-1)$
in the nucleus = $Z +$ charges (as required).

Since the beta ray is an electron, the existence of electrons in the nucleus seemed to explain beta decay as well. However, there were problems with having the electron in the nucleus.

VANISHING MAGNETIC FIELD

The electron rotates (spins) either clockwise or counterclockwise about a *vertical* axis (Chapter 15). However, the motion of charged particles always produces a magnetic field. For example, if you put a compass near a current-carrying wire, you will see the needle align with the magnetic field. So too, the spinning electron produces a magnetic field and, in fact, that field is very similar to a bar magnet's. Now, the proton spins about a vertical axis, either clockwise or counterclockwise, too, and produces a magnetic field similar to a bar magnet's. But there is one important difference. It turns out that the magnetic field of the spinning particle is inversely proportional to its mass. Therefore, the magnetic field generated by the spinning electron is about 2000 times *larger* than that due to the proton. The magnetic field derived from the nucleus should then be due primarily to the *A* minus *Z* electrons in the nucleus. But experimental measurements of the magnetic field of the nucleus showed that it was on the order of the *proton's* magnetic field, not the electron's. This was hard to understand. How could the electron's magnetic field vanish when it was inside the nucleus?

CAN THE ELECTRON BE BOUND TO THE NUCLEUS?

Before answering, let's consider a familiar example first. In the hydrogen atom, an electron with $n = 1$ has an energy E equal to -13.6 eV. The orbiting electron will remain in this orbit *forever* unless disturbed by some outside influence. For instance, if a photon having an energy of 15 eV is absorbed by the hydrogen atom, then the electron, freed from its attractive bonds to the nucleus, escapes with a kinetic energy of 1.4 eV. In this case, the electron's total energy E is $+1.4$ eV. From this example, we conclude:

1. When the electron's total energy E has a negative value, the electron is bound to the nucleus and will remain so unless disturbed.
2. When the electron's total energy E has a positive value, the electron cannot be bound to the nucleus.

Now, consider the beta ray, or an electron, *inside* the nucleus. Here, too, the electron is attracted to the positive charge of the nucleus, but this force of attraction is much larger than

in the hydrogen atom since the charges are separated by such a short distance. However, the same two conclusions apply to the binding of the electron inside the nucleus.

Experimentally, we know that a beta ray spontaneously (that is, without outside influences) escapes from the nucleus with several MeVs of kinetic energy. Therefore, inside the nucleus, the beta ray must have a total energy E of several MeV, where E is a *positive* number. According to our second conclusion, the beta ray should not be bound to the nucleus at all, not even for a second. However, experimentally we know that the beta ray *is* bound to the nucleus for some time before the nucleus disintegrates. Surely all of this is extremely puzzling!

Beta Decay and Energy Conservation

On top of these two problems was another, even more worrisome: Energy did not appear to be conserved in beta decay. Rutherford comments:

The complexity of the (beta) radiation has been shown very clearly by Becquerel in the following way. An uncovered photographic plate, with the film upwards, was placed horizontally in the horizontal uniform magnetic field of an electromagnet. A small, open, lead box containing the radioactive matter, was placed in the centre of the field, on the photographic plate. . . . The whole apparatus [see Fig. 11–7, p. 267] was placed in a dark room. . . . A diffuse impression is observed on the plate giving, so to speak, a continuous spectrum of the rays and showing that the radiation is composed of (beta) rays of widely different velocities.[7]

In 1914, James Chadwick used a Geiger counter to observe the relative number of beta rays at various radii (Fig. 16–10). Since each radius corresponds to a different speed and, hence, to a different kinetic energy, Chadwick

THE NUCLEUS

379

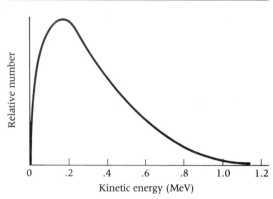

FIG. 16–11 The energy distribution of beta rays from bismuth-210.

TABLE 16–5 EXPECTED KE OF BETA RAY

	KE (amu)	ME (amu)	TE (amu)
Before	0	210.04975 − 83 m_e	210.04975 − 83 m_e
After	KE of beta ray	(210.04850 − 84 m_e) + m_e	210.04975 − 83 m_e

obtained the energy spectrum of beta rays emitted by the source. Figure 16–11 shows such an energy spectrum for the beta rays emitted by $^{210}_{83}$Bi. This is quite different from alpha decay, where *all* alpha particles have the *same* kinetic energy.

According to Rutherford and Soddy's second displacement law, the new element has the same atomic mass number A as its parent, but an atomic number Z greater by 1. Therefore, when a bismuth nucleus emits a beta ray, the following occurs:

$$^{210}_{83}\text{Bi} \rightarrow \beta^- + {}^{210}_{84}\text{Po}. \qquad (6)$$

The bismuth nucleus has 83 + charges, and 83 + charges must remain after the decay. Thus, 84 + plus 1 − equals 83 +. The beta ray carries off little mass, so the nucleus remaining still has 210 units of mass.

Now, we arrive at the crux of the problem. According to Einstein's equivalence of mass and energy, the radioactive decay occurs because the mass afterwards is less than before, with the difference converted to the kinetic energy of the beta ray (primarily). The mass difference obviously has a single value, so the *kinetic energy of the beta rays is expected to have only one value.* That the beta rays have a spread in energy appears to violate conservation of energy. Let's consider the beta decay of bismuth-210 and find out what kinetic energy is expected.

The masses in Table 16–1 are the masses of *atoms,* not nuclei. Since Equation 6 refers to the transformation of the bismuth nucleus, we need nuclear masses. An atom has Z electrons, so the mass of its nucleus is given by:

Mass of nucleus = mass of atom − Zm_e,

where m_e is the mass of an electron. The expected kinetic energy can be found from Table 16–5. The kinetic energy of the beta ray is

$$(210.04975 - 83\,m_e) - (210.04850 - 83\,m_e)$$
$$= 0.00125 \text{ amu}.$$

In general, the kinetic energy of a beta ray is equal to the difference in *atomic* masses of the two isotopes. Since 1 amu corresponds to 931 MeV, the kinetic energy of the beta ray is

Expected
KE of
beta ray = (0.00125 amu)(931 MeV)/amu)
= 1.16 MeV.

In Fig. 16–11, we see perhaps a few beta rays have this kinetic energy, but the vast majority have much less energy. In fact, the *average* energy of the beta rays in Fig. 16–11 is only 0.34 MeV. What happens to the rest of the energy?

The problem of beta decay was still unsolved in 1927, when C. D. Ellis and W. A. Wooster[8] set out to measure the disintegration energy in another way. They placed a bismuth-210 source in a **calorimeter** (an insulated vessel containing water) and measured the temperature rise. Bismuth-210 emits no gamma rays, so they did not have to worry about penetrating gamma rays escaping from the calorimeter and carrying away some energy. When the beta rays (or possibly any other radiation) were stopped in the calorimeter, the beta rays' kinetic energy (or that of any other radiation) was transformed into heat. By comparing the total heat generated by a known number of disintegration, Ellis and Wooster found the average energy per disintegration to be 0.35 ± 0.04 MeV. This is the same as the average kinetic energy of the beta rays.

One theory held that beta rays emerged from the nucleus with a kinetic energy of 1.2 MeV and then interacted with orbiting electrons, transferring some energy to them. The beta ray would emerge from the atom with a reduced energy. If this were the case, then the water would be heated by the beta rays coming to a stop *and* also by the energy transferred to orbiting electrons. Each disintegration should supply an energy of 1.2 MeV to heat the water. The calorimeter experiment showed that each disintegration supplied an energy of only 0.35 MeV. These results ruled out this theory. There was no escape: The beta rays emerge from the nucleus with a distribution of kinetic energies.

There was still no solution to the missing energy problem.

Chadwick Discovers the Neutron

In 1920, Rutherford first suggested the possible existence of a neutral particle: the **neutron.** While Rutherford and Chadwick carried out their experiments on artificial disintegration, looking at flashes on a fluorescent screen in a darkened room, Rutherford "expounded to me [Chadwick] at length his views on the problems of nuclear structure, and in particular on the difficulty in seeing how complex nuclei could possibly build up if the only elementary particles available were the proton and the electron, and the need therefore to invoke the aid of the neutron."[9] In fact, they carried out several unsuccessful attempts to find the neutron.

When alpha particles strike a beryllium target, very penetrating radiation emerges from the target. Irene Curie-Joliot (Marie and Pierre Curie's daughter) and Frederick Joliot found that the penetrating radiation knocked protons from paraffin and suggested that this penetrating radiation consisted of gamma rays. Chadwick describes Rutherford's reaction to this gamma-ray suggestion and how he began his experiment that led to the discovery of the neutron:

Then one morning I read the communication of the Curie-Joliots in the Comptes Rendus, *in which they reported a still more surprising property of the radiation from beryllium—a most startling property—that of ejecting protons from matter containing hydrogen. . . . A little later that morning I told Rutherford. It was a custom of long standing that I should visit him about 11 A.M. to tell him my news of interest and to discuss the work in progress in the laboratory. As I told him about the Curie-*

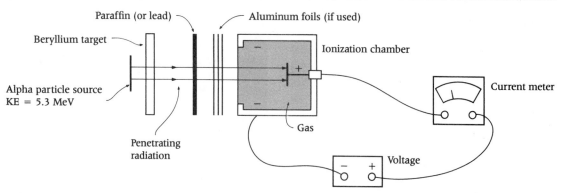

FIG. 16–12   A diagram of Chadwick's apparatus.

Joliot observation and their views on it, I saw his growing amazement: and finally he burst out: "I don't believe it." Such an impatient remark was utterly out of character, and in my long association with him I recall no similar occasion. I mention it to emphasize the electrifying effect of the Curie-Joliot report. Of course, Rutherford agreed that one must believe the observation [knocking protons from paraffin]; *the explanation* [that the radiation was gamma rays] *was quite another matter.*

It so happened that I was just ready to begin experimentation, for I had prepared a beautiful source of polonium.[10]

Figure 16–12 shows a schematic diagram of Chadwick's apparatus for investigating the penetrating radiation produced when alpha particles strike beryllium. The radiation entering the ionization chamber ionizes gas atoms, and the current measures the number of ions produced. Thus, the ionization chamber registers the number of current surges—counts, *as well as their strength.* (A Geiger counter only measures the number of counts.)

When Chadwick placed 2 cm of lead in front of the counter, the number of counts did not change. Indeed, the radiation from beryllium must be *extremely* penetrating to pass through that much lead!

Next, Chadwick placed a sheet of paraffin 2 mm thick in front of the ionization chamber, and the number of counts greatly increased. The strength of the current surges indicated that these particles were protons. The penetrating radiation striking the paraffin, which contains hydrogen, had knocked out hydrogen nuclei—protons, just as the Curie-Joliots had reported. Then Chadwick placed aluminum foils between the paraffin and the chamber and noted the thickness of aluminum needed to absorb the protons. From the results of previous absorption measurements with protons, he deduced that the speed of the protons ejected from the paraffin was 3.3×10^7 m/sec.

CLOUD CHAMBER EXPERIMENTS

Chadwick collaborated with Norman Feather, who had carried out many experiments with a cloud chamber. In this case, the cloud chamber contained water vapor and nitrogen gas. When the penetrating radiation from beryllium entered the chamber and collided with a nitrogen atom, this nitrogen atom was set into motion, producing a track in the chamber. The length of this track was an indication of the initial speed of the struck nitrogen atom. In this way, they found that its speed was 0.44×10^7 m/sec.

COULD THE PENETRATING RADIATION BE GAMMA RAYS?

That the penetrating radiation could *not* be charged particles was obvious. Charged particles would easily be stopped in traveling through even a small thickness of lead. When Irene Curie-Joliot and Frederick Joliot found that the radiation ejected protons from paraffin, they suggested that the radiation consisted of gamma rays with an energy of 50 MeV. We know, for example, that X-rays and gamma rays can set electrons in motion. But protons are nearly 2000 times more massive. Hence, the Curie-Joliots *calculated* that a large energy of 50 MeV was required. We have just learned about Chadwick's and Rutherford's reaction to this suggestion—nonacceptance, and with good reason. Gamma rays from radioactive sources have an energy usually less than 5 MeV. An energy of 50 MeV was hard to reconcile with this evidence.

Chadwick found still another difficulty with the gamma ray theory. Suppose that the beryllium nucleus absorbed the alpha particle to form a nucleus of carbon-13. This carbon-13 nucleus would have excess energy and emit a gamma ray. Using Einstein's mass-energy relationship, Chadwick found that the gamma ray would have an energy of only 16 MeV, nowhere near the required 50 MeV.

THE NEUTRON HYPOTHESIS

Armed with the experimental data and the many inconsistencies of the gamma ray theory, Chadwick proposed:

If we suppose that the radiation is not a quantum of radiation (gamma rays), but consists of particles of mass very nearly equal to that of the protons, all the difficulties connected with the collisions disappear . . . In order to explain the great penetrating power of the radiation we must further assume that the particle has no net charge . . . We must suppose that an alpha particle is captured by a ^{9}Be nucleus with the formation of a ^{12}C nucleus and the emission of a neutron.[11]

The reaction equation would be

$$^4_2\text{He} + {}^9_4\text{Be} \rightarrow {}^{12}_6\text{C} + {}^1_0\text{n}. \qquad (7)$$

A neutral particle would be very penetrating because it simply would not interact with the negatively charged electrons.

When a neutron struck a hydrogen nucleus, it was set into motion with a speed of 3.3×10^7 m/sec and, similarly, when a neutron struck a nitrogen nucleus, the nitrogen nucleus obtained a speed of 0.44×10^7 seconds. Chadwick used this data and the laws of conservation of momentum and energy to show that the neutron's mass was equal to the proton's mass. (See p. 124 for these calculations.)

The discovery of the neutron answered the question posed earlier: Does the nucleus contain electrons? No, it does not. Nor is the neutron the combination of a proton and an electron. In that case, we don't have to worry about the electron's vanishing magnetism in the nucleus, nor the electron's binding to the nucleus. The nucleus is composed of protons and neutrons: Z protons account for the Z-positive charges and Z units of mass, while A minus Z neutrons account for the rest of the mass.

In 1935, Chadwick received the Nobel Prize in physics because his discovery had such a profound impact on our understanding of the nucleus. Even so, several questions remained unanswered: How *are* beta rays emitted by the nucleus? What happens to the missing energy?

Neutrinos Are the Answer

In 1930, Wolfgang Pauli proposed a radical solution to this dilemma: Another particle, called the **neutrino,** is also emitted during beta decay, and the neutrino carries off the missing energy. This particle must be intensely penetrating, which would explain the heat measurements associated with beta decay. If the neutrino were not stopped in the calorimeter, its energy, obviously, would not be transformed into heat. The neutrino mass must be very small. Let's see why by looking at the beta decay of bismuth-210.

Earlier, we saw that the kinetic energy of the beta ray was expected to be 1.16 MeV. According to the neutrino theory, this disintegration energy is distributed in the following way:

$$1.16 \text{ MeV} = \text{KE of beta} \atop + \text{KE of neutrino} + m_v c^2, \tag{8}$$

where m_v is the neutrino mass. Here, we ignore the recoil kinetic energy of the polonium-210 nucleus, since it is so small. When the kinetic energy of the neutrino is zero, Equation 8 becomes

$$1.16 \text{ MeV} = \text{KE}_{max} \text{ of beta} + m_v c^2. \tag{9}$$

Figure 16–11 shows that the maximum kinetic energy of the beta ray is certainly very close to 1.16 MeV. In that case, Equation 9 shows that the neutrino mass is zero. Precise experiments demonstrated that this was so. Like photons of light that have no mass, the neutrinos travel at the speed of light.

In 1934, the Italian-born physicist Enrico Fermi (1901–1954) followed Pauli's suggestion and developed a comprehensive theory for beta decay. The electron does not exist in the nucleus, but is formed at the instant the decay takes place when a neutron transforms into a proton by emitting a beta ray and a neutrino:

$$n \rightarrow p + \beta^- + v. \tag{10}$$

Fermi showed that the shape of the beta-ray energy spectrum (Fig. 16–11) is governed by statistics. The chance that the beta ray will have a certain energy, and the chance that the neutrino will have another energy, are not independent. Rather, they are linked because their total energy must remain constant.

In fact, the neutrino was not detected until nearly 30 years later by two American physicists, F. Reines (b. 1918) and C. C. Cowan (1919–1974). The difficulty was that the neutrino barely interacts with matter. A neutrino travels through the massive sun without being absorbed. Even so, indirect evidence of the neutrino's existence accumulated.

Scientists used the law of conservation of momentum to confirm the existence of a third particle in beta decay. Before beta decay, the momentum of the nucleus is zero. After the decay, the total momentum must likewise be zero. By measuring the momentum of the beta ray as well as the momentum of the remaining nucleus, they saw that momentum was not conserved unless a third particle also had momentum. Figure 16–13 shows such an example. Both the recoil momentum of the polonium-210 nucleus and the momentum of the beta rays are directed toward the right and do not total zero. In order to conserve momentum, a third particle traveling to the left is required. This particle was taken to be the neutrino.

Gamma Rays and Quantized Energy Levels in the Nucleus

So far, we have considered that *all* of the alpha rays emitted by a given radioactive source have the same energy. When S. Rosenblum,[12] in 1929, measured the kinetic energy of alpha rays from bismuth-212, he perhaps anticipated that they would have only one value:

$$^{212}_{83}\text{Bi} \rightarrow ^4_2\text{He} + ^{208}_{81}\text{Tl} \tag{11}$$

In Equation 11, the mass on the left side is larger than the total on the right side. The difference in mass is 0.00666 amu, and the kinetic energy of the alpha particle, therefore, would be

$$\text{KE} = (0.00666 \text{ amu})(931 \text{ MeV/amu}) \atop = 6.200 \text{ MeV.}$$

Rosenblum subjected the alpha rays to a magnetic field, causing them to travel in a circle and strike a photographic plate. His results (see Fig. 16–14) show that there are several different kinetic energies. The alpha particles of smaller speed are more easily bent by a magnetic field and travel in a path of smaller radius. The radius and, therefore, position on the photographic plate, indicate the kinetic energy of the alpha particles.

Gamma rays also accompany the alpha decay of bismuth-212. George Gamow realized that there was a connection between these different alpha particle groups and the gamma rays. The thallium-208 nucleus retains some of the 6.200 MeV disintegration energy, and the remainder goes to the alpha particle. Then, the thallium-208 nucleus gets rid of the excess energy by emitting a gamma ray. The disintegration energy—6.200 MeV—is divided in the following way:

Disintegration
energy, 6.200 MeV
  = thallium nucleus in excited state   (12)
    + KE of alpha particle

Thallium nucleus
in excited state
  = thallium nucleus in ground state   (13)
    + gamma ray.

For example, when an alpha particle has a kinetic energy of only 5.87 MeV (α_3), Equation 12 shows that the thallium nucleus must be in an excited state that has an energy of 6.200 MeV minus 5.87 MeV, or 0.33 MeV, above the ground, or normal, state. The thallium nucleus gets rid of this excess energy by emitting a 0.33-MeV gamma ray. Figure 16–15 shows the quantized energy levels, called the **excited states,** and the gamma rays accompanying the de-excitation.

We can see a similarity here to the light emitted by atoms. An atom emits a photon of light when the electron jumps to a lower-energy level. The photon energies are measured in electron-volts. The nucleus, too, has quantized energy levels, and a gamma ray carries off its excess energy, but the gamma ray energies are in the

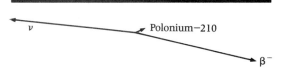

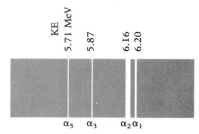

FIG. 16–13   The momentum of the neutrino is needed to conserve momentum.

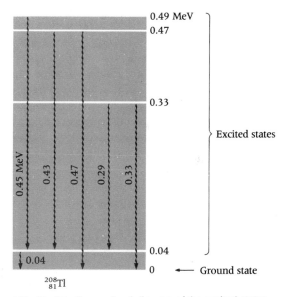

FIG. 16–14   This artist's rendition of the photographic plate shows several groups of alpha particles having differing values of kinetic energy.

FIG. 16–15   Energy-level diagram of the excited states in the thallium-208 nucleus. Gamma rays are emitted when the nucleus gets rid of excess energy. The gamma ray energy is equal to the difference in energy of the two states of the nucleus.

million electron volt range. This high energy has an important application in the field of medicine: Today cobalt-60, not radium, is used in treating cancer because its gamma rays are much more energetic.

Summary

Rutherford and Soddy learned that thorium emitted a fourth kind of radiation—the emanation—which they found to be a penetrating, radioactive, inert gas. A chemical separation procedure showed that thorium contains an impurity that emits the emanation. Several weeks after the separation, thorium regained the ability to emit the emanation, leading to the conclusion that the impurity element is *generated* in the thorium. From many studies, these researchers concluded that radioactivity occurs because the atom disintegrates into two parts: One part is the ray, and the other is the remaining atom of a different element. The disintegration continues in many steps until a stable atom is reached.

Isotopes are several groups of atoms of the *same* element having different atomic masses, the same chemical properties, and, if radioactive, different half-lives. During one half-life, the number of radioactive nuclei in the source and the activity (rays per second) of the source drop by a factor of 2.

Nuclear energy results from the transformation of mass into energy as the nucleus disintegrates (1 amu = 931 MeV).

Natural radioactivity occurs because the nucleus can disintegrate into a smaller total mass; otherwise, the nucleus is stable.

Rutherford placed an alpha source in a chamber of nitrogen gas and detected particles emerging at one end as they hit a fluorescent screen. These particles, so penetrating that they could not be alpha particles, are protons (nuclei of hydrogen atoms) that were released during the collision. Nuclei contain protons.

Penetrating radiation (able to pass through 2 cm of lead) is produced when alpha particles strike a beryllium target. When this radiation strikes paraffin, protons are knocked out, which suggested to the Curie-Joliots that it consists of 50-MeV gamma rays. Chadwick measured the speed of the protons by noting the thickness of material required to stop them. When the penetrating radiation struck nitrogen atoms in a cloud chamber, the length of the track indicated the speed of the struck nitrogen atoms. Using conservation of momentum and energy, Chadwick found that the penetrating radiation consisted of neutral particles—neutrons—having about the same mass as a proton. There are Z protons and A minus Z neutrons in the nucleus.

Electrons are not found in the nucleus. When the nucleus emits a beta ray (electron), one of its neutrons decays into a proton, beta ray, and neutrino. The disintegration energy is shared by the beta ray and neutrino, which accounts for the continuous distribution of beta ray energies.

Gamma rays accompany alpha decay (and beta decay). After the decay, the remaining nucleus has an excess energy, which it gets rid of by emitting one or more gamma rays. The disintegration energy is shared by the alpha particles and the gamma rays.

True or False Questions

Indicate whether the following statements are true or false. Change all of the false statements so that they read correctly.

16-1 The following statements deal with the experiments of Rutherford and Soddy that uncovered the nature of radioactivity:

    a. When Rutherford placed thorium on the bottom plate of the electrical apparatus for studying radioactivity, he found the electrical current to be erratic.

b. The electrical current was erratic because a draft of air carried off the alpha and beta rays.

c. Thorium emits a fourth type of radiation, called the emanation, which is easily stopped by paper.

d. Soddy found that the emanation reacted strongly with sulfuric acid.

e. The emanation is an inert gas.

f. The emanation emits beta rays and has a half-life of 4 days.

g. Soddy found a chemical impurity in the thorium sample.

h. Immediately after the chemical separation of the impurity from thorium, the impurity element emitted the emanation, but the pure thorium did not.

i. Several weeks after the chemical separation of the impurity from thorium, the impurity element emitted the emanation, but the pure thorium did not.

j. As a result of the observations indicated in **h** and **i,** Rutherford and Soddy concluded that the impurity element had been generated in the *initially pure* thorium sample.

k. Several weeks after the chemical separation of the impurity from thorium, the initially pure thorium sample did not show any trace of the impurity element.

l. Because the impurity element emits alpha particles and the emanation at the same rate, Rutherford and Soddy concluded that an impurity atom breaks up into an alpha particle and an atom of the emanation.

m. A thorium atom disintegrates into an alpha particle and an impurity atom, which explains how the impurity is generated in an initially pure thorium sample.

16–2 Two radioactive samples have identical chemical properties, but different half-lives.

a. Both samples have the same atomic number Z.

b. Both samples have the same atomic mass number A.

c. If these two samples are mixed together, it is possible to separate them using chemical separation techniques.

d. The two samples emit the same spectrum of light.

16–3 A sample of a radioactive isotope contains 1 million atoms, and initially emits alpha particles at the rate of 50 alpha particles per second. The half-life of the radioactive isotope is 4 hours.

a. After 8 hours, the sample contains 250,000 atoms that have not yet emitted an alpha particle.

b. After 8 hours, the sample still emits 50 alpha particles per second.

16–4 The isotope polonium-201 is radioactive and decays as follows:

$$^{201}_{84}Po \rightarrow {}^{4}_{2}He + {}^{197}_{82}Pb.$$

(Pb is the chemical symbol for lead.)

a. The isotope polonium-201 is radioactive because the mass of the polonium atom is greater than the sum of the masses of a helium atom and a lead atom.

16–5 The isotope neon-20 is stable and the following reaction does *not* happen in nature:

$$^{20}_{10}Ne \rightarrow {}^{4}_{2}He + {}^{16}_{8}O.$$

a. The isotope neon-20 does not emit alpha particles because the atomic mass number of neon is equal to the sum of the atomic mass numbers of helium and oxygen.

16–6 In a cloud chamber,
a. the air becomes supersaturated with water vapor when the piston compresses the air.

b. a track is produced when water vapor condenses along the path of a particle.

THE NUCLEUS

16-7 a. Experiment shows that the magnetic field produced by a nitrogen nucleus is close to the magnetic field produced by a single proton.

b. Due to its larger mass, a proton generates a larger magnetic field than an electron.

c. If a nitrogen nucleus ($A = 14$, $Z = 7$) contained only protons and electrons, scientists thought that there should be 14 protons and 7 electrons.

d. If two bar magnets of equal strength are aligned, so that the north pole of one is adjacent to the south pole of the other, the *net* magnetic field produced by the two magnets is nearly zero.

e. If 2 protons are aligned (1 with spin up and the other with spin down), so that the north pole produced by 1 proton is adjacent to the south pole produced by the other, the *net* magnetic field of the 2 protons is nearly zero.

f. With proper alignment, the magnetic field produced by 14 protons and 6 electrons should be nearly zero.

g. With proper alignment, the magnetic field produced by 14 protons and 7 electrons (thought to be in the nitrogen nucleus) should be close to that produced by a single electron.

h. The nitrogen nucleus cannot contain electrons because the magnetic field produced by the nitrogen nucleus is much smaller than that produced by a single electron.

16-8 When beta rays from a radioactive source are subjected to a magnetic field, all of the beta rays travel in a circular path of the same radius, showing that the beta rays all have the same speed.

16-9 When alpha particles strike beryllium, penetrating rays are produced.

a. The Curie-Joliots thought that this penetrating radiation, sufficiently energetic to knock protons from paraffin, must be beta rays having an energy of 50 MeV.

b. Chadwick showed that the penetrating radiation was due to particles called neutrons, which have a mass of about 1 amu.

c. When the neutrons entered a cloud chamber containing nitrogen, the struck nitrogen nuclei had a larger speed than the protons ejected from paraffin.

16-10 When the nucleus emits a beta ray,

a. one of the protons in the nucleus decays into a neutron, a beta ray, and a neutrino.

b. the disintegration energy is shared by the beta ray and the neutrino.

Questions for Thought

16-1 Rutherford measured the activity of uranium and thorium using the electrical method.

a. What differences did he observe?

b. How did the motion of air around the equipment affect his measurements?

16-2 Why did Rutherford theorize that thorium emitted a fourth type of radiation, called the emanation?

16-3 a. Describe how Rutherford measured the activity of the emanation.

b. What was the half-life of the emanation?

c. List several other properties of the emanation.

16-4 Soddy found an impurity element in the thorium sample.

a. Immediately after chemical separation, was the emanation emitted by thorium or by the impurity element?

b. What happened several weeks after the chemical separation had been completed?

c. What did Rutherford and Soddy conclude?

16-5 a. What was the half-life of the impurity element?

b. How does the rate at which the impurity element emits the emanation change?

16-6 Rutherford and Soddy concluded that the thorium atoms disintegrate and this disintegration continues. Explain how experimental results support their conclusions.

16-7 Use Rutherford and Soddy's displacement laws and compare the atomic mass number and atomic number of thorium and the impurity element.

16-8 Explain what led Soddy to propose the existence of isotopes.

16-9 Two radioactive samples have identical chemical properties, but different half-lives. Do atoms of both samples have the same atomic number Z? The same atomic mass?

16-10 What is a mass spectrograph?

16-11 What criterion determines whether the atoms of an isotope will be radioactive or stable? Why doesn't oxygen-16 emit alpha particles?

16-12 a. Briefly describe how Rutherford discovered the proton.

b. Write the nuclear reaction equation showing how a proton is formed.

16-13 Describe why a track is produced by a charged particle through a cloud chamber. Why doesn't a neutral particle produce a track?

16-14 Which produces a larger magnetic field—a spinning electron or a spinning nucleus?

Why does your answer present a problem for having electrons in the nucleus?

16-15 Alpha particles from a radioactive source have a definite speed, while beta rays have a continuous distribution of speeds. Explain how, at first glance, it might appear that energy is not conserved.

16-16 a. What kind of radiation is produced when alpha particles strike beryllium?

b. What happens when this radiation strikes paraffin?

c. Is this radiation absorbed by 2 cm of lead?

d. Could this radiation consist of charged particles?

e. What did the Curie-Joliots think the radiation consisted of?

f. Why did Rutherford and Chadwick think that the Curie-Joliots' interpretation was wrong?

16-17 Briefly describe the experiment that Chadwick carried out to find the mass of the neutron.

16-18 An atom of $^7_3$Li has _____ orbital electrons, _____ protons, and _____ neutrons. Another isotope of lithium will have a different number of _____ .

16-19 The nucleus does not contain electrons. What happens in the nucleus when it emits a beta ray (which is another name for an electron)?

16-20 How do neutrinos account for the continuous distribution of beta ray energies?

Questions for Calculation

16-1 Of the atoms of chlorine, 76 percent have an atomic mass of 35 amu and 24 percent have an atomic mass of 37 amu. What is the average atomic mass of chlorine? Check your answer with the value given in Fig. 1-3 (p. 6).

16-2 a. A ball is thrown vertically into the air with a speed of 8 m/sec. Can you determine how high the ball will rise and how long this will take? Will you find

these results every time you repeat the toss?

b. Radium has a half-life of 1620 years. Suppose that you consider 1 atom of radium. Can you predict when it will

---

decay? If you have another atom of radium, will it decay after the same time as the first atom?

c. Explain why your answers to **a** and **b** are so different. What laws cover these two situations?

†**16-3** Uranium-238 emits alpha particles.

    a. What nucleus remains after the decay?

    b. How much nuclear energy is released during the decay? (See Table 16-1.)

    c. Neglecting the small recoil momentum of the larger nucleus, what is the kinetic energy of the alpha particle?

16-4 Polonium-201 has a half-life of 18 minutes, and decays as follows:

$$^{201}_{84}Po \rightarrow {}^4_2He + {}^{197}_{82}Pb.$$

                     (Pb is the chemical symbol for lead.)

If there are 500,000 atoms of polonium initially, then there are _____ atoms of polonium and _____ atoms of lead after 18 minutes; _____ atoms of polonium and _____ atoms of lead after 36 minutes; and _____ atoms of polonium and _____ atoms of lead after 54 minutes.

16-5 Write nuclear reaction equations showing how alpha particles transform fluorine-19 into neon, and sodium-23 into magnesium.

†**16-6 a.** Write a nuclear reaction equation that shows how a proton is formed when an alpha particle strikes an oxygen-16 nucleus.

    b. Rutherford also used oxygen gas in the apparatus shown in Fig. 1-22 (p. 24). Did he detect protons? Assume that he used the same source and construct a ME-KE table. (The kinetic energy of the alpha particles from this source is 7.7 MeV.)

c. A beam of 10-MeV alpha particles from a particle accelerator enters a container filled with oxygen-16. Will protons be formed? If so, find their kinetic energy.

16-7 When alpha particles (kinetic energy = 7.7 MeV) strike an aluminum target ($Z = 13$, $A = 27$), protons are formed. Construct a ME-KE table to show how atomic energy is released. In what form is this energy?

16-8 Beta rays are emitted by atoms of carbon-14.

    a. What is the maximum possible kinetic energy of the beta rays?

    b. Do all of the electrons have this energy?

16-9 Rewrite Table 16-4 using nuclear masses instead of atomic masses in the ME column. Show that the kinetic energy of the proton and oxygen nucleus is the same as found before.

16-10 $^{226}_{88}Ra$ is formed when $^{238}_{92}U$ disintegrates with the successive emission of an alpha particle, a beta ray, a beta ray, an alpha particle, and an alpha particle. Write five nuclear reaction equations to show how radium is formed.

†**16-11** Chadwick thought that possibly gamma rays might be produced, when alpha particles struck beryllium, as follows:

$$^4_2He + {}^9_4Be \rightarrow {}^{13}_6C + \gamma.$$

The alpha particles from the source have a kinetic energy of 5.3 MeV.

    a. Construct a ME-KE table and find the energy of gamma rays produced in this way.

    b. Would these gamma rays have sufficient energy to set protons into motion?

16-12 Chadwick used the following reaction to determine the mass of the neutron precisely:

$$^{11}_5B + {}^4_2He \rightarrow {}^{14}_7N + {}^1_0n.$$

The kinetic energy of the alpha particle, expressed in mass units, was 0.00565 amu. The kinetic energy of the neutron was observed to be 0.0035 amu, and the kinetic energy of the nitrogen nucleus was observed to be 0.00061 amu. The masses obtained from Aston's mass spectrograph measurements were: boron-11, 11.0825 amu; helium-4, 4.00106 amu; and nitrogen-14, 14.0042 amu. Find the mass of the neutron using a ME-KE table and compare it with the presently accepted value of 1.008665 amu.

16–13 a. Write a nuclear reaction equation that shows correctly the beta decay of bismuth-210.

b. If a beta ray emitted by bismuth-210 has a kinetic energy of 0.5 MeV, the neutrino has an energy of _____ . If a beta ray has a kinetic energy of 0.8 MeV, the neutrino has an energy of

_____ .

16–14 An alpha particle emitted by a bismuth-212 nucleus has a kinetic energy of 5.87 MeV. What energy gamma ray(s) can accompany this decay? (See Fig. 16–15.)

Footnotes

[1]E. Rutherford, *Philosophical Magazine*, 47, Ser. 5(1899):109.

[2]E. Rutherford, *Philosophical Magazine*, 49, Ser. 5(1900):1.

[3]Ibid, p. 2.

[4]Alfred Romer, *The Restless Atom*, (Garden City, N.Y.: Anchor Books, 1960), p. 67.

[5]E. Rutherford, *Philosophical Magazine*, 37, Ser. 6(1919):581.

[6]E. Rutherford and J. Chadwick, *Philosophical Magazine*, 42, Ser. 6(1921):809.

[7]E. Rutherford, *Radioactive Substances and their Radiation*, (London: Cambridge University Press, 1913), p. 196.

[8]C. D. Ellis and W. A. Wooster, *Proceedings of the Royal Society*, 117 A, (1927), p. 109.

[9]J. Chadwick, *Adventures in Experimental Physics*, Volume Beta (1972), p. 193.

[10]Ibid, p. 193.

[11]J. Chadwick, *Proceedings of the Royal Society*, 136 A, (1932), p. 692.

[12]S. Rosenblum, *Journ. de Phys.*, 1(1930), p. 438.

Ernest O. Lawrence

CHAPTER 17

Cosmic Rays
to Cyclotrons

Introduction

This chapter continues our study of the nucleus, but from a quite different perspective. In Chapter 16, we saw the scientists *observe* nature at work. Here, the scientists *intervene* and actually cause elements to become radioactive in order to probe still further into the workings of the nucleus.

We'll begin with the study of cosmic rays, which leads to the discovery of a new particle: the positron. The positron has the same mass as an electron, but one unit of positive charge. In 1934, Irene Curie-Joliot and Frederick Joliot discovered that positrons are emitted by a boron target bombarded by alpha particles. In short, they had discovered—*and explained*—induced radioactivity, a phenomenon that had so intrigued Marie and Pierre Curie.

The advent of particle accelerators greatly advanced the study of induced radioactivity. In 1932, the first accelerator was designed by J. Cockcroft and E. T. S. Walton in England. In their first experiments, they proved that mass and energy are equivalent—$E = mc^2$—as Einstein had postulated in 1905. Within a few months, the cyclotron at the University of California at Berkeley signaled American's dramatic entry into the field of nuclear physics. Since this story of discovery is such a fascinating one, we shall consider the experiments with the cyclotron in some detail.

Cosmic Rays

Charge an electroscope and it will gradually discharge, showing that air conducts a slight current. After the discovery of radioactivity, it was assumed that rays from radioactive substances were responsible. And, indeed, *most* of the discharging is due to this. But, in 1911, the Austrian physicist V. F. Hess and W. Kolhoerster (1883–1964) carried an electroscope aloft in a balloon to heights up to 9000 m. They found that, above a few hundred meters, the rate of discharge increased. At 9000 m, it was 7 times greater than at the earth's surface. Hess suggested that penetrating radiation falling upon the earth from outside was the cause—**cosmic rays.**

The cosmic radiation striking the upper reaches of the earth's atmosphere consists of about 90 percent protons, 9 percent alpha particles, and 1 percent heavier nuclei. These particles have a *giga*ntic energy, with their average energy being 1 GeV, where 1 GeV equals 1000 MeV. (The G stands for giga, as in gigantic.) These primary particles strike atoms in the atmosphere and produce secondary radiation, which includes neutrons, gamma rays, electrons, and positrons. Many other kinds of so-called elementary particles have also been observed in cosmic rays. We shall have more to say about this later.

Dirac's Prediction of the Positron

The English physicist P. A. M. Dirac (b. 1902) had predicted the existence of the positron in 1930. He had developed a wave equation for electrons that included the effects of relativity, something Schroedinger did not do. Out of this theory, the notion of positrons emerged, as well as the concept of electron spin, which was also missing in Schroedinger's theory.

Discovery of the Positron

In 1932, C. D. Anderson identified a **positron** (a particle having the same mass as an electron, but positively charged) in cloud chamber photographs of cosmic rays. He relates how this discovery came about:

It has often been stated in the literature that the discovery of the positron was a consequence of its theoretical prediction by Dirac, but this is not true. The discovery of the positron was wholly accidental. . . . (Dirac's theory) played no part whatsoever in the discovery of the poistron.[1]

When gamma rays in cosmic rays strike electrons (in atoms of the earth's atmosphere), some of the energy of the gamma ray is transferred to the electron, and the gamma ray has a correspondingly smaller energy. This process is called the *Compton effect* (see Chapter 10, p. 242), and the struck electrons are called **Compton electrons.** Anderson set out to measure the percentage of electrons having a given energy. He used a cloud chamber and operated it in a strong magnetic field, which caused the electrons to travel in a circular path. The radius and length of the track in the cloud chamber yielded the kinetic energy of the electron. Then Anderson observed something unexpected:

. . . about half of the high-energy cosmic ray particles observed were positively charged and therefore could not represent Compton electrons. At the time they were presumed to be protons. . . . It was, of course, important to provide unambiguous identification of these unexpected particles of positive charge.

. . . a lead plate was inserted across the center of the chamber. . . . It was not long after the insertion of the plate that a fine example [see Fig. 17–1] was obtained. . . . Ionization and curvature measurements clearly showed this particle to have a mass much smaller than that of a proton, and indeed, a mass entirely consistent with an electron mass.[2]

In the middle of the cloud chamber, Anderson placed a 6-mm lead plate. When a charged particle passes through the lead, it loses an amount of energy that depends upon its mass and its charge. As a result of its lower energy, the track has a smaller radius because the magnetic field bends it more easily. In Fig. 17–1, the particle is traveling upward because its radius is smaller above the lead plate. This was really quite a fluke because most cosmic rays would be traveling downward. In this magnetic field, a charged particle traveling upward would bend to the left if positively charged and to the right if negatively charged. The track definitely shows

COSMIC RAYS TO CYCLOTRONS

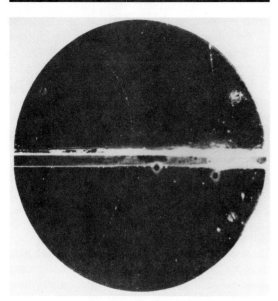

FIG. 17–1 Anderson's cloud chamber photograph showing the track made by a positron. (AIP Niels Bohr Library. Courtesy C. D. Anderson)

that it is a positively charged particle. A proton is ruled out because it would curve very sharply after passing through the lead plate, and such a sharp curvature was not observed. The curvature of the tracks before and after passing through the lead plate, and the particle's energy loss there, led Anderson to conclude that the particle was a positron:

After the existence of the positron was clearly indicated, the question naturally arose as to how they came into being. . . . Did positrons somehow acquire their positive charge from the nucleus? Could they be ejected from the nucleus when there were presumably no positrons present in the nucleus? The idea that they were created out of the radiation itself did not occur to me at that time, and it was not until several months later when Blackett and Occhialini suggested the pair-creation hypothesis that this seemed the obvious answer to the production of positrons in the cosmic radiation.[3]

Pair Production

According to the theory (now confirmed) of P. M. S. Blackett (1897–1974) and G. Occhialini (b. 1907), **pair production** occurs when a gamma ray in the cosmic rays interacts with a nucleus (in the cloud chamber apparatus) causing a positron (β^+) and an electron (β^-) to be formed. Pair production:

$$\gamma \rightarrow \beta^+ + \beta^- \qquad (1)$$

Pair production only occurs when the gamma ray encounters a nucleus because the forward motion of the struck nucleus is needed to conserve momentum. Since the electron mass is equivalent to 0.511 MeV, the gamma ray must have sufficient energy to be transformed to 2 electron masses, or have an energy equal to, or greater than, 1.022 MeV.

The positron will not travel very far before it encounters another electron and, when it does, the positron and electron will **annihilate,** forming two gamma rays. Two gamma rays are required to conserve momentum. Anderson

FIG. 17—2 Frederick Joliot and Irene Curie-Joliot in their laboratory in 1934. (AIP Niels Bohr Library. Courtesy Radium Institute)

observed the positron before it encountered an electron. Annihilation:

$$\beta^+ + \beta^- \rightarrow \gamma + \gamma \qquad (2)$$

Pair production also cleared up another question: How are gamma rays absorbed in passing through matter? Before this discovery, two ways were known: (1) The gamma ray could be absorbed by an atom and its energy transformed to an orbital electron, knocking it out of the atom (the photoelectric effect); and (2) the gamma ray could scatter from an orbital electron in an atom and lose some of its energy (the Compton effect). The problem was that these two processes did not seem to account for the strong absorption of gamma rays. The additional absorption was due to pair production (Equation 1).

Induced Radioactivity

We return now to a phenomenon that so puzzled the Curies: **induced radioactivity.** They found that *any* substance placed near a tube of radium *became* radioactive, even though it was not ordinarily so. This radioactivity, however, quickly decreased. For example, the radioactivity induced in glass had a half-life of four days. It seems appropriate that their daughter Irene Curie-Joliot and her husband Frederick Joliot (Fig. 17—2) should be the ones to provide the basic understanding of induced radioactivity.

COSMIC RAYS TO CYCLOTRONS

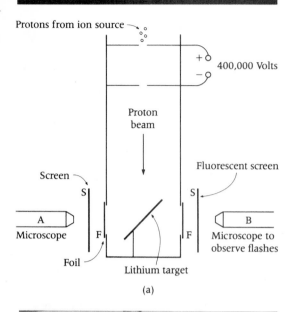

Protons from ion source

+ 400,000 Volts
−

Proton beam

Screen

Fluorescent screen

S

S

A
Microscope

F

F

B
Microscope to observe flashes

Foil

Lithium target

(a)

(b)

FIG. 17–3 (a) A diagram of the Cockcroft-Walton accelerator. (b) Crockcroft and Walton's electrostatic accelerator, with Cockcroft underneath, taking measurements. (Cavendish Laboratory)

In 1934, they observed that boron emits positrons while being bombarded by alpha particles. The positrons were observed in a cloud chamber. At first, they proposed—incorrectly—that the reaction occurred in a single step, as follows:

$$^{10}_{5}B + ^{4}_{2}He \rightarrow ^{1}_{0}n + \beta^{+} + ^{13}_{6}C. \qquad (3)$$

Carbon-13 is a stable isotope. Thus, this view had to be modified when they learned that positrons *continued to be emitted even after the alpha particle source was removed.* The emission of positrons followed the well-known decay law for radioactivity, with the half-life determined to be about 10 minutes. If Equation 3 were correct, a positron would be emitted *promptly only when* the boron nucleus had absorbed an alpha particle. Equation 3 could not explain the delayed emission of positrons after the alpha particle source had been removed.

To solve this dilemma, Curie and Joliot proposed a two-step process. First, the isotope nitrogen-13 was produced according to the following reaction:

$$^{10}_{5}B + ^{4}_{2}He \rightarrow ^{1}_{0}n + ^{13}_{7}N. \qquad (4)$$

Nitrogen-13 is an isotope not found in nature. For example, when a sample of nitrogen gas is analyzed in a mass spectrograph, only two isotopes of nitrogen are found: nitrogen-14 and nitrogen-15. It seemed reasonable to suggest that nitrogen-13 would be radioactive and would decay to a stable isotope by emitting a positron:

$$^{13}_{7}N \rightarrow \beta^{+} + ^{13}_{6}C + \nu. \qquad (5)$$

Carbon-13 is an isotope of carbon that is found in nature. A neutrino (ν) is also emitted. A neutrino has no mass and no charge, but does have energy. We see that there are a total of 7 *positive* charges on both sides of Equation 5. Positrons are not contained inside the nucleus. When a nucleus emits a positron, a proton in the nucleus breaks up into a neutron, positron, and a neutrino. The neutron remains in the nucleus and the positron and neutrino are ejected:

$$p \rightarrow n + \beta^+ + \nu. \qquad (6)$$

Curie and Joliot continued their study of induced radioactivity by bombarding targets of aluminum ($^{27}_{13}$Al) and magnesium ($^{24}_{12}$Mg) with alpha particles. They had solved the puzzle of *induced radioactivity*. It was due to the formation of isotopes—radioactive isotopes—not found in nature. These isotopes then decayed to stable naturally occurring isotopes. For this work, Irene Curie-Joliot and Frederick Joliot received the Nobel Prize for chemistry in 1935.

The study of induced radioactivity advanced quickly with the use of particle accelerators to produce energetic beams of particles. In 1932, J. D. Cockcroft (1897–1967) and E. T. S. Walton (b. 1903) of the Cavendish Laboratory designed an accelerator capable of producing protons having a kinetic energy of 700,000 electron volts. Also, in 1932, E. O. Lawrence (1901–1958) and M. S. Livingston (b. 1905) at the University of California designed a cyclotron capable of producing energetic beams of particles.

Cockcroft-Walton Accelerator

In the **Cockcroft-Walton accelerator** (Fig. 17–3a), protons from an ion source (see thought question 7–22) are accelerated by a voltage of up to 400,000 volts. The proton beam travels down an evacuated tube and strikes a target.

In their first experiment, Cockcroft and Walton used a 0.25-MeV proton beam aimed at a lithium target. The particles produced by the collision pass through the thin foil F and strike a fluorescent screen S, producing flashes of light. Cockcroft and Walton observed these flashes through the microscopes A and B. The brightness of the flashes indicated that they were alpha particles. Further, whenever one experimenter observed a flash in A, the other observed a flash

TABLE 17–1 PROTONS FROM ACCELERATOR STRIKE LITHIUM, PRODUCING TWO ALPHA PARTICLES

	KE (amu)	ME (amu)	TE (amu)
Before collision	0.000268	7.01600 + 1.007825 = 8.023835	8.024103
After collision	KE of 2 alpha particles	4.00260 + 4.00260 = 8.00520	8.024103

in B. This suggested that the alpha particles were produced in pairs, according to the following:

$$^{7}_{3}\text{Li} + ^{1}_{1}\text{H} \rightarrow ^{4}_{2}\text{He} + ^{4}_{2}\text{He}. \qquad (7)$$

They measured the kinetic energy of the alpha particles by placing absorbers between F and S and finding the thickness needed to stop the alpha particles. Since they knew the relationship between absorber thickness and initial kinetic energy, they found the kinetic energy of each alpha particle to be 8.75 MeV. Kinetic energy was being created in the reaction. According to Einstein's mass-energy relationship, mass is being converted to kinetic energy.

Furthermore, Cockcroft and Walton were able to test this relationship because the atomic masses of lithium, hydrogen, and helium were known from mass-spectrograph measurements. From the mass-energy table (Table 17–1), we can find the kinetic energy expected from the conversion of mass.

Before the collision, only the proton has kinetic energy. Since 1 amu is equivalent to 931 MeV, 0.25 MeV equals (0.25 MeV)/(931 MeV/amu), or 0.000268 amu. The total energy before collision is therefore 8.024103 amu and must be equal to the total energy after collision. The total kinetic energy of the two alpha particles is given by

FIG. 17–4 The "old" Radiation Laboratory at the University of California, Berkeley, 1934. M. Stanley Livingston (left) and Ernest O. Lawrence with the 27-inch cyclotron. (Courtesy Lawrence Radiation Laboratory, University of California, Berkeley)

KE of 2 alpha particles

$$= 8.024103 - 8.00520$$
$$= 0.018903 \, amu$$
$$= (0.018903 \, amu)(931 \, MeV/amu)$$
$$= 17.60 \, MeV.$$

Cockcroft and Walton measured the kinetic energy of both alpha particles to be 17.50 MeV. Considering the uncertainty associated with the measurement, Einstein's theory and these experimental results are in excellent agreement.

The Cyclotron

M. Stanley Livingston was a graduate student at the University of California in Berkeley when, during the summer of 1930, Professor E. O. Lawrence suggested the development of the **cyclotron** as a suitable investigation for a Ph.D. dissertation. Under Lawrence's direction, Livingston designed and built the first cyclotron, which produced protons having an energy of 1.2 MeV. (1 MeV = 1 million electron-volts.) Livingston relates:

This small and relatively inexpensive machine could split atoms! This was Lawrence's goal. This is why Lawrence literally danced with glee when, watching over my shoulder as I turned the magnet through resonance, the galvanometer (current meter) spot swung across the scale indicating that 1,000,000-volt ions (protons) were reaching the collector. The story quickly spread around the laboratory and we were busy all that day demonstrating million-volt protons to eager viewers.[4]

Figure 17–5 is a schematic diagram of a cyclotron. An ion source is located between two *dees*—appropriately named short, hollow, half-cylinders. The region surrounding the dees is evacuated, so that the protons will not collide with air molecules and be stopped. A powerful electromagnet produces a magnetic field perpendicular to the plane of this paper. The two dees are connected to an alternating high voltage, V_g, so that for a period of time (T_{AV}) D1 is positively charged and D2, negatively charged. The polarity of the voltage then switches and D1 is negatively charged and D2, positively charged for the same time T_{AV}. Each time the proton crosses the gap, it faces a negatively charged dee, and its kinetic energy increases by V_g electron volts. After it passes through the gap, the magnetic field causes the proton to travel in a circular path in the hollow region of the dee. It then comes to the gap again. By this time, the voltage has switched, so that the proton once again faces an accelerating voltage. The beauty of the cyclotron is that the proton requires the same amount of time, T_{AV}, to travel around a half-circle of *any size*. This ensures that the proton is always accelerated, and not decelerated, when it crosses the gap. Let's see why.

The force on a proton in a magnetic field of strength B is given by qvB, where the force is perpendicular to both the speed and the direction of the magnetic field. As a result, the proton travels in a circle and has acceleration v^2/r. Newton's second law gives:

$$F = ma$$
$$qvB = \frac{mv^2}{r}$$
$$\frac{r}{v} = \frac{m}{qB} .$$

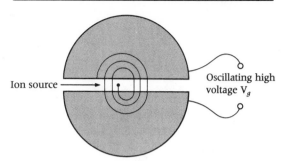

Ion source → Oscillating high voltage V_g

Magnetic field perpendicular to paper

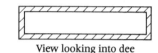

View looking into dee

FIG. 17–5   Diagram of a cyclotron.

Multiplying by π gives

$$\frac{\pi r}{v} = \frac{\pi m}{qB}$$

$$T_{AV} = \frac{\pi m}{qB}, \qquad (8)$$

where πr is the distance around a half circle. This distance divided by speed is the time, T_{AV}, required to travel around the half circle. Since the right side of Equation 8 contains only constants, the time required for a proton to travel around a half circle is the same for *all* values of the radius, and protons are always accelerated as they cross the gap.

The cyclotron was only 11 inches in diameter and the voltage V_g was 2000 volts. In order to produce 1.2 MeV protons, the protons had to cross the gap 600 times.

Livingston continues his story:

Within a few months after hearing the news from Cambridge (about Cockcroft and Walton's experiment) we were ready to try for ourselves. Targets of various elements were mounted on a movable stem which could be swung into the beam of ions (hydrogen ions or protons). The counters clicked, and we were observing disintegrations![5]

Lawrence and Livingston bombarded a lithium target with 0.7-MeV protons and observed the alpha particles from the disintegration. Twice as many disintegrations occurred at this energy compared to Cockcroft and Walton's 0.5 MeV. They also observed the disintegration of boron bombarded by protons:

$$^1_1H + {}^{11}_5B \rightarrow 3\,{}^4_2He.$$

Long before I had completed the 11-inch machine as a working accelerator, Lawrence was planning

the next step . . . a 27-inch cyclotron.*. . . One of the exciting periods was our first use of deuterons (nuclei of the hydrogen isotope ^2_1H containing one proton and one neutron) in the cyclotron.[6]

The deuterons bombarded a target of lithium-6 and the following reaction occurred:

$$^6_3Li + {}^2_1H \rightarrow 2\,{}^4_2He.$$

The alpha particles produced in this reaction had an energy of 12.1 MeV, an energy greater than those found in natural radioactivities. Livingston tells us more:

We also had many successful and exciting moments. I recall the day early in 1934 (February 24) when Lawrence came racing into the lab waving a copy of the Comptes Rendus *and excitedly told us of the discovery of induced radioactivity by Curie and Joliot in Paris, using natural alpha particles on boron and other light elements. They predicted that the same activities could be produced by deuterons on other targets, such as carbon. Now it just so happened that we had a wheel of target inside the cyclotron which could be turned into the beam. . . . We quickly . . . turned the target wheel to carbon, adjusted the counter circuits, and then bombarded the target for 5 minutes. (Then) . . . the counter was turned on, and click-click-click-click-click. We were observing induced radioactivity within less than a half-hour after hearing the Curie-Joliot results.*[7]

Their experiments continued and they bombarded 14 targets with 1.5-MeV protons and 3-MeV deuterons. They found no induced activity using protons, but every target bombarded with deuterons became radioactive and emitted positrons and gamma rays.

In one (of many) experiments, they bombarded nitrogen gas with deuterons.[8,9] The resulting radioactive substance emitted positrons and had a half-life of 126 seconds. To find the energy of the positrons, they put aluminum foils in front of the Geiger counter. From the thickness of aluminum needed to stop the pos-

*The 27-inch cyclotron was developed during 1932 and 1933.

itrons, they determined the positron energy to be 1.7 MeV.

What isotope was emitting positrons? Lawrence and Livingston could not immediately perform a chemical separation of the bombarded nitrogen gas, because the number of radioactive atoms was just too small for that. To answer this question, they had to resort to a bit of chemistry—and ingenuity. To the sample of nitrogen gas that had been bombarded with deuterons, the experimenters added oxygen and hydrogen gas. This mixture was then passed over a heated surface, producing water, which was collected. When this water sample was tested with a Geiger counter, it was found to be radioactive. Lawrence and Livingston concluded that, when nitrogen is bombarded with deuterons, a radioactive isotope of oxygen is produced. Their water sample primarily contained molecules formed with the stable oxygen-16 isotope and a small percentage with the radioactive isotope oxygen-15. With this information, they concluded that the following nuclear reactions took place:

$$^{14}_{7}\text{N} + ^{2}_{1}\text{H} \rightarrow ^{15}_{8}\text{O} + ^{1}_{0}\text{n}$$

$$^{15}_{8}\text{O} \rightarrow ^{14}_{7}\text{N} + \beta^{+} + \nu.$$

In 1934, Livingston left the University of California and went to Cornell University. In the same year, Edwin McMillan came to Berkeley as a research associate and became a professor of physics there in 1946. McMillan was awarded the Nobel Prize for chemistry in 1951. Upon the death of Lawrence in 1958, McMillan became the director of the Lawrence Berkeley Laboratory.

McMillan relates:

At this time (1935) biological experiments were started. . . . The first one that I recall, and I think the first use anywhere of an artificially produced radioisotope in human beings, was an early experiment of Joseph Hamilton in which he measured the circulation time of the blood by a very primitive method. The experimental subject takes some radioactive sodium dissolved in water in the form of sodium chloride, drinks it, and then has a Geiger counter

which he holds in his hand, so that when the radioactive sodium reaches the hand, it starts to register. His hand is in a lead box so that the stuff that's just in his body doesn't affect the counter by gamma rays.[10]

Radioactive sodium is produced in the following reaction:

$$^{2}_{1}\text{H} + ^{23}_{11}\text{Na} \rightarrow ^{24}_{11}\text{Na} + ^{1}_{1}\text{H}.$$

The radioactive sodium then decays as follows:

$$^{24}_{11}\text{Na} \rightarrow ^{24}_{12}\text{Mg} + \beta^{-} + \nu + \gamma.$$

Sodium $-$ 24 has a half-life of 15 hours.

Today, the use of tracers plays an important role in medicine, biology, agriculture, and industry.

Another highlight from 1936 was the first time that anyone tried to make artificially a naturally occurring radionuclide*. . . . This, I think, was a fairly classical experiment because there were then some people who didn't quite believe that artificial radioactive materials were on the same status as the naturally occurring ones.[11]

Bismuth was bombarded with 6-MeV deuterons and the following reaction took place:

$$^{209}_{83}\text{Bi} + ^{2}_{1}\text{H} \rightarrow ^{1}_{1}\text{H} + ^{210}_{83}\text{Bi}.$$

Bismuth-210 was radioactive and emitted beta rays:

$$^{210}_{83}\text{Bi} \rightarrow \beta^{-} + ^{210}_{84}\text{Po} + \nu.$$

The half-lives and energies were identical to those of naturally occurring bismuth-210 and polon-

---

*A radioactive nucleus.

ium-210. (This is the isotope of polonium that Marie Curie found in uranium.)

In 1937, the cyclotron was upgraded to 37 inches, producing deuterons up to an energy of 8 MeV. In 1938, Emilio Segre and C. Perrier discovered the first artificial element, **technetium,** in a molybdenum strip from the cyclotron that had received bombardment by the beam. Technetium has an atomic number Z of 43, but does *not* occur naturally (see Fig. 1–3). Its name is taken from the Greek word *technetos,* which means "artificial." The deuteron bombardment of the molybdenum produced technetium in the following way:

$$^2_1H + ^A_{42}Mo \rightarrow ^{A+1}_{43}Tc + ^1_0n.$$

The isotope technetium-95 is useful for tracer work because it has a half-life of 60 days and produces energetic gamma rays.

In 1938, the 60-inch cyclotron was constructed. In 1939, Ernest Lawrence was awarded the Nobel Prize for his development of the cyclotron and studies of induced radioactivity.

McMillan continues:

Now Ernest Lawrence was never a man who wanted to rest on achievement; he always wanted to go a step farther. I think it was this forward-looking spirit, and his ability to communicate it to others, that was his true greatness. So, even though the 60-inch cyclotron was a beautiful machine, was running fine, and was doing a great deal of important work, he had this dream of 100 million volts.[12]

In 1940, Edwin McMillan and Philip Abelson discovered the first **transuranic element:** an element with atomic number Z greater than 92. The deuteron beam from the cyclotron was aimed at a target producing neutrons, which were then directed onto a uranium target. A nucleus of uranium-238 absorbs a neutron to form uranium-239, which then emits a beta ray to form the element McMillan called **neptunium.**

$$^{238}_{92}U + ^1_0n \rightarrow ^{239}_{92}U$$
$$^{239}_{92}U \rightarrow \beta^- + ^{239}_{93}Np + \nu.$$

Neptunium-239 also emits a beta ray to form plutonium-239.

$$^{239}_{93}Np \rightarrow ^{239}_{94}Pu + \beta^- + \nu.$$

McMillan goes on:

But then the war came along and the whole effort of the Laboratory was diverted to other things. The magnet for this cyclotron was used for research on the electromagnetic isotope separation process (to separate fissionable uranium-235 from uranium-238) and it wasn't until quite a while later that it came back to use as a cyclotron.[13]

The 184-inch cyclotron at the University of California Lawrence Radiation Laboratory produces particles having an energy of 730 MeV—an energy even exceeding Lawrence's dream.

Summary

Anderson discovered positrons (particles having the same mass as electrons, but positively charged) in cloud chamber photographs from the way in which they deflected in a magnetic field. When a gamma ray strikes a nucleus, the gamma ray changes into an electron and a positron. The positron is easily absorbed by matter because a positron and an electron annihilate each other, forming two gamma rays.

After bombarding a boron target with alpha particles, the Curie-Joliots found that positrons continued to be emitted even after they removed the alpha particle source. They had produced an isotope not found in nature and it was radioactive, decaying with a characteristic half-life. When a nucleus emits a positron, a proton changes into a positron, neutron, and neutrino.

Cockcroft and Walton used protons from their accelerator (up to 0.4 MeV) to bombard a lithium-7 target, forming two alpha particles, whose kinetic energy they measured. Since they knew the masses of all particles, they predicted, according to Einstein's theory, the kinetic energy of the alpha particles and compared it with the experimental values. Agreement was excellent. They were the first to verify that mass and energy are interchangeable.

Lawrence and Livingston designed and built the first cyclotron, which produced 1.2-MeV protons. The protons travel in a circular path inside of two evacuated dees due to a perpendicular magnetic field. The polarity of the voltage across the dees alternates periodically, so that a proton is accelerated each time it crosses the gap.

Physicists used the cyclotron to: (1) produce energetic alpha particles by bombarding various targets with protons; (2) study induced radioactivity; (3) produce artificial radioactive isotopes by bombarding targets with deuterons (nuclei of ^2_1H atoms)—for example, radioactive sodium-24 (which emits a beta ray) is produced when deuterons bombard sodium-23; (4) pioneer the use of tracers in medical applications—for example, sodium-24 was used to study the circulation of the blood; (5) discover the artificial element technetium, useful as a tracer; (6) produce artificially radioactive isotopes with Z greater than 80; and (7) discover the transuranic elements (elements with Z greater than 92), neptunium-239 and plutonium-239. These are just some of the important research purposes to which the cyclotron has been put.

The 184-inch cyclotron at the Lawrence Radiation Laboratory produces protons having an energy of 730 MeV—an energy even exceeding Lawrence's dream of 100 MeV.

True or False Questions

Indicate whether the following statements are true or false. Change all of the false statements so that they read correctly.

17–1 When a charged electroscope is carried aloft in a balloon, the rate at which the electroscope discharges is slower than at the earth's surface.

17–2 Dirac predicted the existence of the positron in 1930 and, in 1932, Anderson set out to prove its existence experimentally.

17–3 In Anderson's cloud-chamber experiment,

    a. the positron has a smaller speed after passing through the lead plate (see Fig. 17–1).

    b. the positron travels in a circular path of larger radius after passing through the lead plate (see Fig. 17–1).

    c. the positron in Fig. 17–1 travels counterclockwise around the circular path.

    d. an electron would travel clockwise around the circular path.

    e. a proton would make a track of much larger radius than a positron because the massive proton is harder to bend.

17–4 A beam of gamma rays having an energy of 1.7 MeV strikes an aluminum target.

    a. Positrons are found to come from the aluminum target.

    b. When a gamma ray interacts with an aluminum nucleus, the gamma ray disappears, but 2 positrons are produced.

    c. If the gamma rays have an energy of 0.5 MeV, positrons are still found to come from the aluminum target.

17–5 When a positron interacts with an electron, both disappear, and two gamma rays are produced.

17–6 When Irene Curie-Joliot and Frederick Joliot bombarded a boron target with alpha particles from a radioactive source, they found

---

a. that positrons came from the boron target.

b. that positrons were no longer emitted when the radioactive source was removed.

17–7 When a target is bombarded by energetic particles, induced radioactivity occurs because an isotope not found in nature is produced.

17–8 An isotope not found in nature is always radioactive.

17–9 When a nucleus emits a positron, a proton in the nucleus breaks up into a neutron, a positron, and a neutrino.

17–10 In a cyclotron,
a. the protons travel in a circular path due to the voltage connecting the two dees.

b. the protons are accelerated by the magnetic field.

c. a proton traveling in a circular path of 3-inch radius takes the same amount of time for one revolution as a proton traveling in a circular path of 6-inch radius.

Questions for Thought

17–1 What is a positron?

17–2 Anderson discovered the positron in cloud chamber photographs of cosmic rays.

a. What was the original objective of Anderson's experiment? What did he find?

b. Anderson observed a circular track in a cloud-chamber photograph, but didn't know whether it had been traveling clockwise or counterclockwise. Could he tell whether it was a positively or negatively charged particle?

c. Explain how the difficulty in **b** was resolved when Anderson placed a lead plate in the center of the chamber.

d. Explain how Anderson identified a positron in one of his cloud-chamber photographs.

17–3 What is pair production?

17–4 What happens when a positron collides with an electron?

17–5 The Curie-Joliots observed that positrons are produced when alpha particles bombard a boron target.

a. How did they discover induced activity?

b. How did they explain induced activity?

17–6 What happens in the nucleus when it emits a positron?

17–7 A sample of manganese, analyzed using a mass spectrograph, consists *only* of the isotope manganese-55. This information means that atoms of manganese-54:

a. are stable.

b. are radioactive.

c. will never be formed in any manner.

d. Not enough information provided.

17–8 Describe the basic operation of a cyclotron.

17–9 Nuclear physicists learn a great deal about the forces between protons and neutrons in the nucleus by determining the quantized energy levels, such as that shown in Fig. 16–15 (p. 385) for thallium-208. One simple way to do this is to scatter protons from a target. Some of the protons will scatter from the target nuclei just like a billiard ball collision and have almost the same energy as that of the incident beam. Other protons will have a smaller energy. For example, if 18-MeV protons are scattered from yttrium-89, some protons have an energy of almost 18 MeV, while other groups of protons have an energy of 17.09, 16.49, and 16.25 MeV (Ph.D. research dissertation M. Stautberg Greenwood).

    a. What happened to the rest of the protons' energy?

b. How does this information give us information about the yttrium-89 nucleus?

c. Could gamma rays be found when protons bombard the yttrium-89 target?

Questions for Calculation

17–1 When alpha particles strike an aluminum-27 target, neutrons and a radioactive isotope can sometimes be formed. The radioactive isotope then emits positrons, according to a 2.5 minute half-life.

    a. Write the two nuclear reaction equations.

    b. Suppose that 1 million radioactive nuclei are formed and the alpha particle source is then removed. After 10 minutes, how many radioactive nuclei remain, and how many stable nuclei have been formed?

17–2 In Cockcroft and Walton's experiment, protons bombard a lithium-7 target, forming two alpha particles. Some alpha particles enter a microscope placed at 90° with respect to the proton beam (Fig. 17–3a). Particles having an angle between 85° and 95° will be observed in the microscope. Suppose that the alpha particle entering microscope A has exactly a 90° angle. Use conservation of momentum arguments to explain which one of the following diagrams shows how the other alpha particle enters microscope B (see Fig. 17–6).

17–3 Cockcroft and Walton also bombarded fluorine-19 and sodium-23 targets with protons. In each collision, *one* alpha particle was produced along with another isotope. Write the nuclear reaction equations.

†17–4 The first cyclotron was 11 inches in diameter and produced protons having a kinetic energy of 1.2 MeV.

    a. Find the speed of 1.2-MeV protons. (Mass of proton $= 1.67 \times 10^{-27}$ kg. 1 MeV $= 1.6 \times 10^{-13}$ joules).

    b. Find the time T_{AV} for the proton to

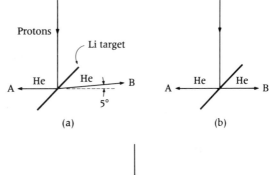

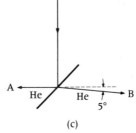

FIG. 17–6   Diagram for calculation question 17–2.

travel around a half circle of radius 5.5 inches, or 0.14 m.

    c. What is the time T_{AV} for a proton to travel around a half circle of radius 0.07 m.

    d. Find the speed of a proton traveling in a circular path of 0.07-m radius.

    e. Find the kinetic energy of a proton in MeV traveling in a circular path of 0.07-m radius.

    f. If the voltage across the gap V_g was 2000 volts, how many times has the proton in **e** crossed the gap?

17–5 To produce protons having an energy greater than about 10 MeV, the design of the conventional cyclotron must be modified in order to take into account relativistic effects.

    a. From the definition of momentum at low and at high speeds, show that we can think of an effective mass defined as

$$m_{eff} = \frac{m_0}{\sqrt{1 - v^2/c^2}}.$$

    b. How does the effective mass in **a** affect the time T_{AV} given in Equation 8?

    c. Explain why a cyclotron of conventional design cannot acceleration protons to speeds near that of light.

17–6 Identify the particle or isotope in the following nuclear reactions:

    a. $^{24}_{12}Mg + ^{4}_{2}He \rightarrow ? + ^{27}_{14}Si.$

    $^{27}_{14}Si \rightarrow ? + ^{27}_{13}Al.$

    b. $^{27}_{13}Al + ^{2}_{1}H \rightarrow ^{25}_{12}Mg + ?$

    c. $^{27}_{13}Al + ^{2}_{1}H \rightarrow ^{28}_{13}Al + ?$

    $^{28}_{13}Al \rightarrow ? + ^{28}_{14}Si.$

17–7 Nuclear physicists often use the following reactions to see what happens when a neutron in a nucleus is replaced by a proton. Identify the missing part of the reaction.

    a. $^{41}_{19}K + ^{3}_{2}He \rightarrow ? + ^{3}_{1}H.$

    b. $^{89}_{39}Y + ^{1}_{1}H \rightarrow ? + ^{1}_{0}n.$

To add a proton to a nucleus:

    c. $^{88}_{38}Sr + ^{3}_{2}He \rightarrow ? + ^{2}_{1}H.$

17–8 In calculation question 17–7, the $^{3}_{1}H$, $^{1}_{0}n$, and $^{2}_{1}H$ are found to have groups of different energies. In each case, how does this give us information about the quantized energy level? For what nucleus?

17–9 A fission reaction occurs when a slow-moving neutron strikes a uranium-235 nucleus, breaking it apart into two nuclei and several neutrons. The following is one possibility:

$$^{235}_{92}U + ^{1}_{0}n \rightarrow ^{143}_{56}Ba + ^{90}_{36}Kr + 3^{1}_{0}n.$$

    a. How much energy is released in this fission reaction?

    b. What form does this energy take? (Assume that the kinetic energy of the incident neutron is very close to zero. Use the masses in Table 16–1.)

Footnotes

[1] C. D. Anderson, *American Journal of Physics*, 29(1961), p. 825.

[2] Ibid, p. 825.

[3] Ibid, p. 825.

[4] M. S. Livingston, "History of the Cyclotron, Part I," *Physics Today*, 10(October, 1959), p. 21

[5] Ibid, p. 21.

[6] Ibid, p. 22.

[7] Ibid, p. 23.

[8] M. S. Livingston and E. McMillan, *Physical Review*, 46(1934), p. 437.

[9] E. McMillan and M. S. Livingston, *Physical Review*, 47(1935), p. 452.

[10] Edwin McMillan, "History of the Cyclotron, Part II," *Physics Today*. 10(October, 1959), p. 26.

[11] Ibid, p. 26.

[12] Ibid, p. 32.

[13] Ibid, p. 33.

Introduction

On December 2, 1942, the first self-sustaining chain reaction was achieved at the University of Chicago. This chapter tells that story, which begins not in Chicago, not even in the United States, but in Italy. It was in Rome where several developments spurred the discovery of fission.

At the University of Rome, Enrico Fermi and his students found that targets bombarded with neutrons become radioactive. Amazingly, this induced activity increases dramatically if the neutrons are slowed down before striking the target. Although uranium ($Z = 92$) becomes very radioactive when bombarded with neutrons, Fermi could not determine the element responsible for this activity. He cautiously suggested that perhaps an element with atomic number Z equal to 93 had been produced in the uranium. (Recall that uranium is the last naturally occurring element in the periodic table.) This suggestion created quite a controversy in the scientific community. The stage had been set for the discovery of fission.

In Berlin, Otto Hahn and Fritz Strassmann repeated Fermi's experiment with uranium and showed that a radioactive isotope of barium ($Z = 56$) is produced. They concluded that, when a neutron strikes a uranium nucleus, the nucleus breaks apart into large fragments, one of which is barium. They had discovered fission. Hahn and Strassmann sent a letter to Lise Meitner, their colleague, who had sought refuge from the Nazis in Stockholm in 1938. When the letter arrived, she discussed it with Otto Frisch, a physicist visiting her during Christmas vacation. They concluded that, when a uranium nucleus is struck by a neutron, the uranium nucleus fissions into a barium nucleus and a krypton nucleus. Most significantly, since these fragments contain many positive charges, the force of repulsion between them will be very great and the particles will move apart at great speeds. That is, fission is accompanied by a large energy release: about 40 times that occurring in natural radioactivity. Frisch returned to

Enrico Fermi

CHAPTER 18
Fission and the Chain Reaction

FIG. 18–1 Enrico Fermi. (Courtesy of Argonne National Laboratory)

Copenhagen and told Niels Bohr about the discovery of fission just as Bohr was about to leave for the United States. Within a few days, Frisch carried out an experiment that confirmed the large energy release accompanying fission.

Bohr visited Fermi, who had just recently come to Columbia University, and told him of fission. Immediately, Fermi and Herbert Anderson began an experiment to look for a large energy release accompanying fission. (Word of Frisch's experiment had not yet reached the United States.) Meanwhile, Fermi attended a physics conference in Washington, D.C. When Anderson confirmed the large energy release, he telegraphed Fermi. At this conference, Fermi suggested that additional neutrons might be released during fission. There was a possibility of a chain reaction: Neutrons from one fission reaction might cause another fission. Fermi rushed back to Columbia and began an experiment to see whether neutrons are released during fission. They are!

The remainder of this chapter centers on the exciting story (presented usually in first-hand accounts by the researchers themselves) of the developments of the first chain-reacting pile. Unlike scientific discoveries before it, the chain reaction depended on the work of many people, rather than just a few. "Big Science," with support from the government, had its beginning here.

Rome: Experiments with Neutrons

Enrico Fermi (1901–1954), shown in Fig. 18–1, was appointed professor of theoretical physics at the University of Rome at the young age of 25. Until 1934, his work had been mainly theoretical. Now, he and his students had decided to undertake an experimental project. They were trying to decide on a direction for their work when artificial or induced radioactivity was discovered by Irene Curie-Joliot and Frederick Joliot in 1934 (see Chapter 17). The Joliots had found that targets bombarded by alpha particles became radioactive and remained radioactive for a short time *even after* the source of alpha particles was removed. This radioactivity was due to the emission of positrons. Many scientists, using alpha particles, began to study the exciting phenomenon of induced radioactivity.

Fermi decided to try a new approach and see if neutrons could induce radioactivity in the target. In this experiment, the target would be bombarded with neutrons for some time, the neutron source removed, and the number of particles per minute emitted by the target (that is, its activity) would be counted by a Geiger counter. The neutrons would be produced by alpha particles from a radioactive source striking a beryllium target. However, for every 100,000 alpha particles, *only 1 neutron was produced!* For this reason, most scientists felt that the induced radioactivity, if any, would be too small to be easily measured. Fermi, however, realized that neutrons also had a distinct advantage. Because a neutron does not experience Coulomb repulsion as it approached a nucleus, it could get very close to the nucleus, much closer than an alpha particle with its +2 charge. This advantage might compensate for the *extremely* small number of neutrons.

The targets were chosen in order of increasing atomic number Z: hydrogen first, followed by lithium, beryllium, boron, carbon, nitrogen, and oxygen ($Z = 8$). In these targets, no induced radioactivity was found. Fermi was just about ready to give up when induced radioactivity was observed in fluorine ($Z = 9$) and in elements of larger atomic number.

For example, when iron ($Z = 26$) was bombarded with neutrons, it became radioactive. What element was responsible for this radioactivity? An element close to iron in the periodic table was suspected. However, because the radioactive atoms were so very few in number, they could not be chemically separated from the iron directly. Laura Fermi, Enrico's wife, relates how this difficulty was cleverly eliminated:

Accordingly, they dissolved the activated iron in nitric acid and added to the solution small amounts of chromium ($Z = 24$), manganese ($Z = 25$), and

cobalt ($Z = 27$). Usual methods of chemical separation were then followed, and the separated elements were tested with Geiger counters. The activity accompanied manganese, and the physicists could then assume that when iron is bombarded with neutrons, it transforms into manganese.[1]

These results suggested that the following reactions had occurred:

$$^{56}_{26}\text{Fe} + ^{1}_{0}\text{n} \rightarrow ^{56}_{25}\text{Mn} + ^{1}_{1}\text{H}. \qquad (1)$$

Naturally occurring manganese occurs only with $A = 55$ and therefore, manganese-56 (Mn-56) is radioactive. It emits a beta ray in order to reach a stable isotope:

$$^{56}_{25}\text{Mn} \rightarrow \beta^{-} + ^{56}_{26}\text{Fe} + \nu. \qquad (2)$$

These beta rays are counted by the Geiger counter.

Chemical separation techniques can only separate elements, not isotopes, and the separated manganese sample contains both the stable Mn-55 and the radioactive Mn-56.

In some cases the nucleus of the target (A, Z) simply absorbs a neutron, forming the radioactive isotope ($A + 1$, Z), which emits beta rays. Silver is an example of this effect.

This procedure—adding elements to the dissolved irradiated targets and noting with which element the radioactivity resides—is used to determine the radioactive element. In the next section, we shall see how this procedure was used in discovering fission.

URANIUM

Finally, uranium ($Z = 92$), the last naturally occurring element in the periodic table, was bombarded with neutrons and it, too, became radioactive. Using a method similar to that for iron, Fermi's researchers found that the radioactive element(s) was (were) not uranium or thorium because the radioactivity did not reside with either uranium or with thorium ($Z = 90$) in the chemical separation. They did not have a sample of protactinium ($Z = 91$). Cautiously, Fermi *suggested* that a new element $Z = 93$ *might* result if uranium-238 (U-238) absorbed a neutron to form U-239, which then emitted a beta ray to form an isotope having $Z = 93$ and $A = 239$.

Much to Fermi's dismay, the Fascist press made much ado about Fermi's *discovery* of a new element. In the scientific community, much controversy reigned over this experiment, and some scientists suggested that the induced radioactivity was due to protactinium ($Z = 91$). We shall see shortly that this controversy spurred the discovery of fission.

SLOW NEUTRONS

In the course of their research, Fermi and his students noticed that the radioactivity induced in the silver varied greatly. For example, the amount of radioactivity was greater if the apparatus was placed on a wooden table than on a piece of metal. Some neutrons might strike the wood (or metal) first, scatter, and then strike the silver. What is the effect of different materials on the neutrons? To study this, they placed different materials between the neutron source and the silver target.

Laura Fermi relates:

A plate of lead made the activity increase slightly. Lead is a heavy substance. "Let's try a light one next," Fermi said, "for instance, paraffin." The experiment with paraffin was performed on the morning of October 22.

They took a big block of paraffin, dug a cavity in it, put the neutron source inside the cavity, irradiated the silver cylinder, and brought it to a Geiger counter to measure its activity. The counter clicked madly. The halls of the physics building resounded with loud exclamations: "Fantastic! Incredible! Black magic!" Paraffin increased the artificially induced radioactivity of silver up to one hundred times.

At noon the group parted reluctantly for the usual lunch recess, which generally lasted a good couple of hours. . . . By the time he [Fermi] went back to the laboratory he had a theory worked out to explain the strange action of paraffin.

Paraffin contains a great deal of hydrogen. Hydrogen nuclei are protons, particles having the same mass as neutrons. When the source is enclosed in a paraffin block, the neutrons hit the protons in the paraffin before reaching the silver nuclei. In the collision with a proton, a neutron loses part of its energy, in the same manner as a billiard ball is slowed down when it hits a ball of its same size. Before emerging from the paraffin, a neutron will have collided with many protons in succession, and its velocity will be greatly reduced. This slow neutron will have a much better chance of being captured by a silver nucleus than by a fast one, much as a slow golf ball has a better chance of making a hole than one which zooms fast and may bypass it.[2]

They found that water, like paraffin, also increased the induced radioactivity.

This idea—slow neutrons are more easily captured by nuclei than fast neutrons—was essential in accomplishing a self-sustaining chain reaction. (This was discussed briefly in Chapter 4.)

They continued their experiments with neutrons until 1938. On December 10, 1938, Fermi received the Nobel Prize for his study of slow-neutron reactions. He traveled with his family to Stockholm to receive the prize, and they never returned to Rome, for they could no longer tolerate the Fascist regime. Instead, they sought refuge in America. Fermi had obtained a professorship at Columbia University.

Berlin: The Discovery of Fission

The next important event in the development of fission took place at the Kaiser Wilhelm Institute of Chemistry in the laboratory of Otto Hahn

(1879–1968) and Lise Meitner (1878–1968), who knew about the controversy over Fermi's neutron experiments with uranium. Otto Hahn relates:

Naturally Dr. Meitner and I were keenly interested, for we knew a good deal about the chemical properties of protactinium, our discovery [in 1917]. We decided to repeat Fermi's experiments to try to determine whether his substance was or was not protactinium.[3]

They had a sample of protactinium, which was added to the dissolved neutron-irradiated uranium target. Their results showed that the radioactivity was not due to protactinium. They also showed that the radioactive elements were not thorium or actinium. They did not immediately consider looking for lighter elements. Time passed. They were joined in their research by Fritz Strassmann (b. 1902).

In July, 1938, Lise Meitner fled Nazi Germany and sought refuge in Stockholm. Hahn and Strassmann continued. In the course of their complex chemical separation experiments, they added (stable) barium to the uranium solution. After the chemical separation of the mixture into elements, they found the barium to be radioactive. The radioactivity could not be separated from the barium by *any* method. Only one conclusion was possible: The radioactivity must be due to a radioactive isotope of barium that was mixed with the stable barium. At first, Hahn and Strassmann could not understand these curious results and performed many additional chemical separations. Barium had an atomic number Z of only 56. How could it be obtained from uranium ($Z = 92$)? They could find no fault with their experiments—a radioactive isotope of barium had indeed been produced when uranium was bombarded with neutrons. Otto Hahn describes the discovery of **fission:**

In January, 1939, we published an account of these experiments that are at variance with all previous experiences in nuclear physics. . . . We did speak of the "bursting" of uranium, as we called the surprising process that had yielded barium, far down

in the periodic table. In this first paper we also speculated on what the other partner of the splitting of the uranium atom might be.[4]

Sweden: Energy Is Released by Fission

That energy release accompanies fission was the contribution of Lise Meitner and Otto Frisch (Fig. 18–2). Otto Frisch (b. 1904) relates how they came to this conclusion:

This is where I came in because Lise Meitner was lonely in Sweden and, as her faithful nephew, I went to visit her at Christmas. There, in a small hotel in Kungälv near Göteborg I found her at breakfast brooding over a letter from Hahn. I was skeptical about the contents—that barium was formed from uranium by neutrons—but she kept on with it. We walked up and down in the snow, I on skis and she on foot (she said and proved that she could get along just as fast that way), and gradually the idea took shape that this was no chipping or cracking of the nucleus but rather a process to be explained by Bohr's idea that the nucleus was like a liquid drop; such a drop might elongate and divide itself.[5]

LIQUID-DROP MODEL

Bohr suggested that the nucleus behaves in many ways like a drop of liquid. Let's investigate this analogy.

In a drop of liquid, the attractive forces between the molecules cause the drop to assume a spherical shape. For example, molecules on the surface cannot interact with as many neighboring molecules as those on the interior. In order to maximize the attraction between the molecules, the surface area must be minimized. For a given volume of liquid, the sphere has the smallest surface area. This effect is called the **surface tension.** The most stable shape for a drop of liquid is a sphere. Similarly, in the

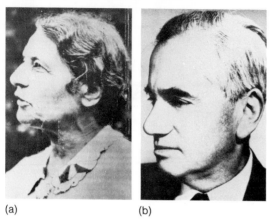

(a)   (b)

FIG. 18-2   (a) Lise Meitner. (Courtesy of National Archives, U.S.A.) (b) Otto Frisch. (Courtesy of Argonne National Laboratory)

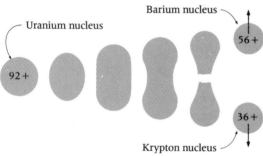

FIG. 18-3   Fission of uranium nucleus into barium and krypton fragments.

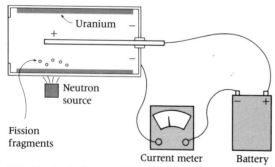

FIG. 18-4   A diagram of Frisch's apparatus for measuring the kinetic energy of fission fragments.

nucleus, the attractive forces between the protons and neutrons cause the nucleus to assume a spherical shape, so that the attraction between photons and neutrons is maximized.

Meitner and Frisch discussed the fission process:

If the movement [of the drop] *is made sufficiently violent by adding energy, such a drop may divide itself into two smaller drops. . . . If one of the parts is an isotope of barium (Z = 56) the other will be krypton (Z = 92 − 56).*[6]

Figure 18-3 illustrates this fission process. Energy is added to the drop by absorbing the kinetic energy of the neutron and by the conversion of mass energy into oscillations of the drop. For example, when U-235 absorbs a neutron to form U-236, an energy of 6.47 MeV is released because the total mass of a U-235 nucleus and a neutron is larger than that of a U-236 nucleus. In 1939, Niels Bohr and John A. Wheeler examined the fission process in detail.

The barium and krypton fragments are formed very close together and both are positively charged. "These two nuclei will repel each other and should gain a total kinetic energy of about 200 MeV."[7] This is an extraordinarily large amount of energy. (Remember that an alpha particle emitted from a radioactive source has kinetic energy of only 5 MeV.)

Using the masses of all nuclei and calculating the difference in mass before and after fission, about 0.2 amu of mass is converted to kinetic energy, an energy of about 200 MeV. (See calculation question 17-9.)

Frisch continues his story:

We spent only two or three days together that Christmas. Then I went back to Copenhagen and just managed to tell Bohr about the idea as he was catching his boat to the U.S. I remember how he struck his head after I had barely started to speak and said: "Oh, what fools we have been! We ought to have seen that before." But he had not—nobody had.[8]

FRISCH'S EXPERIMENT

Frisch performed an experiment to find out if an energy release accompanied fission.[9] He used a device called an **ionization chamber** (see Fig. 18–4). This chamber is similar to a Geiger counter, with one important exception. In the Geiger counter, a particle is detected because of a large burst of current, but the value of the current cannot be measured. In an ionization chamber, a meaningful value of the current can be obtained and indicates the kinetic energy of the particle. Frisch lined the cylinder of an ionization chamber with uranium and brought a neutron source up to it. The neutrons passed through the wall of the cylinder and struck the uranium, causing fission, and the fission fragments (Kr and Ba) caused ionization of the gas in the cylinder. The size of the pulses indicated a kinetic energy of 70 MeV for *one* fragment, in close agreement with the 100 MeV indicated by their theory (for one of the two fragments). This was conclusive evidence that fission (the breaking up of the uranium nucleus into two parts of comparable size) does indeed occur with a release of energy. These results were obtained on January 16, 1939.

Columbia University: Possibility of a Chain Reaction

On January 9, 1939, Fermi arrived with his family in New York and began his work at Columbia University. Bohr arrived in New York one week later and, a few days after he was settled in Princeton, he came to Columbia, bursting with the news of the discovery of fission. Immediately, Fermi and a graduate student, Herbert L. Anderson (b. 1914), planned an experiment. On January 25, Anderson performed the same experiment with the ionization chamber that Frisch had carried out only 10 days earlier in Copenhagen. Anderson, too, observed the large energy release accompanying fission. Fermi left for a theoretical physics conference in Washington, D.C., before he learned of these results, which were then telegraphed to him. Anderson writes:

When the meeting opened the next day both Bohr and Fermi talked about the fission problem. Fermi was able to speak with the conviction of personal experience. Then he mentioned the possibility that neutrons might be emitted during the splitting. After all, the fission products would have a large neutron excess. Although this was only a guess, its implication for a chain reaction produced a great deal of excitement. Physicists at the meeting rushed to call their laboratories, and soon there were confirmations from a number of places throughout the country.[10]

A CHAIN REACTION

Two terms—the *neutron excess* and the *chain reaction*—need clarification. In order to see why Fermi suggested that more neutrons might accompany fission, let us consider the following fission reaction for U-235:

$$^{235}_{92}\text{U} + ^{1}_{0}\text{n} \rightarrow ^{236}_{92}\text{U} \rightarrow ^{144}_{56}\text{Ba} + ^{92}_{36}\text{Kr}. \quad (3)$$

Since, for naturally occurring barium, $A = 137$, and, for krypton, $A = 84$, the radioactive isotopes formed from fission have an *extremely* large number of neutrons, or **neutron excess,** compared to the stable isotopes. For this reason, the radioactive isotopes produced from fission may emit neutrons (as well as beta rays) as they decay to stable isotopes. Suppose that 3 neutrons are released in this way. A chain reaction occurs when one or more of these neutrons causes an additional fission reaction. This is shown in Fig. 18–5, where 2 of the 3 neutrons cause fission and the other one is "lost." (The reason for the "lost" neutrons will be discussed shortly.) A sustained **chain reaction** occurs when one neutron from each fission reaction causes a subsequent fission reaction. If, on the average, less than one neutron produces a fission reaction, the chain reaction will die out. Since an energy of 200 MeV is released with each fission reaction, a chain reaction can release a lot of

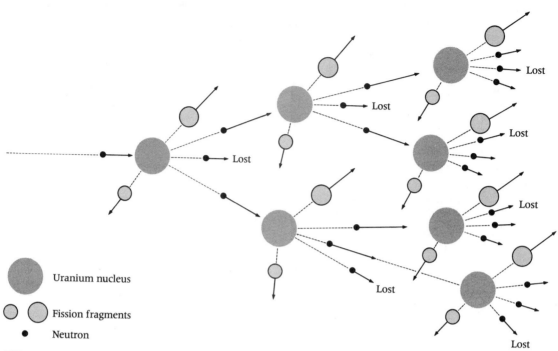

Uranium nucleus

Fission fragments

Neutron

FIG. 18–5   A chain reaction.

energy. For example, in Fig. 18–5 the number of fissions that occur is 1, 2, 4, 8, 16, etc. After 100 generations, there would be 2^{100}, or 1.27×10^{30} fissions and an energy release of 252×10^{30} MeV! Certainly, a cause for great excitement.

AN EXPERIMENT TO TEST NEUTRON PRODUCTION

Fermi did not wait for the conference to end, but rushed back to Columbia and began making plans for experiments. Figure 18–6 depicts the first experiment done to see whether additional neutrons accompanied fission. A neutron source (in a spherical metal container) was placed at the center of a large tank of water 90 cm in diameter and 90 cm high, and a rhodium foil was placed at some distance from the neutron source.[11] As Fermi knew from the Rome experiments, the neutrons would be slowed down as they traveled through the water and would cause sizable induced activity in the rhodium foil. This activity was measured with a Geiger counter. The experiment was then repeated with uranium surrounding the neutron source. If additional neutrons accompanied fission, then the induced activity in the foil should be larger with the uranium surrounding the source. This is exactly what they found. By comparing the induced activity of the foil with and without the uranium, at various distances from the neutron source, the Fermi team deduced that about 2 neutrons were produced for each neutron captured by a uranium nucleus. More sophisticated experiments followed to determine this number more accurately, but these experiments showed that they were on the right track.

In their experiments, they used *natural* uranium, which consists of 99.3 percent U-238 and only 0.7 percent U-235. They did not know whether fission was due to one or to both isotopes of uranium. Based on the liquid-drop model of the nucleus, Bohr advanced the idea that slow neutrons caused the fission of U-235 and *did not* cause the fission of U-238. If this were so, then, by using natural uranium, only 1 nucleus in 140 had the possibility of undergoing fission. If a sample containing *only* U-235 were used, then a chain reaction would be much more likely than it was with natural uranium. Furthermore, the possibility of making an extremely powerful bomb would be great. The isotope of uranium responsible for fission had to be determined immediately.

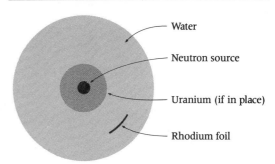

FIG. 18–6 Experimental apparatus for testing whether additional neutrons accompanied fission.

SLOW NEUTRONS CAUSE FISSION OF U-235

This problem was investigated by A. O. Nier at the University of Minnesota and E. T. Booth, J. R. Dunning and A. V. Grosse at Columbia University.[12] Nier used a mass spectrometer to separate U-235 and U-238.* After operating the mass spectrometer for 11 hours, only 2.1×10^{-9} g of U-235 and 2.9×10^{-7} g of U-238 were obtained. These numbers give us some idea of how difficult it would be to obtain kilogram amounts of U-235.

At Columbia, each separated sample was placed inside an ionization chamber and then bombarded with slow neutrons. (Frisch's experiment was essentially repeated, using separated isotopes of uranium.) Very few counts or fissions were observed with the U-238 sample, but a *sizable* number were observed with the U-235 sample. This showed conclusively that

---

*Briefly, positively charged ions of U-235 and U-238 enter a perpendicular magnetic field and travel in a circular path. Because all of the ions have the same charge and the same speed, the forces on the ions due to the magnetic field ($F = qvB$), are identical. Because the U-238 ions are more massive and therefore harder to bend, the paths of the U-238 ions have a larger radius than those of the U-235 ions. It is therefore possible to separate and collect each isotope. The operation of a mass spectrometer is indicated in thought question 7–25.

the U-235 isotope was responsible for fission with slow neutrons.

Today, we know that there are many ways in which fission can occur. The following are examples:

$$^{235}_{92}U + n \rightarrow {}^{236}_{92}U \rightarrow {}^{152}_{60}Nd + {}^{80}_{32}Ge \ + 4\,n \quad (4)$$

$$\rightarrow {}^{149}_{57}La \ + {}^{85}_{35}Br \ + 2\,n \quad (5)$$

$$\rightarrow {}^{139}_{56}Ba \ + {}^{95}_{36}Kr \ + 2\,n \quad (6)$$

$$\rightarrow {}^{132}_{51}Sb \ + {}^{100}_{41}Nb \ + 4\,n \quad (7)$$

$$\rightarrow {}^{130}_{50}Sn \ + {}^{103}_{42}Mo + 3\,n. \quad (8)$$

When the nucleus fissions into two parts, the two most probable values for the atomic mass number A are 141 and 95. Additional neutrons are promptly emitted, as shown in Equations 4 through 8. In their later work, Hahn and Strassman identified over 100 different elements, or fission fragments. Considering all types of fission reactions that occur, 2.5 neutrons are produced *on the average* for each fission reaction.

Feasibility of a Chain Reaction

Even though Fermi knew that U-235 was responsible for fission with slow neutrons, he still decided to use natural uranium. The only alternative was to launch a massive effort to separate U-235 from U-238. Heretofore, only minute quantities of the separated isotopes had been obtained.

From his experiments in Rome, Fermi knew fission of uranium nuclei was about 500 times more probable with *slow* neutrons than with fast neutrons. Neutrons can be slowed down by colliding with lightweight atoms. (This was discussed from the viewpoint of conservation of momentum in Chapter 4.) For example, paraffin or water effectively slows down neutrons due to the scattering from hydrogen.

MODERATOR

A material that slows down neutrons is called a **moderator.** Water seemed a most likely candidate for the moderator. However, experiments showed that water absorbed neutrons in the following way:

$$^{1}_{1}H + {}^{1}_{0}n \rightarrow {}^{2}_{1}H. \quad (9)$$

Neutrons absorbed in this way are "lost" to the fission process.

Instead, graphite (a form of carbon) was chosen because it was plentiful and easily machined. At the time, graphite seemed the only reasonable choice.

No one knew how far Germany had proceeded with uranium research. Fermi relates:

Everybody was conscious early in 1939 of the imminence of a war of annihilation. There was well-founded fear that the tremendous military potentialities that were latent in the new scientific developments might be reduced to practice first by the Nazis. Nobody at that time had any basis for predicting the size of the effort that would be needed, and it well may be that civilization owes its survival to the fact that the development of atomic bombs requires an industrial effort of which no belligerent except the United States would have been capable in time of war. The political situation of the moment had a strange effect on the behavior of scientists. Contrary to their traditions, they set up a voluntary censorship and treated the matter as confidential long before its importance was recognized by the governments and secrecy became mandatory. [13]

Due to this impetus, plans for a chain reaction using a graphite moderator were begun immediately. Fortunately, graphite was an excellent choice for a moderator because it absorbed very few neutrons, as experiments showed. In the fall of 1939, Einstein wrote his famous letter to

President Roosevelt, advising him of the possibility of a chain reaction and the possibility of extremely powerful bombs.[14] As a result, the government granted $6000 for the purchase of tons of graphite.

BASIC DESIGN OF PILE

The basic idea for the chain-reacting **pile** is to construct a lattice of uranium and graphite blocks. A cross section is shown in Fig. 18–7, where lumps of uranium are surrounded by graphite. A neutron leaving one lump of uranium is slowed down as it travels through the graphite and has a high probability of causing fission when it strikes another lump of uranium. On the average, 2.5 neutrons are produced for each fission reaction. The **reproduction factor k** is defined as the number of neutrons from one fission reaction that causes another fission. For a sustaining chain reaction k must be equal to 1. As described in the following list, 1.5 neutrons can be "lost."

A neutron will be "lost" and cannot cause fission if:

1. The neutron is absorbed by U-238 to form U-239, or by U-235 to form U-236 (without fission).

2. The neutron is absorbed by carbon, or by an impurity element in the moderator.

3. The neutron is simply scattered by the uranium and carbon atoms (much like the collision of billiard balls) and arrives at the surface of the pile, where it escapes.

The following section illustrates some of these effects.

EFFECTS OF NEUTRONS STRIKING URANIUM: AN EXAMPLE

In order to illustrate the many effects that occur when neutrons strike uranium, values known accurately today will be used.

Suppose that 10,000 neutrons strike a cube of *natural* uranium in a direction perpendicular to one face of a cube, 1 cm on a side. For slow

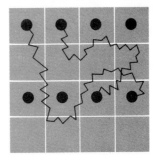

FIG. 18–7 A lattice of graphite blocks, with uranium spheres in some blocks. A fast neutron emerging from one uranium sphere slows down as it travels through graphite. This neutron has a high probability of causing fission when it strikes another uranium sphere.

neutrons having a speed of 2200 m/sec, we find the following:

1950 n	cause fission of U-235 nuclei
360 n	are captured to form U-236, but do not cause fission
30 n	are scattered like billiard balls by U-235 nuclei
3950 n	are scattered by U-238 nuclei
1680 n	are captured to form U-239
2030 n	pass through and do not interact with any nucleus

Total 10,000 n

Since each fission of U-235 produces 2.5 neutrons, there will be 1950×2.5, or 4875 neutrons produced in the sample, compared to $1680 + 360$, or 2040 neutrons that are captured. A total of 10,000 neutrons enter the uranium cube and $4875 + 30 + 3950 + 2030$, or 10,885, leave. For a chain reaction to be feasible, the number of neutrons leaving the uranium lump must be larger than the number entering it.

For fast neutrons having a speed of 2×10^7 m/sec, only 4 out of 10,000 will cause fission. Slow neutrons are indeed much more effective than fast neutrons in causing fission.

CRITICAL SIZE OF PILE

By increasing the size of the pile, the neutron loss can be minimized. As the pile is made larger, a neutron has a greater chance of striking a uranium lump before it reaches the surface. The number of fissions is therefore proportional to the *volume* of the pile. It is also important to know that more neutrons are produced in the uranium lump than are absorbed by U-235 and U-238. This was illustrated in a preceding example. When a neutron reaches the surface, it escapes into the air and is lost. The number of neutrons lost is proportional to the *surface*

area of the pile. First, let us suppose that the graphite does not absorb any neutrons. By increasing the size of the pile, the number of fissions increases *more* than the number of neutrons lost at the surface. If the size of the pile is increased so that these two effects are equal, a sustaining chain reaction occurs. For example, if the length of a cube is doubled, the volume is increased by a factor of 2^3, or 8, while the surface area is increased by a factor of 2^2, or 4. In this case, the number of neutrons produced by fission increases by a factor of 8, while those lost at the surface increase only by a factor of 4.

In practice, some neutrons *are* absorbed by the graphite and impurities, and the number of neutrons lost in this way is also proportional to the volume. For this reason, it seemed impractical to start constructing a pile of uranium and graphite and keep making it larger until a sustained chain reaction occurred. Maybe it would be too large to be feasible. Fermi suggested a trial experiment, called an **intermediate pile.**

INTERMEDIATE PILE

A uranium and graphite pile was arranged, as shown in Fig. 18−8. A neutron source was placed at the bottom.

Anderson recalls the construction of this pile:

We were faced with a lot of hard and dirty work. The black uranium oxide powder had to be packed in cubical tin cans 8 inches on a side. The uranium had to be heated to drive off undesired moisture and then packed hot in the containers and soldered shut. To get the required density, the filling was done on a shaking table. Our little group, which by that time included Bernard Feld, George Weil, and Walter Zinn, looked at the heavy task before us with little enthusiasm. It would be exhausting work. Fortunately, Fermi managed to recruit some members of the Columbia football squad to assist in this. In those days football players were expected to earn some of their support by doing useful work for the University. It was a pleasure to see them work; they made it seem easy. Fermi tried to do his share of the work; he donned a lab coat and pitched in to do his stint with the football men, but it was clear that he was out of his class. The rest of us found a lot to keep us busy with measurements and calibrations that suddenly seemed to require exceptional care and precision.[15]

The intensity of the neutrons throughout the pile was measured by the radioactivity induced in foils placed in the slots. The purpose of this experiment was to mimic the behavior of a very much larger pile of the same design *without* the neutron source and to obtain the value of the reproduction factor k. Was k equal to 1?

In the intermediate pile experiment, the neutrons arise from two effects: (1) neutrons coming directly from the source and (2) neutrons resulting from the fission of uranium. By comparing the induced radioactivity with that expected from the first effect, the experimenters were able to calculate the reproduction factor k, although the analysis is quite complicated and beyond our present discussion. In the first experiment, k was found to be 0.87, which is appreciably less than 1. The value of k became closer to 1 when impurities in the uranium and graphite were eliminated and the dimensions of the uranium and graphite lattice were improved. More about this shortly.

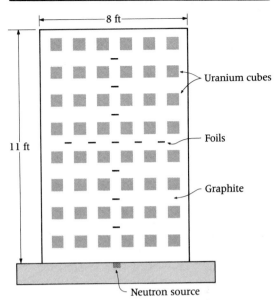

FIG. 18–8   An intermediate pile.

PRODUCTION OF PLUTONIUM-239

A neutron absorbed by U-238 forms Pu-239 in the following way:

$$^{238}_{92}U + ^{1}_{0}n \rightarrow ^{239}_{92}U \rightarrow ^{239}_{93}Np + \beta^- + \nu \quad (10)$$

$$^{239}_{93}Np \rightarrow ^{239}_{94}Pu + \beta^- + \nu. \quad (11)$$

In Fermi's experiment in Rome, the formation of new elements neptunium and plutonium *did* take place. However, the radioactivity that Fermi measured was due primarily to the radioactivity of the fission fragments from the fission of U-235.

According to Bohr and Wheeler's theory of fission, Pu-239 should also be fissionable. When Pu-239 absorbed a neutron, the difference in mass is converted to the oscillations of the "liquid" drop and should be sufficient to cause fis-

FISSION AND THE CHAIN REACTION

sion. Thus, it might be possible to breed Pu-239 at the same time U-235 was being consumed. The Pu-239 could be separated by *chemical* methods from uranium. Thus, in order to obtain kilogram amounts of fissionable material for a bomb, it might be simpler to obtain Pu-239 than to separate U-235 from U-238. The production of Pu-239 was one reason for the urgency of a sustaining chain reaction.

Interestingly, the probability that a neutron will be absorbed by U-238 is larger for *fast* neutrons than for slow neutrons. This is exactly opposite to the fission of U-235, which occurs more readily with slow neutrons. A slow neutron entering a uranium lump has a smaller chance of being absorbed than a fast neutron resulting from fission in that uranium lump. By lumping the uranium, it was hoped that the proper amount of absorption—not too much and not too little—could be achieved.

The Metallurgical Laboratory

The attack on Pearl Harbor occurred on December 7, 1941. The Metallurgical Laboratory was organized, under the direction of Arthur H. Compton (1892–1962) of the University of Chicago, during January 1942. Its mission was to develop the chain reaction with natural uranium and use it to produce plutonium-239. Compton decided that the work on the chain reaction should be concentrated at the University of Chicago. The Columbia group moved to Chicago during the spring of 1942. Fermi describes their progress during that year:

During 1942 some twenty or thirty exponential experiments (or intermediate pile experiments) were carried out at Chicago in the attempt to improve on the condition of the first experiment. Two different types of improvements were pursued. One consisted in a better adjustment of the dimensions of the lattice and the other in the use of better materials. Impurities had to be eliminated to a surprisingly high extent from both uranium and graphite since the parasitic absorption due to elements appearing as common impurities in uranium and graphite was responsible for a loss of an appreciable fraction of the neutrons. The problem was tackled to organize large-scale production of many tons of graphite and uranium of an unprecedented purity. Also the production of uranium in metallic form was vigorously pursued. Up to 1941 uranium metal had been produced only in very small amounts, often of questionable purity. . . .

Toward the fall of 1942 the situation as to the production of materials gradually improved. Through the joint efforts of the staff of the Metallurgical Laboratory and of several industrial firms, better and better graphite was obtained. Industrial production of practically pure uranium oxide was organized and some amount of cast uranium metal was produced. The results of the exponential (intermediate pile) experiments improved correspondingly (i.e., the reproduction factor k was very close to 1) to the point that the indications were that a chain reacting unit could be built using these better brands of materials.[16]

SPONTANEOUS FISSION

In a chain-reacting pile, a neutron source is not needed. Neutrons are produced in uranium because a small percentage of the uranium nuclei *spontaneously* fissions without being hit by a neutron.

In one gram of uranium, about five uranium nuclei per second undergo spontaneous fission. Let us suppose that 10,000 neutrons per second occur from spontaneous fission and that the reproduction factor k for the pile is 0.9. (Remember that k is the number of neutrons from one fission reaction causing another fission reaction.) Therefore, the number remaining after each generation is: 10,000, 9000, 8100, 7290, 6561, or $10,000 \times (0.9)^n$ for the nth generation. After 87 generations, only 1 neutron remains.* The spontaneous fission continues to supply more neutrons. As the size of the pile increases, and k increases, the neutrons from the spontaneous fission "live longer" and the density of neutrons in the pile increases.

Fermi continues:

The actual erection of the first chain reacting unit [see Fig. 18–9] was initiated in October 1942. It was planned to build a lattice structure in the form of a huge sphere supported by a wooden structure. The structure was to be erected in a Squash Court on the campus of the University of Chicago. . . . It took a little over one month to build the structure. A large number of physicists, among them W. H. Zinn, H. L. Anderson, and W. C. Wilson, collaborated in the construction. During this time the approach to the chain reacting conditions was followed day by day by measuring the neutron intensity building up inside the pile. . . . By watching the rise of the neutron density, one obtains, therefore, a positive method for extrapolating to the critical size.

Appreciably before the dimensions originally planned for the structure were reached, the measurements of the neutron density inside the structure indicated that the critical size would soon be attained. From this time on work was continued under careful supervision so as to make sure that criticality would not be inadvertently reached without proper precautions. Long cadmium strips were inserted in slots that had been left for this purpose in the structure. Cadmium is one of the most powerful absorbers of neutrons and the absorption of these strips was large enough to make sure that no chain reaction could take place while they were inside the pile. Each

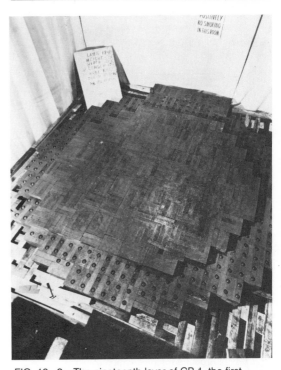

FIG. 18–9   The nineteenth layer of CP-1, the first nuclear reactor, was made up of graphite blocks. In the partially shown eighteenth layer, the graphite blocks contained spheres of uranium oxide. The first chain-reacting pile was moved to Argonne National Laboratory, where this photograph was taken. (Courtesy of Argonne National Laboratory)

*The value of 87 is determined in the following way:

$$10,000 \times (0.9)^n = 1$$

$$(0.9)^n = \frac{1}{10,000} = 10^{-4}$$

$$\log (0.9)^n = \log 10^{-4} = -4$$

$$n \log 0.9 = -4$$

$$n = \frac{-4}{\log 0.9} = \frac{-4}{-0.04576}$$

$$n = 87.4$$

FIG. 18–10 The first nuclear reactor (CP-1) at the University of Chicago as its chain reaction became self-sustaining on December 2, 1942. This is a painting by Gary Sheahan, at the time a *Chicago Tribune* staff artist, who reconstructed the scene from drawings, notes, and conversations with many of the persons who were present. (Courtesy of Argonne National Laboratory)

morning the cadmium strips were slowly removed, one by one, and a determination of the neutron intensity was carried out in order to estimate how far we were from the critical conditions.

On the morning of December 2, 1942, the indications were that the critical dimension had been slightly exceeded and that the system did not chain react only because of the absorption of the cadmium strips.[17]

The drama of this moment in history (Fig. 18–10) is recalled by H. L. Anderson:

When the cadmium rod was pulled out to the position he asked for next, the increase in neutron intensity was noticeably quickened. At first you could hear the sound of the neutron counter, clickety-clack, click-ety-clack. Then the clicks came more and more rapidly, and after a while they began to merge into a roar; the counter couldn't follow anymore. That was the moment to switch to a chart recorder. But when the switch was made, everyone watched in the sudden silence the mounting deflection of the recorder's pen. It was an awesome silence. Everyone realized the significance of that switch; we were in the high intensity regime* and the counters were unable to cope with the situation anymore. Again and again, the scale of the recorder had to be changed to accommodate the neutron intensity which was increasing more and more rapidly. Suddenly Fermi raised his hand: "The pile has gone critical," he announced. No one present had any doubt about it. Then everyone began to wonder why he didn't shut the pile off. But Fermi was completely calm. He waited another

*On December 2, the chain-reacting pile produced about 1/2 watts of power and, later, never exceeded 200 watts because there was no protective shielding against the radiation and no way to remove the heat from the pile.

minute, then another, and then when it seemed that the anxiety was too much to bear, he ordered "Zip in." Zinn released his rope (holding vertical control rod called Zip) and there was a sigh of relief when the intensity dropped abruptly and obediently to a more modest level. It was a dramatic demonstration that the chain reaction worked.

No cheer went up, but everyone had a sense of excitement. They had been witness to a great moment in history. Wigner was prepared with a bottle of Chianti wine to celebrate the occasion. We drank from paper cups and then began to say things to one another. But there were no words that could express adequately just what we felt.[18]

Arthur Compton placed a long-distance call to Mr. Conant of the Office of Scientific Research and Development at Harvard:

"The Italian Navigator has reached the New World," said Compton as soon as he got Conant on the line.
"And how did he find the natives?"
"Very friendly."[19]

Summary

Fermi found that targets bombarded by neutrons became radioactive. The element responsible for the radioactivity was determined by adding *stable* elements to the dissolved neutron-irradiated target and noting with which element the radioactivity resided. Fermi also found that the induced radioactivity may be greatly increased, in some cases nearly a 100-fold, if slow neutrons rather than fast neutrons bombard the target. Slow neutrons are obtained by passing the neutrons through a material containing hydrogen, such as paraffin or water. In the collision of a neutron and a hydrogen atom, some of the neutron's kinetic energy is transferred to the hydrogen atom. Curious results were obtained when uranium ($Z = 92$), the last naturally occurring element in the periodic table, was bombarded with neutrons. Hahn and Strassmann repeated the experiment with uranium. Fission was discovered when they showed that the radioactivity of the uranium target could not be separated from barium, which had been added to the dissolved uranium target.

Bohr theorized that the nucleus behaved like a drop of liquid. Lise Meitner and Otto Frisch realized that an energy release accompanied fission. When the uranium nucleus absorbed a neutron, the nucleus would vibrate, elongate, and then break apart—fission—into two pieces: two *positively* charged pieces, which would fly apart due to Coulomb repulsion. Frisch lined the cylinder of an ionization chamber with uranium and brought a neutron source up to it. Huge pulses, corresponding to an energy of about 100 MeV for one fission fragment, were observed. Fission releases a total energy of about 200 MeV.

The possibility of a chain reaction existed if more neutrons were released during fission. These neutrons could then cause additional fission and so on. At Columbia University, the neutron production was verified. A neutron source was placed in a tank of water and caused radioactivity in a foil placed in the tank. When the neutron source was surrounded by uranium, the induced radioactivity of the foil was larger, indicating that more neutrons were produced as a result of fission.

A mass spectrometer was used to obtain a minute sample of U-235. Experiments using the U-235 sample inside an ionization chamber showed that U-235, and not U-238, fissions with slow neutrons. U-235 can fission in many possible ways. The most probable values for the atomic mass number A are 95 and 141. On the average, 2.5 neutrons are produced for each fission.

In order to test the possibility of chain reaction, Fermi decided to use lumps of uranium surrounded by a graphite moderator. Neutrons slowed or moderated by the graphite would have a larger chance of causing fission than fast neutrons. The design of the pile was tested with an intermediate pile experiment: a small pile of graphite and uranium *with* a neutron source to determine the reproduction factor k—the number of neutrons from one fission reaction, causing another fission. In the final design for a large chain-reaction pile, the chain reaction would be controlled by strips of cadmium (which readily absorbs neutrons) placed into the pile of graphite and uranium. By making the pile sufficiently large, the neutrons lost at the surface would be balanced by the neutron production inside. The minimum size necessary for a sustaining chain reaction is called the critical size.

Spontaneous fission in uranium produces a few neutrons to begin the chain reaction. Neutrons are absorbed by U-238 to form U-239, which emits two beta rays in sequence to form plutonium-239 ($Z = 94$). Pu-239 was also found to be fissionable. The first sustaining chain reaction occurred at the University of Chicago on December 2, 1942.

True or False Questions

Indicate whether the following statements are true or false. Change all of the false statements so that they read correctly.

18–1 Fermi and his students at the University of Rome found that targets bombarded with neutrons became radioactive.

a. Other scientists had not tried this experiment because the chance that a neutron would directly hit a nucleus in the target was so small.

b. The neutrons were produced in the following way:

$$\,^4_2\text{He} + \,^9_4\text{Be} \rightarrow \,^1_0\text{n} + \,^{12}_6\text{C}.$$

c. For every 100 alpha particles that struck the beryllium target, 1 neutron was produced.

d. The induced radioactivity can be greatly increased if slow neutrons, rather than fast neutrons, strike the target.

e. Slow neutrons are obtained by placing a lead absorber between the neutron source and the target.

f. The element causing induced radioactivity in the target can be determined by adding *stable* elements to the dissolved neutron-irradiated target and noting with which element the radioactivity resides after the chemical separation.

g. Fermi suggested that uranium becomes radioactive when bombarded with neutrons due to the following reaction:

$$\,^{238}_{92}\text{U} + \,^1_0\text{n} \rightarrow \,^{239}_{93}\text{Np} + \beta^- + \nu.$$

18–2 Hahn and Strassmann bombarded uranium with neutrons. Then they dissolved the target in acid, added stable barium, and performed a chemical separation procedure.

a. After the chemical separation procedure, the pure barium sample was, as expected, not radioactive.

18–3 a. Meitner and Frisch concluded that fission occurs in the following way:

$$\,_{92}\text{U} + \,^1_0\text{n} \rightarrow \,_{56}\text{Ba} + \,_{36}\text{K}.$$

b. It was only after the discovery of fission that Niels Bohr developed a theory to explain how fission could occur.

c. Frisch lined the cylinder of a Geiger counter with uranium and irradiated it with neutrons. The large pulses in the Geiger counter showed that a large energy release accompanied fission.

18–4 A neutron source is placed in a tank of water, causing a foil submerged in the water to become radioactive.

a. When the neutron source is surrounded by uranium, the activity of the foil does not change.

18–5 The term *neutron excess* refers to the fact that, on the average, 2.5 neutrons are produced for each fission reaction in uranium.

18–6 a. When Frisch's experiment was repeated using separated isotopes of U-235 and U-238, the researchers found that fission with slow neutrons occurs in both U-235 and U-238.

b. The separated isotopes for Frisch's experiments were obtained using chemical separation techniques.

c. In natural uranium ore, only 1 atom in 140 is of U-235.

18–7 In the chain-reacting pile,

a. the purpose of the moderator is to absorb the heat produced by fission.

b. a fast neutron striking a hydrogen nucleus loses more kinetic energy than a fast neutron striking a carbon nucleus.

c. water was not used in the first chain-reacting pile because too many neutrons would be lost when hydrogen absorbed a neutron to form an isotope of hydrogen, ^2_1H.

d. cadmium strips served as a moderator.

18–8 In the chain-reacting pile, the first fission reaction was triggered by a neutron source located within the pile.

18–9 If the reproduction factor k for a certain pile is 0.98,

a. the neutron loss for this pile exceeds the number of neutrons generated by fission.

b. a self-sustaining chain reaction occurs in this pile.

c. the size of the pile must be increased in order for a self-sustaining chain reaction to occur.

18–10 The production of plutonium-239 begins when a nucleus of uranium-235 absorbs a neutron and that absorption is followed by two subsequent beta decays.

Questions for Thought

18–1 Define the term *induced activity.*

18–2 Fermi and his students bombarded targets with neutrons.

a. What was the object of these experiments?

b. Write a nuclear reaction equation showing how the neutrons were produced.

c. What were the advantages and disadvantages of using neutrons rather than alpha particles?

d. Briefly describe their chemical separation technique. What was its objective?

18–3 Fermi and his students bombarded a silver target with neutrons.

a. What happened when they placed a neutron source inside a block of paraffin?

b. How did Fermi explain these results?

18–4 After bombarding a uranium target with neutrons, Hahn and Strassmann dissolved the target in acid, added barium, and performed a chemical separation process.

a. What did they observe then?

b. What were their conclusions?

18–5 How did Meitner and Frisch use Bohr's liquid-drop model of the nucleus to explain fission?

18–6 a. Describe Frisch's experiment, which showed that an energy release does accompany fission.

b. Use conservation of momentum to show that only *one* fission fragment contributed to the pulse in the ionization chamber.

18–7 Define the terms *neutron excess* and *chain reaction*. What is the connection between them?

18–8 Describe the Fermi team experiment that showed that neutrons are released during fission.

18–9 a. Briefly describe how a mass spectrometer was used to separate an isotope of U-235 and U-238 from natural uranium. How much was produced?

b. Describe the experiment that showed that slow neutrons cause fission of U-235 and not of U-238.

18–10 When a nucleus of U-235 absorbs a neutron, a nucleus of _____ is formed. Fission can occur in about (choose one: 1, 10, 50, or 100) _____ way(s), and the two most probable values for the atomic mass number of the fission fragments are _____ and _____ . On the average, _____ neu-trons are produced during each fission reaction. Explain whether a chain reaction would be possible if 0.8 neutron was produced for each fission reaction.

18–11 a. Define the term *moderator.*

b. Would water make a good moderator? Why and/or why not?

c. Why did Fermi choose graphite, a form of carbon, as the best moderator to use with natural uranium?

18–12 Define the term *reproduction factor.*

18–13 Describe an intermediate pile. What was its purpose?

18–14 Why is it easier to obtain a sample of pure plutonium-239 than one of pure uranium-235?

18–15 What is spontaneous fission?

Questions for Calculation

18–1 Complete the following:

$$^{107}_{47}Ag + ^{1}_{0}n \rightarrow ^{108}_{47}Ag$$

$$^{108}_{47}Ag \rightarrow \beta^- + ?$$

18–2 Write two nuclear reaction equations showing how plutonium - 239 is formed when uranium - 238 absorbs a neutron.

18–3 A spherical drop of water with a radius of 1 cm has a volume $(4/3 \ \pi r^3)$ of 4.19 cm^3 and a surface area $(4\pi r^2)$ of 12.6 cm^2. Show that a cube of water of the same *volume* has a side of length 1.61 cm. Find the surface area of this cube.

18–4 Let a circle about three-eights of an inch in diameter represent a water molecule. Draw several rows, one right next to the other, of such circles.

a. With how many nearest neighboring molecules does a molecule on the sur-face interact? With how many does a molecule in the interior interact?

b. Use the results of calculation question 18–3 to explain why the spherical drop is more stable (or the molecules bound more tightly) than a cubic drop.

18–5 A U-235 nucleus absorbs a (slow) neutron to form U-236.

a. Show that an energy of 6.4 MeV is released in this process. (U-235 has a mass of 235.0439 amu; U-236, 236.0457 amu; and a neutron, 1.0087 amu. 1 amu = 931 MeV.)

b. According to Meitner and Frisch, what form did this energy take?

18–6 Use information from the section "Effects of Neutrons Striking Uranium: An Example" to show that fission with slow neutrons is about 500 times more probable than that with fast neutrons.

†18–7 Suppose that the number of neutrons per second causing fission F in a cubic pile (side L in meters) is given by F equals (1 million neutrons per sec/m^3) L^3, and the number of neutrons per second lost at the surface S is given by S equals (4 million neutrons per sec/m^2)L^2.

    a. Make a table showing the values of F and S for values of L ranging from 1 m to 6 m.

    b. Find the critical size of the pile where a sustained reaction will just take place.

    c. What happens when L is larger than the critical size? Smaller than the critical size?

Footnotes

[1] Laura Fermi, *Atoms in the Family: My Life with Enrico Fermi*: (Chicago: The University of Chicago Press, 1954), p. 90. (Reprinted from *Atoms in the Family* by Laura Fermi by permission of the University of Chicago Press. © by The University of Chicago Press, 1954.)

[2] Ibid, p. 98. (Reprinted from *Atoms in the Family*/by Laura Fermi by permission of the University of Chicago Press. © by The University of Chicago Press, 1954.)

[3] Otto Hahn, "The Discovery of Fission," *Scientific American* 198(February, 1958):76.

[4] Ibid, p. 82.

[5] Otto R. Frisch, "How It All Began," *Physics Today* 20(November, 1967):47.

[6] Lise Meitner and O. R. Frisch, "Disintegration of Uranium by Neutrons: A New Type of Nuclear Reaction," *Nature* 143(1939):239.

[7] Ibid, p. 239.

[8] Frisch, *Physics Today,* p. 47.

[9] O. R. Frisch, "Physical Evidence for the Division of Heavy nuclei under Neutron Bombardment," *Nature* 143(1939): 276.

[10] Herbert L. Anderson, "The First Chain Reaction," in Jane Wilson, ed., *All in Our Time: The Reminiscences of Twelve Nuclear Pioneers* (Chicago: Bulletin of the Atomic Scientists, 1975), p. 73.

[11] H. L. Anderson, E. Fermi, and H. B. Hanstein, "Production of Neutrons in Uranium Bombarded by Neutrons," *Physical Review* 55(1939):797.

[12] A. O. Nier, E. T. Booth, J. R. Dunning, and A. V. Grosse, *Physical Review* 57 (1940):546 and 748.

[13] Enrico Fermi, *Collected Works*, Vol. II (Chicago: The University of Chicago Press, 1962), p. 543. (Reprinted from *Collected Works* by Enrico Fermi by permission of the University of Chicago Press. © 1962 by The University of Chicago Press.)

[14] Jerry B. Marion, *A Universe of Physics: A Book of Readings* (New York: John Wiley, 1970), p. 213. (Einstein's letter is reproduced here.)

[15] H. L. Anderson, "The First Chain Reaction," p. 86.

[16] Fermi, *Collected Works*, p. 547. (Reprinted from *Collected Works* by Enrico Fermi by permission of the University of Chicago Press. © 1962 by The University of Chicago Press.)

[17] Enrico Fermi, *The Development of the First Chain Reacting Pile, Collected Works*, pp. 547–48. (Reprinted from *Collected Works* by Enrico Fermi by permission of the University of Chicago Press. © 1962 by The University of Chicago Press.)

[18] Herbert L. Anderson, "The First Chain Reaction," p. 95.

[19] Laura Fermi, *Atoms in the Family*, p. 198. (Reprinted from *Atoms in the Family* by Laura Fermi by permission of the University of Chicago Press. © by The University of Chicago Press, 1954.)

J. Robert Oppenheimer

CHAPTER 19

Nuclear Energy from Fission and Fusion

Introduction

In this chapter, we'll immediately consider two extremely important technological applications of nuclear energy: the atomic bomb and the nuclear reactor. In keeping with our historical approach, we'll talk about the atomic bomb first. Estimates showed that a single bomb would require between 1 and 100 kilograms of pure fissionable material. There were two choices: uranium-235 (U-235) and plutonium-239 (Pu-239). When Bohr considered the prospect of separating U-235 from U-238, he remarked in a conversation with J. A. Wheeler, "It would take the entire efforts of a country to make a bomb."[1] How prophetic were these words!

The powers-that-be ultimately decided to "leave no stone unturned" and to proceed with both projects: to separate U-235 from U-238 and to produce Pu-239 in a chain-reacting pile. We'll start by examining these processes, continue with the development of the atomic bomb, and end our discussion of fission on the subjects of the nuclear reactor and the breeder reactor.

Next, we'll introduce fusion. Remember (Chapter 6) how we saw that the present age of the sun (4.5 billion years) prohibited the conclusion that gravitational contraction could be the cause of the sun's energy. In the 1920s, scientists realized that the conversion of mass into energy just may be the process that fitted the bill, that caused the energy of the sun. Exactly *how* it occurred, though, was not known. We'll look into this energy process of fusion, in the sun and other stars as well, and we shall see how elements are formed in stars. Finally, we'll learn about the hydrogen bomb and see how the fusion reactor may one day supply energy for humankind.

The Production of Plutonium

Extensive research had been carried out at the Metallurgical Laboratory on the chemistry of plutonium and the method of chemically separating it from uranium. In these initial studies, only microgram (a millionth of a gram) amounts were available, but, even so, separation on a large scale did appear possible. The objective was to obtain kilogram amounts of plutonium. In order to grasp the magnitude of the problem, consider the following, which is an example of the power required to generate 1 kg of plutonium per month.

Suppose that, in a fission of a U-235 nucleus releasing an energy of 200 MeV, 2.5 neutrons are produced: one is needed to maintain the chain reaction, one is captured by U-238, ultimately forming Pu-239 (see Chapter 18, Equations 10 and 11), and 0.5 neutrons per fission are lost.

We know that the atomic mass of an element in kilograms contains 6×10^{26} atoms. (Chapter 1, p. 8.) Therefore, 1 kg of plutonium contains $6 \times 10^{26}/239$, or 2.5×10^{24} atoms. Since an energy release of 200 MeV accompanies the formation of one atom of plutonium, then the total amount of energy produced in 30 days is given by

$$\text{Energy} = 2.5 \times 10^{24} \text{ atoms} \times \frac{200 \text{ MeV}}{\text{atom}}$$
$$= 5 \times 10^{26} \text{ MeV}.$$

Since 1 MeV equals 1.6×10^{-13} joules, this energy becomes

$$\text{Energy} = 5 \times 10^{26} \text{ MeV} \times \frac{1.6 \times 10^{-13} \text{ joules}}{\text{MeV}}$$
$$= 8 \times 10^{13} \text{ joules}.$$

Power is defined as the amount of energy per time, and 1 watt of power is a joule/sec. Since 30 days is equal to 2.592×10^6 seconds, the power required would be

$$\text{Power} = \frac{\text{energy}}{\text{time}}$$
$$= \frac{8.0 \times 10^{13} \text{ joules}}{2.592 \times 10^6 \text{ sec}}$$
$$= 3.1 \times 10^7 \text{ watts}$$
$$= 31,000 \text{ kilowatts}.$$

The maximum power generated by the first pile was only 200 watts, or 0.2 kilowatts. Exactly 31 piles having a power of 1000 kilowatts would be needed to produce 1 kg of plutonium per month! Building a reactor with a power of 1000 kilowatts based on experience derived from a small model was indeed a giant step. So, too, was designing a plant capable of extracting grams of plutonium from tons of uranium, from experience with only micrograms of plutonium. But the pressures of wartime and the possibility of achieving tremendously important results provided the impetus.

THE ARGONNE SITE

In 1943, the chain-reacting pile was moved to a site in Argonne, Illinois, about 35 miles west of Chicago, now called Argonne National Laboratory. The chain-reacting piles for the production of plutonium were designed at Argonne. These piles were built at Hanford, Washington. Designs for the radiation shielding of the piles were also carried out at Argonne. Many physics experiments, using the neutrons from the pile, were also performed. Today, Argonne National Laboratory is a leader in reactor development.

CLINTON ENGINEER WORKS IN OAK RIDGE, TENNESSEE

The E. I. du Pont de Nemours and Company was selected to design and construct the pilot plant for the production of plutonium in Tennessee, as well as the production plant at Hanford, Washington. The pilot pile was designed to produce a power of 1000 kilowatts. In order to facilitate the removal of uranium, rods of uranium imbedded in the graphite moderator were used, rather than lumps of uranium. The pile was water-cooled. The pilot plant provided large amounts of plutonium for study of its chemical properties. Oak Ridge National Laboratory now occupies this site.

NUCLEAR ENERGY FROM FISSION AND FUSION

HANFORD ENGINEER WORKS

Construction of the main production plant at Hanford, Washington, was begun in 1943. This site was chosen because of its remoteness and its location on the Columbia River, which provides cooling water. Three piles and plutonium separation plants were constructed. The first large pile began to operate in September, 1944, and the entire plant was in operation by the summer of 1945.

Separation of U-235

Two types of separation plants were constructed at the Clinton Engineer Works beginning in 1943. One utilized the technique of gaseous diffusion, and the other, the technique of electromagnetic separation.

GASEOUS DIFFUSION

For this method, the uranium must be in the form of a gas, uranium hexafluoride (UF_6). At a given temperature, *all* molecules of the gas (those with U-235 atoms *and* those with U-238 atoms) will have the same average kinetic energy. Since the kinetic energy is given by $1/2mv^2$, the lighter molecule will have a larger average speed. Because the molecules differ only slightly in their mass, their speeds likewise differ only slightly. The speed of the $^{235}UF_6$ molecule is only 0.4 percent larger than the $^{238}UF_6$ molecule. If the UF_6 gas is directed toward a porous barrier, the molecules containing U-235 will pass through the barrier at a faster rate than those containing U-238. Thus, on the other side of the porous barrier, the ratio of U-235 to U-238 will be increased, but just by an extremely small amount. If this "enriched" gas is passed through another porous barrier, the enrichment can be increased still further.

About 2000 barriers would theoretically be required to produce uranium enriched to 99 percent U-235, but about 5000 barriers were actually used. The porous barriers were made from thin sheets of metal etched with acid to form very small holes only about 1×10^{-8} m in diameter. The plant—a major engineering feat of our time—was put into successful operation before the summer of 1945.

ELECTROMAGNETIC SEPARATION

In a mass spectrometer, U-235 can be separated from U-238, but this, we have seen, is an exceedingly slow process. In November, 1941, E. O. Lawrence at the University of California assembled a team to convert the 37-inch cyclotron magnet into a mass spectrometer. Many modifications, including a more intense ion source, were made. This method seemed promising. Construction was begun in Tennessee in March of 1943. Oak Ridge, built in about two and a half years under strict security, had a population of 75,000 by 1945.

The Atomic Bomb

ENTER OPPENHEIMER

When fission was discovered in 1939, J. Robert Oppenheimer (1904–1967) was a well-established theoretical physicist and professor of physics at both the University of California, Berkeley, and the California Institute of Technology. Recall (Chapter 18) that the developments leading to the first sustained chain reaction in 1942 were swift. Meanwhile, at Berkeley, Oppenheimer led a group of theoretical physicists who were considering the problem of the atomic bomb. The U.S. Army had appointed General Leslie Groves to direct the atomic energy project that went under the famous code name "Manhattan District." Oppenheimer recommended to Groves that the development of the atomic bomb be concentrated in a single laboratory. Groves thought that an excellent idea, mainly because it would provide greater security. Following Oppenheimer's advice, Groves chose a boys' school in a remote section of New Mexico. In 1943, Site Y (now the Los Alamos Scientific Laboratory) was set up. Because of his work at Berkeley and his magnetic personality, Oppenheimer was selected by Groves to head the laboratory. Even though he had no prior experience with administration, Oppenheimer assembled a first-rate team of scientists and progress flourished under his skillful leadership.

CRITICAL MASS OF THE BOMB

The basic idea behind nuclear bombs is that a sustaining chain reaction will occur when a sufficient mass, called the **critical mass,** has been assembled in one piece. If the mass of fissionable material is less than the critical mass, the chain reaction will die out. To see why, let's consider a neutron produced by spontaneous fission in a sphere of fissionable material. This neutron may make many collisions with atoms, bouncing like billiard balls, before it either causes the fission of an atom, or reaches the surface where it can escape. Thus, if the radius of the sphere is increased, the neutron can make more collisions with atoms before it reaches the surface, and the probability of fission is, therefore, increased.

When the radius of the sphere is increased, the *surface area* of the sphere is likewise increased and, because of this, the probability that the neutron will escape from the surface becomes greater. Thus, there are two competing effects. Increasing the volume raises the probability of fission, while increasing the surface area raises the probability of escape. Since volume is proportional to r^3 and the surface area to r^2, the volume *increases* more rapidly than the surface area, with increasing values of r. For one value of r, the two effects are equal, and the neutrons causing fission just equal those lost at the surface.

A sustaining chain reaction occurs when one neutron from each fission reaction produces another fission reaction. For a smaller value of r, the number of neutrons escaping is greater than the number causing fission, and a chain reaction does not take place. For a large value of r, the number causing fission is greater than those lost at the surface, and a sustaining chain reaction does take place. The following is an example that illustrates these effects.

Suppose that the loss of neutrons at the surface is represented by the simple equation $L = 4r^2$, and the production of neutrons by fission is represented by $P = 2r^3$ in some set of units. When $r = 2$, both L and P are equal to 16. For all values of r less than 2, L is greater than P. When $r = 1$, $P = 2$ and $L = 4$. For all values of r greater than 2, P is greater than L. When $r = 3$, $P = 54$ and $L = 36$.

One of the theoretical problems concerned calculating the critical mass. Scientists could not obtain a precise value because needed constants were not accurately known. In Chapter 18 (p. 418) we saw that, of 10,000 neutrons striking a 1-cm cube of natural uranium, 1950 neutrons caused the fission of U-235 nuclei and 1680 neutrons were captured by U-238 nuclei. Such information was not *accurately* known then for either uranium or plutonium. Of the 10,000 neutrons, 3950 were scattered from the uranium nuclei. How many would be scattered at 10° from their initial direction? At 50°? At 120°? This, too, was not accurately known. Using the information available, the theoretical physicists calculated that the critical mass was somewhere between 1 kg and 100 kg. The work of the experimental physicists involved determining the needed constants more accurately, so that the critical mass calculations could be refined.

When sufficient U-235 and Pu-239 became available, it was also possible to determine experimentally the critical mass and to test the calculations by building near spherical assemblies out of small bricks. Otto Frisch tells of this procedure:

One of the more risky proposals I made was accepted, rather to my surprise. Richard Feynman said it was like tickling the tail of a sleeping dragon, and so it became known as the dragon experiment. It consisted in setting up an assembly made from a hydride of uranium-235 big enough to explode but with the central core missing so that it was safe. That central core was then allowed to fall through the hole so that for a split second the conditions for a (rather mild) nuclear explosion existed. Of course there were smooth guide rails and many safeguards against the core getting stuck! We got quite large bursts of neutrons which behaved exactly as foreseen and which convinced us—if any of us doubted it—that there was no significant delay in the emissions of neutrons from a nuclear fission. It was as close as we could get to an atomic bomb without actually being blown up.[2]

DESIGN OF THE BOMB

In essence, the atomic bomb is a device to assemble the critical mass of fissionable material. The first design considered by the scientists at Los Alamos was a Gun-type assembly.

Gun-Type Bomb Before detonation, the fissionable material is separated into two parts, with each part having less than the critical mass. Using ordinary explosives, the two parts are driven together with high speed so that their total mass exceeds the critical mass (see Fig. 19–1).

One difficulty is that the two parts must be put together *very rapidly.* Otherwise, the chain reaction might begin prematurely, and generate enough energy to blow the device apart. In this case, the chain reaction would stop, and only a small explosion would have occurred.

Only one atomic bomb used the Gun-type assembly with U-235, and it was dropped on Hiroshima. The rest were of the implosion type.

Implosion-Type Bomb In an implosion-type bomb, a sphere of plutonium is surrounded by a shell of ordinary explosives. When the explo-

sives are detonated, the force of the explosion compresses the plutonium sphere *instantaneously* by an appreciable amount. Since the nuclei are now closer together, the probability that a neutron will collide with a nucleus is greater, and the probability that it will escape at the surface is less. This compression permits a sustained chain reaction to occur. The compression must occur quickly, so that the device does not blow apart prematurely, as mentioned earlier. A schematic diagram of the atomic bomb is shown in Fig. 19–2.

The ordinary explosives are formed in a special shape, called *lenses,* so that the pressure is the same everywhere over the surface of the plutonium sphere. The U-238 tamper has two purposes: (1) When the chain reaction begins, the dense U-238 tamper, having a large inertia, holds the plutonium core together; and (2) when neutrons reach the surface of the plutonium sphere, they strike atoms of U-238 and, like the collision between two balls, the neutrons are "reflected" back into the plutonium core. The initiator releases neutrons to start the reaction.

The implosion bomb using plutonium was tested on July 16, 1945, at a site called Trinity, near Alamagordo, New Mexico. (The uranium gun-type bomb was never tested, because only uranium for one bomb was available at that time.) The implosion bomb was detonated atop a 100-foot tower. The test was a success; the bomb released an energy equivalent to the explosion of 20,000 tons of TNT.

EFFECTS OF ATOMIC BOMB EXPLOSION

The energy of the atomic bomb explosion is released in several ways. The extremely rapid expansion of the fireball causes high-pressure air to move away from it. This phenomenon, called a **blast wave,** accounts for about 50 percent of the energy of the explosion. Heat radiation accounts for about 35 percent, and radioactivity for about 15 percent. As a result of the test at Trinity, the scientists found that the blast damage could be increased, and the effects of radioactivity reduced, if the bomb were detonated at a higher altitude. The radioactivity would be spread over a much larger area. On August 6, 1945, a uranium bomb equivalent to 20,000 tons of TNT was dropped on Hiroshima, Japan, at a height of 1850 feet. On August 9, 1945, a plutonium bomb was dropped on the population of Nagasaki, Japan.

Nuclear Reactors*

DESIGN

A diagram of a **nuclear reactor** is shown in Fig. 19−3. The water serves as a moderator and as a coolant. Because water absorbs neutrons (Chapter 18, Equation 9), the uranium rods are enriched to 3 percent uranium-235. The **control rods,** which absorb neutrons, control the power of the reactor. The heat generated by the reactor is used to create steam to drive the turbines of an electrical generator.

SAFETY FEATURES

The safety features of a reactor include the thin metal *cladding* around the uranium rods, the *containment vessel* surrounding the core, and the *hemispherical dome* of steel-reinforced concrete.

HEAT AND POWER PRODUCTION

When a fission reaction occurs, the energy of 200 MeV is shared by the kinetic energy of the fission fragments (about 90 percent) and the kinetic energy of the neutrons (about 10 percent). When these energetic fission fragments collide with the uranium atoms, they transfer some of their kinetic energy to those atoms. This increased motion of the uranium atoms is perceived as heat. In addition, these fission fragments are radioactive and emit neutrons, gamma rays, and beta rays until a stable isotope is formed. When the beta rays travel through the uranium and finally stop, their initial kinetic

*Much of the information in this section is based on the source *Nuclear Power and the Environment: Questions and Answers* (Hinsdale, Ill.: American Nuclear Society, 1976).

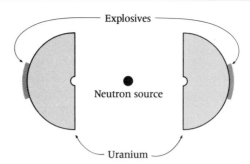

FIG. 19−1   The gun-type assembly atomic bomb.

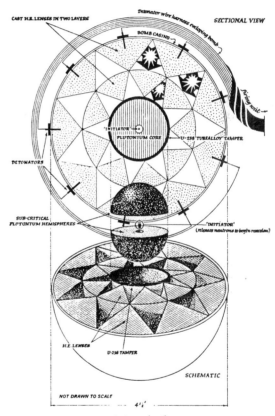

The Gadget

FIG. 19−2   A diagram of an implosion-type atomic bomb. [From Lansing Lamont, *Day of Trinity*, Copyright © 1965 by Lansing Lamont. (New York: Atheneum, 1965) Reprinted by arrangement with Atheneum Publishers.]

NUCLEAR ENERGY FROM FISSION AND FUSION

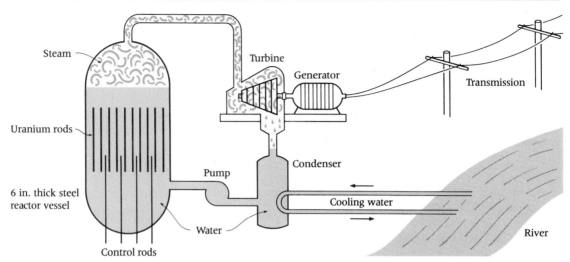

FIG. 19–3 A diagram of boiling-water reactor. A containment building constructed of steel and reinforced concrete surrounds the entire reactor system.

energy is also transformed into heat. The heat produced in the reactor core is transferred to the water passing through the core and is then converted into electricity.

STORAGE OF RADIOACTIVE WASTES

The uranium fuel elements become extremely radioactive. As a fission fragment decays, it can form either a stable isotope or an isotope having a long half-life. For example, if the fission fragment $^{95}_{38}Sr$ is formed, it decays by emitting a neutron to form $^{94}_{38}Sr$ and then two beta rays in succession to form $^{94}_{40}Zr$, which is stable. Often, as a fission fragment decays, it forms an isotope with a long half-life. Examples are: strontium-90 with a half-life of 28 years; cesium-137 with a half-life of 30 years, and technetium-99 with a half-life of a million years. For example, the radioactivity of an isotope with a 30-year half-life is reduced by a factor of 2^4, or 16, after 4 half-lives, or 120 years.

The uranium rods are left in the reactor until about two-thirds of the U-235 is consumed. They must then be replaced. The spent uranium rods are stored under 20 feet of water at the nuclear power plant. A reprocessing plant for commercial power reactors is nearly complete at Barnwell, South Carolina, to separate the valuable uranium and plutonium from the spent fuel rods. Three currently operating government reprocessing plants process fuel from military programs and special reactors. The continued storage and/or disposal of radioactive waste poses serious technical and social problems.

The National Academy of Science has recommended that solidified nuclear wastes be permanently disposed of by deep underground burial in rock-salt formations.[3] These rock-salt formations have not been disrupted by earthquakes for millions of years. Processes for incorporating radioactive wastes into either glass-base or cement materials, which are impervious to water, have been developed and successfully tested.

RADIATION

Radiation is all around us and is part of our natural environment. There is radiation from cosmic rays. There is natural radioactivity in the soil, in the houses we live in, in the food we eat, and in the air we breathe. Some of this natural radiation is due to uranium, thorium, radium, potassium-40, hydrogen-3, carbon-14, and radon gas. Then there is the radiation we are exposed to from certain medical procedures. The unit of radiation, which takes into account its biological effects, is the **millirem** (abbreviation: mrem). We shall not define this unit precisely, but, rather, compare the amounts of radiation from these various effects. Table 19–1 displays the amount of radiation received by the average United States citizen during one year. The total is approximately 200 mrem.

The amount of radiation emitted by a nuclear reactor operating normally is indeed quite small. For example, a person living at the boundary of a nuclear reactor site for one year, 24 hours per day, receives about 5 mrem. At a distance of one mile from a nuclear reactor site, the dose is reduced to 0.5 mrem. Beyond five miles, there is no radiation from the reactor. Compare these values with the radiation of 20 mrem from one dental X-ray.

The radioactivity from a reactor is due, in part, to the gaseous fission fragments xenon and krypton that occasionally escape into the reactor coolant through pinhole defects in the metal cladding around the uranium rods. Small amounts of gaseous wastes are stored to allow some decay of the radioactivity and then discharged to the atmosphere in accordance with regulations.

A REACTOR IS NOT A BOMB

It is simply not possible for a reactor to explode like an atomic bomb. First of all, the uranium is enriched to only 3 percent in U-235, while a bomb contains 100 percent fissionable material. Secondly, the large explosive power of the atomic bomb is due to the *very rapid* assembly

TABLE 19–1   AVERAGE RADIATION DOSE IN ONE YEAR

SOURCE	AVERAGE AMOUNT OF RADIATION IN ONE YEAR (mrem)
Cosmic radiation at sea level (Add 1 mrem for each 100 ft above sea level.)	40
Living in a brick house (75 percent of time)	45
From the ground	15
Water, food, and air	25
Medical procedures (A chest X-ray averages between 100 to 200 mrem; a dental X-ray, 20 mrem; and a gastrointestinal tract X-ray, 2000 mrem.)	61
Total	186 mrem

of the fissionable material into a critical mass. This element is, of course, absent in a nuclear reactor.

MELTDOWN

The most serious accident that could possibly happen in a nuclear reactor is a **meltdown.** Such an accident has never occurred. It could occur if, for some reason, the cooling water did not reach the reactor core. Even if the reactor had been completely shut down because the control rods had been inserted, the intense radioactivity in the uranium rods would be sufficient to melt the core within a few minutes for a 1000-megawatt reactor. (A megawatt is 1 million watts.) The emergency core cooling would have to reach the core within this short time. Following a complete meltdown due to failure of the emergency core cooling system, the molten mass would melt its way through the containment and into the earth. Since it would be melting its way toward China, this effect is called the "China syndrome." The gaseous fission products would escape into the atmosphere, but the rest would remain. The groundwater could be contaminated.

THE ENERGY CRISIS

The term *energy crisis* should, by now, be familiar to everyone. The fact that the energy demands of the United States increase by about the same *percentage* from year to year, rather than the same *amount,* is at the heart of the problem. For example, in 1974, the total amount of all types of energy consumed in the United States was equivalent to 35.4 million barrels of oil per day. This was 3.8 percent more than in 1973. If our energy demands increase by 3.8 percent each year due to a rising population, greater productivity, and an enhanced standard of living, then, in 1993, our energy demand will be $35.4 \times (1.038)^{19}$, which equals 70.8 million barrels of oil per day. Our energy demands will double every 19 years!

If the energy demands of the United States increase by a *larger* percentage each year, the energy demand will double sooner. At present, the total energy needs of the United States are supplied by oil (46 percent), natural gas (31 percent), coal (18 percent), hydroelectricity (4 percent), and nuclear reactors (1 percent). Coal is America's most plentiful fossil fuel. Assessments indicate that our coal reserves may last about 200 years, and natural gas and oil much less than this, perhaps 50 to 100 years. These are, of course, estimated projections and may be pessimistic in outlook.

ENERGY ALTERNATIVES

There are some alternatives to the use of oil as an energy source. **Geothermal power**, which utilizes the natural steam or hot water trapped in the earth, is one possibility. Some geothermal plants are being built, but their output is small—50 to 100 megawatts—compared to fossil fuel at a nuclear plant of 1000 megawatts. **Solar energy,** under development today, is another strong contender, but the technology for a large power plant is still in the research stage. In principle, we know how to convert sunlight to electricity, but making the process *economically competitive* with fossil fuel and nuclear plants presents a barrier not yet overcome. A **fusion reactor** looms on the distant horizon as another alternate possibility, but a fusion reactor producing as much power as it consumes has not yet been achieved. Generally, it takes about 25 years, having a working model, to engineer a competitive power plant.

The U.S. government is committed to a policy of energy independence. This can be achieved by the expansion of nuclear power, economically feasible solar energy, increased use of coal, conservation, and the elimination of energy waste.

The Breeder Reactor

A **breeder reactor** is designed to convert the nonfissionable U-238 into the fissionable Pu-239 and to produce more Pu-239 than it consumes of U-235. Production of Pu-239 occurs when a U-238 nucleus captures a neutron and then emits two beta rays in succession. The probability that a neutron will be captured by a U-238 nucleus is larger for *fast* neutrons and, for this reason, the breeder reactor does not contain a moderating material to slow down the neutrons. However, the probability that a fast neutron will cause the fission of a U-235 nucleus is greatly reduced (by a factor of about 500) compared to slow neutrons. To counter this effect, the uranium fuel is enriched to a larger percentage of U-235 than in slow neutron reactors.

In some designs, the initial uranium fuel is concentrated to nearly 50 percent U-235. On the average, 2.5 neutrons accompany the fission of U-235 nuclei. One neutron is needed to sustain the reaction. If less than 0.5 neutrons of the remaining 1.5 are lost, then more than 1 neutron can be absorbed by U-238 to produce Pu-239. Hence, the breeder reactor produces more fissionable material, Pu-239, than it consumes of U-235. Pu-239 is produced in the uranium rods and also in a blanket of U-238 surrounding the reactor core.

A liquid metal, such as liquid sodium, is used as a coolant in many designs. Because liquid metal readily transfers heat, and because it is molten at a high temperature (boiling point of sodium is 892°C), the conversion of heat to electricity is more efficient with its use. The reactor is controlled by removing or inserting the uranium fuel rods. As the world's supply of uranium dwindles, the breeder reactor can supply the needed fissionable material.

Figure 19−4 shows the experimental breeder reactor, EBR-II, at Argonne's site in Idaho. Since it began operation in the mid-1960s, this facility has provided valuable experimental data and, through 1980, over 1 billion kilowatt hours of electricity.

Fusion: Hydrogen Burning

PROTON-PROTON CHAIN

The center of the sun has a temperature of 15 million °K (degrees Kelvin).* At this temperature, the atoms have been ionized so that the gas consists of free nuclei and electrons. The protons have a kinetic energy of 0.002 MeV. The process of **fusion** begins when two protons collide head-on and slow down. At such a small kinetic energy, there is only an exceedingly small chance that the two nuclei will fuse together

*To find a temperature in °C, subtract 273 from a temperature specified in °K. That is, °C = °K − 273.

rather than retreat because of the Coulomb repulsion between two like charges. A given proton will take about 7 billion years[4] before it fuses with another proton, according to

$$\ce{^1_1H + ^1_1H -> ^2_1H + \beta^+ + \nu}. \tag{1}$$

After about 4 seconds, the deuteron (a nucleus of the $\ce{^2_1H}$ isotope, called deuterium) collides with another proton, according to

$$\ce{^2_1H + ^1_1H -> ^3_2He + \gamma}. \tag{2}$$

After about 400,000 years, the helium-3 nucleus collides with another one, yielding

$$\ce{^3_2He + ^3_2He -> ^4_2He + 2 ^1_1H + \gamma}. \tag{3}$$

Since two helium-3 nuclei are needed, the net effect can be obtained from the following

2 (Equation 1) + 2 (Equation 2) + Equation 3.

This yields

$$\ce{4 ^1_1H -> ^4_2He + 2\beta^+ + 2\nu + 3\gamma}. \tag{4}$$

Thus, the net effect is to take 4 protons and fuse them into 1 helium nucleus with emission of positrons, neutrinos, and gamma rays. Energy is released during fusion because the mass on the left side of Equation 4 is larger than the total on the right, as shown in the following:

4(1.007825 amu)
  = 4.00260 amu + 2(0.00055 amu) + energy
4.03130 amu = 4.00370 amu + energy
Energy = 0.02760 amu
  = (0.02760 amu)(931.0 MeV/amu)
  = 25.7 MeV.

FIG. 19–4 (a) The experimental breeder reactor-II (EBR-II). (b) A cutaway drawing of EBR-II. (Courtesy of Argonne National Laboratory)

(a)

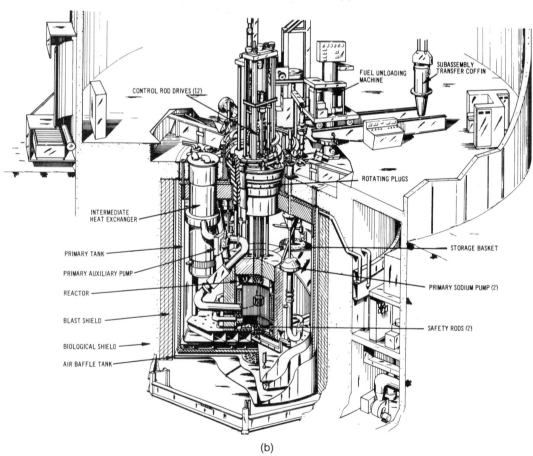

CONTROL ROD DRIVES (12)

FUEL UNLOADING MACHINE

SUBASSEMBLY TRANSFER COFFIN

ROTATING PLUGS

INTERMEDIATE HEAT EXCHANGER

STORAGE BASKET

PRIMARY TANK

PRIMARY AUXILIARY PUMP

PRIMARY SODIUM PUMP (2)

REACTOR

BLAST SHIELD

SAFETY RODS (2)

BIOLOGICAL SHIELD

AIR BAFFLE TANK

(b)

The energy release occurs in the form of the energy of the neutrinos, the kinetic energy of the positrons, and the energy of several gamma rays.

THE CARBON-NITROGEN CYCLE

In 1938, the German-born physicist Hans Bethe (b. 1906) showed that another process, called the **carbon-nitrogen cycle,** also releases energy:

$$^{12}_{6}C + {}^{1}_{1}H \rightarrow {}^{13}_{7}N + \gamma \qquad (5)$$

$$^{13}_{7}N \rightarrow {}^{13}_{6}C + \beta^{+} + \nu \qquad (6)$$

$$^{13}_{6}C + {}^{1}_{1}H \rightarrow {}^{14}_{7}N + \gamma \qquad (7)$$

$$^{14}_{7}N + {}^{1}_{1}H \rightarrow {}^{15}_{8}O + \gamma \qquad (8)$$

$$^{15}_{8}O \rightarrow {}^{15}_{7}N + \beta^{+} + \nu \qquad (9)$$

$$^{15}_{7}N + {}^{1}_{1}H \rightarrow {}^{12}_{6}C + {}^{4}_{2}He + \gamma. \qquad (10)$$

In this series of reactions, the carbon acts as a catalyst because, in the final step, the carbon is again returned. The net reaction, found by adding Equations 5 through 10, is

$$4{}^{1}_{1}H \rightarrow {}^{4}_{2}He + 2\beta^{+} + 4\gamma + 2\nu. \qquad (11)$$

The result is the same: 4 hydrogen atoms have been fused into 1 helium atom with the release of 2 positrons, 2 neutrinos, and several gamma rays.

In the sun, about 95 percent of the energy is due to the proton-proton chain, and about 5 percent from the carbon-nitrogen cycle. In more massive stars, the carbon-nitrogen cycle predominates.

Because hydrogen is "used up" in the process of obtaining energy, the proton-proton chain and the carbon-nitrogen cycle are often referred to as **hydrogen-burning.** In the sun, 700 million tons of hydrogen are converted into helium every second!

Formation of Elements

HELIUM-BURNING

When a new star is formed by the contraction of a gas cloud, the star becomes heated (see Chapter 6, p. 121). Because of continued contraction of the gas, the star eventually heats up to the point where the hydrogen-burning process can begin and, this becomes more frequent as the temperature increases. The hydrogen-burning creates an internal pressure in the gas. Finally, an equilibrium is reached where this internal pressure is just balanced by the gravitational force tending to contract the gas. Thus, the contraction stops. The hydrogen-burning continues and the **helium ash** (so named because it is inactive in supplying energy) accumulates in the core of the star. After about 10 percent of the hydrogen has been consumed, the inert helium core begins to contract because no nuclear reactions are taking place within it, and the star evolves to what astronomers call a **red giant.** Radiation of light accompanies this contraction and causes the hydrogen around the helium core to heat and expand. Because of this expansion, the outer layer becomes cool and red, hence the name red giant. The helium core keeps contracting and when the temperature reaches 100 million °K, the helium ceases to be inert and helium-burning begins, according to

$$^{4}_{2}He + {}^{4}_{2}He \rightarrow {}^{8}_{4}Be$$

$$^{8}_{4}Be + {}^{4}_{2}He \rightarrow {}^{12}_{6}C.$$

The higher temperature is needed because the alpha particles (He-4 nuclei) must have sufficient kinetic energy to overcome the large Coulomb repulsion between particles having *2 positive charges.* Alpha particles can be added to form $^{16}_{8}O$ (oxygen-16), $^{20}_{10}Ne$ (neon-20), and, perhaps, $^{24}_{12}Mg$ (magnesium-24). The produc-

---

NUCLEAR ENERGY FROM FISSION AND FUSION

tion of yet heavier elements due to helium-burning stops here because the Coulomb repulsion becomes too large.

Helium-burning takes place in the core, while hydrogen-burning continues in the shell around the core.

THE ALPHA PROCESS

The **alpha process**[5] supplies more energetic alpha particles to offset the Coulomb repulsion and form elements heavier than $^{20}_{10}\text{Ne}$. When the temperature reaches 1 billion °K due to the contraction of the helium core, the gamma rays have sufficient energy to break apart the neon-20 nucleus, according to

$$^{20}_{10}\text{Ne} + \gamma \rightarrow {}^{4}_{2}\text{He} + {}^{16}_{8}\text{O}.$$

For example, if the gamma ray has an energy of 6 MeV, an energy of 4.75 MeV must be converted into the larger mass of helium-4 and oxygen-16 compared to neon-20. An energy of 1.25 MeV is then shared by the kinetic energy of the helium-4 and oxygen-16, with the larger fraction going to the helium-4 nucleus. In contrast, the average kinetic energy of the helium nucleus due to a temperature of 1 billion °K is only 0.1 MeV.

The energetic alpha particles can then fuse with neon-20 to form magnesium-24, according to

$$^{20}_{10}\text{Ne} + {}^{4}_{2}\text{He} \rightarrow {}^{24}_{12}\text{Mg} + \gamma.$$

The addition of energetic alpha particles yields $^{28}_{14}\text{Si}$ (silicon-28), $^{32}_{16}\text{S}$ (sulfur-32), $^{36}_{18}\text{Ar}$ (argon-36), $^{40}_{20}\text{Ca}$ (calcium-40), and probably $^{44}_{20}\text{Ca}$ [resulting from the decay, in several steps, of $^{44}_{22}\text{Ti}$ (titanium-44)] and $^{48}_{22}\text{Ti}$.

Finally, at still higher temperatures, subsequent addition of energetic alpha particles yields $^{52}_{24}\text{Cr}$ (chromium-56) and $^{56}_{26}\text{Fe}$ (iron-56). The addition of energetic alpha particles stops with the production of Fe-56.

NEUTRON CAPTURE

Elements beyond iron in the periodic table are formed by the capture of neutrons. The capture of one or more neutrons is followed by beta decay because neutron-rich isotopes are not stable. The capture of one neutron increases the atomic mass A by one unit, while beta decay increases the atomic number Z by one unit. For example, when $^{56}_{26}\text{Fe}$ absorbs 3 neutrons, $^{59}_{26}\text{Fe}$ is produced, which emits a beta ray to form stable $^{59}_{27}\text{Co}$ (cobalt-59). In turn, $^{59}_{27}\text{Co}$ can absorb 1 neutron to form $^{60}_{27}\text{Co}$, which emits a beta ray to form $^{60}_{28}\text{Ni}$ (nickel-60). These processes continue with the formation of elements having a mass number over 200. The buildup stops due to alpha decay and neutron-induced fission of the heavy elements.

DEATH OF A STAR

The final step in the evolution of a star depends upon its mass. A star having a mass about the mass of the sun becomes a **white dwarf.** When the hydrogen- and helium-burning have stopped, the star contracts due to gravitation until the atoms resist further contraction. This star is now small—a dwarf—and white hot. It cools for billions of years. A star somewhat larger than the mass of the sun has sufficient gravitation to squeeze the electrons into nuclei-forming neutrons and becomes a **neutron star.**

Stars having a mass greater than 1.4 times that of the sun may end their existence in two dramatic ways: in a *supernova explosion,* or as a *black hole.* In a **supernova explosion,** the star explodes and again becomes a cloud of gas, expanding at two-and-a-half million miles per hour. This cloud contains the elements formed in the star. From this cloud, new stars will form. Our sun must have been formed from the remnants of a supernova explosion because it contains not only hydrogen and helium, but a total of 65 elements.

A star having a mass about 3 times that of the sun can evolve into a **black hole.** Here, the gravitational force is so strong that the contraction continues until the mass of the star concentrates into a sphere of infinitesimally small size—nearly a point mass. Such a phenomenon is called a black hole because its gravity is so strong that light cannot escape from it; hence, it appears black. Astronomers search for black holes by scanning the cosmos for the X-rays produced when gas from a nearby star is sucked into a black hole. Because X-rays are absorbed by the earth's atmosphere, the X-ray detectors are placed aboard NASA satellites. The star Cygnus X-1 has been identified as a possible black hole.

The Hydrogen Bomb

The hydrogen bomb makes use of the following fusion reaction:

$$D + T \rightarrow {}_2^4He + {}_0^1n, \qquad (12)$$

where D is the symbol for the isotope of hydrogen ${}_1^2H$ called *deuterium* and T symbolizes the isotope of hydrogen ${}_1^3H$ called *tritium*. The kinetic energies of the He-4 and the neutron share the energy release of 17.6 MeV.

The hydrogen bomb consists of a small atomic bomb placed near the chemical lithium hydride (LiH). The lithium is enriched in the isotope ${}^6Li$ and the hydrogen is the isotope deuterium. The atomic bomb has two purposes: to supply neutrons, and to supply heat for the tremendous temperature needed for the fusion reaction in Equation 12. The neutrons from the atomic bomb form tritium, according to

$$ {}_3^6Li + {}_0^1n \rightarrow {}_2^4He + T. \qquad (13)$$

These neutrons release an energy of 5 MeV. The tritium nuclei produced in this way then undergo the fusion reaction in Equation 12.

Fusion Reactor

The fusion reactor may be the ultimate energy resource for humankind, but there are many problems that must be overcome before this reactor can be a reality.

The most promising fusion reaction is the fusion of deuterium and tritium, according to Equation 12.

Fusion reactions with protons are responsible for the energy release in the sun. However, the fusion of 2 protons occurs with such a small probability that it is not practicable in a fusion reactor. While only 1 atom in over 6000 atoms of hydrogen is deuterium, large amounts of deuterium could be obtained from ocean water. Tritium is radioactive with a half-life of 12 years, and can be produced in fission reactors.

In order for the fusion reaction in Equation 12 to occur at an appreciable rate, the mixture of deuterium and tritium gases must be heated to a temperature over 50 million °K! The temperature at the center of the sun is only about 20 million °K. How can the gas mixture be heated to this incredible temperature, and what kind of container can hold it? Scientists have found some answers to these questions, but the crux of the problem is to get *more* energy out of the reactor than has been put in. So far, this has not been achieved, but great strides toward this goal have been made in recent years.

TOKAMAK

One of the most promising developments has been the **Tokamak,** originally developed in the USSR. Figure 19−5 shows the Tokamak Fusion Test Reactor now operating at Princeton Plasma Physics Laboratory. At such a high temperature, the gas atoms are all ionized, consisting of positively charged nuclei and electrons. This ionized gas is called a **plasma.** Because a magnetic field exerts a force on a charged particle, a mag-

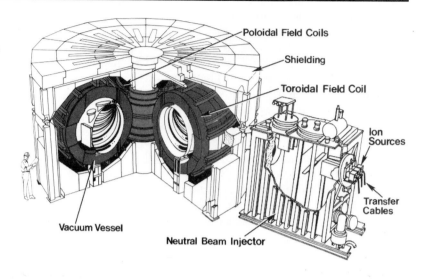

FIG. 19–5 The Tokamak Fusion Test Reactor. (Courtesy of Princeton Plasma Physics Laboratory)

Poloidal Field Coils

Shielding

Toroidal Field Coil

Ion Sources

Transfer Cables

Neutral Beam Injector

Vacuum Vessel

netic field can be used to confine a plasma. There are several ways to do this, but, in a Tokamak, the plasma is contained in a doughnut-shaped ring by a magnetic field, which is designed so that the plasma does not strike the walls for a period of time, which is called the **confinement time.** More fusion reactions can obviously occur if the confinement time is increased, and this is one area in which research with Tokamaks is proceeding. More fusion reactions can also occur if the density of ions (number of ions per cm^3) is increased. In order for the energy output of the reactor to *just equal* the energy input, calculations show that the density of ions multiplied by the confinement time must equal 10^{14} sec/cm^3. The Tokamak at Princeton has achieved a product of 1.5×10^{13} sec/cm^3, with a confinement time of 0.025 seconds; the maximum temperature of the plasma was 82 million °K.

LASER FUSION

One of the newest ideas is to use lasers to heat and increase the density of a frozen deuterium-tritium pellet. No magnetic fields would be used. Instead, enough fusion reactions must take place before the "pellet" expands. The lasers must be exceptionally powerful, perhaps as much as 1 million joules, and they must deliver a pulse of light energy in only one-billionth of a second. Electrical energy must be converted to laser light, and laser light energy has to be converted to heat in the plasma. There are losses at each conversion. Only further research can determine whether this idea is feasible.

It is not possible to predict when, or even if, fusion power will become a reality. Much has been accomplished in the last 20 years, but much more remains to be done. It seems safe to say that fusion power will not be a reality before the year 2000.

Summary

Pure fissionable material is needed to produce an atomic bomb. For the first atomic bomb, both U-235 *and* Pu-239 were obtained. Pu-239 was obtained by *chemically* separating the plutonium produced in a chain-reacting pile (now called a reactor) from the uranium. These piles were designed at Argonne, Illinois; a pilot plant was built in Tennessee; and three piles were constructed at Hanford, Washington.

The separation of U-235 from U-238 was accomplished in Oak Ridge, Tennessee, by two means: gaseous diffusion and electromagnetic separation. With the gaseous diffusion technique, the gaseous uranium molecules, all having the same kinetic energy, are passed through a porous barrier. Because of their smaller mass, U-235 molecules have a larger speed than U-238 molecules. The U-235 molecules thus pass through the barrier more rapidly. About 5000 barriers are needed to achieve 99 percent U-235. In the electromagnetic separation technique, uranium ions travel through a magnetic field and, because the U-235 ions have a smaller radius path than U-238 ions, separation occurs.

The first atomic bomb was designed at Los Alamos, New Mexico. The principle behind the bomb is: a chain reaction will occur when a critical mass of fissionable material is assembled in one piece. An implosion-type atomic bomb consists of a sphere of plutonium surrounded by a shell of ordinary explosives which, when detonated, compress the uranium sphere. Because the uranium nuclei are closer together, a chain reaction can occur.

In a nuclear power reactor, water serves as a coolant and moderator. Because water absorbs neutrons, the fuel elements are enriched to 3 percent U-235. The power of the reactor is controlled by control rods, which absorb neutrons. Heat generated by the reactor is used to produce electricity. A breeder reactor is designed to convert the nonfissionable U-238 into Pu-239.

As a star is formed by the gravitational contraction of a cloud of hydrogen gas, the gas heats up, allowing protons to have sufficient energy so that 2 protons can fuse together to form hydrogen-2. A third proton fuses to hydrogen-2 to form helium-3. The final step in the fusion process occurs when two helium-3 nuclei fuse to form helium-4. In essence, 4 protons fuse to form helium-4, with an energy release of about 25 MeV. In massive stars, the carbon-nitrogen cycle predominates. In this process, a proton fuses with carbon-12. After a sequence of several steps, essentially 4 protons fuse to form helium-4. As this hydrogen-buring process continues, the helium ash (inert) accumulates in the center of the star and then begins to contract due to gravity. Helium-burning begins and 2 helium-4 nuclei fuse to form beryllium-8, to which another helium-4 is added to form carbon-12, then oxygen-16, and, finally, neon-20. Helium-burning cannot form heavier elements because the Coulomb repulsion is too large. The alpha process is responsible for the formation of elements from neon-20 to iron-56. A gamma ray breaks up a neon-20 nucleus into oxygen-16 and a *very energetic* alpha particle (helium-4), which has sufficient energy to overcome Coulomb repulsion and fuse with neon-20 to form magnesium-24, silicon-28, . . . , and finally iron-56. Elements with larger atomic numbers A are formed by capturing a neutron, increasing A by one unit. Beta decay increases the atomic number Z by one unit. Stars evolve to either a white dwarf, a neutron star, a black hole, or end their existence in a supernova explosion that produces a cloud of gas from which new stars form.

The hydrogen bomb consists of a small atomic bomb placed near the chemical lithium hydride, which is enriched in the isotope lithium-6 and deuterium (hydrogen-2). Neutrons from the atomic bomb are captured by lithium-6 to form tritium (hydrogen-3). Fusion of the deuterium and tritium, isotopes of hydrogen, releases an energy of 17.6 MeV.

The fusion reactor also utilizes the fusion of deuterium and tritium. The ionized gas, or plasma, is confined by a magnetic field. Fusion research is addressed to increasing the confinement time of the plasma (before it strikes the walls) and the density of the plasma. One of the most promising designs for a fusion reactor is the Tokamak. The use of an intense laser beam to heat a deuterium-tritium pellet is a recent idea for generating fusion without employing magnetic fields.

True or False Questions

Indicate whether the following statements are true or false. Change all of the false statements so that they read correctly.

19–1 Plutonium can be separated from uranium by chemical separation techniques.

19–2 A vessel contains the gas uranium hexafluoride (UF_6).

    a. Molecules with U-238 have the same speed as those with U-235.

    b. Molecules with U-238 have the same kinetic energy as those with U-235.

    c. Molecules with U-235 hit the walls of the vessel more often than those with U-238.

    d. If one wall of the vessel were porous, molecules of U-238 would escape more easily.

19–3 Two identical cubes of plutonium are separated by 25 cm, and a self-sustaining chain reaction does not occur in either one.

    a. The mass of a single cube is less than the critical mass.

    b. If the two cubes are moved side by side, a self-sustaining chain reaction will never occur.

19–4 a. The density of a sphere of jello the size of a golf ball cannot be increased by squeezing it in the palm of your hand.

    b. The density of a sphere of plutonium can be increased only if the pressure exerted on the surface of the sphere is the same everywhere over the surface.

19–5 a. In a breeder reactor, liquid sodium serves as a moderator.

    b. In a breeder reactor, fission occurs with slow neutrons.

19–6 Because the temperature in the center of the sun is 15 million °K, a proton will take about seven years to fuse with another proton.

19–7 Hydrogen-burning in a star refers to the transformation of helium to hydrogen.

19–8 a. Elements between helium and iron in the periodic table are produced by helium-burning.

    b. Elements beyond iron in the periodic table are produced by neutron capture and subsequent beta decay.

19–9 Our sun will eventually become a black hole.

19–10 The same nuclear reactions will occur in a fusion reactor that occur in the sun.

Questions for Thought

19–1 What was the objective of the Argonne site? Of Clinton Engineer Works in Oak Ridge, Tennessee? Of Hanford Engineer Works? Of Site Y at Los Alamos, New Mexico?

19–2 What two methods were used to obtain uranium-235?

19–3 What is meant by the critical mass of an atomic bomb?

19–4 Describe two types of atomic bombs.

19–5 Describe the basic operation of a nuclear reactor.

19–6 Explain how heat is produced in a reactor core.

19–7 Explain why a reactor cannot explode like an atomic bomb.

19–8 Describe why a meltdown of a reactor core would occur if the flow of cooling water to the core were blocked.

19–9 a. Describe how a breeder reactor produces more fuel than it uses.

b. Why must a breeder reactor use fast neutrons?

c. Why must a breeder reactor use uranium greatly enriched in U-235?

19–10 In the sun, _____ million tons of hydrogen are converted to helium every second.

19–11 Define:
a. hydrogen-burning
b. proton-proton chain
c. carbon-nitrogen cycle
d. helium-burning
e. helium ash
f. red giant
g. white dwarf
h. neutron star
i. black hole

19–12 By what process are the following elements produced in stars: Helium-4? Oxygen-16? Sulfur-32? Gold-197?

Questions for Calculation

†19–1 Suppose that the energy demands of the United States increase by 5 percent each year. Use a hand calculator (or logarithms) to find the total increase after 10 years. How many years will it take for our energy demands to double?

19–2 Complete the following:

$$^1_1H + ^1_1H \rightarrow ? + \beta^+ + \nu$$

$$^2_1H + ^1_1H \rightarrow ? + \gamma$$

$$2\,? \rightarrow ^4_2He + 2\,^1_1H + \gamma.$$

19–3 Show that Equation 4 results from Equations 1, 2, and 3.

19–4 Show that Equation 11 results from Equations 5 through 10.

†19–5 Protons at the center of the sun where the temperature is 15 million °K have a kinetic energy of 0.002 MeV. Suppose that 2 protons approach each other head on.

a. Using conservation of energy, determine how far these protons are when they come to a stop. (1.6×10^{-13} joules $= 1$ MeV, $e = 1.6 \times 10^{-19}$ coulombs, $K = 9 \times 10^9$ newtons-m^2/coulomb^2, PE $= KQ_1Q_2/d$.)

b. The diameter of the deuterium nucleus is about 10^{-15} m. According to the classical calculation carried out in part **a**, will these two protons get close enough to fuse together?

c. Recall that alpha decay was explained (Chapter 15) in terms of barrier penetration. Make a similar explanation here for the difficulties observed in part **b**.

Footnotes

[1] John A. Wheeler, "Mechanism of Fission," *Physics Today* 20(November, 1967):52.

[2] Otto R. Frisch, "Investigating Fission," in Jane Wilson (ed.), *All in Our Time: The Reminiscences of Twelve Nuclear Pioneers* (Chicago: Bulletin of the Atomic Scientists, 1975), p. 62.

[3] Bernard L. Cohen, "The Disposal of Radioactive Wastes from Fission Reactors," *Scientific American* 236(June, 1977):21.

[4] David Bergamini, *The Universe* (New York: Time-Life Books, 1966), p. 86.

[5] E. Margaret Burbidge, G. R. Burbidge, William A. Fowler, and F. Hoyle, "Synthesis of the Elements in Stars," *Reviews of Modern Physics* 29(1957):547.

NUCLEAR ENERGY FROM FISSION AND FUSION

Hideki Yukawa

CHAPTER 20
Elementary Particles

Introduction

We have, by now, learned a great deal about the structure of the nucleus and the myriad ways in which the nucleus breaks apart. So far, however, we've avoided an extremely important question: What holds the nucleus together? The force of repulsion between the protons is huge at such short distances. Why, then, don't the protons move apart? Further, the neutrons have no charge. What, then, holds the neutrons in the nucleus? In the first section of this chapter, we shall see how, in 1934, a young Japanese physicist named Hideki Yukawa tackled this problem. Yukawa proposed the existence of a meson (from the Greek word *mesos,* meaning "intermediate"), a particle having a mass about 200 times that of an electron.

The first elementary particles were discovered in nature's laboratory—in cosmic rays: the muon in 1936; Yukawa's meson, the pi-meson, in 1947; and the lambda-naught also in 1947. The year 1948 marked a turning point because a particle accelerator was used to produce pi-mesons. Because accelerators could produce copious quantities of elementary particles in contrast to cosmic rays, great strides were made in the study of elementary particles. For instance, Dirac's theory, which predicted the existence of the positron, also predicted the existence of the antiproton, a particle with the same mass as a proton, but having one negative charge. The Bevatron at the University of California was designed to have sufficient energy (6200 MeV) to produce the antiproton. In 1955, experiments confirmed the existence of the antiproton. Nature exhibits a pleasing symmetry.

The invention of the hydrogen bubble chamber was another milestone in the progress of elementary particle physics. Particles from an accelerator entered a chamber filled with liquid hydrogen and collided with hydrogen nuclei. The particles emerging from the collision created tracks (made bubbles) as the hydrogen liquid began to boil along the path of the particles. The curvature of the tracks due to a magnetic field enables the researchers to identify particles, as had been previously done with cloud chambers.

Symmetry considerations indicated the existence of the omega-minus particle and 2 bubble-chamber photographs out of 50,000 photographs demonstrated its existence.

In 1964, Gell-Mann and Zweig proposed that elementary particles are composed of quarks. The proton, for example, is thought to be composed of three quarks. When the high-energy electron beam from the Stanford Linear Accelerator was directed onto a hydrogen target, the number of electrons scattered to various angles was larger than expected. Just as Rutherford's scattering experiments with alpha particles indicated the existence of the nucleus within the atom, these electron-scattering experiments indicated the existence of small objects (or particles) within the proton. However, despite exhaustive searches, no free quarks have ever been detected.

At Fermi National Accelerator Laboratory in Batavia, Illinois, protons are accelerated to an energy of 500 GeV (1 GeV = 1 billion eV). The upsilon, a particle having a mass about 10 times that of a proton, was discovered at this laboratory in 1977. Surely many exciting discoveries will be made when the beam energy is upgraded to 1000 GeV. The new facility to accomplish this will begin operation in 1983.

Yukawa and the Nuclear Force

Hideki Yukawa was graduated from the Kyoto University in 1929. In 1933, he was appointed lecturer at Osaka University. He describes his fascination with the nuclear force:

In the fall of 1934 when my second son was born, I became so engrossed in thinking about the nuclear forces, that I began to find it difficult to fall asleep at night. . . . my ideas began to crystallize and so I was able in October to present my ideas at a seminar. I concluded that in this new field the rest mass of the new quanta should be about two hundred times the electron mass . . ."[1]

In 1949, Yukawa was awarded the Nobel Prize. In his acceptance speech, he explains how he arrived at his theory:

. . . Fermi developed a theory of beta-decay based on the hypothesis by Pauli, according to which a neutron, for instance, could decay into a proton, an electron, and a neutrino, which was supposed to be a very penetrating neutral particle with a very small mass. [See Chapter 16, Equation 10.] This gave rise, in turn, to the expectation that nuclear forces could be reduced to the exchange of a pair of an electron and a neutrino between two nucleons,* just as electromagnetic forces were regarded as due to the exchange of photons between charged particles.† It turned out, however, that the nuclear forces thus obtained were much too small, because the beta-decay was a very slow process compared with the supposed rapid exchange of the electric charge responsible for the actual nuclear forces. The idea of the meson field was introduced in 1935 in order to make up this gap.[2]

Yukawa proposed that one nucleon (a proton or a neutron) emits a meson, which is then absorbed by the other nucleon. This rapid exchange of mesons between two nucleons is responsible for the nuclear force between them: a force that is strong and attractive when the nucleons are separated by a distance of 10^{-15} m or less, but zero for large separation distances. This attractive force between nucleons holds the nucleus together. The nuclear force between 2 protons (and between 2 neutrons) occurs by the exchange of a neutral **pi-meson,**

---

*A term referring to either a proton or a neutron.

†In Chapter 3, we learned that Coulomb's law described the attraction or repulsion between charged particles. Modern physics has arrived at a more fundamental understanding of this force. One of the charged particles emits a photon of light that is absorbed by the other particle. This rapid exchange of photons between the two charged particles is responsible for the mutual force between them.

denoted by the symbol π^0, between the two nucleons. The nuclear force between a proton and a neutron occurs in two ways. The first is when a proton throws out a positively charged pi-meson (π^+), and the proton transforms into a neutron.

$$p \rightarrow n + \pi^+. \tag{1}$$

This meson is caught by the neutron, transforming it into a proton.

$$n + \pi^+ \rightarrow p. \tag{2}$$

This process is shown in Fig. 20–1.

The second way is when a neutron throws out a negatively charged pi-meson (π^-), and the neutron transforms into a proton. This meson is caught by the proton, transforming it into a neutron.

$$n \rightarrow p + \pi^- \tag{3}$$
$$p + \pi^- \rightarrow n. \tag{4}$$

In Equation 1, we see that there is more mass on the right side than on the left. How can the proton *spontaneously* break up into something *heavier*? Doesn't this violate conservation of energy? To see why this can occur, we must consider Heisenberg's uncertainty principle, given by

$$\Delta p \, \Delta x \geqslant \frac{h}{2\pi}.$$

If the particle's momentum is uncertain, then its kinetic energy is likewise uncertain. If a particle travels between two places, but the distance between them is uncertain by an amount Δx, then the time to travel between the two places is uncertain by an amount Δt. An alter-

nate form* of Heisenberg's uncertainty principle is

$$\Delta E \, \Delta t \geqslant \frac{h}{2\pi}. \tag{5}$$

If the energy on the right side of Equation 1 is *uncertain* by an amount equal to the mass energy

*The kinetic energy of a particle, $\frac{1}{2}mv^2$, can also be written in terms of its momentum p (where $p = mv$), as KE equals $p^2/2m$. If the particle's momentum is uncertain, then its energy is likewise uncertain. If the momentum is $p \pm \Delta p$, then the kinetic energy is

$$K \pm \Delta E = \frac{(p \pm \Delta p)^2}{2m}$$

$$= \frac{p^2}{2m} \pm \frac{2p \, \Delta p}{2m} + \frac{(\Delta p)^2}{2m}.$$

We shall neglect the last term because it is so much smaller than the other two terms. We can then identify the uncertainty in the kinetic energy as

$$\Delta E = \frac{p}{m} \, \Delta p$$

$$= v \, \Delta p,$$

or

$$\Delta p = \frac{\Delta E}{v}. \tag{a}$$

If a particle travels between two points, but the distance between them is uncertain by the amount Δx, then the time to travel between them is uncertain by the amount Δt:

$$\Delta t = \frac{\Delta x}{v}$$

or

$$\Delta x = v \, \Delta t. \tag{b}$$

Using Equation (a), Equation (b), and Heisenberg's uncertainty relationship

$$\Delta p \, \Delta x \geqslant \frac{h}{2\pi},$$

we find,

$$\left(\frac{\Delta E}{v}\right)(v \Delta t) \geqslant \frac{h}{2\pi},$$

or

$$\Delta E \, \Delta t \geqslant \frac{h}{2\pi}. \tag{c}$$

of the pi-meson, then conservation of energy still holds. Therefore,

$$\Delta E = m_\pi c^2.$$

Since the range of the nuclear force is about 10^{-15} m, the pi-meson can travel no farther than this. If its speed is nearly equal to the speed of light, then the uncertainty in its time of travel is

$$\Delta t = \frac{10^{-15}\,\text{m}}{c}.$$

Heisenberg's uncertainty principle then becomes

$$\Delta E = \frac{h}{2\pi\,\Delta t}$$

$$m_\pi c^2 = \frac{hc}{2\pi(10^{-15}\,\text{m})}$$

$$= \frac{(4.13 \times 10^{-15}\,\text{eV-sec})(3 \times 10^{8}\,\text{m/sec})}{2\pi(10^{-15}\,\text{m})}$$

$$= 1.97 \times 10^{8}\,\text{eV}$$

$$= 197\,\text{MeV}.$$

Since 1 amu equals 931 MeV, the mass of the pi-meson is

$$m_\pi = 197\,\text{MeV} \times \frac{1\,\text{amu}}{931\,\text{MeV}}$$

$$= 0.21\,\text{amu}.$$

The masses of the proton and neutron are 1 amu, and the mass of the pi-meson is about one-fifth that of a proton or a neutron.

Figures 20-2 and 20-3 show some interesting analogies to the nuclear force.

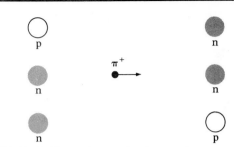

FIG. 20–1   The exchange of a pi-meson between a proton and a neutron.

Discovery of the Muon

Examining cloud-chamber photographs of cosmic rays in 1936, Carl Anderson and S. H. Neddermeyer saw evidence of a particle having

FIG. 20—2 Using this graphic illustration, George Gamow colorfully explains the exchange of a neutral meson between a proton and a neutron: "Probably the best way to picture a force of attraction between two bodies caused by the presence of a third body is to imagine two hungry dogs who come into possession of a juicy bone and are grabbing it from each other to take a bite. The tasty bone is continuously passing from the jaws of one of them into the jaws of the other, and in the resulting struggle, the two dogs become inseparably locked." [G. Gamow, *The Atom and Its Nucleus* (Englewood Cliffs, New Jersey: Prentice-Hall, 1961), p. 137.]

FIG. 20—3 De Paul University teammates, Teddy Grubbs (left) and Clyde Bradshaw (right) are linked together as they pass the ball. (Courtesy De Paul Sports Information Office)

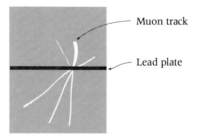

FIG. 20—4 Drawing of a cloud-chamber photograph showing the existence of a muon. [*Source:* C. D. Anderson and S. H. Neddermeyer, *Physical Review* 50(1936): 270.]

a mass between that of an electron and proton. Carl Anderson explains:

We know that the muon was the first particle to be discovered which has a mass between that of an electron and a proton. . . . The discovery of the muon, unlike that of the positron, was not sudden and unexpected. Its discovery resulted from a two-year series of careful, systematic investigations all arranged to follow certain clues and to resolve some prominent paradoxes which were present in cosmic rays.

The gist of the matter was as follows. Professor Seth H. Neddermeyer and I were continuing the study of cosmic-ray particles using the same magnet cloud chamber in which the positron was discovered. In these experiments it was found that most of the cosmic-ray particles at sea level were highly penetrating in the sense that they could transverse large thickness of heavy materials like lead and lose energy only by directly produced ionization which amounted to something like 20 million eV per cm of lead. A principal aim of the experiments was to identify these penetrating cosmic-ray particles. They had unit electric charge and were therefore presumably either positive or negative electrons or protons, the only singly charged particles known at that time.

. . . To interpret these particles as protons would mean assuming the existence of protons of negative charge since these sea-level particles occurred equally divided between negative and positive charges, and at that time there was no evidence for the existence of protons of negative charge.[3]

When high-energy electrons strike a lead plate they lose energy, predominantly (1) by ionizing lead atoms and (2) by radiation. The second process occurs when the electron closely passes by a lead nucleus and experiences a *large* acceleration. Accelerating charged particles emit electromagnetic radiation and, hence, lose energy. (Remember, for instance, that electrons emit X-rays when they strike the anode of an X-ray tube and come to a stop.) This process is important only when the electrons have a speed near that of light. Furthermore, a proton, for example, will not lose much energy by radiation when it strikes a lead plate because its mass is so much larger than an electron's, and its acceleration, consequently, so much less.

As Anderson mentions, the penetrating cosmic-ray particles striking lead lost energy *only by ionization,* not by radiation. Thus, it was difficult to interpret the penetrating cosmic-ray particles as electrons. Anderson continues:

This then was the situation in 1934 in which the sea-level penetrating particles had this paradoxical behavior. They seemed to be neither electrons nor protons. . . .

Evidence of an entirely new type was soon obtained. In experiments carried out on the summit of Pikes Peak in 1935 a number of cases of cosmic-ray produced nuclear disintegrations were observed from which many protons were ejected, but showing also in a few cases particles which, from ionization and curvature measurements, were lighter than protons and heavier than electrons.[4]

Figure 20−4 shows one such photograph.

Neutral particles or photons do not cause ionization of the air in the cloud chamber and, therefore, leave no tracks. Apparently, such a nonionizing ray struck the lead plate and produced the six tracks that emerged from it. One track is much denser (that is, produces more ionization) than the others. The density of this track, the one in question, is greater than would be produced by an electron or a positron. Could the track, then, be due to a proton? (A proton's track would be denser than an electron's.) The dense track is about 4 cm long[5] and is bent by the magnetic field in a circle having a radius of 7 cm.

A proton with a speed of 1.7×10^7 m/sec would come to a stop after traveling 4 cm in air. When a particle is subjected to a magnetic field, the radius of its path can be found as follows:

$$F = ma$$

$$qvB = \frac{mv^2}{r}$$

$$r = \frac{mv}{qB}. \tag{6}$$

The experimenters knew all the terms on the right side of Equation 6.

The way in which the track curves in the magnetic field indicates whether it is negatively or positively charged. If the particle were a proton, the track was calculated to have a radius of about 20 cm, not 7 cm. This track must, therefore, be due to a particle with a mass smaller than a proton's, but larger than an electron's. (The magnetic field would bend such a particle more easily than a proton.) Equation 6 shows that a smaller mass gives a smaller radius. Today, this particle is called a **muon.**

A muon is a much more penetrating particle than an electron because, with its larger mass, the muon will suffer little energy loss by emitting radiation.

Anderson continues:

Yukawa estimated from the known range of nuclear forces that this carrier (of the nuclear force) should have a rest mass about 200 times that of an electron.

This novel suggestion of Yukawa was unknown to the workers engaged in the experiments on the muon until after the muon's existence was established. Although Yukawa's suggestion preceded the experimental discovery of the muon, he published it in a Japanese journal which did not have a general circulation in this country. It is interesting to speculate on just how much Yukawa's suggestion, had it been known, would have influenced the program of the experimental work on the muon. My own opinion is that this influence would have been considerable. . . . My reason for believing this is

that for a period of almost two years there was strong and accumulating evidence for the muon's existence, and it was only the caution of the experimental workers that prevented an earlier announcement of its existence. [6]

For a while, scientists identified the muon as Yukawa's particle. However, there was one urgent problem. The chief property of the muon is its penetrability. Yukawa's meson, on the other hand, should interact strongly with protons and neutrons and, hence, should be readily absorbed. This mystery was not cleared up until 1947 with the discovery of another particle—which turned out to be Yukawa's meson.

Discovery of the Pi-Meson

At the University of Bristol, the English physicist C. F. Powell (1903–1969) and his group pioneered the use of photographic emulsions to trace the passage of charged particles, producing photographs much like those from cloud chambers. The emulsions, greatly enriched in a silver halide compound, were on the order of one-tenth of a millimeter thick. Because energetic charged particles will not be stopped in passing through one emulsion, the researchers used a stack of emulsions stripped from their glass backings. After development, they examined the tiny tracks (at most about a millimeter long) through a microscope and traced the track produced by a charged particle from one emulsion to the next.

Figure 20–5 displays one such mosaic-type photograph produced by exposing a stack of emulsions to cosmic rays at high altitudes. First of all, notice the curvature of both tracks. These tracks could not be due to protons because, from experiments, it was known that proton tracks are straight. A charged particle, entering the emulsion, would feel a force of repulsion (or attraction) as it passes close to a nucleus. The amount that a particle deflects depends upon its mass. A proton, for example, would deflect only a little. In these emulsions, the passage of electrons (or positrons) could not be detected because their tracks are simply not dense enough. Therefore, the particles producing the tracks in Fig. 20–5 must have a mass intermediate between that of an electron and a proton. From the density of the tracks and their range in the emulsion, the Powell group estimated that the primary particle (labeled π) had a mass about 300 times that of an electron, and the secondary particle (labeled μ) had a mass about 200 times that of an electron.

The track of the primary particle thickens as it slows down because the particle produces more ionization *per unit length* at a smaller speed. This track comes to a stop at the lower left of Fig. 20–5 and an energetic particle, indicated by its thin track, emerges and travels toward the right. This secondary track also thickens as the particle slows down, suggesting that the primary particle transforms into the secondary particle with the difference in mass converted into the kinetic energy of the secondary particle. The researchers identified the primary particle as Yukawa's particle—the **pi-meson**—and the secondary particle—the **muon**—the penetrating particle found at sea level. Later, using a more sensitive emulsion, they found that the muon, when stopped in the emulsion, produced an extremely light track, identified as being due to an electron. Today, we know that the following transformations* occur:

$\pi^{\pm} \rightarrow \mu^{\pm} + \nu$ (primary decay mode)   (7)

$\rightarrow e^{\pm} + \nu$ (occurs about 10^{-2} percent of the time).   (8)

$\pi^{0} \rightarrow \gamma + \gamma$ (98 percent)   (9)

$\rightarrow e^{+} + e^{-} + \gamma$ (1.2 percent)   (10)

$\mu^{\pm} \rightarrow e^{\pm} + 2\nu.$ (11)

Table 20–1 lists the masses of these particles.

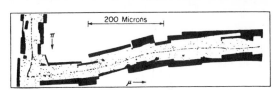

FIG. 20–5 Photograph showing the transformation of a pi-meson into a muon (1 micron = 10^{-6} m). [*Source:* C. M. G. Lattes, H. Muirhead, G. P. S. Occhialini, and C. F. Powell, *Nature* 159(1947):694.]

TABLE 20–1   SOME ELEMENTARY PARTICLES

NAME OF PARTICLE	MASS-ENERGY (MeV)	MASS IN UNITS OF ELECTRON MASS
Ξ-hyperon	1318	2579
Σ-hyperon	1193	2335
Λ^{0}-hyperon	1115	2182
neutron	939.6	1839
proton	938.3	1836
$K^{\pm}$ and K^{0} meson	496	970.6
$\pi^{\pm}$ and π^{0} meson	138	270.1
$\mu^{\pm}$ (or muon)	106	207.4
$e^{\pm}$	0.511	1

*In this chapter, the symbols e^{-} abd e^{+} will be used, respectively, to denote an electron (beta ray) and a positron to conform to current usage in high-energy physics.

Artificial Production of Pi-Mesons

In 1948, Eugene Gardner and C. M. G. Lattes were the first to produce mesons artificially. Using the Berkeley 184-inch cyclotron, these researchers bombarded a carbon target with 380-MeV alpha particles. Some of the alpha particles' kinetic energy was transformed into the mass of meson, having a mass-energy of 140 MeV. Figure 20–6 shows the experimental setup used. The perpendicular magnetic field causes the negative pi-mesons to travel in a circle and strike the stack of photographic emulsions. Gardner and Lattes determined the radius of the circle by locating where the mesons struck the emulsion. Using Equation 6, they determined the meson's momentum (mv). By measuring the meson's range in the emulsion, they obtained its kinetic energy. Values for the momentum and kinetic energy then yielded the mass of the meson. In only 10 minutes, 50 meson tracks were produced on each emulsion. Gardner and Lattes comment:

. . . it seems certain that this marks the beginning of meson study under controllable laboratory conditions. The large intensities, approximately 10^8 times those available in cosmic rays, mean that the rate of progress in this field can be greatly accelerated.[7]

Still More Particles

Experiments with cosmic rays revealed still more particles. In 1947, G. D. Rochester and C. C. Butler obtained a cloud chamber photograph that showed two particles emerging from a point in the gas. There was, however, no track *approaching* this point. Speculation was that a neutral cosmic-ray particle, which would not leave a track, decayed into two charged particles. From energy and momentum considerations, Rochester and Butler concluded that one

track was due to a proton and the other was due to a pi-meson. This neutral particle, called **lambda-naught** (symbol: Λ^0) has to be more massive than a proton in order for such a decay to occur. This difference in masses of the particles on the left and right side of Equation 12 is converted into the kinetic energy of the proton and pi-meson:

$$\Lambda^0 \rightarrow p + \pi^-. \qquad (12)$$

A particle that has a mass greater than that of a neutron is called a **hyperon.**

Discoveries of other particles soon followed: the **sigma hyperons** (symbol: Σ) and another group of mesons, called the **kaons** (symbol: K). Table 20–1 lists these particles.

The Cosmotron

The Cosmotron, completed in 1952 at Brookhaven National Laboratory in Long Island, is called a **synchrotron,** a type of particle accelerator that accelerates protons to an energy of 3000 MeV. (Thought question 7–23 outlines the operation of a synchrotron.) In this process, the protons are injected into a circular ring with a diameter of 75 ft. A perpendicular magnetic field keeps the protons traveling in a circular path. The protons gain about 3000 eV energy per revolution and make about 1 million revolutions to achieve an energy of 3000 MeV. As the proton's energy increases, the magnetic field must likewise increase, so that the radius of the path does not change. That is, the increasing magnetic field is synchronized with the increasing energy (which explains the name *synchrotron*). This process is repeated to produce a beam consisting of periodic bursts of protons.

In one of the first experiments carried out at the Cosmotron, the energetic protons were directed onto a target. As a result, a neutron beam was produced. This was then directed toward the cloud chamber. A neutron interacted with a nucleus in the wall of the cloud chamber and a Λ^0-hyperon was produced. Alan

M. Thorndike (b. 1918) describes the discovery of the Λ^0-hyperion in the cloud-chamber photographs:

With everything in operation, the day's run was finally really under way. Apart from minor interruptions for checking and servicing equipment, the run continued until midnight. By that time eleven 100-foot rolls of 35 mm film had been filled with cloud-chamber photographs—about 4000 in all. Several additional days' runs followed soon afterwards. Altogether some 20,000 neutron beam pictures were taken.

At the end of the afternoon of April 3 the four physicists (Fowler, Shutt, Thorndike and Whittemore) were checking over the new events found in the pictures that had been looked at during the day. There seemed to be a considerable number of "good" events, so it was clear that the next day would be a busy one. Noticing that it was just five o'clock, the punctual Whittemore left to pick up his wife, who worked in another building, and the group broke up. Thorndike lingered on, feeling the universal curiosity about what will happen next, and idly turned ahead to look at the next few pictures, as one often looks ahead a few pages before putting down a fast-moving book. He did not turn far, though. On the first new picture a track junction caught his eye. He looked at it for a minute—certainly that characteristic V-shape pair of tracks could be nothing else but a Λ^0 hyperon! This case . . . [see Fig. 20–7] was the first definite example of an artificially produced hyperon.[8]

The Antiproton

We have talked about how C. D. Anderson discovered the anti-electron (called the positron) quite accidentally from cloud-chamber photographs (Chapter 17). The discovery of the positron, though, was predicted by P. A. M. Dirac, who had developed a wave equation that included relativistic effects. Dirac's equation, however, required that both negative *and* positive electrons exist. Since both the electron and proton have spin ½ (both have spin "up" and

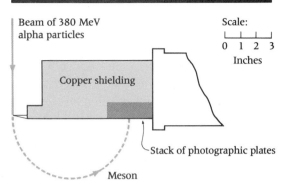

FIG. 20–6 The experimental setup used for producing pi-mesons and detecting them in a stack of photographic emulsions.

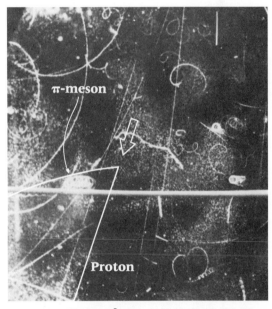

FIG. 20–7 The first Λ^0-hyperon produced artificially. This diffusion cloud-chamber photograph shows that the Λ^0 decays below the point of the arrow into a π^- meson and a proton in the process $\Lambda^0 \rightarrow \pi^- + $ p. The Λ^0 was presumably produced in a collision between a neutron from the Cosmotron and a proton or neutron in an iron nucleus in the wall of the chamber above the area shown, but the tracks observed in the chamber do not give any information about this collision. (Courtesy of Brookhaven National Laboratory)

"down"), Dirac's equation should apply to protons as well and, in fact, predicts the existence of the **antiproton,** a negative particle having the same mass as a proton.

In contrast to the accidental discovery of the positron, four researchers, O. Chamberlain, E. Segre, C. Wiegand, and I. Ypsilantis, made a concerted effort to track down the antiproton, using the then newly constructed (1954) proton accelerator, the Bevatron, at the University of California Radiation Laboratory in Berkeley. When these energetic protons struck a target, the researchers hoped that an incident proton would interact with a proton in the target, producing an antiproton, as follows:

$$p + p \rightarrow p + p + p + \bar{p}. \qquad (13)$$

Note that charge is conserved in this reaction, but there is another conservation law evident here that we have not yet discussed: **conservation of nucleons.** This conservation law, so far obeyed in all nuclear reactions, states that the number of nucleons always remains unchanged in any interaction. Equation 13 shows that there are 2 protons (2 nucleons) before the interaction. There must then be 2 nucleons afterwards as well. The antiproton on the right side of Equation 13 can be viewed as "cancelling out," with 1 proton leaving 2 protons remaining. Hence, the nucleon number is conserved.

The Bevatron, in fact, was designed to generate sufficient energy to produce antiprotons. How much energy is this? The mass of 2 protons must be *created* (Equation 13), corresponding to an energy of 2(938 MeV), or 1876 MeV. This means that 1876 MeV of the proton's kinetic energy must be transformed into the masses of a proton and an antiproton. Conservation of momentum plays an important role here. The proton beam must have a *minimum* kinetic

energy equal to 5628 MeV—not 1876 MeV! Figure 20−8a shows the reason for this. Suppose that a proton had only a kinetic energy of 1876 MeV. All of this energy would be used up in creating a proton and an antiproton, and the (hypothetical) four particles would be at rest afterwards. But this violates conservation of momentum and, hence, simply doesn't happen.

Figure 20−8(b) shows that both momentum and energy are conserved. The incident proton has a kinetic energy of 5628 MeV and a momentum denoted by the symbol p. Afterward, each proton has a momentum $\frac{1}{4}p$ and a kinetic energy of 938 MeV. The *total* momentum is equal to p, both before and afterward. The difference in kinetic energy (5628 − 3752 = 1876 MeV) is just that needed to create the mass of the proton and antiproton. Relativistic formulas show that the *minimum* beam energy required to produce antiprotons must be 5628 MeV. (See Chapter 9, p. 220.) The Bevatron accelerated protons to an energy of 6200 MeV, or 6.2 BeV, and, hence, the name Bevatron. Today, 10^9 eV is designated as 1 GeV (where, remember, G stands for giga, as in gigantic).

Figure 20−9 shows the experimental setup for observing antiprotons. The 6.2-GeV protons bombard a copper target, and many kinds of particles are produced besides the rare antiprotons. These particles then pass through a magnetic field, oriented so that only negatively charged particles emerged. But, in addition, only those emerged that had a certain radius in the magnetic field, and, thus, the magnet selected the momentum of the emerging particles. When p equals mv is used, Equation 6 becomes

$$p = qBr. \qquad (14)$$

At this momentum, π^- mesons have a speed of $0.99c$, and the antiprotons have a speed of $0.78c$, where c is the speed of light—3×10^8 m/sec.

When the speed of a particle is much less than the speed of light, the momentum p is given by p equals mv. However, when the speed of the particle is close to the speed of light, its relativistic momentum is given by

$$p = \frac{m_0 v}{\sqrt{1 - v^2/c^2}} \,. \qquad (15)$$

(See Equation 14 of Chapter 9, p. 217.)

The momentum has already been determined by the particle's radius in a magnetic field. Equation 15 shows that, if p and v are known, the mass m_0 can be determined. The basic object of the experiment was to identify the antiproton, measure its speed, and, in this way, determine its mass.

Segre and Wiegand describe the problem:

. . . It turned out that there were about 40,000 mesons for each rare antiproton in the stream of emerging particles focused by our magnets. The mesons follow exactly the same trajectory as the antiprotons, but they are lighter and travel with a velocity practically identical to that of light whereas the heavier antiproton moves with 78 percent of the velocity of light. The problem was to pick out of the stream the occasional heavy particle (one in 40,000) moving with the right velocity to be an antiproton.[9]

The basic idea is to measure the antiproton's speed by recording its travel time between two detectors D_1 and D_2, separated by a distance of 12 m. The detectors are made of a plastic material that emits light when a charged particle passes through it. The light triggers electronic equipment that yields the travel time. If the time is 51×10^{-9} seconds, the particle is an antiproton and is counted, but if the time is 40×10^{-9} seconds, the particle is a π^- meson, and is rejected.

Segre and Wiegand continue:

When the discovery of the antiproton was announced last October (1955), 60 of them had been recorded, at an average rate of about four to each hour of operation of the Bevatron.

An interesting subject for contemplation is the possible existence of an "anti-world." This would be a world in which all particles are opposite in charge to our one: the hydrogen atom, for instance, would have an antiproton as its nucleus and a positron in place of the electron. We know of no method by which we could recognize the existence of such a universe by astronomical observation. But if antimatter exists

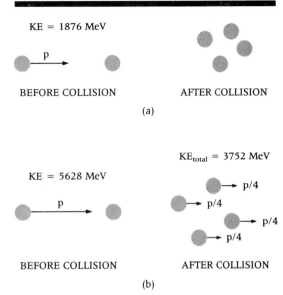

FIG. 20–8 (a) This reaction does not occur because momentum is not conserved. (b) This reaction occurs because momentum and energy are both conserved.

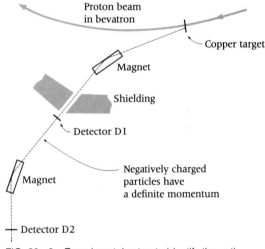

FIG. 20–9 Experimental setup to identify the antiproton and measure its speed.

and if it should come into contact somewhere with ordinary matter as we know it, the two forms of matter would annihilate each other with a huge release of energy, mostly as mesons. . . .[10]

A proton and an antiproton annihilate in one of the following ways:

$$p + \bar{p} \rightarrow \pi^+ + \pi^- \qquad (16)$$
$$\rightarrow \pi^+ + \pi^- + \pi^+ + \pi^-.$$

In 1959, E. Segre and O. Chamberlain were awarded the Nobel Prize for their discovery of the antiproton.

The Overthrow of Parity

A principle in physics, called **conservation of parity,** states that it is not possible to distinguish between a real object or event and its mirror image. Until 1956, this conservation law was just as revered as the conservation laws of energy and momentum. Even so, T. D. Lee (b. 1926) of Columbia University and C. N. Yang (b. 1922) of the Institute for Advanced Study at Princeton University boldly suggested that it might be possible to determine which is the mirror image. If one could determine the mirror image—if parity were not conserved—a serious problem concerning two elementary particles, called the **theta** (symbol: θ) and **tau** (symbol: τ) **mesons,** would be resolved. These particles decay as follows:

$$\theta^+ \rightarrow \pi^+ + \pi^0. \qquad (17)$$
$$\tau^+ \rightarrow \pi^+ + \pi^+ + \pi^-. \qquad (18)$$

Yang explains:

In the years 1954–1956 a puzzle called the θ - τ puzzle developed. The θ and τ mesons are today known to be the same particle which is usually called

K. *In those years, however, one only knew that there were particles that disintegrate into two π mesons and particles that disintegrate into three π mesons. They were called respectively θ's and τ's, the τ being the name given to it by Powell in 1949. As time went on, measurements became more accurate and the increasing accuracy brought out more and more clearly a puzzlement. On the one hand it was clear that θ and τ had very accurately the same mass. They were also found to behave identically in other respects. So it looked as if θ and τ were really the same particle disintegrating in two different ways. On the other hand, increasing accurate experiments showed that θ and τ did not have the same parity and could not therefore be the same particle.*[11]

If parity, however, were not conserved, then θ and τ could be the same particle. Lee and Yang searched the literature to see if any experiments concerning beta decay had ever shown that parity must be conserved. They could find no such references, and described several experiments to test the conservation of parity.

One was carried out by C. S. Wu (b. 1913) and her co-workers studying the beta decay of cobalt-60:

$$^{60}_{27}\text{Co} \rightarrow {}^{60}_{28}\text{Ni} + \beta^- + \nu.$$

The cobalt nucleus and the nickel nucleus spin about their axes much like the earth daily rotates about its axis. This rotating charge produces a magnetic field, whose orientation is shown in Fig. 20–11.

In cobalt, at room temperature, the axes of rotation are oriented at random. By placing the cobalt in a region of a strong, external magnetic field, the little nuclear magnets will align with this strong magnetic field. By cooling the cobalt to less than one degree above absolute zero, the vibration of the atoms is greatly reduced and the nuclear magnets stay aligned. The object of the experiment was to compare the number of electrons emitted from the south end with those emitted from the north end.

According to mirror symmetry, or conservation of parity, the same number should be emitted from each end as shown in Fig. 20–12.

DEVELOPMENT OF MODERN PHYSICS

To determine if two objects on a desktop are identical, you can turn them around. Similarly, to determine if the mirror image in part (b) is identical to part (a), you can turn (b) upside down, as illustrated in part (c). Parts (a) and (c) are identical—it is impossible to distinguish the beta decay event from its mirror image. Thus, according to conservation of parity, the same number of electrons should be emitted from both ends.

Conservation of parity breaks down if more electrons are emitted from the south end, as shown in Fig. 20–13. The mirror image is shown in part (b) of this figure. But now turn part (b) upside down. Parts (a) and (c) show that the directions of rotation are the same, but the directions in which more electrons are emitted do not agree. Clearly, you can pick out part (b) as being the mirror image.

Wu and her co-workers found that more electrons were emitted from the south end of the aligned cobalt-60 nuclei, as shown in Fig. 20–13, and parity was not conserved.

In 1957, Lee and Yang were awarded the Nobel Prize for their prediction.

The Bubble Chamber

In 1952, the American physicist Donald A. Glaser (b. 1926) invented the **bubble chamber,** an apparatus that has become so significant in studying elementary particle events. The *Scientific American* describes what prompted Glaser to begin his research on the bubble chamber:

He [Glaser] *went to do his graduate work at the California Institute of Technology under Carl D. Anderson, and for his Ph.D. thesis investigated high-energy cosmic rays. Glaser claims that the encouragement to try his bubble-chamber idea, which he had already been mulling over, came one night during a beer session with some physicist friends: "After several pitchers of beer we began to wax philosophical about physics. One of the boys, looking dreamily into the pitcher of beer before him, saw the usual streamers of bubbles and remarked, 'Nuclear physics should be easy. You can see tracks in nearly every-*

FIG. 20–10   Chiens Shiung Wu. (AIP Niels Bohr Library)

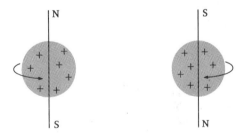

FIG. 20–11   The magnetic field produced by a nucleus rotating about its axis.

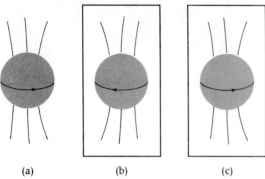

(a)   (b)   (c)

FIG. 20–12   Predictions of conservation of parity. (a) An equal number of electrons emitted from both ends of aligned cobalt nuclei. For simplicity, only one nucleus is shown. (b) A mirror image of part (a). (c) Part (b) turned around for comparison with part (a).

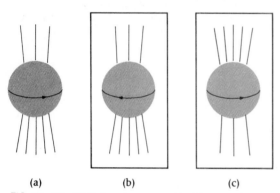

(a)   (b)   (c)

FIG. 20–13   Parity is not conserved in this example (a) More electrons are emitted by the south end of the aligned cobalt nuclei. (b) A mirror image of part (a). (c) Part (b) turned around for comparison with part (a).

thing.' Just for fun I actually exposed some beer to gamma rays the next day in the laboratory. Nothing happened." But . . . he went on to more serious tests which succeeded.[12]

The basic idea of the bubble chamber rests on the fact that the boiling point of a liquid depends upon pressure. For example, in a pressure cooker, the boiling point of water is greater than 100°C because of the increased pressure. Suppose now that we have a chamber filled with liquid hydrogen. At atmospheric pressure, liquid hydrogen boils at a temperature of −253°C. By subjecting the chamber to an increased pressure, the boiling point of the liquid hydrogen can be raised to −248°C. When the pressure is reduced once again, the liquid hydrogen will soon start to boil. If, at that instant, the chamber is subjected to charged particles, bubbles will form along the path of the particle and nowhere else; that path can be photographed through windows in the chamber. With the application of a magnetic field, the path of the charged particles becomes curved, and information about them can be obtained. Figure 20–14 is an example of a hydrogen bubble-chamber photograph.

Donald Glaser was awarded the Nobel Prize in 1960.

The Omega-Minus: Prediction and Discovery

So many elementary particles! What does it all mean? Just as Dimitri Mendeleev sought order in constructing the periodic table of the elements, scientists sought for some link connecting properties of the particles. In 1961, Murry Gell-Mann (b. 1929), an American physicist at the California Institute of Technology, and the Israeli physicist Yuval Ne'eman (b. 1925) of Tel-Aviv University independently proposed a solution.

The four delta particles—Δ^-, Δ^0, Δ^+, and Δ^{++}—with a mass of 1238 MeV, and the three sigma particles—Σ^-, Σ^0, Σ^+—with a mass of

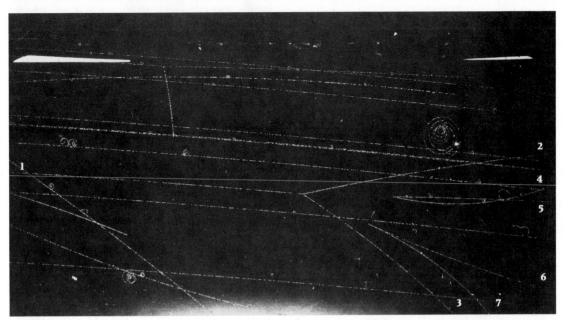

FIG. 20–14   A 2.85-GeV proton (1) from the Cosmotron collides with a proton in the liquid hydrogen bubble chamber. A proton (2) and π^+ meson (3) are emitted and, in addition, two neutral particles. The first neutral particle is a Λ^0-hyperon, which disintegrates to give a π^- meson (4) and a proton (5). The second neutral particle is a K^0 meson, which disintegrates to give a π^- meson (6) and a π^+ meson (7). [Courtesy of Brookhaven National Laboratory. Description obtained from David H. Frisch and Alan M. Thorndike, *Elementary Particles* (Princeton, N.J.: D. Van Nostrand, 1964, Plate V).]

1385 MeV, had been discovered. The difference in their masses was 147 MeV, which suggested that two elementary particles might be found that have a mass of 147 MeV larger than that of the sigmas, or a mass of 1532 MeV. The *Scientific American* describes the events leading to the discovery of the omega-minus:

At the biennial International Conference on High Energy Physics in Geneva in July 1962, however, experimeters reported the discovery of a pair of xi particles (xi-minus and xi-zero) of mass 1530 MeV. Gell-Mann, who was present at the conference, immediately saw the connection between this almost routine announcement and the eightfold way (as Gell-Mann's theory is called from mathematical considerations). The new xi particles qualified for membership in the family with the delta and sigma particles by their mass (very close to the predicted 1,532

MeV). . . . It was now possible to predict with reasonable confidence that the pyramid [see Fig. 20–15] must be crowned by a singlet at the apex—a particle . . . (having) a mass about 1676 MeV (146 MeV higher than that of the xi particles). A determined, full-scale search for this particle, already named omega-minus, was therefore launched almost immediately. . . . The Brookhaven National Laboratory, with its 33-GeV . . . accelerator and a new 80-inch liquid hydrogen bubble chamber, was prepared for the opportunity, and welcomed the task of searching for the omega-minus.[13]

At the Brookhaven synchroton, 33-GeV protons struck a tungsten target, producing a ple-

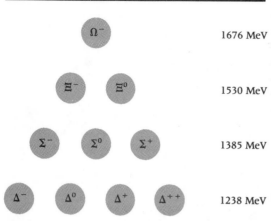

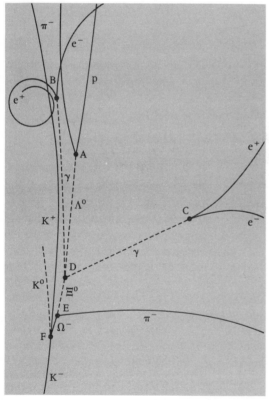

FIG. 20–15 A diagram predicting the existence of the Ω^- particle.

FIG. 20–16 Diagram of bubble-chamber photograph. [*Source:* W. B. Fowler and N. P. Samios, "The Omega-Minus Experiment," *Scientific American* (October, 1964): p. 39. Copyright © 1964 by Scientific American, Inc. All rights reserved.]

thora of particles, among them the desired K-minus particles (symbol: K^-). To obtain a beam of *pure* K^- particles, a multiparticle beam is passed through several magnets, which allow only the K^- particles to pass through. This K^- beam, having a kinetic energy of 5000 MeV, is then directed into a hydrogen bubble chamber. Doing this, the researchers hoped, as their calculations suggested, that the omega-minus (symbol: Ω^-) might be produced when the K^- interacted with a hydrogen nucleus in the following way:

$$K^- + p^+ \rightarrow \Omega^- + K^+ + K^0. \quad (19)$$

The researchers planned to identify the Ω^- from its decay, which the scientists predicted to be as follows:

$$\Omega^- \rightarrow \Xi^- + \pi^0 \quad (20)$$
$$\rightarrow \Xi^0 + \pi^- \quad (21)$$
$$\rightarrow \Lambda^0 + K^-. \quad (22)$$

The *Scientific American* continues:

After six weeks of arduous labor in perfecting the operation of the beam equipment to obtain a sufficiently pure and intense beam of K-minus mesons, photographs of the bombardment of the bubble chamber by the beam began on December 14, 1963. In the ensuing months the entire system—the accelerator, the beam equipment, the bubble chamber and the camera—was operated on an around the clock basis. By January 30, 50,000 good photographs had been obtained.[14]

On January 31, the researchers found the photograph corresponding to the decay given by Equation 21 (see Fig. 20–16).

One month later, they found another photograph showing the decay given by Equation 22. In 50,000 photographs, only two showed the Ω^-. In fact, the researchers considered themselves lucky in finding two within a month.

The researchers used conservation of momentum and energy to determine the mass of the Ω^-. For example, the photon that leaves no track decays at point C into an electron-positron pair. The radius of curvature in a magnetic field gives the momentum and energy of each particle. This, in turn, gives the energy of the gamma ray. Continuing this process gives the mass of the Ω^-, which they determined to be "between 1668 and 1686 MeV (allowing for the uncertainties in the calculation). This is precisely on the mark for the predicted mass of the omega-minus particle—1676 MeV."[15]

Quarks

In 1964, Gell-Mann and Zweig proposed that elementary particles are different arrangements of particles called **quarks,** whose properties are presented in Table 20–2. Massive particles, called **baryons,** consist of 3 quarks, and their antiparticles consist of 3 antiquarks. Mesons result from the combination of a quark with an antiquark. Table 20–3 shows the construction of baryons and mesons from quarks.

Despite an exhaustive search for free quarks, they have never been found. Physicists are divided upon the question of whether quarks really exist, or whether they provide a way to make mathematical relationships more easily visualized. There is some convincing evidence that the proton and neutron do have internal structure. Let's turn our attention to this evidence.

The Stanford Linear Accelerator

Stanford Linear Accelerator Center (SLAC), completed in 1966, houses the world's largest **linear accelerator.** In this process, the electron is raised to a maximum energy of 21 GeV as it travels down a two-mile evacuated pipe. At this energy, an electron travels at a speed greater than 99.9999999 percent of the speed of light.

You may wonder why such a large energy and such a large speed are necessary. An electron traveling at 99.99 percent of the speed of

TABLE 20–2 PROPERTIES OF QUARKS

QUARK		MASS (MeV)	CHARGE
u	(up)	336	$+\frac{2}{3}$
d	(down)	338	$-\frac{1}{3}$
s	(sideways or strange)	540	$-\frac{1}{3}$
ANTIQUARK			
$\bar{u}$		336	$-\frac{2}{3}$
$\bar{d}$		338	$+\frac{1}{3}$
$\bar{s}$		540	$+\frac{1}{3}$

TABLE 20–3 CONSTRUCTION OF BARYONS AND MESONS FROM QUARKS

BARYONS	QUARK TRIPLET
proton	$u\,u\,d$
neutron	$d\,d\,u$
Δ^-	$d\,d\,d$
Δ^0	$d\,d\,u$
Δ^+	$u\,u\,d$
Δ^{++}	$u\,u\,u$
Σ^-	$d\,d\,s$
Σ^0	$u\,d\,s$
Σ^+	$u\,u\,s$
Ξ^-	$d\,s\,s$
Ξ^0	$u\,s\,s$
Ω^-	$s\,s\,s$
Λ^0	$u\,d\,s$
antiproton	$\bar{u}\,\bar{u}\,\bar{d}$

MESONS	QUARK + ANTIQUARK
π^+	$u\,\bar{d}$
π^0	$u\,\bar{u}$ or $d\,\bar{d}$
π^-	$d\,\bar{u}$
K^+	$u\,\bar{s}$
K^0	$d\,\bar{s}$
K^-	$s\,\bar{u}$

light has an energy of 36 MeV. Why does such a small change in speed result in such a large change in energy? At these speeds, relativistic formulas for energy and momentum must be used. The formula for energy is given by

$$E = \frac{m_0 c^2}{\sqrt{1 - v^2/c^2}}, \qquad (23)$$

and the formula for momentum is given by

$$p = \frac{m_0 v}{\sqrt{1 - v^2/c^2}}. \qquad (24)$$

Here, small changes in v result in large changes for E and p. One purpose of an accelerator is to create new particles. This is accomplished when, in a collision, the kinetic energy of the electron is transformed into the mass of a new particle. Obviously, the larger the kinetic energy of the electron beam, the more massive are the particles created. Another purpose is to scatter electrons from a target and to study the structure of the proton and neutron; that is, to use the electron beam as a probe. Let's consider this latter purpose now.

The proton has a diameter of 1.6×10^{-15} m. Suppose that we aim a beam of electrons at the target protons to see if the proton has some internal structure. What energy electron beam is needed to carry out this mission?

Electrons, recall (Chapter 15), also have a wave nature and that wave nature determines the electron's behavior. The de Broglie wavelength is given by

$$\lambda = \frac{h}{p}. \qquad (25)$$

Here p, given by Equation 24, must be used. If the wavelength of the electron is about 10^{-15} m, the diffraction effects will be very great because the wavelength and "obstacle" are nearly

the same. In that case, it will not be possible to probe the internal structure at all. To reduce diffraction, the wavelength must be smaller than the proton "obstacle." At an energy of 21 GeV, an electron has a wavelength of 0.6×10^{-16} m, obtained by using Equations 24 and 25.

Before the scattering experiment began, theorists calculated (assuming that the proton was diffuse and homogenous) the relative number of electrons that should be scattered to various angles. When the scattering experiment was performed, the actual number of electrons scattered to various angles exceeded the theorists' expectation by as much as a factor of 40.[16] Just as the backward-scattering of alpha particles by a gold atom implied that the atom had within it a small positive nucleus, these results implied that the proton had embedded within it objects whose diameters are no more than one-fiftieth of the proton as a whole. Richard P. Feynman of the California Institute of Technology called these objects **partons.** There was a natural tendency for scientists to identify partons as quarks.

Discovery of Neutral Massive Mesons

THE J PARTICLE

In Fig. 20–17, the American physicist Samuel Ting (b. 1936) holds a graph showing evidence for the J particle, which he and his co-workers discovered in 1974 using Brookhaven National Laboratory's alternating gradient synchrotron. In their experiment, 30-GeV protons were aimed at a beryllium target. At this energy, many kinds of particles are produced, but Ting was interested in one particular reaction—that in which electron-positron pairs are produced:

$$p + \text{Be} \rightarrow e^+ + e^- + x. \qquad (26)$$

The x stands for the other particles in which he had no interest.

Suppose, now, that, for a brief period of time,

a *J* particle exists that subsequently decays into an electron-positron pair, as follows:

$$p + \text{Be} \rightarrow J + x \qquad (27)$$
$$J \rightarrow e^+ + e^-.$$

The question is, How can you tell the difference in an experiment between Equations 26 and 27? To do this, the Ting group placed two elaborate detectors at a fixed angle θ from the incident beam direction, as shown in Fig. 20–18. These detectors identified the particle, its mass, and measured its momentum. Events were observed in which an electron entered one detector and, simultaneously, a positron entered the other. The question is, How can the researchers distinguish between Equations 26 and 27 by knowing the momentum of the electron and the momentum of the positron?

Figure 20–19 shows the decay of a fast-moving *J* particle into an electron and positron pair. Momentum is conserved—always! Therefore, if you measure p_{e^-} and p_{e^+}, and know the angle θ (fixed by the experimental setup), you can determine the total momentum afterwards. This total momentum, of course, must be equal to the total momentum before, or p_J. Thus, by their measurement, and some calculations, the researchers were able to determine p_J.

At these high speeds, relativistic formulas must be used, but the nonrelativistic formulation,

$$\text{KE} = \frac{1}{2}mv^2 = \frac{p^2}{2m},$$

shows that the kinetic energy does depend upon both mass and momentum of the particle. Similarly, in the relativistic formulation, the kinetic energy depends upon the mass of the particle and its momentum. (See calculation question 9–13.) To determine the mass of the *J* particle, we use conservation of energy. There, the total energy *E* of a particle is given by

$$E = \text{KE} + m_0 c^2,$$

FIG. 20–17 The combined effort of the Massachusetts Institute of Technology (MIT) and Brookhaven National Laboratory (BNL) brought about the physics discovery of the decade made at the alternating gradient synchrotron at Brookhaven National Laboratory—the discovery of the *J* particle. The MIT/BNL group that discovered the *J* particle is seen with the graph of the events, which is held by Samuel Ting. (Courtesy of Brookhaven National Laboratory)

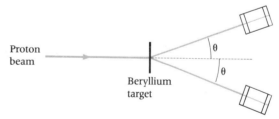

FIG. 20–18 A diagram of the Ting-group experiment.

ELEMENTARY PARTICLES

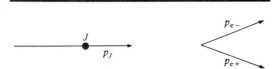

FIG. 20–19 The total momentum of the electron and positron equals the momentum of the J particle.

where m_0 is the mass of the particle at rest and c is the speed of light (see Equation 16 of Chapter 9, p. 218).

Using conservation of energy, we find:

$$TE_{before} = TE_{after}$$

$$KE_J + m_J c^2 = KE_{e-} + m_e c^2 \qquad (28)$$
$$+ KE_{e+} + m_e c^2.$$

We know all terms on the right side, and KE_J depends upon p_J (which we know) and m_J. Therefore, the only unknown in Equation 28 is m_J and we can solve for m_J.

In summary, the momenta of the electron and positron are measured in the experiment and, from conservation of momentum and energy, the mass of the hypothetical J particle can be determined.

On a graph, Ting plotted the mass of the J particle (obtained from his calculations) along the x-axis, and the number of times he saw that value of the mass along the y-axis. He observed a peak in the graph at a mass of 3.1 GeV. This is exactly what you would expect to find: a *single value* for the mass. The J particle has a mass *3.3 times that of the proton!*

If nature had instead conformed to Equation 26, the graph of number of events versus mass would have been a *smooth* curve. This would show that there is no correlation between the electron's momentum and energy and those of the positron. The momentum and energy of the electron and positron would have been distributed randomly, and the resulting mass would have been random as well. This would result in a smooth curve.

THE Ψ PARTICLE

Almost simultaneously with the Ting group, other physicists—Burton Richter, Gerson Goldhaber, and their numerous collaborators—used the Stanford Positron-Electron Accelerating Ring (SPEAR) and discovered the same particle as Ting's group, calling it the Ψ particle.

At SPEAR, energetic electrons and positrons

travel in a circular ring and then are made to collide *head-on*. Sometimes, the electrons and positrons entering the collision region do not interact at all; sometimes, they simply scatter from each other; and, rarely, their collision results in the production of **hadrons**—a collective term referring to protons, neutrons, mesons, hyperons, and their antiparticles. Hadrons are strongly interacting particles, in contrast to electrons and muons, which interact only weakly. In their experiment, the researchers counted the events (using elaborate detectors and an on-line computer) and determined the following probability:

$$\text{Probability} = \frac{\left(\begin{array}{l}\text{number of events resulting}\\ \text{in hadron production}\end{array}\right)}{\left(\begin{array}{l}\text{number of positrons and}\\ \text{electrons entering collision}\\ \text{region}\end{array}\right)}.$$

The object of the experiment was to see how this probability changed as the energy of the electrons and positrons changed (with the restriction that the energy of the positrons equals that of the electrons). When the electrons and positrons had a kinetic energy of 1.55 GeV, the probability increased tremendously. On a graph of probability (*y*-axis) versus kinetic energy (*x*-axis), a narrow peak centered on an energy of 1.55 GeV. This group of physicists at Stanford had discovered a new elementary particle.

Goldhaber describes their discovery:

Later that morning when I returned to the SPEAR control room, a number of extra runs had been taken at various energies below and above 1.56 GeV. . . . At about 9:00 A.M. we began to zero in on trying to find where the peak might be. Suddenly, at about noon, we found an energy where the cross section (probability) had risen by a factor of 7! By this time our excitement rose to a fever pitch. Having been nurtured on strong interactions, I had hardly ever seen an elementary-particle cross-section that rose by a factor of 7 above the background level. At this point the results were so convincing that I suggested to some of my colleagues (Burt Richter, Roy Schwit-

ters, William Chinowsky, and others) that we should sit down and write a paper. They agreed with me and Burt led me to a side room where I could quietly start this task. I picked up a piece of computer output from the floor to write on, and in about one hour I had written the first draft for our paper—an activity, incidentally, I had never done before: writing a paper on-line, so to speak.

At that point Willy Chinowsky burst into the room where I was working and told me very excitedly— and I have rarely seen Willy so excited—that the cross-section had risen by another factor of 10! With this news I jumped up and ran to the one-event display in the control room, and there indeed we were seeing hadron events coming in one after the other. . . . It was absolutely unbelievable.

Were I to try to express my feelings as I saw the data coming in on the cathode ray tube, I would say that this was perhaps the highest level of excitement I have ever experienced—in a laboratory, that is.

. . . I was obliged to go and delete all the numbers I had tentatively written in the paper and up everything by a factor of 10! . . . The news spread like wildfire.

A typical conversation went something like this:

"There really is a peak in the e^+e^- cross section. How high above background do you think it goes?"

Answers that came back were:

"50%?"

"A factor of 2?"

"A factor of 5?"

I would finally say:

"A factor of 70!" And then I always had to add, "That is, seven-zero!!"

I wish I had been able to record the expressions of amazement that resounded in my ears.

The next few hours were spent talking and speculating what new quantum numbers or what new selection rule must be involved to inhibit the decay of our resonance (peak) and make it so narrow.[17]

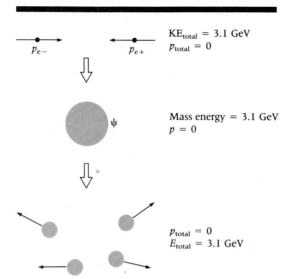

$KE_{total} = 3.1$ GeV
$p_{total} = 0$

Mass energy = 3.1 GeV
$p = 0$

ψ

$p_{total} = 0$
$E_{total} = 3.1$ GeV

FIG. 20–20   A positron and an electron collide head-on. Since they have the same speeds, the total momentum before collision is zero. Their kinetic energy is converted into the Ψ particle, which then disintegrates into hadrons.

Let's take a closer look at this team's data and see why they really had discovered a new particle. Figure 20–20 displays an interpretation of their results.

Before the collision, the total momentum is zero because the electron and positron travel in opposite directions with the same speed. Because the total momentum is zero, the entire energy before collision—3.1 GeV—can be converted into the mass of the Ψ particle, which, since the momentum is also zero, is at rest. The mass-energy of the Ψ particle is 3.1 GeV, but lives only for a short time before decaying into hadrons. This process is akin to a backward-run motion picture of the reaction used by Ting.

The simultaneous discoveries of the J/Ψ particle were published side by side in Volume 33 of the *Physical Review Letters,* 1974. In 1976, Samuel Ting and Burton Richter shared the Nobel Prize in physics for their discovery.

Rapidly, the group at SPEAR discovered yet another particle (symbol: Ψ') having a mass of 3.7 GeV.

These new particles "upset the apple cart," to be sure, for they did not fit into the 3-quark classification scheme. For one thing, the new particles are so massive. The combined mass of the 3 quarks is only 1.2 GeV. Thus, if the J/Ψ particle is composed of quarks, there must be a more massive fourth quark. These discoveries created a great deal of excitement for, once again, theories had to be reformulated. Physicists now think that there is a fourth quark, called the **charmed quark,** having a mass of approximately 1.6 GeV, and a charge of $+\frac{2}{3}$. The J/Ψ particle results from the combination of a charmed quark c with a charmed antiquark $\bar{c}$; the Ψ' particle is an excited state of $c\bar{c}$.

The combination of 4 quarks predicts the existence of still other particles, and physicists are currently searching for them—with some successes and some failures.

Fermi National Accelerator Laboratory

Fermilab at Batavia, Illinois, about 35 miles west of Chicago (see Fig. 20–21), was completed in 1972. Here, protons are accelerated to a maxi-

FIG. 20–21   An aerial view of the principal components of the accelerator system at the Fermi National Accelerator Laboratory, Batavia, Illinois. The largest circle is the main accelerator, which is four miles in circumference. The smaller circle, with a cooling-water pond in the center, is the booster accelerator. Experimental areas lie at a tangent to the left. (Courtesy of Fermi National Accelerator Laboratory)

mum energy of 500 GeV (1 GeV = 1 billion electron-volts) in four stages (see Fig. 20–22). A Cockcroft-Walton generator accelerates protons to 0.75 MeV; a linear accelerator, or *linac*, to 200 MeV; a booster synchrotron to 8 GeV; and the main ring synchrotron to an energy of 500 GeV. In the 145-m length of the linac, many electrodes within set up electric fields to boost the energy of the protons. In both synchrotrons (descendants of the Cosmotron and Bevatron), magnetic fields force the protons around the circular path. As the speed of the protons increases, the strength of the magnetic field also increases. There are 954 magnets placed around the 6.3-km circumference of the main ring. To achieve an energy of 500 GeV, the protons make about 176,000 revolutions around the ring in 4 seconds, receiving an energy from radio-frequency cavities of 2.8 MeV on each turn. It's interesting to note that the protons enter the main ring with a speed of 99.45 percent of the speed of light and emerge with a speed of 99.9998 percent of the speed of light. The kinetic energy is given by

$$\mathrm{KE} = \frac{m_0 c^2}{\sqrt{1 - v^2/c^2}} - m_0 c^2.$$

Thus, at speeds near that of light, a small increase in the speed results in an extremely large increase in kinetic energy.

After the protons are extracted from the main ring, they are guided by some 100 magnets to one of the three experimental areas shown in Fig. 20–22. Copious quantities of pi-mesons are produced when protons strike a target. Experiments can be carried out with a secondary beam of pi-mesons. Since the pi-mesons decay into muons and neutrinos, experiments can be performed using muon or neutrino beams.

In the first experiment at Fermilab, protons were aimed at target protons. Analysis of the scattering of the protons yielded the radius of the proton.

ELEMENTARY PARTICLES

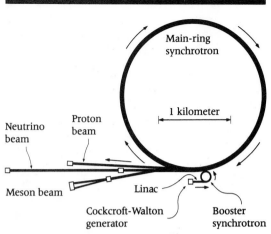

FIG. 20–22 A diagram of the Fermilab accelerator and the experimental areas.

In the meson area, researchers carried out several experiments to detect isolated quarks, but the elusive quarks were not to be found. No research to date has been successful in this area.

Earlier, we learned that the proton has structure. In the manner of Rutherford's experiments, electrons striking protons were scattered to larger angles more frequently than would occur if the proton were a homogenous sphere. At Fermilab, experiments exhibit, this feature, too. R. R. Wilson, the first director of Fermilab, describes the results of these experiments:

By analogy with Rutherford's finding, the new observation that high-energy particles are emitted at large angles from the proton implies that parts of the proton are "hard." The observation holds true whether the proton is bombarded with neutrinos, with muons, or with other protons, and whether the emerging particles are muons, pions, photons or other protons. In short, there is little doubt that we are beginning to perceive an inner structure of the proton.[18]

In 1977, a massive new particle, called the **upsilon,** was discovered by a group led by Leon Lederman, the present director of Fermilab.[19] In principle, the Lederman-group experiment is quite similar to Ting's experiment. In the Lederman method, protons are directed toward a target and two detectors are set up (see Fig. 20–18) to observe simultaneously a muon and its positively charged antiparticle. The group analyzed the data to see if a particle existed for a short period of time before it decayed into these two muons. They discovered the upsilon particle (symbol: Υ) that has a mass of 9.5 GeV, a mass that is about 10 times that of a proton. Another particle (symbol: Υ') has also been discovered and it has a mass of 10.0 GeV. To "explain" the existence of these particles, elementary particle physicists invoke the existence of yet a fifth quark, called the **bottom quark.** Just as the Ψ/J particle and the Ψ' particle result from the combination of a charmed quark c and a charmed antiquark $\bar{c}$, so the Υ and the Υ' result from the combination of the bottom quark b and an antibottom quark $\bar{b}$. The bottom quark has a charge of $-\frac{1}{3}$. Theorists also predict the existence of a **top quark,** the counterpart of the bottom quark, with a charge of $+\frac{2}{3}$. So far, no new particles have been discovered that imply the existence of this sixth quark, the top quark.

THE TEVATRON

The Tevatron[20] will raise the proton energy at Fermilab to 1 TeV (1000 GeV or 1 trillion eV). It is scheduled to begin operation with an energy greater than 500 GeV in May, 1983. Here, in the main ring beneath the original magnets, 1000 superconducting magnets are being installed. These magnets have windings made from a niobium-titanium alloy embedded in copper. When cooled to the temperature of liquid helium (4.5 °K), the resistance of the magnet wire to electrical current is nearly zero. The maximum current will be about 4600 amperes. These magnets will produce a magnetic field twice that of the original ones, so that the maximum energy is doubled to 1 TeV. To maintain this low temperature, a central helium liquifier will produce about 4000 liters per hour. This new facility will be used in two ways: (1) to deliver protons having a maximum energy of 1 TeV to a fixed target; and (2) to collide beams of protons and antiprotons.

For the fixed-target experiments, the protons are first accelerated to an energy of 500 GeV, guided by the original magnets. Then they are switched to the superconducting set, where they are accelerated to 1 TeV.

To see why the colliding-beam experiments offer such promise, review Fig. 20−8. Here, we see that, of the 5.6-GeV incident energy, 3.8 GeV appears after the collision as kinetic energy. Only 1.8 GeV remains for the creation of new particles. Similarly, when 500-GeV protons strike stationary protons, 469 GeV appears as kinetic energy afterwards, with only 31 GeV available for the creation of new particles. In the colliding-beam experiments, the momentum before collision is zero. Therefore, after collision, the 2 protons can be at rest, and the total momentum is still zero. Thus, if both the proton and anti-

proton have an energy of 1 TeV, then 2 TeV is available for the creation of new particles.

In the colliding-beam mode, protons would be accelerated to 100 GeV, extracted from the main ring, and aimed at a target, producing antiprotons. After they are formed into a well-collimated beam, the antiprotons would be returned to the main ring. Then, protons and antiprotons would be accelerated simultaneously to 1 TeV. Since they are oppositely charged, the protons and antiprotons would travel in opposite directions around the ring and collide head-on at several points in the main ring.

When the new facility is completed, researchers will surely look for particles that are composed of top and bottom quarks, and will certainly try to find isolated quarks. Perhaps, with so much energy available, quarks can be torn from the nucleus.

At this higher energy, the W particle, the so-called **weakon,** may also be detected. In the first part of this chapter, we saw that Yukawa described the nuclear force as originating with the exchange of a pi-meson between a proton and neutron. Similarly, in beta decay, the weak interaction (in comparison to the nuclear force) is thought to be mediated by the W particle. For example, a neutron emits a W^- particle and transforms into a proton. The W^- particle travels a short distance and decays into an electron and a neutrino. The weakon comes in three varieties, the W^-, the W^+, and the Z^0. These particles are expected to have a mass of about 100 GeV, or 100 times that of a proton.

The Future

Physicists will surely continue their search for the weakon and for particles that indicate the existence of the top quark. As accelerators reach ever higher energy, researchers will probe the substructure of the proton and neutron with greater resolution.

Some theoretical calculations suggest that the proton may be unstable, with an extremely long half-life, on the order of 10^{32} years. To shield the detectors from cosmic rays, experiments that look for proton decay are being carried out in mines. Since proton decay is such a rare event, researchers may detect less than about 40 decays in one *year*.[21]

Another fascinating speculation is that the neutrino may have an exceedingly small mass. To test this hypothesis, physicists can study beta decay and measure, with great precision, the maximum kinetic energy of the beta ray. (See Chapter 16, Equation 8.)

These are just some of the challenging questions that confront researchers in elementary particle physics. Without doubt, as research in *all* fields of physics progresses, we shall witness the *excitement of discovery*.

Summary

In 1935, Yukawa suggested that the attractive force between two nucleons (a term for either the proton or neutron) results from the exchange of a pi-meson ($m_\pi \cong 200\ m_e$) between them. In 1936, Anderson and Neddermeyer discovered in cosmic rays a *penetrating* particle, today called a muon, which has a mass between an electron's and a proton's. The pi-meson is easily absorbed by matter because it interacts strongly with protons and neutrons, while a muon is more penetrating than an electron.

The pi-meson was discovered in 1947 when researchers exposed a stack of photographic emulsions (not sensitive to electrons) to cosmic rays. From the density of the track and its length, the researchers estimated that the pi-meson had a mass of about 200 m_e. A charged pi-meson decays into a muon and a neutrino; a neutral pi-meson decays into two gamma rays. In 1948, pi-mesons were produced artificially when 380-MeV alpha particles from the Berkeley cyclotron struck a target. During the collision, 140 MeV of kinetic energy was converted into the mass of the pi-meson.

Still more particles were discovered in cosmic rays: hyperons (particles more massive than a neutron)—Ξ, Σ, Λ^0, and more mesons (particles that interact strongly with the nucleus), such as kaons. The lambda-naught hyperon (Λ^0) decays into a proton and negative pi-meson. Brookhaven's Cosmotron produced the first *artificial* hyperon—a Λ^0 hyperon—when neutrons entered a hydrogen-filled cloud chamber.

In 1954, scientists at the Berkeley Bevatron demonstrated the existence of the antiproton, a particle having the same mass as a proton but negative charge. A 6.2-GeV proton (1 GeV = 1000 MeV) beam hit a target, and an additional proton and antiproton were produced in the collision. A magnetic field selected the momentum of the negatively charged particles leaving it. The antiprotons were distinguished from the plentiful pi-mesons by measuring the transit time between two detectors.

The principle of conservation of parity states that it's impossible to distinguish between a real object or event and its mirror image. In 1956, Lee and Yang suggested that parity might not be conserved in beta decay. Wu and her co-workers studied the beta decay of cobalt-60, by aligning the nuclear "magnets" (produced by the rotating nuclear charge) at low temperature in a strong magnetic field. They found that more beta rays emerged from the south pole of the nucleus, which was in sharp contradiction to the equal amounts that conservation of parity predicted.

Theorists had predicted the existence of an omega-minus particle, having a mass of about 1532 MeV. K-minus particles from the Brookhaven synchrotron entered a hydrogen bubble chamber (invented in 1952). The omega-minus particle was identified in bubble-chamber photographs by using the principles of conservation of momentum and energy.

In 1964, Gell-Mann and Zweig suggested that elementary particles are themselves composed of three kinds of quarks, with nucleons and hyperons containing 3 quarks and a meson, a quark and an antiquark. Despite exhaustive searches, individual quarks have never been detected.

Physicists used the 21-GeV electron beam at the Stanford Linear Accelerator Laboratory to probe the structure of the proton and neutron. At this energy, the position of the electron can be identified as close as 10^{-17} m. Therefore, such an energetic electron can effectively sample the internal structure of the proton (diameter = 1.6×10^{-15} m). The scattering at some angles was much larger than expected if the proton were a homogenous sphere. These results indicated that the proton is composed of point-like objects, initially called partons and, now, some say, quarks.

In 1974, two groups of researchers simultaneously discovered a neutral massive meson, called the J/Ψ particle, which has a mass 3.3 times that of a proton. Theorists explain its existence by introducing a fourth quark, called the charmed quark; the J/Ψ particle allegedly consists of a charmed quark and its antiquark.

In 1977, the Υ and the Υ' particles were discovered. The idea of a fifth quark, called the bottom quark, was introduced, and these new particles theoretically result from the combination of this bottom quark and its antiquark.

At the Fermi National Accelerator, protons are accelerated to 500 GeV, and a new facility, the Tevatron, will soon boost the proton energy to 1000 GeV. We can expect a plethora of new particles to be discovered at this higher energy.

True or False Questions

Indicate whether the following statements are true or false. Change all of the false statements so that they read correctly.

20–1 a. The nuclear force between two protons occurs because they exchange a positively charged pi-meson between them.

b. So that conservation of energy is not violated, the proton emitting the pi-meson must have an uncertainty in its energy smaller than the mass-energy of the pi-meson.

c. The uncertainty in the position of the pi-meson is equal to the range of the nuclear force.

d. If the mass of the pi-meson were smaller, then the range of the nuclear force would likewise be smaller.

20–2 In 1936, Anderson and Neddermeyer discovered the muon in cloud-chamber photographs of cosmic rays.

a. The muon has a mass between that of an electron and a proton.

b. In a magnetic field, the track made by a muon has a radius larger than that made by a proton.

c. The muon could not be identified as Yukawa's meson because the muon penetrated matter so easily, while the meson should be readily absorbed.

d. Like an electron, the muon loses energy as it travels through matter by ionization, and by radiation as it passes a nearby nucleus.

20–3 a. The pi-meson was discovered in cloud-chamber photographs of cosmic rays.

b. The positively charged pi-meson decays into a positively charged muon and a neutrino.

ELEMENTARY PARTICLES

20-4 a. The Λ^0 (lambda-naught) decays into a proton and a negatively charged pi-meson.

b. The Λ^0 has a mass greater than the mass of a proton plus the mass of pi-meson.

c. The Λ^0 belongs to a class of particles called hyperons.

20-5 Protons from the Bevatron strike a target and produce antiprotons.

a. The antiprotons cannot be produced in the following way, because it would violate conservation of energy:

$$p + p \rightarrow p + p + \bar{p}.$$

b. The kinetic energy of the incident proton is completely converted into the mass of the newly formed proton and antiproton.

c. In the experiment, the antiproton could be distinguished from the plentiful pi-mesons by passing the beam through a magnetic field.

20-6 The beta decay of cobalt-60 nuclei, aligned with a magnetic field, showed that parity was not conserved because

a. the mirror image of the decay was not identical to the decay itself.

b. equal numbers of beta rays were emitted from both ends of the cobalt-60 nuclei.

20-7 a. In a bubble chamber, the boiling point of the liquid does not change when a pressure is exerted upon the liquid hydrogen.

b. A track is produced by bubbles of hydrogen gas forming along the path of the particle as the liquid hydrogen boils.

20-8 A proton is thought to consist of three quarks, and a pi-meson of a quark and an antiquark.

20-9 Electrons from an accelerator can probe the substructure of the proton only if the de Broglie wavelength of the electron is larger than the diameter of the proton.

20-10 The J/Ψ particle is thought to consist of a charmed quark and its antiquark.

Questions for Thought

20-1 Define the term *nucleon*.

20-2 Anderson and Neddermeyer studied the highly penetrating cosmic rays.

a. Explain how electrons lose energy in traveling through matter.

b. In what way was the behavior of the penetrating rays different from that of electrons?

c. Why did they think the rays weren't protons?

20-3 The cosmic rays entering the cloud chamber were subjected to a magnetic field. In one of the Anderson-Neddermeyer cloud-chamber photographs, the researchers found a dense circular track about 4 cm long, with a radius of curvature of 7 cm.

a. Explain why a proton could not have made this track.

b. Why must the particle's mass be less than a proton's mass?

20-4 What was the difficulty in identifying this penetrating cosmic-ray particle as the pi-meson?

20-5 What experimental method did Powell and his group use to detect pi-mesons?

20-6 How do muons and pi-mesons decay?

20-7 Define the term *hyperon*.

20-8 Explain how lambda-naught hyperons were produced artificially at the Cosmotron. How were they identified in cloud-chamber photographs?

20-9 What is an antiproton?

20-10 On what grounds did physicists suspect the existence of an antiproton?

20-11 Why can't the following reaction occur?

$$p + p \rightarrow p + p + \bar{p}.$$

What conservation law is violated?

20-12 A proton beam from the Bevatron has an energy of 6 GeV. Explain why a particle having a mass-energy of 6 GeV cannot be created when this beam strikes a stationary target?

20-13 In the experiment searching for antiprotons, a 6.2-GeV proton beam was directed onto a copper target. Elementary particles entered a magnet.

> **a.** What was the purpose of the magnet?
>
> **b.** How were the antiprotons distinguished from the negatively charged pi-mesons? What was the ratio of the number of mesons to the number of antiprotons?

20-14 What is the speculation about the existence of an antiworld?

20-15 State the principle of conservation of parity.

20-16 How did the θ-τ puzzle lead Yang to propose nonconservation of parity?

20-17 Describe the experiment performed by Wu and her co-workers to look for parity violation in the beta decay of cobalt-60.

20-18 A sample of cobalt-60 at room temperature is placed on a table in a laboratory (and is *not* subjected to an experimental magnetic field). Will the number of beta rays emitted per time be the same in all directions? Why or why not?

20-19 An elementary particle enters a hydrogen bubble chamber. Explain why the liquid hydrogen boils along the path of the particle.

20-20 On what grounds was the existence of the omega-minus predicted?

20-21 a. In order to look for the omega-minus, what kind of beam entered the hydrogen bubble chamber?

> **b.** How was this beam produced by the Brookhaven synchrotron?
>
> **c.** What nuclear reaction produced the omega-minus?
>
> **d.** What reaction identified the omega-minus?
>
> **e.** In 50,000 bubble-chamber photographs, how many showed the existence of the omega-minus?

20-22 List the properties of the up, down, and sideways quarks.

20-23 A proton is considered to be composed of 2 up quarks and 1 down quark. Show that this combination gives a charge of +1. The combined mass of these three quarks is greater than 938 MeV. What happens to the difference in mass?

20-24 What did the 21-GeV electron-scattering experiment show about the structure of the proton? What is the analogy between this experiment and Rutherford's scattering experiment with alpha particles?

20-25 Ting and his group searched for a new particle by bombarding a beryllium target with 30-GeV protons.

> **a.** What observations did they make during the experiment?
>
> **b.** Did the new particle enter their detectors?
>
> **c.** One of their graphs showed a peak. What quantity was plotted on each axis? How did the peak confirm the existence of a new particle?
>
> **d.** How was conservation of energy and momentum used to find the mass of the *J* particle?

20-26 Define the term *hadron*.

20–27 At SPEAR, physicists looked for a new particle by observing what happened when electrons and positrons of the same energy collided head-on.

    a. What observations did they make during the experiment?

    b. How did these observations (and some calculations) show the existence of a new particle?

    c. Use the principle of conservation of nucleons to show why the experimenters must have observed particles (and some antiparticles as well).

20–28 Physicists are very interested in colliding-beam experiments, such as the one at SPEAR, for the production of elementary particles. Explain the reason for their interest.

Questions for Calculation

†20–1 What is the force of repulsion between two protons separated by a distance of 10^{-15} m? ($e = 1.6 \times 10^{-19}$ C, $K = 9 \times 10^9$ N-m^2/C^2.)

20–2 How did Yukawa explain the attractive force between two nucleons?

20–3 A proton and a neutron exchange a positively charged pi-meson, according to Equations 1 and 2.

    a. In order that conservation of energy remains valid, what must be the uncertainty in energy in Equation 1? (Use the known mass of the pi-meson.)

    b. Use Heisenberg's uncertainty principle to find the uncertainty in the time of travel.

    c. Assume that the meson travels close to the speed of light. Find the range of the nuclear force.

    d. Show that if the range of the pi-meson were 10^{-13} m instead of what it is, the pi-meson would have to travel at a speed greater than the speed of light (which is impossible).

†20–4 The range R of the nuclear force is only about 10^{-15} m.

    a. Show that the mass of a pi-meson exchanged by two nucleons is given by

$$m_\pi = \frac{h}{2\pi Rc}$$

    b. If the range R of the pi-meson were 10^{-13} m, find the required mass energy $m_\pi c^2$ of the pi-meson in MeV.

20–5 The force between two charged particles results from the exchange of a photon of electromagnetic radiation. Use the ideas of calculation question 20–4 to show why the range of the force between the two particles approaches infinity.

20–6 A lambda-naught hyperon having a kinetic energy of 100 MeV decays into a proton and a negatively charged pi-meson. What must be the combined kinetic energy of the proton and meson?

20–7 An electron traveling at 99.99 percent of the speed of light has an energy of 36 MeV, while one traveling at 99.9999999 percent of the speed of light has an energy of 21 GeV.

    a. Use Equation 23 to compare the values of $\sqrt{1 - v^2/c^2}$.

    b. The 21-GeV electron has a de Broglie wavelength of 0.6×10^{-16} m. Use the results of **a** to find the de Broglie wavelength of the 36-MeV electron.

    c. Explain why a 36-MeV electron beam cannot be used to probe the structure of the proton.

Footnotes

[1] Hideki Yukawa and Chihiro Kukuchi, *American Journal of Physics* 18(1950):154.

[2] Niels H. de V. Heathcote, *Nobel Prize Winners in Physics 1901–1950,* (New York: Henry Schuman, 1953), p. 448.

[3] C. D. Anderson, *American Journal of Physics* 29(1961):825.

[4] Ibid, p. 825.

[5] J. D. Stranathan, *The Particles of Modern Physics* (Philadelphia: The Blakiston Co., 1942), p. 525.

[6] C. D. Anderson in *American Journal of Physics,* volume 29.

[7] Eugene Gardner and C. M. G. Lattes, *Science* 107(1948):270.

[8] David H. Frisch and Alan M. Thorndike, *Elementary Particles* (Princeton, N.J.: D. Van Nostrand Co., Inc., 1964), p. 70.

[9] From Emilio Segre and Cylde E. Wiegand, "The Antiproton," *Scientific American* 194(June, 1956):37. (Copyright © 1956 by Scientific American, Inc. All rights reserved.)

[10] Ibid, p. 37.

[11] C. N. Yang, *Elementary Particles* (Princeton, N.J.: Princeton University Press, 1961), p. 54. (Reprinted by permission of Princeton University Press.)

[12] *Scientific American,* 192(February, 1955):26.

[13] From W. B. Fowler and N. P. Samios, "The Omega-Minus Experiment," *Scientific American* 211(October, 1964):36. (Copyright © 1964 by Scientific American, Inc. All rights reserved.)

[14] Ibid, p. 36.

[15] Ibid, p. 36.

[16] Henry W. Kendall and Wolfgang K. H. Pahofsky, "The Structure of the Proton and the Neutron," *Scientific American,* (June, 1971), p. 61.

[17] Gerson Goldhaber, "Discovery Story," *Adventures in Experimental Physics,* Epsilon Volume (Princeton, N.J.: World Science Communications), p. 131.

[18] R. R. Wilson, "The Batavia Accelerator," *Scientific American,* 230(February, 1974):72.

[19] James S. Trefil, *From Atoms to Quarks* (New York: Charles Scribner, 1980), p. 182.

[20] R. R. Wilson, "The Next Generation of Particle Accelerators," *Scientific American,* 242(January, 1980):42.

[21] James W. Cronin and Margaret Stautberg Greenwood, "CP Symmetry Violation," *Physics Today,* 35(July, 1982):44.

Large and small numbers may be conveniently represented using the powers of 10, as shown in Table A–1. Arithmetical calculations are also much simpler. The following demonstrates how numbers can be expressed as a power of 10:

$$50,000 = 5 \times 10,000$$
$$= 5 \times 10^4$$
$$0.006 = 6 \times .001$$
$$= 6.0 \times 10^{-3}.$$

Count the number of places you move the decimal point. If to the left, the power of 10 is a positive number; if to the right, it is a negative number.

To multiply, you add the exponents, as the following examples show:

$$5000 \times 40 = 200,000$$
$$(5 \times 10^3)(4 \times 10^1) = 20 \times 10^4,$$
$$\text{or } 2 \times 10^5.$$
$$0.006 \times 200 = 1.2$$
$$(6 \times 10^{-3})(2 \times 10^2) = 12 \times 10^{-1},$$
$$\text{or } 1.2.$$

To divide, you subtract exponents:

$$\frac{4800}{160} = 30$$
$$\frac{4.8 \times 10^3}{1.6 \times 10^2} = 3 \times 10^1 = 30.$$
$$\frac{100}{0.0005} = 200,000$$
$$\frac{1 \times 10^2}{5 \times 10^{-4}} = 0.2 \times 10^{2-(-4)}$$
$$= 0.2 \times 10^6.$$

To add or subtract numbers, both numbers must have the same exponent:

$$20,000 + 300,000 = 320,000$$
$$2 \times 10^4 + 30 \times 10^4 = 32 \times 10^4.$$

APPENDIX A

Powers of 10

TABLE A–1   POWERS OF 10

$10^5 = 10 \times 10 \times 10 \times 10 \times 10 = 100,000$
$10^4 = 10 \times 10 \times 10 \times 10 = 10,000$
$10^3 = 10 \times 10 \times 10 = 1000$
$10^2 = 10 \times 10 = 100$
$10^1 = 10$
$10^0 = 1$
$10^{-1} = 1/10 = 0.1$
$10^{-2} = 1/100 = 0.01$
$10^{-3} = 1/1000 = 0.001$
$10^{-4} = 1/10,000 = 0.0001$
$10^{-5} = 1/100,000 = 0.00001$

To find the square root of a number, express it as an even power of 10 and divide the exponent by 2:

$$\sqrt{2500} = 50$$
$$\sqrt{25 \times 10^2} = 5 \times 10^1.$$

To find the cube root of a number, express it as a power of 10 that is divisible by 3. Divide the exponent by 3:

$$\sqrt[3]{27,000} = 30$$
$$\sqrt[3]{27 \times 10^3} = 3 \times 10^1$$
$$= 30.$$

Exercises

1. Express the following in powers of 10:

  a. 460   *Answers:* **a.** 4.6×10^2
  b. 6,800,000
  c. 2400   **c.** 2.4×10^3
  d. 0.00016
  e. 0.000054   **e.** 5.4×10^{-5}
  f. 0.003

2. Evaluate the following:

  a. $\dfrac{4000 \times 200}{120 \times 0.001}$ *Answers:*   **a.** 6.67×10^6

  b. $\dfrac{6 \times .002}{300 \times 0.2}$

  c. $\dfrac{400 \times 800000}{.001 \times 0.016}$ **c.** 2×10^{13}

  d. $\sqrt{640,000}$

  e. $\sqrt[3]{8,000,000}$

Units

$$1 \text{ km} = 1000 \text{ m}$$
$$1 \text{ m} = 100 \text{ cm}$$
$$1 \text{ cm} = 10 \text{ mm}$$
$$1 \text{ in} = 2.54 \text{ cm} = 0.0254 \text{ m}$$
$$1 \text{ ft} = 0.3048 \text{ m}$$
$$1 \text{ yd} = 0.9144 \text{ m}$$
$$1 \text{ m} = 3.28 \text{ ft} = 39.36 \text{ in}$$
$$1 \text{ mi} = 1609 \text{ m}$$
$$1 \text{ mi} = 5280 \text{ ft}$$
$$1 \text{ lb} = 454 \text{ grams} = 0.454 \text{ kg}$$
$$1 \text{ kg} = 1000 \text{ g}$$
$$1 \text{ kg} = 2.20 \text{ lb}$$

APPENDIX B
Units and Unit Conversion

Unit Conversion

To convert from one set of units to another, simply multiply by "1", which is expressed in terms of units. The undesirable units cancel, while the desirable ones do not. For example, to convert 20 m/sec to mi/hr:

$$20 \text{ m/sec} \times 1 \times 1$$

$$= 20 \frac{\text{m}}{\text{sec}} \times \frac{1 \text{ mi}}{1609 \text{ m}} \times \frac{3600 \text{ sec}}{1 \text{ hr}}$$

$$= 44.7 \text{ mi/hr.}$$

Answers to Selected Questions

Unless otherwise specified, the answers refer to questions for calculation.

1–5 a. 16 g, 18 g, 2, 16 g, 1, 18 g, 1
  b. N_A
  d. 44 amu, 44 g
1–6 a. 8.91 g
  d. 1.06×10^{-22} g or 1.06×10^{-25} kg
2–16 82.32 m/sec, 345.7 m
2–18 a. 17.8 m/sec
  b. 12.9 m
  c. 1.62 sec
2–19 yes, $d = 31.5$ m
3–10 4.9 m/sec²
3–11 a. 12.6 m/sec
  b. 158.8 m/sec²
  c. 63.5 newtons
3–13 a. 3×10^{21} m/sec²
  b. 2.00×10^{-5} newtons
  c. 4.61×10^{-6} newtons
  d. 2.00×10^{-5} newtons
  e. maximum force must be greater than F_{AV}.
4–4 a. polonium-218
  b. 2.95×10^5 m/sec
4–5 a. zero
  b. in the opposite, larger than
  c. 1.57×10^7 m/sec
5–1 $F_{grav} = 3.61 \times 10^{-47}$ newtons,
  $F_{charge} = 8.20 \times 10^{-8}$ newtons,
  F_{charge} is 2.3×10^{39} times larger than F_{grav}.
5–4 4.36 m/sec²
5–5 0.80 years
6–6 a. 4×10^6 joules
  b. 2.4×10^6 joules
  c. 1.6×10^6 joules
  d. 333 m
  e. heat
6–7 $a = v^2/r$
  a. 7.67 m/sec
  b. 2940 joules
  c. 14,700 joules
  d. no; PE and TE at 14 m are only 13,720 joules.
6–9 a. 45 joules
  b. yes; 0.3 m
  c. machines do work by exerting a smaller force through a larger distance
6–19 2.40×10^{-28} kg, 263 m_e
7–20 (thought) 8, 8, 4

7–2 a. $E = \dfrac{V}{S}$

  b. $F = \dfrac{KQ_1 Q_2}{d^2}$

   E due to $Q_1 = \dfrac{KQ_1}{d^2}$,

   E due to $Q_2 = \dfrac{KQ_2}{d^2}$.

7–4 4.80×10^{-17} newtons,
9.60×10^{-19} joules,
6 eV,
9.60×10^{-19} joules,
6 eV,
8 eV

7–5 9.60×10^{-17} newtons,
19.20×10^{-19} joules,
12 eV,
19.20×10^{-19} joules,
12 eV,
14 eV

8–1 **a.** 0.32 m/sec
b. 0.08 m
c. 8 per sec, 4 per sec

8–6 **a.** constructive interference
b. constructive interference
c. destructive interference

8–11 1100 mph, 1000 mph

9–4 **a.** Brian's relative speed is 1 m/sec,
Susan's relative speed is 5 m/sec
b. no; no.
c. 3 m/sec, 3 m/sec; yes; yes

9–6 **a.** 10×10^{-8} sec
b. 6×10^{-8} sec
c. 14.4 m

9–7 **a.** 3.33×10^{-7} sec
b. t_{diff}; events occur at opposite ends of spaceship
c. 2.66×10^{-7} sec
d. 48 m
e. 48 m
f. 10,240 km

9–9 **a.** 4 sec
b. 19.2 m, L = 9.6 m
c. 3 m, 6 m, $x_C = ct$
d. 3 m, 6 m, $x_D = ct$
e. 7.8 m, 6.0 m, $d_L = 9.6 - 0.6\,ct$
f. 11.4 m, 13.2 m, $d_R = 9.6 + 0.6\,ct$
g. 2 sec
h. 8 sec
i. 4 sec, 4 sec, 2 sec, 8 sec

9–10 0.8 c, 10.0 m, 6.0 m

9–11 0.95 c

9–15 a. $E = pc$

10–15 (thought)
a. electrons will not be emitted
b. KE remains unchanged
c. yes
d. f_2 is larger than f_1.
e. B_A is larger than B_B.
f. KE_4 is larger than KE_3.
g. KE_4 is larger than KE_2.

10–2 **a.** 3.10 eV
b. 2.02×10^{20} photons/sec

10–3 $y = 3x - 1, y = 3x - 2.5$;
same coefficient of x.

10–7 **a.** 6.20 eV
b. 4.20 eV

11–3 **a.** 1.43×10^5
b. 4.09×10^4 per cm
c. 2.86×10^4
d. 87.2 per cm

11–4 **a.** 9.4/182
b. $1.43 \times 10^5\,N_a$
c. 872
d. $872\,N_b$
e. $\dfrac{872 N_b}{1.43 \times 10^5 N_a} = \dfrac{9.4}{182}$

$N_b/N_a = 8.5$

11–9 **a.** 3.3×10^{22} for lead,
6.0×10^{22} for A1,
b. 82 for lead,
13 for A1
c. 27.1×10^{23} for lead,
7.8×10^{23} for A1
d. 3.5 times as many electrons in 1 cm³ of lead than
in A1

11–10 **a.** zero
b. 10^{1800}
c. 3×10^{-6} sec

11–11 250 counts per min at 0.4 m,
62.5 counts per min at 0.8 m.

11–12 **a.** 4.46×10^3 curies
b. 140 watts of heat
c. 0.18 g, 1.25 watts
d. longer half-life

12–4 **a.** $25 \times 10^{-6}\,\text{m}^2$
b. circular area of 0.5 r radius,
$8.04 \times 10^{-28}\,\text{m}^2$ for one nucleus,
$2.41 \times 10^{-9}\,\text{m}^2$ for all nuclei
c. 9.64×10^{-5}
d. 1 in 10,373

13–9 2.11 eV

14–2 **c.** maximum frequency is 9.67×10^{18} per sec,
minimum frequency is zero,
minimum wavelength is 0.031 nm

15–3 7.29×10^6 m/sec, 3.99×10^3 m/sec

15–5 **a.** 323
b. 677
c. 438

15–9 **a.** 6.58×10^{-7} MeV-sec/m
b. 197 MeV

15–10 4.60 sec

15–11 when the hills are separated by 5 m, the half-life is
4.60 sec

16−3 **a.** thorium-234

 b. 4.28 MeV

 c. 4.28 MeV

16−6 **a.** H-1 and F-19 are produced by collision.

 b. no protons detected; ME-KE table shows KE equal to -0.00045 amu, which is impossible.

 c. Yes, 1.88 MeV (neglecting recoil KE of F-19)

16−11 **a.** 15.9 MeV

 b. No

17−4 **a.** 1.52×10^7 m/sec

 b. 2.9×10^{-8} sec

 c. 2.9×10^{-8} sec

 d. 7.6×10^6 m/sec

 e. 0.3 MeV

 f. 150 times

18−7 **b.** When F equals S, L = 4.

 c. uncontrolled chain reaction, no sustained chain reaction

19−1 63 percent increase, 14 years

19−5 **a.** 3.6×10^{-13} m

 b. No

 c. small probability that protons will cross potential energy barrier

20−1 230.4 newtons

20−4 **b.** 1.97 MeV

Index

Atom:
    Bohr's model of. *See* Atom, hydrogen, Bohr's model of
        calculating diameter of, 8–11
    calculating mass of, 4–5, 8–10
    Dalton's theory of, 4
    defined, 4
    disintegration of, 368–371
    Nagoaka's model of, 282–283
    Rutherford's model of, 286–291
    Thomson's model of, 282, 284–286
Atom, hydrogen, Bohr's model of, 23, 24, 296, 297–298
    circular motion of electrons, 76–78
    energy for first four allowed orbits, 131–134
    selection of allowed orbits, 94–95
    speed of electrons in first allowed orbit, 78
    and wave nature of electrons, 341–342
The Atom and Its Nucleus, 450
Atomic bomb, 430–433
    compared to nuclear reactor, 435
    critical mass of, 431–432
    design of, 432
    effects of explosion, 432–433
Atomic energy, 126. *See also* Nuclear energy
Atomic mass number, 8
Atomic mass unit, 5
The Atomic Nucleus, 360
Atomic number, 6
    determining significance of, 329–333
Avogadro, Count Amedeo, 8, 27
Avogadro's number, 8
    determining from electrolysis, 152

Backward scattering. *See* Alpha particles, backward
    scattering of
Bainbridge, K. T., 372
Balmer, Johann, 14, 299
Balmer formula, 24, 299, 319
Barium, radioactivity of, 260, 261
Baryons, 463
Battaglia, Nancie, 312
Becquerel, Henri, 3, 17, 27, 254–255, 257, 263, 266,
    270, 379
Becquerel rays, 255
Bell-shaped curve, 286
Beta decay, 379–381, 447
Beta radiation. *See* Beta rays
Beta rays, 19, 265, 266, 269
Bethe, Hans, 439
Bevatron, 220, 221, 456
Bismuth, radioactivity of, 260

Black bodies, 231
Blackett, P. M. S., 378, 394
Black hole, 440, 441
Blast wave, 432
BNL. *See* Brookhaven National Laboratory
Bohr, Aage, 301
Bohr, Niels, 3, 23, 28, 76–78, 94–95, 112, 131, 291,
    296, 297–298, 307, 314, 329, 330, 340, 341–342,
    348–349, 350, 354, 362, 408, 411, 412, 413, 419,
    423, 428
Bohr Institute of Theoretical Physics, 301
Booth, E. T., 415
Born, Max, 339, 343, 348, 354
Bottom quark, 470
Bradshaw, Clyde, 450
Bragg, Sir William Henry, 321, 323–328, 329, 330, 334,
    335
Bragg, Sir William Lawrence, 321, 323–328, 329, 330,
    334, 335
Brahe, Tycho, 98
Bravais, Auguste, 322, 323
Bravais lattices, 323
Breeder reactor, 436–437
Brookhaven National Laboratory, 461, 464, 465
Bubble chamber, 459–460
Bunsen, Robert, 308
Bunsen burner, 308
Butler, C. C., 454

Calcium atom, structure of, 333
Calculus, 105
Caloric theory of heat, 117
Calorie, defined, 117, 270
Calorimeter, 381
Cancer:
    cobalt-60 to treat, 386
    radiation to treat, 19, 263, 271
Capacitor, 146–148
Carbon-nitrogen cycle, 439
Cathode, defined, 12
Cathode rays, 14–15, 145
Cathode ray tube, 12, 14, 15
Cavendish, Henry, 100, 102, 103, 107
Cavendish Laboratory, 102, 154
Center of gravity, 101
Centrifugal force, and circular motion, 67, 68–69
    example of, 69–70
Chadwick, James, 3, 25, 28, 112, 124, 126, 127, 134,
    367, 376, 377, 379, 381–383, 386
Chain reaction, in fission, 413–423
Chamberlain, O., 456, 458
Charge, electric, 10–11, 146–148
    on drop of water, measuring, 156–158
    granular nature of, 151
Charmed quark, 468
"China syndrome," 436
Chinowsky, William, 467

INDEX